Handbook of Experimental Pharmacology

Volume 115

Toxicology of Metals

Biochemical Aspects

Contributors

N. Ballatori, M.G. Cherian, D.C. Dawson, M. Delnomdedieu
P. Druet, B.R. Fisher, P. Goering, R.A. Goyer, S. Himeno
N. Imura, E.H. Jeffrey, M.M. Jones, J.H.R. Kägi, S. Kawanishi
C.D. Klaassen, J. Koropatnick, J.S. Lazo, M.E.I. Leibbrandt
K. Miura, A. Naganuma, E.J. O'Flaherty, L. Pelletier
T.G. Rossman, S. Silver, M. Styblo, D.M. Templeton, D.J. Thomas
M.P. Waalkes, M. Walderhaug, J.S. Woods, J. Zeng

Editors

Robert A. Goyer and M. George Cherian

Springer-Verlag
Berlin Heidelberg New York London Paris
Tokyo Hong Kong Barcelona Budapest

Robert A. Goyer, M.D.
Professor Emeritus
Department of Pathology
University of Western Ontario
Health Sciences Centre
London, Ontario, Canada N6A 5C1

M. George Cherian, Ph.D.
Professor
Department of Pathology
University of Western Ontario
Health Sciences Centre
London, Ontario, Canada N6A 5C1

With 37 Figures and 17 Tables

ISBN-13:978-3-642-79164-2 e-ISBN-13:978-3-642-79162-8
DOI: 10.1007/978-3-642-79162-8

Library of Congress Cataloging-in-Publication Data. Toxicology of metals: biochemical aspects/contributors N. Ballatori . . . [et al.]; editors, Robert A. Goyer and M. George Cherian. p. cm. – (Handbook of experimental pharmacology; v. 115) Includes bibliographical references and index. ISBN-13:978-3-642-79164-2 1. Metals – Toxicology. I. Ballatori, N. II. Goyer, Robert. III. Cherian, M. George, 1941– . IV. Series. [DNLM: 1. Metals – toxicity. 2. Maximum Permissible Exposure Level. W1 HA51L v. 115 1995/QV 290 T755 1995] QP905.H3 vol. 115 [RA1231.M52] 615'. 1 s–dc20 [615.9'253] DNLM/DL-C for Library of Congress 94-26484

Softcover reprint of the hardcover 1st edition 1995

Typesetting: Best-set Typesetter Ltd., Hong Kong

SPIN: 10096516 27/3130/SPS – 5 4 3 2 1 0 – Printed on acid-free paper

List of Contributors

Ballatori, N., Department of Environmental Medicine, University of Rochester School of Medicine, Rochester, NY 14642, USA

Cherian, M.G., Department of Pathology, University of Western Ontario, Health Sciences Centre, London, Ontario, Canada N6A 5C1

Dawson, D.C., Department of Physiology, University of Michigan Medical School, 6811 Med.Sci. 2, Ann Arbor, MI 48109-0622, USA

Delnomdedieu, M., Center for Environmental Medicine and Lung Biology, University of North Carolina at Chapel Hill, Chapel Hill, NC 27514, USA

Druet, P., Institut National de la Santé et de la Recherche Médicale (INSERM U28), Hôpital Broussais, 96 rue Didot, F-75674 Paris, Cedex 14, France

Fisher, B.R., Health Sciences Branch, Division of Life Sciences, Office of Science and Technology, Center for Devices and Radiological Health, Food and Drug Administration, 12709 Twinbrook Parkway, Rockville, MD 20857, USA

Goering, P.L., Health Sciences Branch, Division of Life Sciences, Office of Science and Technology, Center for Devices and Radiological Health, Food and Drug Administration (HFZ-112), 12709 Twinbrook Parkway, Rockville, MD 20857, USA

Goyer, R.A., Department of Pathology, University of Western Ontario, Health Sciences Centre, London, Ontario, Canada N6A 5C1

Himeno, S., Department of Public Health and Molecular Toxicology, School of Pharmaceutical Sciences, Kitasato University, Minato-ku, Tokyo 108, Japan

Imura, N., Department of Public Health and Molecular Toxicology, School of Pharmaceutical Sciences, Kitasato University, Minato-ku, Tokyo 108, Japan

Jeffrey, E.H., Institute for Environmental Studies, University of Illinois at Urbana-Champaign, 1101 W. Peabody Drive, Urbana, IL 61801, USA

JONES, M.M., Department of Chemistry and Center in Toxicology, Vanderbilt University, P.O. Box 1583, Station B, Nashville, TN 37235, USA

KÄGI, J.H.R., Biochemistry Institute of the University of Zürich, Winterthurerstraße 190, CH-8057 Zürich, Switzerland

KAWANISHI, S., Department of Public Health, Faculty of Medicine, Kyoto University, Kyoto 606, Japan

KLAASSEN, C.D., Department of Pharmacology, Toxicology, and Therapeutics, University of Kansas Medical Center, 39th and Rainbow, Kansas City, KS 66103, USA

KOROPATNICK, J., Department of Oncology, London Regional Cancer Centre, University of Western Ontario, 790 Commissioners Road E., London, Ontario, Canada N64 4L6

LAZO, J.S., Department of Pharmacology, University of Pittsburgh, Experimental Therapeutics Program, Pittsburgh Cancer Institute, E-1340 Biomedical Science Tower, Pittsburgh, PA 15261, USA

LEIBBRANDT, M.E.I., Parke-Davis Pharmaceutical Research, General Toxicology, 2800 Plymouth Road, Ann Arbor, MI 48105, USA

MIURA, K., Department of Environmental Sciences, Wako University, Machida-shi, Tokyo, 104-01, Japan

NAGANUMA, A., Department of Public Health and Molecular Toxicology, School of Pharmaceutical Sciences, Kitasato University, Minato-ku, Tokyo 108, Japan

O'FLAHERTY, E.J., Department of Environmental Health, University of Cincinnati College of Medicine, 3223 Eden Ave., Cincinnati, OH 45267-0056, USA

PELLETIER, L., Institut National de la Santé et de la Recherche Médicale (INSERM U28), Hôpital Broussais, 96 rue Didot, F-75674 Paris, Cedex 14, France

ROSSMAN, T.G., Nelson Institute of Environmental Medicine and Kaplan Cancer Center, New York University Medical Center, 550 First Avenue, New York, NY 10016, USA

SILVER, S., Department of Microbiology and Immunology, University of Illinois College of Medicine, M/C 790, Room 703, 835 South Wolcott Ave., Chicago, IL 60612-7344, USA

STYBLO, M., Curriculum in Toxicology, University of North Carolina at Chapel Hill, Chapel Hill, NC 27514, USA

TEMPLETON, D.M., Department of Clinical Biochemistry, University of Toronto, Banting Institute, 100 College Street, Toronto, Ontario, Canada M5G 1L5, and Division of Biochemistry, The Research Institute, Hospital for Sick Children, 555 University Ave., Toronto, Ontario, Canada M5G 1X8

THOMAS, D.J., Health Effects Research Laboratory, MD-74, U.S. Environmental Protection Agency, Research Triangle Park, NC 27711, USA

WAALKES, M.P., Inorganic Carcinogenesis Section, Laboratory of Comparative Carcinogenesis, National Cancer Institute, Frederick, MD 21702, USA

WALDERHAUG, M., Division of Microbiological Studies, Microbial Ecology Branch, Food and Drug Administration, CFSAN HFS-517, 200 C. Street, SW, Washington, DC 20204, USA

WOODS, J.S., Battelle Seattle Research Center, 400 N.E. 41st Street, Seattle, WA 98105, USA

ZENG, J., Biochemistry Institute of the University of Zürich, Winterthurerstraße 190, CH-8057 Zürich, Switzerland. Present address: Laboratory of Biochemistry, Bldg. 37, Rm. 4A13, National Cancer Institute, National Institutes of Health, Bethesda, MD 20892, USA

Preface

The toxicology of metals has been concerned in the past with effects that produced clinical signs and symptoms. However, this view of metal toxicology has expanded in recent years due principally to two advances. There has been a considerable increase in our knowledge of the biochemical effects of metals. In addition, biomarkers of toxicity can now be recognized that identify toxicity at levels of exposure that do not produce overt clinical effects. Thus, the toxicology of metals is now focused on nonclinical events that reflect adverse health effects. This new awareness has produced the challenge of determining the lowest adverse level of exposure. With increasing analytical sensitivity and methodologies to detect small changes at the molecular level, the lowest level of exposure of some toxic metals, like lead, is very small. Indeed, for metals in which there is no biologic requirement, it may be questioned whether there is a level of exposure that does not produce some degree of toxicity. For essential metals, the question is being asked as to the levels at which exposure exceeds biologic requirements and excess exposure becomes toxic. The appropriateness of health decisions and the formation of public policy are dependent on the availability of current scientific information that addresses these questions.

The information in this volume is intended to be a resource for this purpose as well as a reference for students of toxicology and other health professionals. Specific topics of current relevance are addressed in chapters on aluminum, cadmium, chromium, lead, and mercury. A number of chapters are concerned with metal–protein interactions, including effects on heat shock proteins and on metallothionein, a protein important in the metabolism of both essential and toxic metals. Zinc-containing finger proteins are related to gene expression, as are chapters on the mutagenicity of metals, plasmid-mediated metal resistance, drug resistance, and the effects of oxygen radicals in the production of DNA damage. Some metals may become involved in basic physiologic processes within cells including transport at the level of the cell membrane and porphyrin metabolism. These are effects that may produce new biochemical markers that provide early evidence of toxicity. And, finally, the toxicity of low-level exposures is stimulating increased efforts to develop chelators or therapeutic agents that may be used for intervention and prevention of toxicity.

The field of biochemical toxicology is rapidly expanding in terms of the sophistication of the science base and a corresponding public interest. The editors are hopeful that this volume will contribute to these interests.

London, Ontario, Canada
August 1994

M. George Cherian
Robert A. Goyer

Contents

CHAPTER 4

Immunotoxicology of Metals

CHAPTER 5

Effects of Metals on Gene Expression

CHAPTER 6

Metallothionein and Its Interaction with Metals

CHAPTER 7

Biochemical Mechanisms of Aluminum Toxicity

CHAPTER 8

Mercury Toxicity

CHAPTER 9

Toxicology of Cadmium

CHAPTER 11

Metals and Stress Proteins

CHAPTER 12

Metals and Anticancer Drug Resistance

CHAPTER 13

CHAPTER 18

Biological Mechanisms and Toxicological Consequences of the Methylation of Arsenic

CHAPTER 19

Bacterial Plasmid-Mediated Resistances to Mercury, Cadmium, and Copper

CHAPTER 1

Transplacental Transfer of Lead and Cadmium

R.A. Goyer

A. Introduction

The placenta provides the route of transfer of both essential and nonessential metals from mother to fetus. It is now recognized that for many toxic metals adverse health effects occur at levels not previously thought to be toxic and that the developing fetus and newborn are particularly vulnerable. Emerging questions that must be addressed concern the lowest levels of exposure that produce toxicity. Are there, in fact, thresholds for such effects, or is the failure to detect effects in the newborn merely a reflection of the inability to recognize or detect subtle effects? Are such effects of longer term consequence, that is, are they reversible? And, finally, what are some of the avenues for intervention including nutrition?

Concern for fetal effects of toxic metals must begin with the role of the placenta in fetal exposure. The placenta is the interface between the mother and the ambient environment and the fetus. Substances cross the placenta by a number of transport mechanisms as in other body tissues, simple diffusion for small molecules, active transport for larger molecules but with a molecular weight of less than 40000, and pinocytosis for macromolecules. The principal function of the placenta, of course, is to provide a conduit for the nourishment of the fetus. The placenta has mechanisms that enhance the transport of those substances that are needed and restricts entry of those substances that are toxic or otherwise harmful. These mechanisms are subject to hormonal influences, some oxidative reactions which are mostly nonenzymatic, and, most importantly, to umbilical cord and fetal blood flow. Not all of the activities of the placenta are critical to transfer of toxic metals, but nevertheless there has been relatively little study to date regarding mechanisms for transport of cations apart from the essential metals, calcium and iron.

The purpose of this chapter is provide an overview of the mechanisms for transport of two toxic metals, lead and cadmium, and to identify factors that may influence this process. The mechanisms for transfer of these two metals differ in that lead is readily diffusible across the placenta. There is no placental barrier for lead, whereas the placenta provides a protective barrier to fetal exposure to cadmium. Consideration of the placental transport of these two metals is prefaced by a few comments on possible experimental models to study placental structure and function. Also some comments are provided on methods for sampling the human placenta. The availability of human fetal tissue is very limited and there is no ideal animal surrogate. The primate might be the best alternative for obvious reasons, but there are constraints here as well. High cost is a major concern and it is difficult to conduct studies with sufficient statistical power to provide definitive conclusions. Rodents provide the most commonly used experimental models so that similarities and differences must be considered in terms of relevance to humans.

I. Comparison of Human and Rodent Fetal-Maternal Blood Barriers

The site of transfer of nutrients from mother to fetus is the fetal-maternal blood barrier. There is somewhat of an analogy between this interface and the blood-brain barrier in the central nervous system. Both must provide mechanisms for transfer of essential nutrients yet provide some protection against transfer of toxic substances. In mammalian systems the interface between maternal and fetal blood is located in the chorionic villus. There

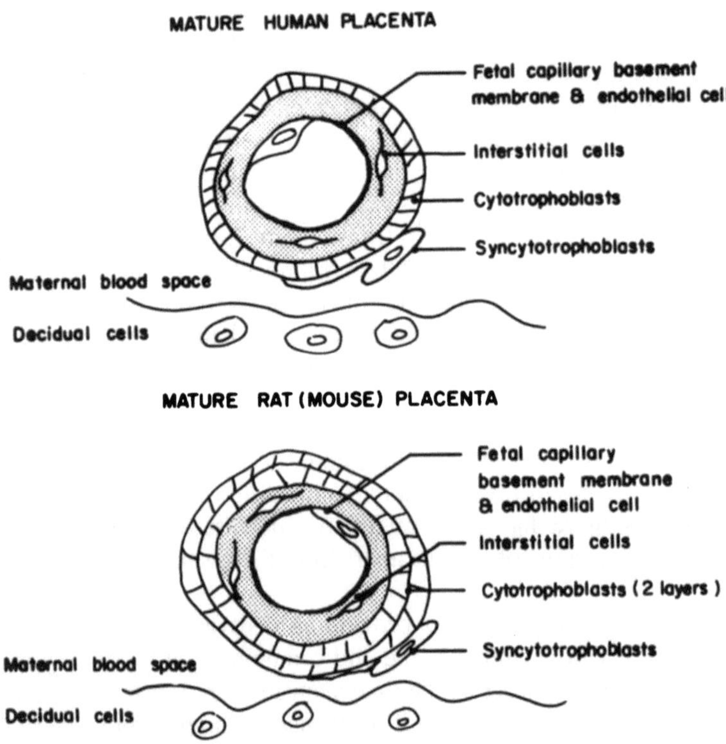

Fig. 1. Comparison of the chorionic villi of mature human placenta and rat placenta. (From GOYER 1990)

are anatomic similarities as well as differences in this interface between various species. These differences and similarities are discussed in reviews by STEVEN (1975) and BECK (1981). The fetal-maternal blood barrier in the mature rat placenta is very similar anatomically to that of the human placenta (Fig. 1) (BECK 1981). Both are hemochorial placentas, that is, the fetal chorionic villi bathe in lacunae of maternal blood. In the human placenta, the villous capillary is separated from maternal blood by three layers of cells, the capillary endothelial cell, an interstitial or connective tissue layer, and a layer of cytotrophoblasts. Syncytial trophoblasts do not form a continuous layer but are intermittent. The layers of the chorionic villus in the rat are similar but differ by the presence of an additional layer of cytotrophoblasts. Electron microscopic studies and immunohistochemical analyses have demonstrated that the cytotrophoblast is primarily the germinative cell layer, whereas the syncytiotrophoblast is the secretory layer as evidenced by hormone content (steroid and peptide hormones) and numerous mitochondria as energy source to support the high level of metabolic activity as well as transport function (YEH and KURMAN 1989).

II. Methods for Sampling the Human Placenta

Because the placenta contains a relatively large amount of blood, some maternal, some fetal, as well as placental tissue that differs in maturity, measurement of placental composition or content is subject to considerable variation. Methods involving a single measurement from only one site are the least reliable. To achieve representative sampling most studies have used random sampling from multiple sites. Alternatively, the entire placenta might be homogenized in order to obtain a homogeneous sample, but this approach is technically cumbersome and subject to contamination if trace metals are to be measured. MANCI and BLACKBORN (1987) proposed sampling from four anatomic zones, one paracentral near the umbilical cord, two from the central portion of the placenta, and a fourth from the periphery. A central basilar sample contains more maternal than fetal hemoglobin and the central chorionic plate region contains more fetal hemoglobin. Using this approach they showed that copper was greatest in fetal membranes, zinc lower in the fetal or chorionic plate area, and calcium greatest in the periphery, as might be predicted because of its greater maturity. There is presently little experience with this approach by others.

B. Placental Transfer of Lead

I. Mechanism of Placental Transfer of Lead

The mechanism for the transport of lead through the placenta is not well defined. There are several reasons for suggesting it may be a matter of simple diffusion from maternal to fetal circulation (GOYER 1990). For one, umbilical cord blood lead levels are generally positively correlated with maternal blood lead levels and are slightly but significantly lower in concentration than maternal levels (LAUWERYS et al. 1978; MOORE 1980; ALEXANDER and DELVES 1981; RABINOWITZ and NEEDLEMAN 1982). This difference seems to be less with higher than with lower blood lead levels and may be closely correlated with blood lead levels in the 10- to 15-μg/dl range (BAGHURST et al. 1992). Secondly, the transfer of lead from mother to fetus has been shown in experimental models (guinea pig) and to be linearly related to umbilical cord blood flow rate (KELMAN and WALTER 1980). There does not seem to be a placental barrier or enhancing mechanism, although fetal blood lead may be slightly lower than maternal blood lead concentration (MOORE 1980). As the fetus grows umbilical cord blood flow increases and transfer of lead increases. An important question is how early in gestation does transfer of lead occur. There are few measurements of lead levels in fetal tissues at different ages of gestation. The largest number of measurements in a single study were reported by BARLTROP (1969). The fetuses were from abortions from mothers in the general population starting with fetuses about 13 weeks of age and extended to term. There is no reason to believe transfer of lead from maternal blood to the fetus does not occur earlier.

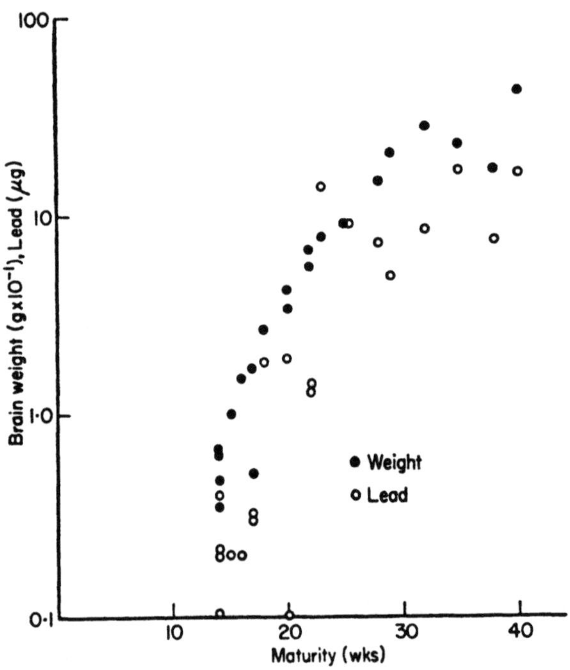

Fig. 2. Brain weight and lead content of 21 fetuses with maturities of 14–40 weeks. (From BARLTROP 1969)

Placental transport of lead may be associated with placental transport of calcium. BARLTROP (1969) observed that femur lead increased rapidly in the third trimester, corresponding to onset of ossification and deposition of calcium. There is a question whether the increased deposition of lead in the femur along with calcium is because of some commonality between lead and calcium metabolism. Figure 2 shows the close correlation between size of fetal brain and increase in lead content as the brain matures. Whether the movement of lead into the brain is a matter of simple diffusion or whether there is some relationship to brain calcium is not known. Calcium deficiency does enhance lead absorption and the pathological effects of lead (MAHAFFEY et al. 1973). The placenta also contains a calcium-binding protein identical to that present in the intestine (VAN DIJK 1981). A comparison of placental transfer of toxic metals by Nakano and Kurosa from the Minamata Institute in Japan (unpublished observation) found that lead levels in the placenta were strongly correlated with calcium, suggesting that lead deposition is associated with calcium deposition or with areas of dystrophic calcification that are present in the mature placenta.

There are several reports of placental lead concentrations in the literature (KARP and ROBERTSON 1981; TSUCHIYA er al. 1984; ERNHART 1986; KORPELA et al. 1986; SCHRAMEL et al. 1988). Since the placenta is mostly blood, maternal and fetal, it might be expected that the lead content would be equivalent to blood levels unless there are foci of concentration. But values in various reports are quite variable, mostly higher than blood values. There are two possible explanations for this. For one, there may, in fact, be a lead-concentrating mechanism in some component of the placenta, in trophoblasts for instance, but this has not been demonstrated, and from what is known about mechanisms for transplacental transport of lead this is unlikely. KHERA and coworkers (1980), in a study of lead in placentas from births from women in lead industries, found that the placental level of lead in normal births was $0.29 \pm 0.9\,\mu g/g$ whereas lead in placentas from stillborns was $0.45 \pm 0.32\,\mu g/g$. Normals were defined as having had no occupational exposure to lead for the 2 years prior to delivery. Nevertheless these values are higher than those for the general population. Stillborns as a group showed even higher levels, perhaps more calcium-lead precipitation in the placenta. Finally, as might be expected, increase in lead in the placenta was related to length of occupational lead exposure and age. Measurements of lead in amniotic fluid and fetal membranes are also quite variable and are, generally, higher than those for the placenta but the significance of this, if any, is not known.

II. Maternal Blood Lead Levels During Pregnancy

Maternal blood lead levels and umbilical cord blood lead levels are the most direct measure of fetal exposure to lead, reflecting recent lead exposure and lead mobilized from tissue stores (MUSHAK 1989). Recent studies of mobilized lead from bone using stable isotope-mixing techniques in nonpregnant women suggest that 40%–70% of blood lead reflects lead from bone (GULSON and MAHAFFEY, unpublished). These observations have not been extended to women during pregnancy but the contribution of bone lead might be expected to be greater than during the nonpregnant state because of comobilization with calcium required for fetal growth. There are limited studies of blood lead during pregnancy, but they have found blood lead levels to be lower than for nonpregnant women (ALEXANDER and DELVES 1981). The reasons for this are not clear but may be in part the hemodilution which occurs during pregnancy, increased urinary excretion, and deposition in fetal tissues.

III. Effect of Maternal Lead on Birth Outcomes

Historical documentation of effects of lead on pregnancy outcomes indicates that lead at high doses may be an abortifacient (OLIVER 1914; LANE 1949). This must be a rare event today and there are no recent reports relating maternal lead exposure to spontaneous abortion or stillbirth. However, epidemiological studies do suggest that maternal blood lead levels may affect duration of pregnancy and fetal birth weight and stature. A study in Port Pirie, Australia (MCMICHAEL et al. 1986), showed that there was an 8.7 times greater risk with maternal blood lead levels of 14 μg/dl or greater than with a maternal blood lead level of 8.0 μg/dl. MOORE and coworkers (1982) showed in a cross-sectional study of 236 mothers and their infants that gestational age was significantly reduced as a function of increasing umbilical cord or maternal blood lead levels. BORSCHEIM et al. (1989) reported that in the Cincinnati Prospective Study there was a ½ week decrease in gestation for a 10-μg/dl increase in maternal blood lead. These investigators also found reduced birth weight and birth length related to maternal blood lead levels.

IV. Effect of Lead on Neurobehavioral and Cognitive Development In Utero

The critical effect of lead on the fetus and developing infant is on the central nervous system with impairment of cognitive and behavioral development. Three prospective studies, two in the United States and one in Australia, have provided data of the effect of intrauterine low-level lead exposure on subsequent neurobehavioral development. The study by BELLINGER and coworkers (1987) followed 249 children from birth to 2 years of age. Prenatal lead exposure was determined by measuring lead levels in umbilical cord blood. Multivariate regression analysis showed an association between cord blood lead and performance in Bailey Mental Developmental Index (MDI) scores. Children with high umbilical cord blood lead levels, between 10 and 25 μg/dl, achieved significantly lower MDI scores through 2 years of age than did infants whose cord blood lead levels were from <3 to 7 μg/dl. Although these early effects are reversible in some children by the age of 5 years, higher prenatal exposure lead exposure is associated with an increased risk of early cognitive deficit, particularly in children with a less than optimal sociodemographic environment. A study of a cohort in Cincinnati (DIETRICH et al. 1990) of infants with mean cord blood lead levels of 8.0 μg/dl found statistically significant relationships between prenatal and neonatal blood lead level and 3- and 6-month covariate-adjusted Bayley Mental and/or Psychomotor Development Index. However, the effects were strongest in male infants and infant from the poorest families. The results suggest that deficits or delays in early mental development appeared to be partly mediated by lead-related lower birth weight and decreased gestational maturity. A prospective study of maternal and infant blood lead levels and neurodevelopment in Port Pirie, Australia, found that for every 10 μg/dl increase in maternal blood lead level MDI scores at 24 months were decreased by about 2 points (MCMICHAEL et al. 1988). In contrast to the Boston and Cincinnati studies, reversibility of early effects was not discernible perhaps because of continuous exposure from the lead smelter in Port Pirie. These studies, particularly the Boston and Cincinnati cohorts, suggest that measurements of neurobehavioral status in early infancy do not have high predictive validity for later intellectual status and give rise to the question of reversibility of intrauterine effects. Alternatively, lead effects per se at this age may not be detectable by available methods of assessment. Blood lead levels at 2 years of age are more predictable of a longer term adverse neurological outcome than umbilical cord blood lead concentration.

V. Mechanisms for the Neurotoxicity of Lead

The fetal brain is more susceptible to the toxic effects of lead than the more mature brain because of the immaturity of the blood-brain barrier in the fetus. ROSSOUW et al. (1987) found that the uptake of lead by the fetal rat brain during gestation is greater than after birth. They found a 6-fold increase in total brain lead prenatally compared to a 3.5-fold increase during weaning and a 2-fold increase postweaning with the same level of lead exposure. Experimental studies suggest that the immature endothelial cells forming the capillaries of the developing brain are less resistant to the effects of lead than capillaries from mature brains and permit fluid and cations including lead to reach newly formed components of the brain, particularly astrocytes and neurons (TOEWS et al. 1978).

Neurodevelopment is highly complex and there are numerous opportunities for lead to interfere with normal development. SILBERGELD (1992) has divided the intrauterine neurological effects of lead into two broad categories, morphological effects and pharmacological effects. A highly significant morphological effect is the result of lead impairment of timed programming of cell: cell connections resulting in modification of neuronal circuitry (BULL et al. 1983). COOKMAN et al. (1987) showed that this lead effect is the result of decreasing the sialyation state of neural cell adhesion molecules producing a failure in synaptic structuring. COOKMAN et al. (1988) found that lead also induces precocious differentiation of the glia upon which cells migrate to their eventual positions during structuring of the central nervous system (COOKMAN et al. 1988).

Lead functions pharmacologically by interfering with synaptic mechanisms of transmitter release. It is suggested that lead can substitute for calcium and possibly zinc in ion-dependent events at the synapse and is possibly responsible for the observed impairment of various neurotransmitter systems (cholinergic, noradrenergic, GABergic, and dopaminergic) (CARROL et al. 1977; SILBERGELD et al. 1979). MARKOVAC and GOLDSTEIN (1988) found that micromolar concentrations of lead activate protein kinase C in brain microvessels from 6-day-old rat pups. This is a calcium phospholipid-dependent kinase which acts as a second messenger in the regulation of cellular metabolism. It is very sensitive to lead and impairment results in a breakdown in the normally tight blood-brain barrier.

Lead may replace calcium in calmodulin-dependent reactions, inhibit membrane-bound Na,K-ATPase, and interfere with mitochondrial release of calcium with impairment of energy metabolism (SIMONS 1986). These effects are potentially reversible if lead can be removed from active sites. Although removal from lead exposure and chelation therapy lower blood lead levels, there is little or no information regarding the removal of lead from these sensitive molecular sites of neural toxicity.

C. Placental Transfer of Cadmium

Studies comparing human maternal and umbilical cord blood cadmium levels are inconsistent, but for most studies fetal umbilical cord blood cadmium levels range from 40% to 70% maternal levels (LAUWERYS et al. 1978; ALESSIO et al. 1984; KORPELA et al. 1986; LAGERKVIST et al. 1992), indicating that the placenta acts as a barrier to fetal exposure to cadmium. The divergent results may reflect methodological problems in measurement of low levels of cadmium. The mechanism and factors involved in the retention of cadmium by the placenta are not well defined. The most plausible explanation is that cadmium is retained by binding to metallothionein and/or other compounds at specific sites in the placenta. The presence of metallothionein has been demonstrated in human placenta and fetal membranes (WAALKES et al. 1984). Synthesis can also be induced in cultured human trophoblasts by both cadmium and zinc salts. The syncytiotrophoblast has been shown to be the site of metallothionein synthesis, a protein that binds cadmium and may also enhance transport of the essential trace metals zinc and copper (MILLER 1986; GOYER and CHERIAN 1992). Metallothionein mRNA is increased in spongiotrophoblasts of mouse placenta following injection of cadmium and zinc (DE et al. 1989). Constitutively high levels of metallothionein and mRNA are also detected in the deciduum following early implantation. Using a cytohistochemistry technique and antibody to metallothionein, GOYER et al. (1992) identified metallothionein in the amniotic membrane, in syncytical trophoblasts, in interstitial cells within the villi, and in maternal decidual cells of human placenta.

I. Cadmium Levels in Human Placenta

Cadmium levels in human placentas are variable. KORPELA et al. (1986) from Finland reported mean placenta cadmium of 20.4 ± 14.4 ng/g. This study included only 19 placentas. Age of mothers ranged from 17 to 42 years. Eight were primigravidal whereas others had given birth to 1 to 13 children. Only two women were smokers. The variance in placenta cadmium levels was very wide. BERLIN et al. (1992) measured cadmium concentration in 27 placentas collected from female battery workers. The mean concentration was 21 ± 2.2 ng/g and ranged from 2 to 95 ng/g. Placenta cadmium concentrations were positively correlated with maternal blood cadmium. Morphological studies including ultrastrutural examination did not reveal any abnormalities. KUHNERT et al. (1988) from Cleveland reported lower values. Their sample included 185 placentas and mean values and SDs are reported in the form of graphs so values can only be estimated. They seem to range between 10 and 16 ng/g for smokers and even lower levels for nonsmokers. In an earlier paper they reported placenta cadmium concentrations of 18.1 ± 7.3 ng/g for smokers and 13.7 ± 6.4 for nonsmokers (KUHNERT et al. 1982). They also found that increased parity was related to increased levels of placental cadmium and reduction of zinc. The results suggest there is depletion of zinc stores with increasing parity and increasing placental nitrogen: cadmium ratios. Experimental studies of BHATTACHARYYA et al. (1988) in mice indicate that maternal renal cadmium increases with each round of pregnancies. These studies suggest that multiparity increases the maternal body burden of cadmium and possibly lowers zinc stores.

GOYER and CHERIAN (1992) measured cadmium, zinc and copper, and metallothionein content of human placentas from 55 uncomplicated, full-term deliveries. The mothers' ages ranged from 22 to 39 years, mean 29 years. All were current nonsmokers, but 16 (30%) acknowledged smoking in the past. None were on medication apart from iron and vitamin supplements. For 43 it was the first delivery, for 9 the second, for 1 the third, and for 2 the fourth. Samples of maternal and fetal blood were not obtained. Metals were measured by atomic absorption spectrophotometry and metallothionein by a silver saturation method (SCHEUHAMMER and CHERIAN 1986). The results are shown in Table 1. Zinc levels of placentas were almost the same as those in the KUHNERT et al. (1988) study, but copper levels were considerably lower. No other comparable measurements of metallothionein were found in the literature. There is a strongly positive correlation between

Table 1. Cadmium (Cd), zinc (Zn), copper (Cu), and metallothionein (Mt) content of 55 human placentas. (From GOYER et al. 1992)

Cd (ng/g)	Zn (μg/g)	Cu (μg/g)	Mt (μg/g)
32.3 ± 16.1	9.1 ± 1.8	0.59 ± 0.18	1.63 ± 0.5

zinc and metallothionein and copper and metallothionein in the placenta, but no relationship between cadmium and metallothionein. These results suggest that zinc and copper are the primary or major determinants of metallothionein levels in the placenta with the low level of exposure from the ambient environment. It should be noted that cadmium is present in only nanogram quantities and is competing with microgram levels of zinc and copper for metallothionein-binding sites so that there must be a considerably larger exposure to cadmium before a cadmium influence on metallothionein synthesis and binding will be detected.

II. Cadmium Effects on Placenta and Fetus

Large doses of cadmium salts administered parenterally during late gestation may produce placental injury and fetal death. Teratogenic effects are more likely when cadmium is administered early in pregnancy. PARIZEK (1964) and PARIZEK et al. (1968) found that a single subcutaneous dose of cadmium chloride, acetate, or lactate (4.5 or 3.3 mg cadmium/kg body weight) administered during day 17–21 of pregnancy was followed by placental necrosis and fetal death. Dose-dependent fetal mortality and teratogenicity, jaw defects, and cleft palate were observed by CHERNOFF (1973) in rats with dose levels of cadmium above 0.3 mg/kg body weight. Daily gastric gavage from day 6 to 19 of high levels of cadmium (20, 40, 60, and 80 mg/kg body weight) produced cardiovascular and hepatic defects (SCHARPF et al. 1972). FERM and CARPENTER (1968) found that zinc and/or selenium supplementation protected from the teratogenic effects of parenteral cadmium administration. HURLEY et al. (1971) reported that maternal zinc deficiency per se may produce congenital malformations. Administration of cadmium to zinc-deficient rats further increased the incidence of malformations. PARZYK et al. (1978) and SAMARAWICKRAMA and WEBB (1979) found that cadmium administration to rats produced zinc deficiency, which was responsible in part for the cadmium-induced abnormalities.

MILLER (1986) have summarized the ultrastructural changes in placenta perfused by cadmium chloride. The mitochondrion appears to be the principal target organelle, resulting in ultrastructural changes similar to that occurring in renal tubular cells of rats exposed to cadmium chloride (GOYER et al. 1984) and cadmium-metallothionein (CHERIAN et al. 1976) and humans with cadmium nephropathy (GARRY et al. 1986).

BOADI and coworkers (1991) observed a 60-fold increase in cytosolic cadmium in isolated placenta perfused with a solution containing 12 μg cadmium/ml. There were no significant changes in oxygen and glucose consumption and placental morphology. Perfusion with a solution containing 24 μg cadmium/ml for 5 h resulted in significant elevation of metallothionein. These authors suggested that the observed elevation in placental metallothionein would provide the fetus some protection from this toxic metal.

LEVIN et al. (1987) found the placenta to be relatively impermeable to a single subcutaneous injection of 40 μmol/kg cadmium chloride on day 18 of pregnancy in Wistar rats. Other experimental studies usually using lower doses of cadmium administered in vivo have shown that transport of cadmium through the placenta is limited. Only a small fraction of cadmium injected into pregnant rats and hamsters reaches the fetus (SONAWANE et al. 1975; AHOKAS and DILTS 1979; DENCKER et al. 1983), and the relationships between zinc and cadmium during pregnancy may have a role in cadmium effects observed during pregnancy and the placental transport of cadmium.

DANIELSSON and DENKER (1984) have suggested that increases in maternal and placental cadmium levels are inversely related to birth weight possibly due to impairment of fetal nutrition. More recently, SORRELL and GRAZIANO (1990) found that there is a relationship between maternal oral cadmium ingestion and decreased zinc stores. Cadmium at four dose levels (0, 5, 50, or 100 ppm) was fed to rats in drinking water on days 6 through 20 of pregnancy. In comparison to controls fetal and maternal weights were slightly reduced in the 50- and 100-ppm groups. Cadmium accumulated in a dose-related manner in maternal and fetal tissues and nitrogen decreased in fetal liver. These authors suggest that impaired fetal growth and zinc deprivation are secondary to maternal zinc retention and also that oral exposure to cadmium may have a more marked effect on zinc than was observed in this experiment, where cadmium was administered intraperitoneally, perhaps through an effect on gastrointestinal metallothionein and absorption of cadmium.

III. Interactions in Placenta Between Cadmium, Zinc and Copper, and Metallothionein

Placental metallothionein binds zinc and copper as well as cadmium. Zinc and copper are essential nutrients for the fetus whereas cadmium is toxic to the fetus and retained rather than transferred to the fetus. There is a question, therefore, as to how the essential metals are preferentially transported to the fetus while the toxic metal, cadmium, is retained. One possibility is that there is a greater sensitivity for zinc and copper metallothionein than cadmium metallothionein to the action of proteolytic enzymes present in trophoblasts. Degradation of the zinc and copper metallothionein complex facilitates the release of these metals to fetal blood, whereas cadmium metallothionein is resistant to this effect.

Feldman et al. (1978) showed in in vitro studies of rat liver that the rate of degradation of zinc metallothionein was twice that of cadmium metallothionein when the two species of metallothionein were incubated with trypsin. No digestion occurred when metal metllaothioneins were incubated with chymotrypsin and rates of degradation were equal when incubated with pronase. Rate of degradation of zinc metallothionein was twice that observed with cadmium metallothionein when proteins were incubated at pH 5.0 with a purified lysosomal extract. The authors concluded that metals stabilize metallothionein and that degradation of metallothionein in vivo is regulated in part by the species of the complexed metal. In cultured He La cells, metallothionein induced by zinc has a much longer half-life than metallothionein induced by dexamethasone (Karin et al. 1981).

In an effort to better define the metabolic relationships between cadmium and metallothionein, Goyer and Cherian (1992) administered cadmium as cadmium chloride at five dose levels to pregnant rats on gestation days 17–19 and found that placental transfer of cadmium and birth weight of pups did not change regardless of dose. The five doses of cadmium showed different levels of response in metallothionein induction and cadmium retention. Placenta showed a dose-related increase in metallothionein and cadmium. Several other studies have demonstrated the presence of metallothionein in rodents (Lucis et al. 1972; Hanlon et al. 1982) and human placenta (Goyer et al. 1992; Boadi et al. 1991), but there is little information regarding factors affecting induction of this protein and interactions with zinc and copper.

Studies on the influence of placental cadmium on birth weight in humans is presently inconclusive. LoIacono et al. (1992) found a higher mean placental cadmium level in cadmium-exposed women than in those who did not smoke. After controlling for potentially confounding variables there was no association between placental cadmium and birth weight. There was no measurement of any parameter of zinc stores. In contrast to these studies, Frery et al. (1993) found that, even at environmental levels, cadmium may induce a decrease in birth weight mediated by cadmium-induced placental calcification. The investigators measured cadmium in placental and hair samples collected from 102 mothers and their newborns and found a significant relationship between decrease in birth weight and increase in newborn hair cadmium levels, which varied in the presence of placental calcifications. In cases of parenchymal calcifications, placental cadmium levels were higher and newborn hair levels were lower than in the absence of calcification. In the absence of calcification, a decrease in birth weight was observed for the upper values of cadmium content in newborn hair. The difference in birth weight between infants in the first and last quadrilles was 472 g in cases of calcification and 122 g in the absence of calcification. These results were significant even after taking into account smoking habits and gestational age.

D. Summary

The placental interface between maternal and fetal blood is in the chorionic villus and consists of capillary endothelial cells and cytotrophoblasts. Human and rodent placentas are hemochorial, where fetal chorionic villi bathe in lacunae of maternal blood. However, they differ in that rat chorionic villi contain a second layer of cytotrophoblasts. Sampling of human placenta should be from multiple sites because of variation in maturity in different areas and foci of degeneration.

Transfer of lead across the placenta is by diffusion. Maternal and fetal blood lead levels are similar although fetal blood lead is usually slightly but significantly lower at low blood lead levels. Umbilical blood lead measurements serve as an estimate of fetal exposure. Diffusion of lead into fetal tissues has been shown to occur as early as the end of the first trimester and may occur earlier. Lead content of fetal tissues tends to parallel that of calcium. Lead in the brain replaces calcium in enzyme systems, interfering with dentritic branching and neurotransmitter activity. Birth weight, birth length, and length of gestation are inversely related to umbilical cord blood lead levels.

Umbilical cord blood levels of cadmium tend to be 40%–70% of maternal blood levels, indicating that the placenta provides an incomplete barrier to fetal cadmium exposure. However, one study has shown a positive correlation between maternal blood cadmium levels and placental cadmium concentration. Human infants are born with very low tissue levels of cadmium.

Experimental studies have shown that cadmium can induce metallothionein synthesis in the placenta, particularly in trophoblasts. Metallothionein in the placenta binds the essential trace metals, zinc and copper. The mechanism whereby essential metals are preferentially transported to the fetus and cadmium retained is not known. Experimental studies have shown that exposure of pregnant animals to high levels of cadmium early in pregnancy is teratogenic. Human studies suggest maternal cadmium exposure may reduce fetal birth weight and that this may be an indirect effect of zinc deprivation that occurs with cadmium exposure.

References

Ahokas RA, Dilts PV (1979) Cadmium uptake by the rat embryo as a function of gestational age. Am J Obstet Gynecol 135:219–222

Alexander FW, Delves HT (1981) Blood lead levels during pregnancy. Int Arch Occup Environ Health 48:35–39

Alessio L, Dell'orto A, Calzaferri G, Buscaglia M, Motta G, Rizzo M (1984) Cadmium concentrations in blood and urine of pregnant women at delivery and their offspring. Sci Total Environ 34:261–266

Baghurst PA, McMichael AJ, Wigg NR, Vimpani GV, Robertson EF, Roberts RJ, Tong S-L (1992) Life-long exposure to environmental lead and children's intelligence at age seven. N Engl J Med 327:1269–1284

Barltrop D (1969) Transfer of lead to the human foetus. In: Barltrop D, Burland W (eds) Mineral metabolism in pediatrics. Blackwell Science, Oxford, pp 135–151
Beck F (1981) Comparative placental morphology and function. In: Kimmel CA, Buelke-Sam J (eds) Developmental, toxicology. Raven, New York, pp 35–54
Bellinger D, Leviton A, Waternaux C, Needleman H, Rabinowitz M (1987) Longitudunal analyses of prenatal and postnatal lead exposure and early cognitive development. N Engl J Med 316:1037–1043
Berlin M, Blanks R, Catton M, Kazantzis G, Mottet NK, Samiullah Y (1992) Birth weight of children and cadmium accumulation in placentas of female nickel-cadmium (long-life) battery workers. In: Nordberg GF, Herber RFM, Alessio L (eds) Cadmium in the human environment: toxicity and carcinogenicity. International Agency for Research on Cancer, Lyon, pp 257–262
Bhattacharyya MH, Whelton BD, Peterson DP, Carnes BW, Guran MS, Moretti ES (1988) Kidney changes in multiparous mice fed a nutrient-sufficient diet containing cadmium. Toxicology 50:205–215
Boadi WY, Yannai S, Urbach J, Brandes JM, Summer KH (1991) Transfer and accumulation of cadmium at the level of metallothionein in the perfused placenta. Arch Toxicol 65:318–323
Borscheim RL, Grote J, Mitchell T, Succop PA, Dietrich KN, Krafft KM, Hammond PB (1989) Effects of prenatal lead exposure on infant size at birth. In: Smith MA, Grant LD, Sors AI (eds) Lead exposure and child development. Kluwer Academic, Boston, pp 307–319
Bull RJ, McCauley PT, Taylor DH, Croften KM (1983) The effects of lead on the developing nervous system of the rat. Neurotoxicology 4:1–18
Carroll PT, Silbergeld EK, Goldberg AM (1977) Alteration of central cholinergic function by chronic lead exposure. Biochem Pharmacol 26:397–402
Cherian MG, Goyer RA, Richardson LS (1976) Cadmium metallothionein-induced nephropathy. Toxicol Appl Pharmacol 38:399–408
Chernoff N (1973) Teratogenic effects of cadmium in rats. Teratology 8:29–32
Cookman GR, King W, Regan CM (1987) Chronic low-level lead exposure impairs embryonic to adult conversion of the neural adhesion molecule. J Neurochem 49:399–403
Cookman GR, Hemmens SE, Keane GJ, King WB, Regan CM (1988) Chronic low level lead exposure precociously induces rat glial development in vitro and in vivo. Neurosci Lett 86:33–37
Danielsson BRG, Denker L (1984) Effects of cadmium on placental uptake and transport to the fetus of nutrients. Biol Res Pregnancy 5:93–101
De SK, McMaster MT, Dey SK, Andrews GK (1989) Cell-specific metallothionein gene expression in mouse decidua and placentae. Development 107:611–621
Dencker L, Daniesson B, Khayat A, Lindgren A (1983) Deposition of metals in the embryo and fetus. In: Clarkson TW, Nordberg GF, Sager PR (eds) Reproductive and developmental toxicity of metals. Plenum, New York, pp 607–632
Dietrich KN, Succop RA, Bornschein RL, Krafft KM, Berger O, Hammond PB, Buncher CR (1990) Lead exposure and neurobehavioral development in later infancy. Environ Health Perspect 89:13–20
Ernhart CB, Wolf AW, Kennard MJ, Erhard P, Filipovich HF, Sokol RJ (1986) Intrauterine exposure to low levels of lead: the status of the neonate. Arch Environ Health 41:287–291
Feldman AL, Failla ML, Cousins RJ (1978) Degradation of rat liver metallothioneins in vitro. Biochim Biophys Acta 544:638–646
Ferm VH, Carpenter SJ (1968) The relationship of cadmium and zinc in experimental mammalian teratogenesis. Lab Invest 18:429–432
Frery N, Nessman C, Girard F, Lafond J, Moreau T, Blot P, Lellouch J, Huel G (1993) Environmental exposure to cadmium and human birthweight. Toxicology 79:109–118

Garry VF, Pohlman BL, Wick MR, Garvey JS, Zeisler R (1986) Chronic cadmium intoxication: tissue response in an occupationally exposed patient. Am J Ind Med 10:153–161
Goyer RA (1990) Transplacental transport of lead. Environ Health Perspect 89:101–105
Goyer RA, Cherian MG (1992) Role of metallothionein in human placenta and rats exposed to cadmium. In: Nordberg GE, Herber RFM, Alessio L (eds) Cadmium in the human environment: toxicity and carcinogenicity. International Agecy for Research on Cancer, Lyon, pp 239–247
Goyer RA, Cherian MG, Delaquerriere-Richardson L (1984) Correlation of parameters of cadmium exposure with onset of cadmium-induced nephropathy in rat. J Environ Pathol Toxicol Oncol 5:89–100
Goyer RA, Haust MD, Cherian MG (1992) Cellular localization of metallothionein in human term placenta. Placenta 13:349–355
Gulson B, Mahaffey KR (1993) Contribution of tissue lead to blood lead of adult female subjects bases on stable lead isotope methods (unpublished)
Hanlon DP, Specht C, Ferm VH (1982) The chemical status of cadmium ion in the placenta. Environ Res 27:89–94
Hurley LS, Gowan J, Swenerton H (1971) Teratogenic effects of short-term and transitory zinc deficiency in rats. Teratology 4:199–204
Karin M, Slater EP, Herschman HR (1981) Regulation of metallothionein synthesis in HeLa cells by heavy metals and glucocorticoids. J Cell Physiol 106:63–74
Karp B, Robertson AF (1981) Cadmium, the placenta, and the infant. In: Nriagu JO (ed) Cadmium in the environment, part II, health effects. Wiley, New York, pp 729–742
Kelman BJ, Walter BK (1980) Transplacental movements of inorganic lead from mother to fetus. Proc Soc Exp Biol Med 163:278–282
Khera AK, Wibberley DG, Dathan JG (1980) Placental and stillbirth tissue lead concentrations in occupationally exposed women. Br J Ind Med 37:394–396
Korpela H, Loueniva R, Yrjanheikki E, Kauppila A (1986) Lead and cadmium concentrations in maternal and umbilical cord blood, amniotic fluid, placenta, and amniotic membranes. Am J Obstet Gynecol 155:1086–1089
Kuhnert BR, Kuhnert PM, Zarlingo TJ (1988) Associations between placental cadmium and zinc and age and parity in pregnant women who smoke. Obstet Gynecol 71:67–70
Kuhnert PM, Kuhnert BR, Bottoms SF, Erhard P (1982) Cadmium levels in maternal blood, fetal cord blood, and placental tissues of pregnant women who smoke. Am J Obstet Gynecol 142:1021–1025
Lagerkvist BJ, Nordberg GF, Soderberg HA, Ekesrydh S, Englyst V, Gustavsson M, Gustavsson NO, Wiklund DE (1992) Placental transfer of cadmium. In: Nordberg GF, Herber RFM, Alessio L (eds) Cadmium in the human environment: toxicity and carcinogenicity. IARC scientific publications, Lyon, pp 287–291
Lane RE (1949) The care of the lead worker. Br J Ind Med 6:125–143
Lauwerys R, Buchet JP, Roels H, Hubermont G (1978) Placental transfer of lead, mercury, cadmium and carbon monoxide in women. I. Comparison of the frequency distributions of the biological indices in maternal and umbilical cord blood. Environ Res 15:278–289
Levin AA, Kilpper RW, Miller RK (1987) Fetal toxicity of cadmium chloride: the pharmacokinetics in the Wistar rat. Teratology 36:163–170
LoIacono NJ, Graziano JH, Kline JK, Popovac D, Ahmedi G, Gashi E, Mehmedi A, Rajovic B (1992) Placental cadmium and birthweight in women living near a lead smelter. Arch Environ Health 47:250–255
Lucis OJ, Lucis R, Shaik ZA (1972) Cadmium and zinc in pregnancy and lactation. Arch Environ Health 25:14–22

Mahaffey KR, Goyer R, Haseman JK (1973) Dose-response to lead ingestion in rats fed low dietary calcium. J Lab Clin Med 82:92–100
Manci EA, Blackburn WR (1987) Regional variations in the levels of zinc, iron, copper, and calcium in the term human placenta. Placenta 8:497–502
Markovac J, Goldstein GW (1988) Lead activates protein kinase C in immature rat brain microvessels. Toxicol Appl Pharmacol 96:14–23
McMichael AJ, Vimpani GV, Robertson EF, Baghurst PA, Clark PD (1986) The Port Pirie study: maternal blood lead and pregnancy outcome. J Epidemiol Community Health 40:18–25
McMichael AJ, Baghurst PA, Wigg NR, Vimpani GV, Robertson EF, Roberts RJ (1988) Port Pirie Cohort study: environmental exposure to lead and children's abilities at the age of four years. N Engl J Med 319:468–475
Miller RK (1986) Placental transfer and function: the interface for drugs and chemicals in the conceptus. In: Fabro S, Scialli AR (eds) Drug and chemical action in pregnancy pharmacological and toxicological principles. Dekker, New York, pp 123–152
Moore MR (1980) Prenatal exposure to lead and mental retardation. In: Needleman ML (ed) Low level lead exposure. Raven, New York, pp 53–65
Moore MR, Goldberg A, Pocock SJ, Mereith PA, Stewart IM, McAnespie H, Lees R, Low A (1982) Some studies of maternal and infant lead exposure in Glasgow. Scott Med J 27:113–122
Mushak P (1989) Biological monitoring of lead exposure in children: overview of selected biokinetic and toxicological issues. In: Smith MA, Grant LD, Sors AI (eds) Lead exposure and child development. Kluwer Academic, Boston, pp 129–147
Oliver T (1914) Lead poisoning: from the industrial, medical and social points of view. Lewis, London, pp 1–294
Parizek J (1964) Vascular changes at sites of oestrogen biosynthesis produced by parenteral injection of cadmium salts: the destruction of placenta by cadmium salts. J Reprod Fertil 7:263–265
Parizek J, Ostadalova I, Benes I, Pitha J (1968) The effect of a subcutaneous injection of cadmium salts on the ovaries of adult rats in persistent oestrus. J Reprod Fertil 17:559–562
Parzyk DC, Shaw SM, Kessler WV, Vetter RJ, Van Sickle DC, Mayes RA (1978) Fetal effects of cadmium in pregnant rats on normal and zinc-deficient diets. Bull Environ Contam Toxicol 19:206–214
Rabinowitz MB, Needleman HL (1982) Temporal trends in the lead concentrations of umbilical cord blood. Science 216:1429–1431
Rossouw J, Offermeier J, Van Rooyen JM (1987) Apparent central neurotransmitter receptor changes induced by low-level lead exposure during different developmental phases in the rat. Toxicol Appl Pharmacol 91:132–139
Samarawickrama GP, Webb M (1979) Acute effects of cadmium on the pregnant rat and embryo-fetal development. Environ Health Perspect 28:245–249
Scharpf LG Jr, Hill ID, Wright PL, Plank JB, Keplinger ML, Calandra JC (1972) Effect of sodium nitrilotriacetate on toxicity, and tissue distribution of cadmium. Nature 239:231–234
Scheuhammer AM, Cherian MG (1986) Quantification of metallothioneins by a silver-saturation method. Toxicol Appl Pharmacol 82:417–425
Schramel P, Hasse S, Ovcar-Pavlu J (1988) Selenium, cadmium, lead and mercury concentrations in human breast milk, in placenta, maternal blood, and the blood of the newborn. Biol Trace Elem Res 15:111–124
Silbergeld EK (1992) Mechanisms of lead neurotoxicity, or looking beyond the lamppost. FASEB J 6:3201–3206
Silbergeld EK, Miller IP, Kennedy S, Eng N (1979) Lead, GABA and seizures: effects of subencephalic lead exposure on seizure sensitivity and GABAergic function. Environ Res 19:371–382

Simons TJB (1986) Cellular interactions between lead and calcium. Br Med Bull 42:431–434
Sonawane BR, Nordberg M, Nordberg GF, Lucier GW (1975) Placental transfer of cadmium in rats: influence of dose and gestational age. Environ Health Perspect 12:97–102
Sorrell TL, Graziano JH (1990) Effect of oral cadmium exposure during pregnancy on maternal and fetal zinc metabolism in the rat. Toxicol Appl Pharmacol 102:537–545
Steven D (1975) Anatomy of the placental barrier. In: Steven DH (ed) Comparative placentation. Essays in structure and function. Academic, pp 25–57
Toews AD, Kolber A, Hayward J, Kingman MR, Morrell P (1978) Experimental lead encephalopathy in the suckling rat: concentration of lead in cellular fractions enriched in brain capillaries. Brain Res 147:131–138
Tsuchiya H, Mitani K, Kodama K, Nakata T (1984) Placental transfer of heavy metals in normal pregnant Japanese women. Arch Environ Health 39:11–17
Van Dijk HP (1981) Active transfer of the plasma bound compounds calcium and iron across the placenta. In: Wallenburg HCS, Van Kreel BK, Van Dijk JK (eds) Transfer across the primate and non-primate placenta. Saunders, London, pp 139–164
Waalkes MP, Poisner AM, Wood GW, Klaassen CD (1984) Metallothionein-like proteins in human placenta and fetal membranes. Toxicol Appl Pharmacol 74:179–184
Yeh I-T, Kurman RJ (1989) Functional and morphologic expressions of trophoblast. Lab Invest 61:1–4

CHAPTER 2

Porphyrin Metabolism as Indicator of Metal Exposure and Toxicity

J.S. Woods

A. Introduction

Numerous studies during the past several decades have demonstrated that porphyrins and other constituents of the heme biosynthetic pathway might serve as sensitive and specific biomarkers of toxic metal exposures in human subjects. Porphyrins (in the reduced form, porphyrinogens) are formed as intermediates of heme biosynthesis in essentially all eukaryotic tissues and are readily measured following extraction in the oxidized form (porphyrins) in blood cells, urine, feces, and other accessible tissues. The utility of porphyrins as biomarkers of metal exposures is based largely on the properties of metals to selectively alter porphyrinogen metabolism in target tissues by mechanisms which lead to metal-specific changes in urinary porphyrin excretion patterns. Of particular importance with respect to the utility of porphyrins as biomarkers of metal effects in target tissues is the property of some specific metals, not only to impair porphyrin(ogen) metabolism, but also to facilitate the oxidation of reduced porphyrins which subsequently accumulate in tissue cells. Evidence indicates that the prooxidant action of metals which promotes porphyrinogen oxidation may also underlie the oxidation of other cellular constituents, such as lipids and proteins, a postulated cause of cell injury. Hence, a common mechanistic etiology underlying the porphyrinogenic and tissue-damaging properties of metals provides the rationale for use of porphyrin measurements as an indicator of metal exposure as well as potential toxicity.

This chapter describes the mechanistic basis of metal-induced alterations in porphyrin metabolism as derived from experimental studies and summarizes the evidence supporting the utility of porphyrin measurements as a biomarker of exposure and effects of specific metals in human subjects, with specific emphasis on lead, mercury, and arsenic. Some perspectives and research needs regarding the use of porphyrin measurements as biomarkers of metal exposures and effects in human populations are also presented.

B. Heme Biosynthesis and Porphyrin Metabolism

The heme biosynthetic pathway, described in Fig. 1, functions in all nucleated cells to provide heme, chlorophyll, and related structures. In mam-

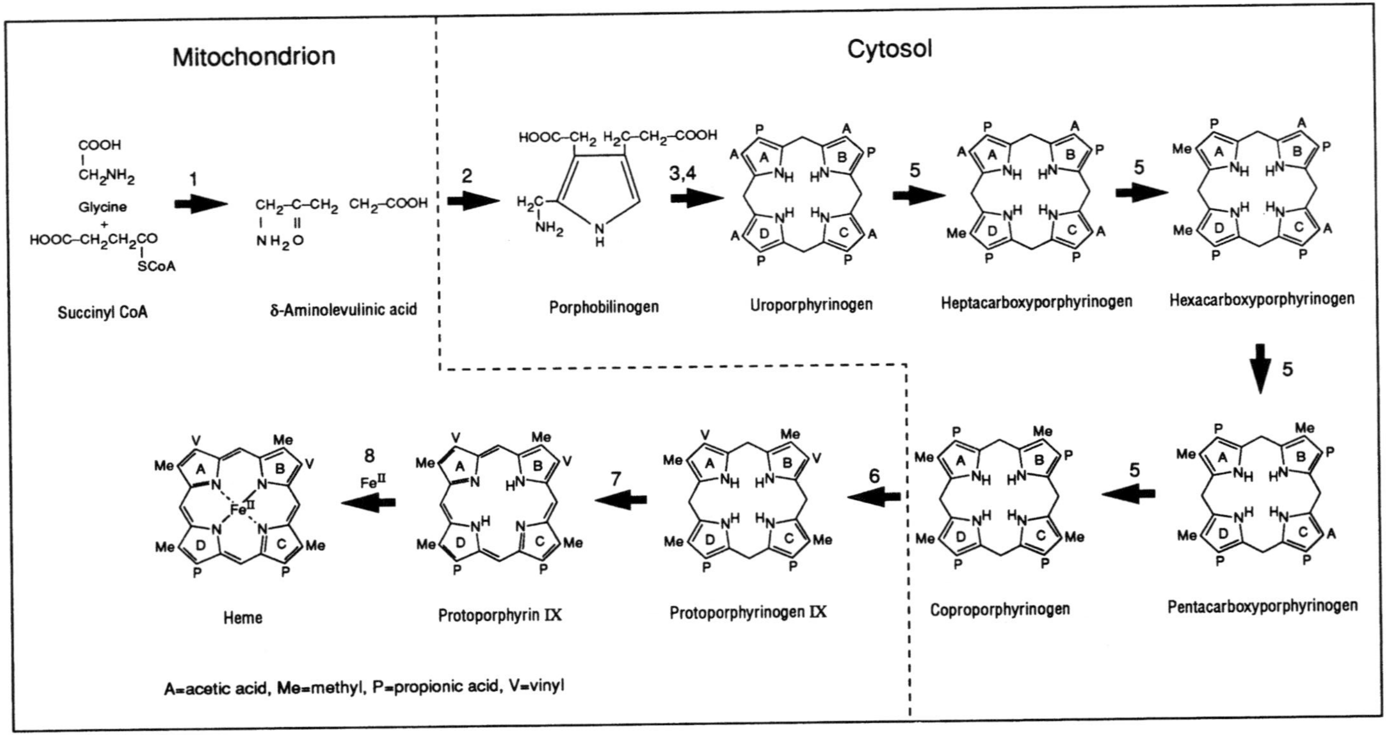

Fig. 1. Heme biosynthetic pathway. Steps are catalyzed by *1*, δ-aminolevulinic acid (ALA) synthetase; *2*, ALA dehydratase; *3*, uroporphyrinogen I synthetase (PBG deaminase); *4*, uroporphyrinogen III cosynthetase; *5*, uroporphyrinogen decarboxylase; *6*, coproporphyrinogen oxidase; *7*, protoporphyrinogen oxidase; and *8*, ferrochelatase (heme synthetase)

malian and avian tissues the principal product of this pathway is heme (protoheme, iron protoporphyrin IX), an essential component of oxygen transport systems, mixed function oxidative reactions, and other oxidative metabolic processes.

The heme biosynthetic pathway utilizes glycine and succinyl coenzyme A as substrates to form the 5-carbon aminoketone, δ-aminolevulinic acid (ALA). This step is catalyzed by ALA synthetase (EC 2.3.1.37), a mitochondrial enzyme which is generally considered rate-limiting in this process in normal adult tissues (GRANICK and SASSA 1971). ALA then passes into the cytoplasm of the cell where two molecules are condensed by ALA dehydratase (porphobilinogen synthase) (EC 4.2.1.24) to form the pyrrole, porphobilinogen (PBG). Two cytoplasmic enzymes, PBG deaminase (uroporphyrinogen I synthetase) (EC 4.3.1.8) and uroporphyrinogen III cosynthetase, then act in concert to convert four molecules of PBG to 8-carboxyl porphyrinogen (uroporphyrinogen). The physiologically relevant isomeric form of 8-carboxyl porphyrinogen is uroporphyrinogen III, in which the propionic acid and acetic acid side chains of the "D" ring are in reversed configuration as compared with the arrangement of these groups on the other three pyrrole rings of the molecule (Fig. 2). The III isomeric form of uroporphyrinogen serves as substrate for another cytoplasmic enzyme, uroporphyrinogen decarboxylase (porphyrinogen carboxylase) (EC 4.1.1.37), which converts the four acetic acid side chains of uroporphyrinogen III to methyl groups, yielding the 4-carboxyl porphyrinogen, coproporphyrinogen III.

The remaining steps of the heme biosynthetic pathway, by which coproporphyrinogen is converted to heme, are mediated by three mitochondrial enzymes. Coproporphyrinogen oxidase (EC 1.3.3.3), which is bound to the mitochondrial inner membrane (ELDER and EVANS 1978), decarboxylates the

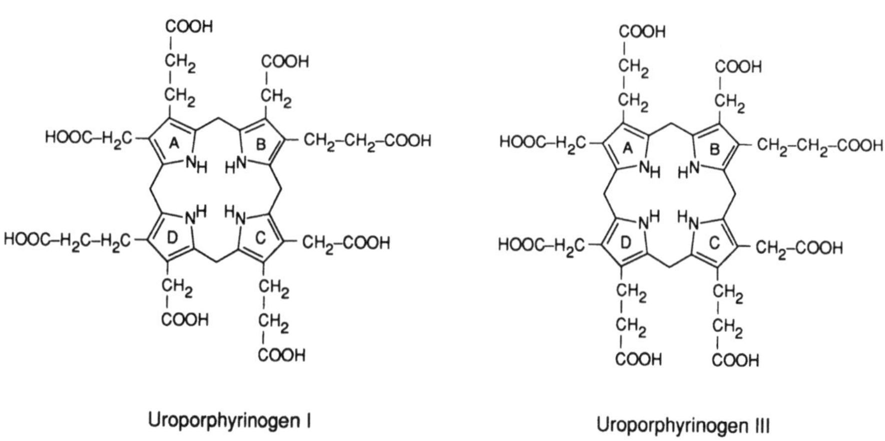

Fig. 2. Structures of the I and III isomers of uroporphyrinogen

two propionic acid side chains of rings "A" and "B" of coproporphyrinogen to vinyl groups, yielding protoporphyrinogen "IX". An intermediate 3-carboxyl porphyrinogen with a single vinyl group in the A ring, harderoporphyrinogen, is formed in the process. Protoporphyrinogen oxidase (EC 1.3.3.4), also associated with the mitochondrial inner membrane (Poulson 1976), then oxidatively cleaves six hydrogen atoms from protoporphyrinogen IX to yield protoporphyrin IX. In the final step, Fe^{2+} is incorporated into the protoporphyrin ring by the inner mitochondrial membranal enzyme, ferrochelatase (heme synthetase) (EC 4.99.1.1) (Jones and Jones 1969).

Porphyrinogen synthesis in most tissues typically occurs in excess of that required for heme biosynthesis, with excess porphyrinogens accumulating in tissue cells. Porphyrinogens are readily oxidized by products of oxidative metabolism to the corresponding porphyrins, and are excreted in urine or feces consistent with their relative hydrophilicity. Thus, porphyrins with 8, 7, 6, and 5 carboxyl groups are excreted largely in the urine, whereas the less polar 2-carboxyl porphyrin (protoporphyrin) is excreted almost exclusively in the feces. 4-Carboxyl porphyrin (coproporphyrin) is found in relatively high concentrations in both urine and feces. In the absence of inherited abnormalities in heme biosynthetic pathway enzyme levels or chemical-induced modification of heme metabolism, porphyrins with 8 through 4 carboxyl groups are excreted into the urine in a well-extablished pattern. In normal human urine coproporphyrin constitutes 70%–80% of total porphyrin content, whereas uroporphyrin makes up 8%–15%. Hepta-, hexa- and pentacarboxyl porphyrins comprise approximately 4%, 1%–2%, and 5% of the total urinary porphyrins, respectively. Total urinary porphyrins have been reported to vary between 20 and 40 μg/l (Borup et al. 1980). Woods et al. (1993) have reported ranges of urinary porphyrin concentrations in both 24-h and spot urine samples collected from healthy adult male subjects. These are presented in Table 1. These values are

Table 1. Normal human urine porphyrin concentrations from 24-h and randomly collected (spot) urine samples

Porphyrin	Urinary porphyrin concentation Means ± SD and (ranges)	
	24-h samples (nmol/24 h)	Random (spot) samples (μg/l)
Uro-	6.0 ± 1.8 (3.2 – 18.5)	3.0 ± 1.0 (1.7 – 19.6)
Hepta-	3.8 ± 1.5 (1.0 – 7.8)	3.2 ± 0.6 (0.6 – 5.4)
Hexa-	1.0 ± 1.0 (0 – 3.5)	1.5 ± 1.2 (0 – 3.7)
Penta-	3.3 ± 3.3 (0.4 – 15.4)	2.2 ± 1.5 (0 – 4.7)
Copro-	52.0 ± 11.4 (29.8 – 92.6)	27.5 ± 16.9 (11–61)

Porphyrin levels as measured in urine samples collected from healthy adult male subjects presenting to this laboratory during the period 1990–1992 for research purposes (Bowers et al. 1992). Values represent the means and standard deviations of 34 individual determinations for 24-h samples and 72 individual determinations for spot samples.

consistent with those employed as reference standards for clinical urinary porphyrin determinations (Mayo Medical Laboratories 1990).

C. Mechanistic Basis of Metal-Induced Porphyria (Porphyrinuria)

Studies by a number of investigators using experimental animal models have demonstrated metal-specific changes in urinary porphyrin excretion patterns during prolonged exposure to metals such as mercury, arsenic, and lead (see reviews by MARKS 1985; WOODS 1989). Investigations on the mechanisms underlying the porphyrinogenic action of these metals have demonstrated that metal-mediated impairment of specific steps of the heme biosynthetic pathway constitutes the principal mechanism underlying their porphyrinogenic effects. In addition, evidence indicates that some metals can promote the oxidation of reduced porphyrins which accumulate in target tissue cells because of their impaired metabolism, accentuating the porphyrinogenic response obtained. The differential properties of metals with respect to their interaction with specific sites of porphyrin metabolism in target tissues underlie the uniqueness of the porphyrinogenic response observed.

I. Metal Effects on Specific Steps of the Heme Biosynthetic Pathway

Metals may impair the metabolism of porphyrins and other heme precursors by direct or indirect compromise of heme pathway enzymes or by reducing the availability of substrates and/or cofactors required for enzyme function.

All eight steps of the heme biosynthetic pathway are catalyzed by enzymes which require functional sulfhydryl (-SH) groups for optimal catalytic activity, either as part of the active site configuration or to maintain their structural integrity. Since most metals have a strong affinity for nucleophilic ligands, each step of the heme biosynthetic pathway is theoretically susceptible to direct inhibition as a result of metal-mercaptide bond formation with functional SH groups. Several heme biosynthetic pathway enzymes have been shown to be particularly susceptible to impairment by specific metals in vitro. Evidence for direct metal-enzyme interactions in vivo, however, is less well substantiated.

ALA dehydratase, the second enzyme of the heme biosynthetic pathway, is inhibited in vitro by metals and other SH-directed agents. Inhibition of ALA dehydratase in liver and erythrocytes in vitro has been observed with silver, iron, manganese, copper, and zinc ions (GIBSON et al. 1955) and in vivo with lead (SASSA 1978) and cobalt (NAKEMURA et al. 1975). The inhibition of ALA dehydratase in erythrocytes by lead in vivo is well established, and this effect is considered a highly sensitive measure of lead exposure in human subjects (considered further in Sect. D.I, below).

Porphobilinogen deaminase, which catalyzes the third step in heme biosynthesis, is also readily inhibited by numerous trace metals in vitro. Lead is a particularly effective inhibitor (PIPER and VAN LIER 1977; ANDERSON and DESNICK 1980). Mercury, cadmium, copper, iron, magnesium, and calcium also impair the purified enzyme, although to a lesser degree than lead (ANDERSON and DESNICK 1980). Studies in vivo have shown that platinum significantly inhibits renal PBG deaminase in kidney (MAINES and KAPPAS 1977). Uroporphyrinogen III cosynthetase has also been shown to be directly inhibited by various metals in vitro, including K^{1+}, Na^{1+}, Mg^{2+}, Ca^{2+}, Zn^{2+}, Cd^{2+}, and Cu^{2+} (PIPER et al. 1983; CLEMENT et al. 1982).

Uroporphyrinogen (urogen) decarboxylase from all tissues thus far studied has an absolute requirement for free SH groups for catalytic activity (ELDER et al. 1983; ELDER and URQUHART 1984), and is readily inhibited by numerous trace metals. Among the most potent of the metals evaluated is Hg^{2+}, which effectively eliminates enzyme activity in vitro in concentrations as low as $100 \mu M$ (WOODS et al. 1984). Urogen decarboxylase from rat kidney appears to be much more readily inhibited by Hg^{2+} than the enzyme from liver, by a factor of at least 10 when measured in crude enzyme preparations from rat tissues (WOODS et al. 1984). However, prolonged exposure of animals to mercury compounds does not elicit uroporphyrinuria, as would be expected if urogen decarboxylase were inhibited by Hg^{2+} in vivo. Of note, however, is the observation that prolonged mercury exposure elicits a porphyrinuria characterized by substantially increased urinary concentrations of 5-carboxyl porphyrin (described further in Sect. D.II, below), suggesting impairment of the fourth decarboxylation site catalyzed by urogen decarboxylase (WOODS et al. 1984).

Coproporphyrinogen (coprogen) oxidase is inhibited by numerous trace metals in vitro (BATLLE et al. 1965; KARDISH et al. 1980; ROSSI et al. 1992). In a study of crude human lymphocyte coprogen oxidase, ROSSI et al. (1992) tested seven metals on enzyme activity at concentrations of $50 \mu M$. Enzyme activity was completely inhibited by Hg^{2+}, whereas Pb^{2+}, Mn^{2+}, Sn^{2+}, Ni^{2+}, and Al^{3+} had no effect. Three organometal compounds were also tested at concentrations of $100 \mu M$; tributyltin and methyl mercury inhibited the enzyme, whereas tetraethyl lead had no effect. Additionally, mercury appears to directly inhibit coprogen oxidase in kidney and other tissue in vivo (WOODS and SOUTHERN 1989; ROSSI et al. 1992), although other studies have suggested that impairment of coproporphyrinogen metabolism may reflect damage to the mitochondrial membrane with which coprogen oxidase is associated in tissue cells, rather than direct inhibition of the enzyme per se (FOWLER and WOODS 1977; WOODS and FOWLER 1987). Alternatively, impaired mitochondrial transport of coproporphyrinogen (ROSSI et al. 1993) or diminished cellular redox status via glutathione depletion (WOODS 1989; LUND et al. 1993) could account for diminished coproporphyrinogen metabolism and consequent coproporphyrinuria.

Ferrochelatase, the final enzyme in heme biosynthesis, has also been shown to be susceptible to direct inhibition by a number of metals in vitro (LABBE and HUBBARD 1961; TEPHLY et al. 1971; TAKETANI and TOKUNAGA 1981; DAILEY and FLEMING 1983). As with coprogen oxidase, however, less evidence is available to support this effect in vivo. It has been noted that, since ferrochelatase, like coprogen oxidase, is an integral constituent of the mitochondrial inner membrane, it is possible that metals may indirectly compromise ferrochelatase activity by impairing the integrity of membranal components with which this enzyme is structurally associated in the intact cell (WOODS 1989). Evidence also supports the view that impairment of ferrochelatase may represent metal-mediated iron deficiency, resulting in decreased availability of Fe^{2+} as substrate for this enzyme (LABBE et al. 1987). This issue is also described in further detail in Sect. D.I, below.

II. Metal-Induced Oxidation of Reduced Porphyrins

Reduced porphyrins (porphyrinogens) are readily oxidized to the corresponding porphyrins by free radicals, including reactive oxygen species (O_2^-, [H_2O_2 + Fe^{2+}], $OH^\bullet$) in vitro (DE MATTEIS 1988; FRANCIS and SMITH 1988; WOODS 1988a; WOODS and CALAS 1989). The reduced and oxidized forms of uroporphyrin are shown in Fig. 3. It has been postulated that chemicals which promote the formation of endogenous reactive oxidants such as H_2O_2 promote porphyria via direct porphyrinogen oxidation in vivo (FERIOLO et al. 1984; SMITH and FRANCIS 1987; BONKOVSKY et al. 1987; MUKERJI and PIMSTONE 1990; SINCLIR et al. 1987; DE MATTEIS et al. 1988). Numerous studies (e.g., SUNDERMAN 1986; WOODS et al. 1990a,b) have shown that porphyrinogenic metals (e.g., Hg^{2+}; As^{3+}, As^{5+}, Pb^{2+}, Fe^{2+}/Fe^{3+}) can

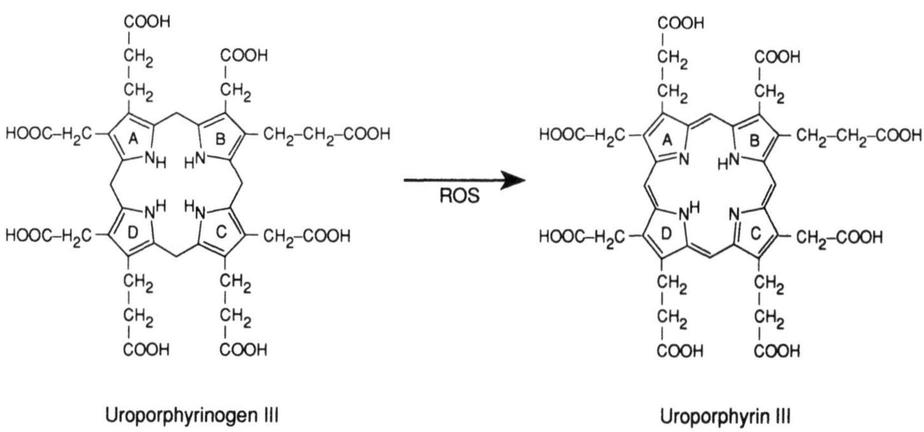

Fig. 3. Structures of uroporphyrinogen III and uroporphyrin III. *ROS*, reactive oxygen species

initiate oxidative events in vivo by uncoupling mitochondrial and microsomal electron transport processes, by increasing production of H_2O_2 and other endogenous reactive oxidants, and by redox cycling. These effects lead to free radical formation with consequent porphyrinogen oxidation and potential damage to cell constituents.

Mercury has been particularly well characterized in this respect (Cantoni et al. 1982; Woods et al. 1990a,b; Miller et al. 1991; Miller and Woods 1993; Lund et al. 1991, 1993). In the kidney, mercury (Hg^{2+}, CH_3Hg^+) both inhibits porphyrinogen metabolism at the levels of coporgen oxidase as well as facilitates the oxidation of the porphyrinogens which subsequently accumulate. This combined action of mercury results in dramatically increased concentrations of 4- and 5-carboxyl porphyrins in the urine (Woods et al. 1991). Similar processes are likely to be involved in the porphyrinogenic action of other metals, including As^{3+}, As^{5+}, Pb^{2+}, and Fe^{2+}/Fe^{3+}, which have also been shown to promote reactive oxidant formation in target tissue cells and the oxidation of cellular constituents via free radical-mediated mechanisms (Sunderman 1986; Quinlan et al. 1988; Hermes-Lima et al. 1991; Yamanaka et al. 1990).

The principal mechanisms underlying metal-induced porphyrinuria are schematically summarized in Fig. 4. Of importance from the biomonitoring

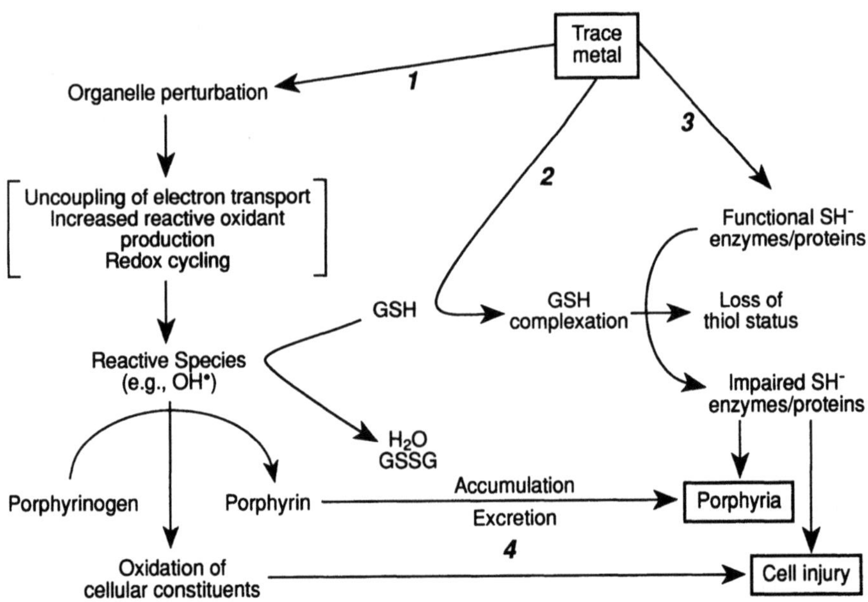

Fig. 4. Mechanisms of trace metal-induced porphyria and cell injury: *1*, metals promote increased reactive oxidant formation; *2*, metals complex with GSH, compromising antioxidant and thiol status; *3*, metals impair SH-dependent enzymes and other proteins via mercaptide bond formation and/or exchange reactions; *4*, metal-induced oxidant stress causes oxidation of reduced porphyrins (porphyrinuria) and other biomolecules (cell injury)

perspective is the concept portrayed in Fig. 4 that the properties of metals which cause the porphyrinogenic response (i.e., their affinity for mercaptic formation with cellular ligands and their pro-oxidant potential) may also underlie concurrent or subsequent oxidative damage to other cellular constituents with resultant toxicity. In this respect, the porphyrinogenic and tissuedamaging properties of metals can be viewed as sharing a common mechanistic etiology, supporting the utility of urinary porphyrin measurements as an indicator of metal exposure as well as of ensuing or prevailing toxicity. Studies supporting this concept are described further in Sect. D.II, below.

D. Metal- and Metalloid-Induced Porphyrinopathies and Porphyrinurias

I. Lead

The diagnostic utility of porphyrins and other heme precursors as indicators of lead exposure has long been recognized (CHISOLM 1971; PIOMELLI and DAVIDOW 1972; FELL 1984), although the greatest use has been restricted to the diagnosis and management of clinical lead poisoning, compatible with blood levels in excess of 50 μg/dl. In recent years, it has been established that potentially adverse effects of lead to the nervous system, especially in undernourished children, may occur during exposures compatible with much lower blood lead levels, perhaps as low as 10 μg/dl (NEEDLEMAN and GATSONIS 1990; BAGHURST et al. 1992; ANGLE 1993; GOYER 1993). These findings support the need for reassessment of established lead biomarkers with respect to their sensitivity in detecting neurologic and other bioeffects at low levels of lead exposure.

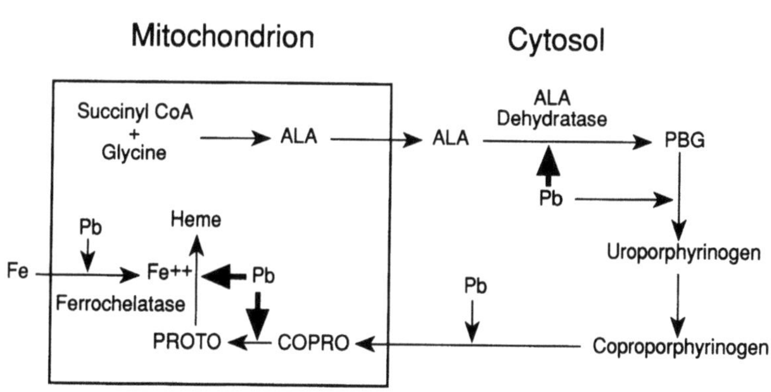

Fig. 5. Sites of lead interaction with the heme biosynthetic pathway. Principal sites are indicated by *heavy arrow*; other sites are indicated by *light arrows*. (Adapted from SASSA 1978)

Lead interacts with the heme biosynthetic pathway at three principal sites, as illustrated in Fig. 5. Additional sites of heme synthesis may also be affected by lead to a lesser degree (PIPER and TEPHLY 1974; FELL 1984). The affinity of lead for erythropoietic tissues predisposes to specific effects of lead at the sites of ALA and protoporphyrin utilization, with effects manifested principally in red blood cells. The preferential uptake of lead also by the kidney predisposes to the inhibition of coproporphyrinogen uptake or utilization by renal mitochondria, with consequent excretion of coproporphyrin in high levels in the urine.

1. Erythrocyte ALA Dehydratase

ALA dehydratase (ALAD) catalyzes the second step in heme biosynthesis in which two molecules of ALA are transformed into the pyrrole, PBG. ALAD is particularly susceptible to inhibition by metals such as lead which readily complex with vicinal dithiol groups at the active site (SHEEHRA et al. 1981). Moreover, the protein structure of ALAD contains 28 SH groups (FARANT and WIGFIELD 1987), enhancing its susceptibility to inhibition by lead through conformational changes. Additionally, ALAD is a Zn-dependent enzyme, and there is evidence that lead may displace Zn from the catalytic site, further compromising enzyme activity (JAFFE et al. 1991).

The inhibition of ALAD in erythrocytes serves as a highly sensitive indicator of environmental lead exposure in humans. As shown in Fig. 6,

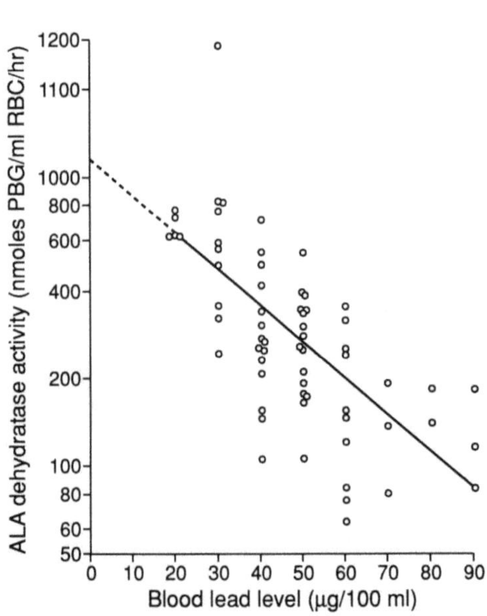

Fig. 6. Relationship between erythrocyte ALA dehydratase activity and blood lead level. (Adapted from SASSA 1978)

erythrocyte ALAD activity is inversely correlated with blood lead over a wide range of concentrations, the log of ALAD activity being linearly related to the blood lead concentration. This relationship was initially established in occupationally exposed workers (HAEGER-ARONSEN 1960; DE BRUIN 1968; HERNBERG et al. 1970), and has subsequently been demonstrated in urban populations as well as in populations residing near lead smelters (HERNBERG and NIKKANEN 1970; NORDMAN et al. 1973). This relationship has also been demonstrated among children living in lead-pulluted areas and ingesting large amounts of lead through pica (MILLAR et al. 1970; WEISBERG et al. 1971; ROELS et al. 1976).

ALAD may not be well suited as a biomarker of lead exposure at blood lead levels greater than 50 μg/dl. The decrease in ALAD activity is accompanied by a concomitant increase in the total amount of ALAD protein in erythrocyte precursors. At blood lead levels of 40–50 μg/dl the amount of ALAD reaches a plateau corresponding to about two times the normal concentration, probably reflecting the maximum capacity of ALAD synthesis (BERNARD and LAUWERYS 1987). The practical consequence of this phenomenon is that assessment of ALAD activity alone does not accurately reflect the extent of enzyme inhibition at higher blood lead levels.

The sensitivity of red cell ALAD at blood levels less than 10 μg/dl has also yet to be well documented. This issue is of considerable importance with regard to the utility of ALAD as a biomarker of lead effects at exposure levels which produce blood lead concentrations below this level. Erythrocyte ALAD activity in human subjects with blood lead levels at 10 μg/dl and below has been reported as ranging from ~75 to 180 U/ml red blood cells (HERNBERG et al. 1970). Given this wide range of interindividual variability among human subjects, a substantial decrease in enzyme activity could be required in order to detect a statistically significant effect of lead on ALAD activity. Kinetic studies have shown that lead inhibition of ALAD is primarily noncompetitive, with the apparent K_i for the inhibition of erythrocyte ALAD by lead being on the order of 0.5 μM or 10 μg/dl blood (SASSA 1978). Thus, a blood lead concentration of 10 μg/dl would produce a 50% reduction in ALAD activity. A reduction of this magnitude would be difficult to detect with a high degree of accuracy unless restricted to assessments of lead exposure among individual subjects or within a selected population in which the range of normal variability in ALAD activity is first well defined.

Genetically linked deficiencies of erythrocyte ALAD in human subjects have been described by a number of investigators (BIRD et al. 1979; DOSS and MULLER 1982; BENKMANN et al. 1983; ASTRIN et al. 1987; WETMUR et al. 1991). In normal individuals, ALAD activity is sufficiently high, even under conditions of moderate lead toxicity, to sustain heme biosynthesis without the prospect of clinical effects related to heme deficiency. In contrast, when ALAD levels are significantly reduced due to an inherited deficiency of this enzyme, affected individuals could be at greater risk of lead toxicity related to impaired heme biosynthesis (Doss and Muller 1982; Astrin et al. 1987). The presence of individual carriers of this deficiency in the general population contributes to the wide range in variability of ALAD activity observed. Identification of such individuals through genetic screening could decrease this variability and also serve to prevent lead toxicity related to impaired ALAD activity.

2. Erythrocyte Zinc-Protoporphyrin

Elevated erythrocyte protoporphyrin concentrations were described as a characteristic of lead exposure as early as 1946 (Watson 1946), and blood protoporphyrin assays have been employed almost since that time as an indicator of lead toxicity (Chisolm 1971; Piomelli et al. 1973). The procedure for measuring blood porphyrins was simplified considerably by the development of a microfluorometric procedure by Piomelli and colleagues in 1972 (Piomelli and Davidow 1972; Piomelli et al. 1972), and this method has remained the most widely utilized in the clinical setting. Initially, it was thought that free-base protoporphyrin was the principal porphyrin constituent extracted from red blood cells, and the term "free erythrocyte protoporphyrin" (FEP) was widely used to describe changes in blood porphyrins due to lead exposure. However, in 1974 Lamola and Yamane (1974) identified the principal porphyrin fluorescence in blood as attributable, not to FEP, but rather to the zinc chelate of protoporphyrin, zinc protoporphyrin (ZPP). Unlike FEP, which can elute from the red blood cell by diffusion through the membrane, ZPP is tightly bound to the heme pocket of the hemoglobin molecule and does not elute from the red cell into the tissues (Lamola et al. 1975). Under steady-state conditions of exposure, the logarithm of ZPP is linearly related with blood lead level (Fig. 7).

The biochemical etiology of ZPP in erythrocytes during lead exposure was held for some time to have its mechanistic basis in the direct inhibition by lead of ferrochelatase. However, evidence that Pb^{2+} directly inhibits ferrochelatase, particularly in vivo, is not conclusive. Labbe and Hubbard (1961) first described the inhibition of ferrochelatase from rat liver by lead in vitro, although at concentrations unlikely to be achieved in reticulocytes in vivo. Similar findings were reported by Tephly et al. (1971). Fowler et al. (1980) reported decreased ferrochelatase activity to 63% of control levels in kidneys of rats with mean blood lead levels of 71 μg/dl following exposure to 250 ppm lead acetate in drinking water for 9 months. In contrast, Henderson and Toothill (1983) reported no decrease in ferrochelatase activity in kidneys of rabbits administered lead acetate at doses sufficient to produce mean blood lead levels of 93 μg/dl. More recently, Piomelli et al. (1987) reported that lead competitively inhibits ferrochelatase in human reticulocytes in vitro by binding to the enzyme in such a way as to modify the conformation of the active site. However, the effect of lead was shown the occur only when iron was not already bound to the active site of the enzyme. These findings suggest that, while lead may directly inhibit fer-

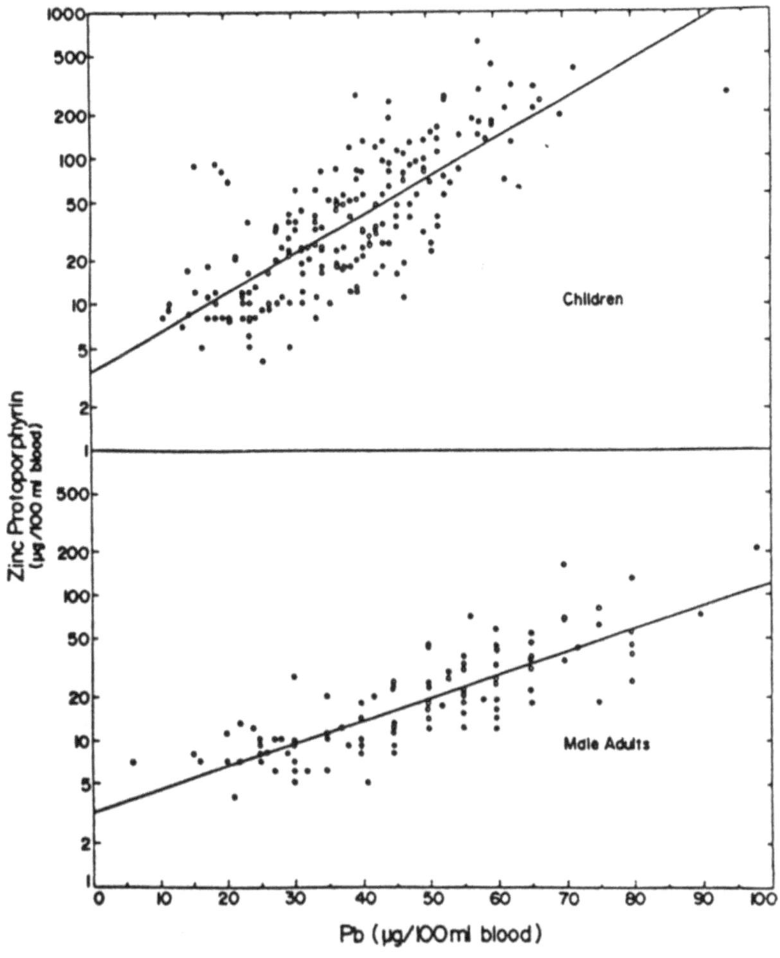

Fig. 7. Relationship between erythrocyte ZPP and blood lead level. (Adapted from LAMOLA et al. 1975)

rochelatase in vivo, this effect is expressed or pronounced only in the case of iron deficiency.

An alternate mechanism to explain the etiology of elevated erythrocyte ZPP during lead exposure has its basis in the relative affinities of iron and zinc as substrates for ferrochelatase. Ferrous iron is the preferential substrate of ferrochelatase and is also an effective inhibitor of zinc utilization by this enzyme. However, when the concentration of iron as Fe^{2+} decreases to suboptimal levels, as in iron deficiency, zinc is utilized by ferrochelatase as a substrate. Studies have suggested that lead decreases the availability of Fe^{2+} as a substrate for ferrochelatase by inhibiting the enzymatic reduction of Fe^{3+} to Fe^{2+} within mitochondria (TAKETANI et al. 1985), as required for use by ferrochelatase. The result of this effect is the increased utilization of zinc as substrate by ferrochelatase, resulting in increased ZPP production. The effects of lead on iron availability and on ferrochelatase may, in fact, be inextricably linked, inasmuch as studies with bovine heart mitochondria have demonstrated that reduction of iron occurs in the same portion of the mitochondrial inner membrane (complex I of the electron transport chain) where ferrochelatase is localized (Taketani et al. 1986). Thus, reduction in the rate of iron use by ferrochelatase could be the result of inhibition of either step by lead. The implication of these findings is that lead mediates elevated erythrocyte ZPP levels primarily by reducing iron availability as a substrate for ferrochelatase, with a direct compromise on ferrochelatase activity in vivo occurring subsequent to iron depletion.

In practical usage, ZPP, expressed as the molar ratio of ZPP to hemoglobin (μmoles ZPP/moles hemoglobin), or, alternatively, as μmoles ZPP/100 ml blood, has been shown to be well correlated with blood lead levels only when the blood lead concentration is more than 40 μg/dl (Haeger-Aronson 1982). Zwennis et al. (1990) described the relationship between the concentration of lead in blood and ZPP, expressed as the ratio to hemoglobin, among men who had been occupationally exposed to lead for more than 1 year. The 95% confidence intervals for blood lead for a given value of ZPP were too large to permit estimation of blood lead on the basis of ZPP measurements, and it was concluded that the ZPP procedure may be applicable only for identifying individuals with blood lead levels in excess of 50 μg/dl for the purpose of further evaluation. The lack of a high correlation between blood lead and ZPP at lower blood lead levels might be explained by the relatively high variability in the kinetics of lead absorption and ZPP formation among individuals, as seen among subjects with blood lead levels in the more commonly observed range of 5–30 μg/dl (Zwennis et al. 1990).

Efforts to improve the sensitivity of ZPP as a measure of low blood lead levels have met with limited success. Labbe et al. (1979) proposed the use of the ZPP/heme molar ratio (μmoles ZPP/moles heme) as a more sensitive and accurate indicator of erythrocyte porphyrin content, obviating the need to compensate for hematocrit variability in computation of ZPP concentrations. Sun et al. (1992) recently investigated the dose-effect relationship between blood and ZPP among lead-exposed smelter workers with blood lead levels between 6 and 153 μg/dl and developed a nonlinear regression model which improved substantially, as compared with a linear model, in estimating the threshold ZPP level at which a response due to lead exposure is detected. Also, Piomelli et al. (1975) have reported a threshold blood lead level between 14 and 17 μg/dl for a significant change in erythrocyte porphyrin content when the analysis was confined to a large population of children with blood lead levels in the range of 5–30 μg/dl. Therefore, as in the case of erythrocyte ALAD activity, ZPP may have improved diagnostic utility at low blood lead levels when performed on subjects from populations in which the underlying variability of erythrocyte ZPP has been initially well characterized, or when employed in a before/after context to determine individual responses to lead exposure.

3. Urinary Coproporphyrin

A third site of the heme biosynthetic pathway which is susceptible to alteration by lead is the site of conversion of coproporphyrinogen to protoporphyrinogen, catalyzed by coprogen oxidase (Fig. 5). The effect of lead at this site is an increase in the concentration of coproporphyrin in the urine. Studies on coprogen oxidase from various sources have characterized it as an SH-containing protein (YOSHINAGA and SANO 1980), although the extent to which functional SH groups are required for catalytic activity is not clear. The possibility exists that lead directly inhibits enzyme activity through mercaptide formation with functional SH groups. However, as in the case of ferrochelatase, discussed above, the evidence for this effect in vivo is inconclusive. BATLLE et al. (1965) showed that purified rat liver coprogen oxidase was inhibited by lead in vitro but only at lead concentrations as high as 20 m*M*. KARDISH et al. (1980) also showed that the activity of coprogen oxidase measured in rat liver mitochondria was reduced only in the presence of lead at concentrations as high as 10 m*M*. More recently, ROSSI et al. (1993) reported the lack of inhibition of coprogen oxidase from human lymphocytes by lead at 50 μM in reaction mixtures. In contrast, in studies in vivo, HENDERSON and TOOTHILL (1983) reported a decrease in coprogen oxidase to 30% of control values when masured in mitochondria from kidneys of rabbits administered lead acetate in drinking water sufficient to achieve blood lead levels of 93 μg/dl and a renal lead concentration of 41 μg/g protein. The latter study is of interest in suggesting that coprogen oxidase may be more readily susceptible to impairment by lead in vivo than in vitro. One explanation for this apparent inconsistency could be that lead indirectly affects coprogen oxidase in cells by compromising the structural integrity of membrane structures with which the enzyme is associated in tissue mitochondria. WOODS and FOWLER (1987) have provided evidence that disruption of membrane integrity characterized by mitochondrial swelling and fragmentation of mitochondrial inner membrane segments occurs at tissue lead concentrations as low as 10 μM, considerably below those which have been shown to inhibit coprogen oxidase activity in vitro. It is possible, therefore, that disruption of the structural association between coprogen oxidase and the mitochondrial membrane could account for the effect of lead on enzyme activity observed at lower lead concentrations.

Other studies have suggested that lead may also impair the transport of coproporphyrinogen into mitochondria subsequent to its synthesis by uroporphyrinogen decarboxylase in the cytoplasmic portion of the cell (ROSSI et al. 1993) (Fig. 5). The processes by which coproporphyrinogen is transported into the mitochondrion or through the outer to inner membrane spaces are not defined. If lead were to interfere with these processes, the oxidation and excretion of accumulated coproporphyrinogen could constitute the source of excess coproporphyrin in the urine.

Studies in experimental animal models have suggested that the effect of lead to promote coproporphyrin excretion occurs predominantly in the kidney (Henderson and Toothill 1983; Haeger-Aronsen 1960; Woods 1988b), a principal target organ for lead compounds. Henderson and Toothill (1983) reported a 7.3-fold increase in renal coproporphyrin (from 32 to 233 pmol/mg protein) and a 37-fold increase in urinary coproporphyrin (from 12 to 443 nmol/24 h) following treatment of New Zealand white rabbits with lead acetate in drinking water sufficient to produce mean blood lead levels of 93 μg/dl. Ichiba and Tomokoni (1987) exposed male Wistar rats to drinking water containing 500 ppm lead acetate and found a progressive increase in urinary coproporphyrin (measured as the predominant III structural isomer form) to 3.4-fold over the first 13 days of exposure. The concentration of coproporphyrin in the urine remained significantly elevated compared with the initial level for at least 13 days following cessation of lead exposure, suggesting the correlation of urinary coproporphyrin with renal lead levels.

In human populations, urinary coproporphyrin analysis was commonly used to monitor occupational lead exposure before the availability of reliable atomic absorption methods for determining blood lead levels. Subsequently, coproporphyrin measurements were largely abandoned owing to the relative difficulty, time, and expense associated with this procedure. However, the more recent development of highly sensitive HPLC spectrofluorometric methods for isolating and quantitating urinary porphyrins in a single procedure (Ford et al. 1981; Lim and Peters 1984; Westerlund et al. 1988; Ho 1990; Woods et al. 1991; Bowers et al. 1992) renews the attractiveness of measuring urinary coproporphyrin as a bioassay of lead exposure. In human urine, coproporphyrin is excreted as a mixture of the I and III structural isomers, and HPLC procedures have been described which permit separation and quantitation of these constituents (Ichiba and Tomkouni 1987). Omae et al. (1988) employed HPLC to measure urinary coproporphyrin isomers in occupationally lead exposed men and reported a sensitivity and specificity exceeding 80% and a false positive rate of <10% when the health-based blood lead limit and screening levels of urinary coproporphyrin (as coproporphyrin III) were previously established. Similar success with this method has been reported by Ichiba and Tomouni (1987) in the evaluation of workers employed in a number of lead industries. From these studies, the urinary level of coproporphyrin III was found to be more highly correlated with blood lead levels than was that of coproporphyrin I. However, since coproporphyrin I constitutes only 10%–15% of the total coproporphyrin excreted in normal urine and is not altered during lead exposure (Ichiba and Tomouni 1987), there is questionable need to separate the III and I isomers when this procedure is employed for lead exposure assessments. Studies in this laboratory involving healthy adult male and female human subjects

(BOWERS et al. 1992) have shown that the levels of total coproporphyrin in either spot urine samples or 24-h collections fall within a relatively well defined range (Table 1). In view of the sensitivity and magnitude of the coproporphyrin response to lead exposure as demonstrated from animal as well as human studies, it is likely that values exceeding the normal range could be readily detected among lead-exposed subjects if employed in populations in which the normal intra- and interindividual variability is first well defined. Under these circumstances the sensitivity of urinary coproporphyrin may equal or exceed that of erythrocyte ALAD as a measure of low level lead exposure.

Finally, experimental studies from this laboratory (SIMMONDS et al. 1994) have shown that urinary coproporphyrin levels are as highly correlated with renal lead levels (r = 0.79) as with blood lead levels (r = 0.75) after 3 weeks following initiation of prolonged lead exposure in rats. These findings suggest that elevated urinary coproporphyrin may represent kidney lead accumulation leading to nephrotoxicity, and, hence, may be useful in diagnostic evaluation. Further studies are required to substantiate this possibility.

a) Confounding Factors in the Assessment of Urinary Coproporphyrin as a Biomarker of Lead Exposure

A number of factors and conditions, in addition to lead exposure, can give rise to excessive amounts of coproporphyrin in the urine. Relatively rare diseases such as hereditary coproporphyrinuria (characterized by an inherited deficiency in coprogen oxidase), hemolytic disorders, and alcoholic or nutritional liver disease may cause increased excretion of coproporphyrin in the urine. So, too, may a number of chemical exposures. Polyhalogenated aromatic hydrocarbons (PHAH), such as hexachlorobenzene, 2,3,7,8-tetrachlorodibenzo-*p*-dioxin (TCDD), and various polychlorinated biphenyls, have been reported to increase coproporphyrin excretion (OCKNER and SCHMID 1961; SAN MARTIN DE VIALE et al. 1970; TALJAARD et al. 1972). However, the principal manifestations of the type of porphyrinuria associated with PHAH exposure are distinct from that caused by lead in that high concentrations of the more highly carboxylated porphyrins, particularly uroporphyrin, are also present. Other trace metals, including arsenic and mercury, have also been found to elicit substantially increased excretion of coproporphyrin in the urine. The changes in the total porphyrin excretion pattern caused by these metals, however, are distinct from that produced by lead, as discussed further in this chapter. From a biomonitoring perspective, the principal concerns associated with urinary coproporphyrin assessments for occupational lead exposure are heavy alcohol use and other metal exposures. The extent to which these factors might confound the use of urinary coproporphyrin assays as a biomarker of lead exposure in human subjects has not been evaluated on a systematic or quantitative basis. There is a need, therefore, to consider these factors on a case by case basis.

II. Mercury

A number of investigators (DE SALAMANCA et al. 1983; GOLDWATER and JOSELOW 1967) have described elevated urinary porphyrin concentrations in workers with high-level occupational mercury exposure. However, the potential utility of urinary porphyrin measurements as a biomarker of exposure to mercury compounds in human subjects has been only recently described (WOODS et al. 1991, 1993; BOWERS et al. 1992). The underlying mechanistic etiology of this effect arises from the high affinity of mercury as Hg^{2+} (or CH_3Hg^+) for kidney proximal tubule cells and the property of mercury as Hg^{2+} to promote the accumulation of 4- and 5-carboxyl porphyrin(ogen)s (WOODS and MILLER 1993) and their subsequent excretion as the oxidized porphyrins in high concentrations in the urine (WOODS et al. 1991, 1993). Two principal actions of Hg^{2+} are involved in this process: (a) impairment of porphyrin metabolism at the level of coproporphyrinogen (coprogen) oxidase in the kidney and (b) facilitated oxidation of reduced porphyrinogens which consequently accumulate in kidney cells. An atypical porphyrin, described below, is also excreted during mercury exposure.

1. Mercury-Directed Alteration of Renal Coproporphyrinogen Metabolism

The mechanism underlying the accumulation of 5- and 4-carboxyl porphyrins observed during mercury exposure has been shown to involve mercuric ion (Hg^{2+}) (or CH_3Hg^+)-directed impairment of renal coprogen metabolism (WOODS and SOUTHERN 1989). Coprogen oxidase from kidney is highly sensitive to inhibition by Hg^{2+} in vitro and in vivo. WOODS and SOUTHERN (1989) assessed the effects of Hg^{2+} at various concentrations on coprogen oxidase in mitochondrial preparations from rat liver and kidney in vitro. The activity of the enzyme from both tissues was inhibited in a dose-related manner, with activity reduced to approximately 50% of control levels by Hg^{2+} at 100 μM in reaction mixtures.

In studies in vivo (WOODS and FOWLER 1977; WOODS and SOUTHERN 1989), treatment of male Sprague Dawley rats with methyl mercury hydroxide (MMH) for 6 weeks at 0.6 or 1.2 mg/kg per day by drinking water produced a pronounced coproporphyrinuria characterized by a dose-related increase in the urinary coproporphyrin concentration up to 20 times control levels. Renal coprogen oxidase activity was depressed by as much as 56% of the control value 3 weeks after initiation of MMH treatment, whereas hepatic coprogen oxidase was not affected. Depression of renal coprogen oxidase activity correlated closely with increasing mercury levels in the kidney, suggesting that the interaction of Hg^{2+} either with the enzyme itself or with structural components of the mitochondrial membrane upon which coprogen oxidase may be dependent for optimal activity was responsible for the decrease in activity observed. Coprogen oxidase from kidney may be particularly sensitive to inhibition by Hg^{2+} in vivo, as compared with that in other tissues, owing to the preferential accumulation of Hg^{2+} by renal mitochondria, and the demonstrated capacity of Hg^{2+} to disrupt mitochondrial membrane structures with which coprogen oxidase is associated in kidney proximal tubule cells (FOWLER and WOODS 1977).

Distinct from the effects of lead, mercury exposure in animals is characterized by the excretion of highly elevated levels of 5-carboxyl porphyrin in addition to coproporphyrin in the urine. The etiology of this effect may lie in the observation that, while both lead and mercury inhibit coprogen oxidase, mercury exposure is also characterized by a two- to threefold elevation in the activity of ALA synthetase, the first and rate-limiting enzyme in heme and porphyrin biosynthesis, in kidney cells (WOODS and FOWLER 1977, 1987). Increased ALA synthetase activity leads to increased porphyrinogen production, which, coupled with significantly impaired porphyrinogen metabolism at the level of coprogen oxidase, results in accumulation of both 5- and 4-carboxyl porphyrinogens in kidney cells. Increased ALA synthetase activity has not been demonstrated in the case of lead exposure. Mercury, also unlike lead, may impair the decarboxylation of 5-carboxyl porphyrinogen by renal uroporphyrinogen decarboxylase in vivo (WOODS et al. 1984), also contributing to 5-carboxyl porphyrinogen accumulation. Additionally, lead may impair the overall porphyrin synthesis through effects at other sites in the heme biosynthetic pathway which are not affected by mercury in vivo (Fig. 5). These differences may account for the distinction in the porphyrin excretion patterns observed in the case of these metals.

Of additional interest with regard to the porphyrinogenic action of mercury is the observation from animal studies that prolonged mercury exposure is characterized by the excretion in the urine of an atypical porphyrin, not normally found in urine of unexposed subjects, which elutes on HPLC approximately mid-way between the 5- and 4-carboxyl porphyrins (WOODS et al. 1991). This porphyrin has been referred to as "precoproporphyrin" in several previous publications (BOWERS et al. 1992; WOODS et al. 1993) and appears to be unique to mercury exposure. Absorption and electrospray ionizing mass spectroscopic analyses have suggested that this atypical porphyrin shares structural properties of a keto derivative of isocoproporphyrin. However, the precise structural characteristics as well as the biochemical etiology of this porphyrin remain to be established. This porphyrin has been detected in urine both of mercury-exposed animals and human subjects (WOODS et al. 1991, 1993).

2. Mercury-Facilitated Porphyrinogen Oxidation

A second property of mercury which accounts for its porphyrinogenic action is the capacity of Hg^{2+} or CH_3Hg^{2+} to promote formation of H_2O_2 and other reactive oxidants (Woods et al. 1990a,b; Lund et al. 1991, 1993; Miller and Woods 1993), which facilitates the conversion of reduced porphyrins which accumulate in kidney cells to the oxidized, readily excreted, porphyrins. Mercury as Hg^{2+} has been shown to dramatically increase the formation of H_2O_2 in rat renal mitochondria both in vitro and in vivo, with an associated increase in the oxidation of mitochondrial constituents (Lund et al. 1993). This effect is exacerbated by Hg^{2+}-mediated depletion of mitochondrial antioxidants, particularly reduced glutathione (GSH), and by the formation of Hg^{2+}-thiol complexes, which themselves possess pro-oxidant activity towards reduced porphyrins (Woods et al. 1990a,b; Miller and Woods 1993).

As demonstrated in Fig. 4, the pro-oxidant properties of mercury appear to contribute to both the porphyrinogenic as well as tissue-damaging effects observed during prolonged mercury exposure. Recent studies have demonstrated significantly elevated rates of formation of thiobarbiturate reactive substances, indicative of lipid peroxidation, in kidneys of Hg^{2+}-treated rats, associated with increased reactive oxidant formation (Lund et al. 1993). Typically, the porphyrinogenic effects of mercury exposure precede to a considerable extent oxidative tissue damage which ensues during prolonged exposure to mercury compounds, although both phenomena are well correlated with mercury concentrations in renal tissue (Woods et al. 1991; Lund et al. 1993). Thus, changes in porphyrin excretion patterns appear to both mechanistically and quantitatively reflect a cellular response to mercury-induced oxidative stress in kidney cells.

Findings supporting the potential utility of urinary porphyrin profiles in the assessment of mercury exposure in human subjects have been derived from recent studies in dentists (Woods et al. 1993). Dentists and other dental professionals incur occupational exposure to mercury as mercury vapor through the use of elemental mercury in the preparation of amalgam tooth restorative materials. Urinary mercury levels of up to $115\,\mu g$ Hg/l among practicing dentists have been reported (Naleway et al. 1991). A study comparing urinary porphyrin and mercury concentrations among dentists who participated in a Health Screening Program at the 1991 and 1992 annual meetings of the American Dental Association found that the mean concentrations of 4- and 5-carboxyl porphyrins, as well as of precoproporphyrin, were significantly elevated among a group of dentists who had urinary mercury in excess of $20\,\mu g$ Hg/l ($20.34-135.65\,\mu g$/l), as compared with a group having no detectable mercury in their urine (Woods et al. 1993). These differences are described in Table 2. No changes were observed in the concentrations of 8-, 7-, or 6-carboxyl porphyrins, similar to results observed in animal studies. While these findings must be viewed as

Table 2. Urinary porphyrin concentrations in dentists

Porphyrin	Urinary porphyrin concentration (μg/l)	
	No urinary mercury	Urinary mercury >20 μg/l
Pentacarboxylporphyrin	0.76 ± 0.81	3.07 ± 1.41[a]
Precoproporphyrin	1.98 ± 1.37	7.58 ± 4.55[a]
Coproporphyrin	22.97 ± 5.80	74.45 ± 8.00[a]

Porphyrin concentrations as measured in spot urine samples obtained from dentists whose urine contained no detectable mercury (n = 37) and whose urine had total mercury concentrations equal to or greater than 20 μg/l (n = 56). Values represent means ± standard deviations of individual determinations.

[a] Significantly different from urinary porphyrin concentrations in samples with no detectable mercury ($p < 0.05$).

preliminary, they are nonetheless encouraging from the biomonitoring perspective in suggesting that changes in the porphyrin excretion pattern comparable to those observed in highly exposed rats may be detectable among human subjects with much lower levels of mercury exposure.

III. Arsenic

The porphyrinogenic action of arsenic compunds has been described by a number of investigators. Woods and Fowler (1978) exposed male Sprague-Dawley rats and C57BL mice to arsenic as sodium arsenate (As^{5+}) at dose levels of 20, 40, or 85 ppm in drinking water for up to 6 weeks and noted the development of a unique porphyrinuria characterized by a 12-fold increase in the urinary concentration of uroporphyrin, a lesser increase in coproporphyrin, but no changes in levels of intermediate carboxylated porphyrins. Similar results were observed by Fowler and Mahaffey (1978) following the feeding of Sprague-Dawley rats a diet containing 50 ppm either sodium arsenate or arsanilic acid for 10 weeks. In other studies, Martinez et al. (1983) administered arsenic as sodium arsenite (As^{3+}) to female Wistar rats at doses of 5, 50, or 100 ppm in drinkig water for 7 weeks and also noted elevated urinary uroporphyrin levels as the predominant indicator of arsenite exposure. Similar results were reported by Webb et al. (1984) in male Fischer-344 rats 14 days after intratracheal instillation of 10, 30, or 100 mg/kg gallium arsenide (GaAs). These investigators found three- to fourfold increased concentrations of urinary uroporphyrin (as compared with a < twofold increase in coproporphyrin) and concluded that urinary uroporphyrin levels may serve as a sensitive indicator of GaAs exposure.

Studies on the mechanisms of arsenic-induced porphyrinuria have demonstrated that both As^{3+} and As^{5+} inhibit urogen decarboxylase and coprogen oxidase from rat liver and kidney (Woods et al. 1981; Woods and Southern 1989). Studies on urogen decarboxylase have shown that the primary focus of this effect appears to be at the enzyme site which catalyzes decarboxylation of the 8- to 7-carboxyl porphyrinogen. This unusual effect is consistent with the urinary porphyrin pattern observed during prolonged exposure to arsenic compounds. However, inhibition of urogen decarboxylase by arsenic is observed in vitro only at relatively high arsenic concentrations (e.g., As^{5+} = 10–50 m*M*). This substantially exceeds tissue concentrations which are observed in vivo during prolonged arsenic treatment regimens, e.g., hepatic or renal total arsenic concentration measured after exposure of rats to sodium arsenate for 6 weeks (150–200 μM). It is possible, therefore, that additional factors, such as the direct oxidation of porphyrinogens by reactive oxidants which could be produced during the interaction of arsenic ions with mitochondrial or microsomal cellular electron transport processes, might be involved in the etiology of arsenic-induced porphyrinuria, similar to that proposed for Hg^{2+}.

The coproporphyrinuria observed during prolonged arsenic exposure of rats may also be accounted for by the inhibition of coprogen oxidase activity by arsenicals, principally in the kidney. Renal coprogen oxidase was readily inhibited by organic (dimethylarsenic acid) and inorganic (sodium arsenate) arsenic in vitro at concentrations as low as 100 μM, although the hepatic enzyme was not affected at arsenic concentrations of less than 10 m*M* in reaction mixtures (WOODS and SOUTHERN 1989). In studies in vivo, prolonged exposure of rats to arsenic as sodium arsenate in concentrations of 40 or 85 ppm in drinking water resulted in a progressive and dose-related increase in urinary coproporphyrin levels, reaching 3.3 and 5.5 times control levels, respectively, at 5 weeks following initiation of exposure. Changes in urinary coproporphyrin content were accompanied by a progressive decline in renal coprogen oxidase activity and an increase in renal cortical arsenic concentrations to 7 and 13 μg/g tissue at the 40 and 85 ppm exposure levels, respectively.

Several studies have reported changes in the urinary porphyrin excretion pattern in arsenic-exposed human subjects, consistent with those observed in animal studies. GARCIA-VARGAS et al. (1991) compared the concentrations of uroporphyrin, coproporphyrin, and total arsenic in urine of 21 individuals exposed to arsenic in drinking water (390 μg arsenic/l) and 19 control subjects having drinking water containing 12 μg arsenic/l. An inversion of the copro-/uroporphyrin ratio was found among most exposed subjects, owing both to a decrease in coproporphyrin excretion and an increase in uroporphyrin excretion. However, no correlation was found between the copro-/uroporphyrin ratio and total urinary arsenic content among exposed subjects in this study. More recently, TELOLAHY et al. (1993) measured uroporphyrin and coproporphyrin levels in urine of 84 smelter workers exposed to arsenic trioxide compared with 22 nonexposed controls. High- and low-exposure groups had urinary arsenic levels of 257 and 129 μg arsenic/g creatinine, respectively, while controls had 9.9 μg/g creatinine. Total coproporphyrin levels increased by two- to threefold in both exposure groups as compared to controls, whereas urinary uroporphyrin levels were not changed. The absence of a strong correlation between changes in the urinary porphyrin profile and urinary arsenic content in either of these studies may suggest that urinary porphyrin excretion is more closely related to arsenic content in target tissues than to the level in urine, as observed in animal studies (Woods et al. 1991). Further studies are required to determine the potential of urinary porphyrins as a measure of arsenic exposure and of the biological effects of arsenic in target tissues.

IV. Other Metals

1. Cadmium

Cadmium is not viewed as a porphyrinogenic metal owing to the absence of measurable effects on porphyrin or heme metabolism in mammalian tissues in vivo and the lack of porphyrinuria during acute or prolonged cadmium exposure (Woods et al. 1981). These results are consistent with the known property of cadmium to induce metallothionein synthesis in vivo and to bind avidly to metallothionein in liver and other tissues (Cherian et al. 1976), effectively preventing binding with functional groups of heme pathway enzymes or other molecules. In contrast, Cd^{2+} has been shown to cause significant inhibition of urogen decarboxylase from rat liver in vitro. Enzyme activity was decreased to 56% and 17% of control levels at Cd^{2+} concentrations in reaction mixtures of $10 \mu M$ and $100 \mu M$, respectively (Woods et al. 1981). Cd^{2+} has also been shown to inhibit coprogen oxidase in crude preparations from rat liver mitochondria in vitro, although only at ten times the concentrations required to significantly inhibit urogen decarboxylase. The effects of Cd^{2+} in vitro are most likely attributable to the strong SH-binding capacity of Cd^{2+}, which, in the absence of induced metallothionein, are directed toward available functional SH groups of the specific enzymes. Although not in itself porphyrinogenic, Cd^{2+} has been shown to alter the porphyrin excretion patterns elicited during exposure to other metals (described in Sect. D.IV.4, below).

2. Platinum

Platinum is of clinical importance as *cis*-dichlorodiammineplatinum (*cis*-platinum), a widely used and effective chemotherapeutic drug against a variety of human cancers. *Cis*-platinum, as well as platinum chloride, has been shown to decrease ALA dehydratase activity in the kidney in vivo (Alexopoulos et al. 1986; Maines 1986). *Cis*-platinum also significantly inhibits PBG deaminase activity in rat prostate in vivo, possibly limiting overall heme biosynthetic capacity in this organ (Iscan and Maines 1990). Maines (1990) examined the effects of *cis*-platinum on heme pathway enzymes in rat liver, kidney, testes, prostate, and adrenals. The kidney was found to be particularly susceptible to the inhibiting effects of this drug, with total renal porphyrin content depleted to 50% of control values 7 days after *cis*-platinum treatment (9 mg/kg, i.v.). In addition to depressed renal porphyrin biosynthesis, an increased utilization of protoporphyrin IX resulting from an accelerated rate of ferrochelatase activity was observed. The effects of *cis*-platinum on the porphyrin-metabolizing enzymes, urogen decarboxylase, coprogen oxidase, and protoporphyrinogen oxidase, have not been described. However, a significant decrease in the concentrations of coproporphyrin and protoporphyrin in erythrocytes, as well as decreased levels of ALA, PBG, uroporphyrin, and coproporphyrin in the urine of *cis*-platinum-treated patients, has been reported (Alexopoulos et al. 1986). It is likely, therefore, that *cis*-platinum inhibits the activities of these enzymes as well. An extensive review of the effects of *cis*-platinum on heme, drug, and steroid metabolism has been recently published (Maines 1990).

3. Aluminum

Potential aluminum intoxication is an important consideration among patients with chronic renal failure (Tsukamoto et al. 1980). Sears and Eales (1973) reported aluminum-induced porphyria in male Wistar rats following i.p. treatment with $Al(OH)_3$ (130 mg/kg), characterized by highly elevated concentrations of total porphyrins in liver, lachrymal gland, bone, teeth, kidney, and other organs, as well as by increased excretion of porphyrins in urine and feces. A marked accumulation of 7- and 8-carboxyl, porphyrins in the livers of Al-treated rats suggests a principal effect of Al on urogen decarboxylase, although no enzyme activities were directly measured. Other investigators (Abdulla et al. 1979) have described inhibition of erythrocyte ALA dehydratase in vitro by aluminum but activation of the enzyme in vivo. Meredith et al. (1974) measured ALA dehydratase activity in rat erythrocytes in vitro and demonstrated enzyme activation by aluminum as either the sulfate or chloride salt to 2.5 times control levels at a concentration of 2 m*M* in reaction mixtures, but inhibition at concentrations greater than 4 m*M*. When Al and Zn (0.18 m*M*) were both present, ALA dehydratase activity was increased to 3.6 times control levels. In studies in vivo, ALA dehydratase activity was increased 1.5-fold with respect to when measured in livers of rats treated with aluminum sulfate (150 mg/kg) 1 h prior to sacrifice. These findings suggest that aluminum at physiologically relevant concentrations may increase ALA dehydratase activity in erythrocytes and other tissues, possibly contributing to the excess accumulation of porphyrins observed in previous studies (Sears and Eales 1973). An inhibitory effect of aluminum on porphyrin-metabolizing enzymes is also suggested by the available findings. The effect of aluminum exposure on porphyrin excretion in humans has, however, not been described.

4. Metal Interactions

Several studies have evaluated the effects of multiple metal exposures on urinary excretion patterns of porphyrins and porphyrin precursors. FOWLER and MAHAFFEY (1978) exposed rats to various combinations of lead, arsenic, and/or cadmium and found distinctive porphyrin excretion patterns in each case. Thus, lead and arsenic elicited porphyrin profile changes comparable to those described above for these metals. Lead and arsenic together produced an exaggerated excretion of coproporphyrin as well as an increase in uroporphyrin, suggesting an additive effect of these metals compared with each given alone. Although cadmium did not produce porphyrinuria alone, administration of cadmium with arsenic and/or lead produced specific changes in the porphyrin excretion pattern as well as changes in the overall magnitude of the porphyrin excretion rate. In another study, WEBB et al. (1984) monitored urinary porphyrin levels in rats following acute intrathecal administration of gallium arsenide to rats and found a porphyrin excretion pattern comparable to that for arsenic, as described above. More recently, GOERING and FOWLER (1987) described increased urinary ALA levels in gallium arsenide treated rats and attributed this effect to inhibition of erythrocyte ALA dehydratase by gallium. Thus, the interaction of gallium and arsenic appears to result in a distinct urinary profile in which the excretion of ALA and uroporphyrin exceed that of coproporphyrin. Various studies (MEREDITH et al. 1974; WOODS and FOWLER 1982; WOODS et al. 1981) have evaluated the effects of multiple metal exposures on specific heme biosythetic pathway enzymes in animal tissues either in vitro or following metal treatment. While the effects of multiple metal exposures on enzyme activities are, in some cases, highly distinct from the effects of metals administered individually, the influence of these changes on urinary excretion patterns of porphyrins and other heme precursors has not been described. The results from the metal-interaction studies described here point, however, to the potential importance of such determinations as indices of metal exposures.

E. Perspectives on the Use of Porphyrins as Biomarkers of Metal Exposure in Human Studies

The distinctive porphyrin patterns associated with exposure of animals to lead, mercury, and arsenic, compared with the normal profile, are depicted in Fig. 8. Changes in urinary porphyrin patterns associated with these and other metals offer promise as biomarkers of metal exposures and potential toxicity in humans from several perspectives. As biological responses to metal effects in target tissues, changes in porphyrin patterns are indicative of the internal concentration of metals in target tissues, and clear dose-response and time-related effects of metals with respect to development of porphyrin profile changes have been demonstrated in animal studies. These findings

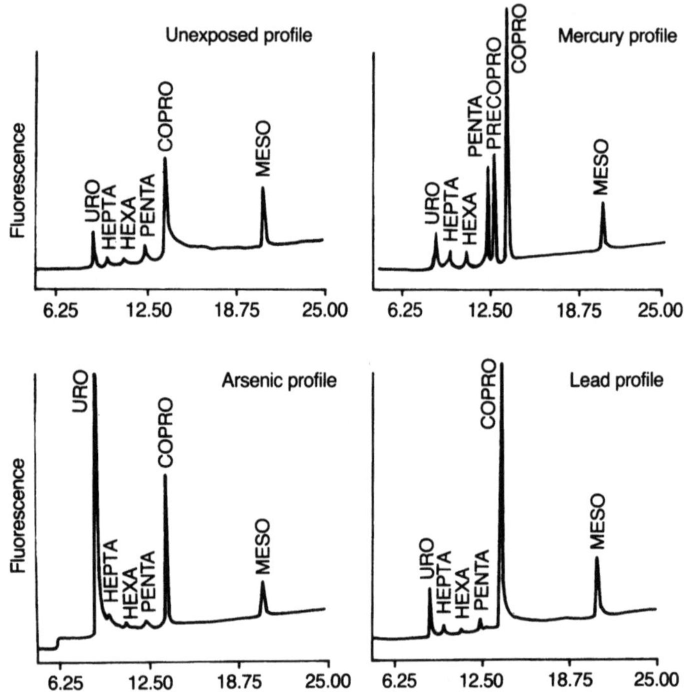

Fig. 8. Urinary porphyrin profiles from Fischer-344 rats showing distinctly altered porphyrin excretion patterns associated with prolonged mercury, arsenic, and lead exposures. *Abscissa* represents HPLC elution time (minutes)

indicate that urinary porphyrin profile changes should reflect metal exposures over a wide range of exposure levels in human subjects, particularly at low metal concentrations. Also, as measures of internal dosage, porphyrin parameters are indicative of cumulative metal exposure and associated health effects; in contrast, measures of blood or urine metal content largely represent recent or current exposures without the necessary mechanistic relationship to biological outcomes or effects. Changes in porphyrin parameters typically appear prior to the manifestation of clinical toxicity and persist after the cessation of metal exposure, consistent with the clearance of metal body burden. These characteristics support the utility of porphyrin parameters both in the predictive assessment of potential metal toxicity as well as in the evaluation of treatment regimens for facilitating metal clearance. Changes in porphyrin parameters are, for the most part, specific to the metal in question; therefore, they offer the diagnostic potential for defining specific metal exposures. Finally, in light of the development and widespread availability of appropriate instrumentation, porphyrin assessments in blood or urine are often more easily and economically accomplished, as compared with blood and urine metal analyses. These features extend the utility of porphyrin profile measurements to a wide range of practical applications.

While porphyrin parameters as biomarkers of metal exposures have been well characterized in animal models, the need remains for their further validation in human subjects and populations. Of primary importance is the determination of the range of normal variability in porphyrin parameters within the human population, with particular attention directed toward establishing the range of variation expressed among those who fall within the "normal ranges" of metal exposures. This issue is of particular importance in terms of defining the diagnostic utility of porphyrin measurements as biomarkers of low-level metal effects either within select populations or among individual human subjects. Further studies to determine threshold porphyrinogenic responses to metals in target tissues, and to define the mechanistic relationship between porphyrinogenic and toxic responses of metals, are also required.

The utility of spot urine samples, versus 24-h urine collections, for measurement of porphyrin excretion patterns as biomarkers of metal exposures in human subjects also requires further evaluation. Spot urine samples may be problematic in this respect, because of factors such as diet, medications, or fluid consumption, which alter porphyrin excretion rates. Also, many metabolites, including proteins and hormones, are known to display cyclical rhythms that can be circadian, monthly, or seasonal in nature. Similar fluctuations may pertain with respect to porphyrin excretion as well. Knowledge of such fluctuations is essential to determination of appropriate urine collection and sampling procedures, as well as to the evaluation of inter- and intraindividual differences in porphyrin excretion patterns subsequent to metal exposures.

Many of these characteristics have been evaluated regarding the use of ALAD, ZPP, and urinary coproporphyrin as biomarkers of occupational lead exposure. However, the sensitivity of these measures at low levels of lead exposure or the physiological relevance of porphyrin changes to lead-induced clinical effects have yet to be well defined. Such efforts might be initiated by establishing the range of variability of a specific biomarker within a selected population in which the extent of metal exposure is first well defined. Evidence of the effectiveness of such an approach has been demonstrated in the determination of ZPP among children from a population of subjects with a well-defined range of "normal" blood lead levels, described by Piomelli et al. (1987).

The efficacy of urinary porphyrin profiles as a biomarker of exposures to metals other than lead in human populations is less well established. Studies on dentists with occupational mercury exposure have provided preliminary evidence of the potential utility of urinary porphyrin changes as a biomarker of low-level mercury exposure in human subjects (Woods et al. 1993), and further studies have demonstrated their efficacy as a measure of cumulative effects of mercury on specific tests of neurobehavioral function (Echeverria et al. 1994). The relationship of mercury-induced changes in urinary porphyrin excretion patterns to damage to renal cell constituents from animal studies (Lund et al. 1991, 1993) also supports the utility of urinary porphyrin profiles as a bioindicator of nephrotoxicity or damage to other target organs. Similar preliminary evidence exists with respect to the efficacy of porphyrin profile measurements as a biomarker of human arsenic exposure.

Much remains to be accomplished with respect to the characterization and validation of porphyrins and other heme pathway parameters as biomarkers of metal exposures and effects in human subjects. Continued advances in the development of porphyrin assessment techniques and in the understanding of the mechanistic relationship of metal-induced porphyrinurias to the biological effects of metal exposures stimulate considerable interest and research in this area. Additional aspects of metal effects on heme and porphyrin metabolism have been described in previous reviews (Maines 1984; Marks 1985; Woods 1988b, 1989).

Acknowledgements. The author gratefully acknowledges the helpful comments and suggestions of Robert F. Labbe, PhD, in the preparation of this chapter. This work was supported by USPHS grants ES03628 and ES04696.

References

Abdulla M, Svensson S, Haeger-Aronsen B (1979) Antagonistic effects of zinc and aluminum on lead inhibition of δ-aminolevulinic acid dehydratase. Arch Environ Health 34:464–469

Alexopoulos CG, Chalevelakis G, Katsoulis C, Pallikaris G (1986) Adverse effects of cis-diamminedichloroplatinum II (CDDP) on porphyrin metabolism in man. Cancer Chemother Pharmacol 17:165–170

Anderson PM, Desnick RJ (1980) Purification and properties of uroporphyrinogen I synthase from human erythrocytes. J Biol Chem 255:1993–1999

Angle CR (1993) Childhood lead poisoning and its treatment. Annu Rev Pharmacol Toxicol 32:409–434

Astrin KH, Bishop DF, Wetmur JG, Kaul B, Davidow B, Desnick RJ (1987) δ-Aminolevulinic acid dehydratase isozymes and lead toxicity. Ann N Y Acad Sci 514:23–29

Baghurst PA, McMichael AJ, Wigg NR, Vimpani GV, Robertson EF, Roberts RJ, Tong S-L (1992) Environmental exposure to lead and children's intelligence at the age of seven years. N Engl J Med 327:1279–1284

Batlle AM del C, Benson A, Rimington C (1965) Purification and properties of coproporphyrinogenase. Biochem J 97:731–740

Benkmann HG, Bogdanski P, Goedda HW (1983) Polymorphism of delta-aminolevulinic acid dehydratase in various populations. Hum Hered 33:61–64

Bernard A, Lauwerys R (1987) Metal-induced alterations of δ-aminolevulinic acid dehydratase. Ann N Y Acad Sci 514:41–47

Bird TD, Hamernynik P, Nutter JY, Labbe RF (1979) Inherited deficiency of delta-aminolevulinic acid dehydratase. Am J Hum Genet 31:662–668

Bonkovsky HL, Sinclair PR, Bement WJ, Lambrecht RW, Sinclair JF (1987) Role of cytochrome P-450 in porphyria caused by halogenated aromatic compounds. Ann N Y Acad Sci 514:96–112

Borup P, Vordac V, Pederson JS, With TK (1980) The porphyrin pattern of normal urine. Int J Biochem 12:1075–1079

Bowers MA, Aicher LD, Davis HA, Woods JS (1992) Quantitative determination of porphyrins in rat and human urine and evaluation of urinary porphyrin profiles during mercury and lead exposures. J Lab Clin Med 120:272–281

Cantoni O, Evans RM, Costa M (1982) Similarity in the acute cytotoxic response of mammalian cells to mercury and X-rays: DNA damage and glutathion depletion. Biochem Biophys Res Commun 108:614–619

Cherian MG, Goyer RA, Delaquerriere-Richardson L (1976) Cadmium-metallothionein-induced nephropathy. Toxicol Appl Pharmacol 38:399–408

Chisolm JJ Jr (1971) Screening techniques for undue lead exposure in children: biological and practical considerations. J Pediatr 79:719–725

Clement RP, Kohashi K, Piper WN (1982) Rat hepatic uroporphyrinogen III cosynthase: purification, properties, and inhibition by metal ions. Arch Biochem Biophys 214:657–667

Dailey HA, Fleming JE (1983) Bovine ferrochelatase. Kinetic analysis of inhibition by *N*-methylprotoporphyrin, manganese and heme. J Biol Chem 258:11453–11459

De Bruin A (1968) Effect of lead exposure on the level of δ-aminolevulinic dehydratase activity. Med Lav 59:411–418

De Matteis F (1988) Role or iron in the hydrogen peroxide-dependent oxidation of hexahydroporphyrins (porphyrinogens): a possible mechanism for the exacerbation by iron of hepatic uroporphyria. Mol Pharmacol 33:463–469

De Matteis F, Harvey C, Reed C, Hempenius R (1988) Increased oxidation of uroporphyrinogen by an inducible liver microsomal system. Biochem J 250: 161–169

De Salamanca RE, Molina C, Olmos A, Chinarro S, Perpina J, Munoz JJ, Pena ML, Valls V (1983) Excretion de porfirinas y precursores en ratas cronicamente intoxicadas por mercurio. Gastroenterol Hepatol 6:20–23

Doss M, Muller WA (1982) Acute lead poisoning in inherited porphobilinogen synthase (aminolevulinic acid dehydrase) deficiency. Blut 45:131–139

Echeverria D, Heyer N, Woods JS, Martin MD, Naleway CA (1994) Effects of low-level exposure to elemental mercury among dentists. Neurotoxicol Teratol (to be published)

Elder GH, Evans JO (1978) Evidence that coproporphyrinogen oxidase activity in rat liver is situated in the intermembrane space of mitochondrion. Biochem J 172:345–351

Elder GH, Urquhart AJ (1984) Human uroporphyrinogen decarboxylase. Do tissue-specific isoenzymes exist? Biochem Soc Trans 12:663–664

Elder GH, Tovey JA, Sheppard DM (1983) Purification of uroporphyrinogen decarboxylase from human erythrocytes. Biochem J 215:45–55

Farant JP, Wigfield DC (1987) Interaction of divalent metal ions with normal and lead-inhibited human erythrocytic porphobilinogen synthetase in vitro. Toxicol Appl Pharmacol 89:9–18

Fell GS (1984) Lead toxicity: problems of definition and laboratory evaluation. Ann Clin Biochem 21:453–460

Ferioli A, Harvey C, De Matteis F (1984) Drug-induced accumulation of uroporphyrin in chicken hepatocyte cultures. Biochem J 224:769–777

Ford RE, Ou CN, Ellefson RD (1981) Liquid-chromatographic analysis for urinary porphyrins. Clin Chem 27:397–401

Fowler BA, Mahaffey KR (1978) Interaction among lead, cadmium and arsenic in relation to porphyrin excretion patterns. Environ Health Perspect 25:87–90

Fowler BA, Woods JS (1977) Ultrastructural and biochemical changes in renal mitochondria following chronic oral methyl mercury exposure: the relationship to renal function. Exp Mol Pathol 27:403–412

Fowler BA, Kimmel CA, Woods JS, McConnell EE, Grant LD (1980) Chronic low level toxicity of lead in the rat. III. An integrated assessment of long-term toxicity with special reference to the kidney. Toxicol Appl Pharmacol 56:59–77

Francis JE, Smith AG (1988) Oxidation of uroporphyrinogen by free radicals. Evidence for nonporphyrin products as potential inhibitors of uroporphyrinogen decarboxylase. FEBS Lett 233:311–314

Garcia-Vargas GG, Garcia-Rangel A, Aguilar-Romo M, Garcia-Salcedo J, Maria del Razo L, Ostrosky-Wegman P, Cortinas de Nava C, Cebrian ME (1991) A pilot study on the urinary excretion of porphyrins in human populations chronically exposed to arsenic in Mexico. Hum Exp Toxicol 10:189–193

Gibson RD, Neuberger A, Scott JJ (1955) The purification and properties of δ-aminolevulinate dehydratase. Biochem J 61:618–629

Goering PL, Fowler BA (1987) Mechanism of urinary excretion of δ-aminolevulinic acid after intrathecal instillation of gallium arsenide. Ann N Y Acad Sci 514: 330–332

Goldwater LJ, Joselow MM (1967) Absorption and excretion of mercury in man. XII. Effects of mercury exposure on urinary excretion of coproporphyrin and delta-aminolevulinic acid. Arch Environ Health 15:327–331

Goyer RA (1993) Lead toxicity: current concerns. Environ Health Perspect 100: 177–187

Granick S, Sassa S (1971) δ-Aminolevulinic acid synthetase and the control of heme and chlorophyll synthesis. In: Vogel HJ (ed) Metabolic pathways, vol V, 3rd edn. Academic, New York, pp 77–141

Haeger-Aronsen B (1960) Studies on urinary excretion of δ-aminolevulinic acid and other haem, precursors in lead workers and lead intoxicated rabbits. Scand J Clin Lab Invest 12 [Suppl 47]:1–28

Haeger-Aronsen B (1982) Why is the patient with lead intoxication not light sensitive? Acta Dermatol 100 [Suppl]:67–71

Henderson MJ, Toothill C (1983) Urinary coproporphyrin in lead intoxication: a study in the rabbit. Clin Sci 65:527–532

Hermes-Lima M, Pereira B, Bechara EJH (1991) Are free radicals involved in lead poisoning? Xenobiotica 8:1095–1090

Hernberg S, Nikkanen J (1970) Enzyme inhibition by lead under normal urban conditions. Lancet i:63–64

Hernberg S, Nikkanen J, Mellin G, Lilius H (1970) δ-Aminolevulinic acid dehydratase as a measure of lead exposure. Arch Environ Health 21:140–145

Ho JW (1990) Determination of porphyrins in human blood by high performance liquid chromatography. J Liquid Chromatog 13:2179–2192

Ichiba M, Tomokuni K (1987) Urinary excretion of 5-hydroxyindoleacetic acid, δ-aminolevulinic acid and coproporphyrin isomers in rats and men exposed to lead. Toxicol Lett 38:91–96

Iscan M, Maines MD (1990) Differential regulation of heme and drug metabolism in rat testis and prostate: response to cis-platinum and human chorionic gonadotropin. J Pharmacol Exp Ther 253:73–79

Jaffe EK, Bagla S, Michini PA (1991) Reevaluation of a sensitive indicator of early lead exposure. Biol Trace Element Res 28:223–231

Jones MS, Jones OTG (1969) The structural organization of heme synthesis in rat liver mitochondria. Biochem J 113:507–514

Kardish R, Fowler BA, Woods JS (1980) Alteration in urinary coproporphyrin and hepatic coproporphyrinogen III oxidase activity following exposure to toxic metals. 19th annual meeting of the Society of Toxicology, abstract A125

Labbe RF, Hubbard N (1961) Metal specificity of the iron-protoporphyrin chelating enzyme. Biochim Biophys Acta 52:131–135

Labbe RF, Finch CA, Smith NJ, Doan RN, Sood SK, Nishi M (1979) Erythrocyte protoporphyrin/heme ratio in the assessment of iron status. Clin Chem 25:87–92

Labbe RF, Rettmer RL, Shah AG, Turnlund JR (1987) Zinc protoporphyrin. Past, present and future. Ann N Y Acad Sci 514:7–14

Lamola AA, Yamane T (1974) Zinc protoporphyrin in the erythrocytes of patients with lead intoxication and iron deficiency anaemia. Science 186:936–938

Lamola AA, Joselow M, Yamane T (1975) Zinc protoporphyrin (ZPP): a simple sensitive fluorometric screening test for lead poisoning. Clin Chem 21:93–97
Lim CK, Peters TP (1984) Urine and faecal porphyrin profiles by reversed-phase high-performance liquid chromatography in the porphyrias. Clin Chem Acta 139:55–63
Lund B, Miller DM, Woods JS (1991) Mercury-induced H_2O_2 production and lipid peroxidation in vitro in rat kidney mitochondria. Biochem Pharmacol 42:S181–S187
Lund BO, Miller DM, Woods JS (1993) Studies on Hg(II)-induced H_2O_2 formation and oxidative stress in vivo and in vitro in rat kidney mitochondria. Biochem Pharmacol 45:2017–2024
Maines MD (1984) New developments in the regulation of heme metabolism and their implications. Crit Rev Toxicol 12:241–314
Maines MD (1986) Differential effect of cis-platinum (cis-diammine-dichloroplatinum) on the regulation of liver and kidney heme and hemoprotein metabolism: possible involvement of γ-glutamyl cycle enzymes. Biochem J 237: 713–721
Maines MD (1990) Effect of cis-platinum on heme, drug, and steroid metabolism pathways: possible involvement in nephrotoxicity and infertility. Crit Rev Toxicol 21:1–20
Maines MD, Kappas A (1977) Enzymes of heme metabolism in the kidney. J Exp Med 146:1286–1293
Marks GS (1985) Exposure to toxic agents: the heme biosynthetic pathway and hemoproteins as indicator. Crit Rev Toxicol 15:151–179
Martinez G, Cebrian M, Chamorro G, Jauge P (1983) Urinary uroporphyrin as an indicator of arsenic exposure in rats. Proc West Pharmacol Soc 26:171
Mayo Medical Laboratories Interpretive Handbood (1990) Mayo Medical Laboratories, Rochester, MN, pp 149–152
Meredith PA, Moore MR, Goldberg A (1974) The effects of aluminum, lead and zinc on δ-aminolevulinic acid dehydratase. Biochem Soc Trans 2:1243–1245
Millar JA, Cumming RL, Battistini V, Cabswell F, Goldberg A (1970) Lead and δ-aminolevulinic acid dehydratase levels in mentally retarded children and in lead-poisoned suckling rats. Lancet ii:695–698
Miller DM, Woods JS (1993) Redox activities of mercury-thiol complexes: Implications for mercury-induced porphyria and toxicity. Chem Biol Interact 88:23–35
Miller DM, Lund B, Woods JS (1991) Reactivity of Hg(II) with superoxide: evidence for the catalytic dismutation of superoxide by Hg(II). J Biochem Toxicol 6:293–298
Mukerji S, Pimstone N (1990) Free radical mechanism of oxidation of uroporphyrinogen in the presence of ferrous iron. Arch Biochem Biophys 281:177–184
Nakemura M, Yasuhochi Y, Minokami S (1975) Effects of cobalt on heme biosynthesis in rat liver and spleen. J Biochem 78:373–380
Naleway C, Chou H-N, Muller T, Dabney J, Roxe D, Siddiqui F (1991) On-site screening for urinary Hg concentrations and correlation with glomerular and renal tubular function. J Public Health Dent 51:12–17
Needleman HL, Gatsonis CA (1990) Low-level lead exposure and the IQ of children. J Am Med Assoc 263:673–678
Nordman Ch, Hernberg S, Nikkanen J, Rykanen A (1973) Blood lead levels and erythrocyte δ-aminolevulinic acid dehydratase activity in people living around a secondary lead smelter. Work Environ Health 10:19–25
Ockner RK, Schmid R (1961) Acquired porphyria in man and rat due to hexachlorobenzene intoxication. Nature 189:499
Omae K, Sakurai H, Higashi T, Hosoda K, Teruya K, Suzuki Y (1988) Reevaluation of urinary excretion of coproporphyrins in lead-exposed workers. Int Arch Occup Environ Health 60:107–110

Piomelli S, Davidow B (1972) Free erythrocyte protoporphyrin concentration: a promising screening test for led poisoning. Pediatr Res 6:366
Piomelli S, Young P, Gay G (1972) A micromethod forree erythrocyte porphyrins: the FEP test. J Lab Clin Med 81:932–940
Piomelli S, Davidow B, Guinee VF, Young P, Gay G (1973) The FEP (free erythrocyte porphyrins) test: a screening micromethod for lead poisoning. Pediatrics 51:254–259
Piomelli S, Lamola AA, Poh-Fitzpatrick MB, Seaman C, Harber L (1975) Erythropoietic protoporphyria and Pb intoxication: the molecular basis for difference in cutaneous sensitivity. I. Different rates of diffusion of protoporphyrin from erythrocytes, both in vivo and in vitro. J Clin Invest 56: 1519–1527
Piomelli S, Seaman C, Kapoor S (1987) Lead-induced abnormalities of porphyrin metabolism. The relationship with iron deficiency. Ann N Y Acad Sci 514:278–288
Piper WN, Tephly TR (1974) Differential inhibition of erythrocyte and hepatic uroporphyrinogen I synthetase activity by lead. Life Sci 14:873–876
Piper WN, van Lier RBL (1977) Pteridine regulation of inhibition of hepatic uroporphyrinogen I synthetase activity by lead chloride. Mol Pharmacol 13: 1126–1135
Piper WN, Tse J, Clement RP, Kohashi M (1983) Evidence for a folate bound to rat hepatic uroporphyrinogen III cosynthase and its role in the biosynthesis of heme. In: Blair IA (ed) Chemistry and biology of pteridines. De Gruyter, Berlin, p 415
Poulson R (1976) The enzymatic conversion of protoporphyrinogen IX to protoporphyrin IX in mammalian mitochondria. J Biol Chem 251:3730–3733
Quinlan GJ, Halliwell B, Moorhouse CP, Gutteridge JMC (1988) Action of lead and aluminum ions on iron-stimulated lipid peroxidation in liposomes, erythrocytes and rat liver microsomal fractions. Biochim Biophys Acta 962:196–200
Roels H, Buchet JP, Lauwerys R, Hubermont G, Bruaux P, Claeys-Thoreau F, Lafontaine A, Van Overschelde J (1976) Impact of air pollution by lead on the heme biosynthesis pathway in school-age children. Arch Environ Health 31: 310–316
Rossi E, Attwood PV, Garcia-Webb P (1992) Inhibition of human coproporphyrinogen oxidase activity by metals, bilirubin and haemin. Biochim Biophys Acta 1135:262–268
Rossi E, Taketani S, Garcia-Webb P (1993) Lead and the terminal mitochondrial enzymes of haem synthesis. Biomed Chromatog 7:1–6
San Martin de Viale LC, Viale AA, Nacht S, Grinstein M (1970) Experimental porphyria induced in rats by hexachlorobenzene. A study of the porphyrins excreted by urine. Clin Chem Acta 28:13–17
Sassa S (1978) Toxic effects of lead, with particular reference to porphyrin and heme metabolism. In: DeMatteis F, Aldridge WN (eds) Heme and hemoproteins, chap 11. Springer, Berlin Heidelberg New York, pp 333–371
Sears WG, Eales L (1973) Aluminum-induced porphyria in the rat. IRCS International Research Communication System J (73-11) 3-10-35
Sheehra JS, Gore MG, Chaudhry AG, Jordan PM (1981) δ-Aminolevulinic acid dehydratase: the role of sulphydryl groups in 5-ALA dehydratase from bovine liver. Eur J Biochem 114:263–269
Simmonds PL, Luckhurst CL, Woods JS (1994) Quantitative evaluation of heme biosynthetic pathway parameters as biomarkers of low level lead exposure in rats. J Toxicol Environ Health (in press)
Sinclair PR, Lambrecht R, Sinclair J (1987) Evidence for cytochrome P-450-mediated oxidation of uroporphyrinogen by cell-free liver extracts from chick embryos treated with 3-methylcholanthrene. Biochem Biophys Res Commun 146:1324–1329

Smith AG, Francis JE (1987) Chemically-induced formation of an inhibitor of hepatic uroporphyrinogen decarboxylase in inbred mice with iron overload. Biochem J 246:221–226
Sun J, Wang J, Liu J (1992) Effects of lead exposure on porphyrin metabolism indicators in smelter workers. Biomed Environ Sci 5:76–85
Sunderman FW Jr (1986) Metals and lipid peroxidation. Acta Pharmacol Toxicol 59 [Suppl 7]:248–255
Taketani S, Tokunaga R (1981) Rat liver ferrochelatase: purification, properties and stimulation by fatty acids. J Biol Chem 256:12748–12753
Taketani S, Tanaka A, Tokunaga R (1985) Reconstitution of heme-synthesizing activity from ferric ion and porphyrins, and the effect of lead on the activity. Arch Biochem Biophys 242:291–296
Taketani S, Tanaka-Yoshioka A, Masaki R, Tashiro Y, Tokunaga R (1986) Association of ferrochelatase with complex I in bovine heart mitochondria. Biochim Biophys Acta 883:227–283
Taljaard JJF, Shanley BC, Deppe WM, Joubert SM (1972) Porphyrin metabolism in experimental hepatic siderosis in the rat. II. Combined effect of iron overload and hexachlorobenzene. Br J Haematol 23:513–517
Telolahy P, Javelaud B, Cluet J, de Ceaurriz J, Boudene C (1993) Urinary excretion of porphyrins by smelter workers chronically exposed to arsenic dust. Toxicol Lett 66:89–95
Tephly TR, Hasegawa D, Baron J (1971) Effects of drugs on heme synthesis in the liver. Metabolism 20:200–210
Tsukamoto Y, Iwanami S, Marumo F (1980) Disturbances of trace element concentrations in plasma of patients with chronic renal failure. Nephron 26:174–179
Watson CJ (1946) Some newer concepts of the natural derivatives of hemoglobin. Blood 1:99–120
Webb DR, Sipes IG, Carter DE (1984) In vitro solubility and in vivo toxicity of gallium arsenide. Toxicol Appl Pharmacol 76:96–104
Weisberg JB, Lipschultz F, Osko FA (1971) δ-Aminolevulinic acid dehydratase activity in circulating blood cells: a sensitive laboratory test for the detection of childhood lead poisoning. N Engl J Med 284:565–569
Westerlund J, Pudek M, Schreiber WE (1988) A rapid and accurate spectrophotometric method for quantification and screening of urinary porphyrins. Clin Chem 34:345–351
Wetmur JG, Lehnert G, Desnick RJ (1991) The δ-aminolevulinate dehydrase polymorphism: higher blood lead levels in lead workers and environmentally exposed children with the 1-2 and 2-2 isozymes. Environ Res 56:109–119
Woods JS (1988a) Attenuation of porphyrinogen oxidation by glutathione and reversal by porphyrinogenic trace metals. Biochem Biophys Res Commun 152: 1428–1434
Woods JS (1988b) Regulation of porphyrin and heme metabolism in the kidney. Semin Hematol 25:336–348
Woods JS (1989) Mechanisms of metal-induced alterations of cellular heme metabolism. Comments Toxicol 3:3–25
Woods JS, Calas CA (1989) Iron stimulation of free radical-mediated porphyrinogen oxidation by hepatic and renal mitochondria. Biochem Biophys Res Comm 160:101–108
Woods JS, Fowler BA (1977) Renal porphyrinuria during chronic methyl mercury exposure. J Lab Clin Med 90:266–272
Woods JS, Fowler BA (1978) Altered regulation of mammalian hepatic heme biosynthesis and uroporphyrin excretion during prolonged exposure to sodium arsenate. Toxicol Appl Pharmacol 43:361–371
Woods JS, Fowler BA (1982) Selective inhibition of delta-aminolevulinic acid dehydratase by indium chloride in rat kidney: biochemical and ultrastructural studies. Exp Mol Pathol 36:306–315

Woods JS, Fowler BA (1987) Metal alteration of uroporphyrinogen decarboxylase and coproporphyrinogen oxidase. Ann N Y Acad Sci 514:55–64
Woods JS, Southern MR (1989) Studies on the etiology of trace metal-induced porphyria: effects of porphyrinogenic metals on coproporphyrinogen oxidase in rat liver and kidney. Toxicol Appl Pharmacol 97:183–190
Woods JS, Miller HD (1993) Quantitative measurement of porphyrins in biological tissues and evaluation of tissue porphyrins during toxicant exposures. Fundam Appl Toxicol 21:291–297
Woods JS, Kardish RM, Fowler BA (1981) Studies on the action of porphyrinogenic trace metals on the activity of hepatic uroporphyrinogen decarboxylase. Biochem Biophys Res Commun 103:264–271
Woods JS, Eaton DL, Lukens CB (1984) Studies on porphyrin metabolism in the kidney. Effects of trace metals and glutathione on renal uroporphyrinogen decarboxylase. Mol Pharmacol 26:366–341
Woods JS, Calas CA, Aicher LD, Robinson BH, Mailer C (1990a) Stimulation of porphyrinogen oxidation by mercuric ion. I. Evidence of free radical formation in the presence of thiols and hydrogen peroxide. Mol Pharmacol 38:253–260
Woods JS, Calas CA, Aicher LD (1990b) Stimulation of porphyrinogen oxidation by mercuric ion. II. Promotion of oxidation from the interaction of mercuric ion, glutathione, and mitochondria-generated hydrogen peroxide. Mol Pharmacol 38:261–266
Woods JS, Bowers MA, Davis HA (1991) Urinary porphyrin profiles as biomarkers of trace metal exposure and toxicity: studies on urinary porphyrin excretion patterns in rats during prolonged exposure to methyl mercury. Toxicol Appl Pharmacol 110:464–476
Woods JS, Martin MD, Naleway CA, Echeverria D (1993) Urinary porphyrin profiles as a biomarker of mercury exposure: studies in dentists with occupational exposure to mercury vapor. J Toxicol Environ Health 40:239–250
Yamanaka K, Hoshino M, Okamoto M, Sawamura R, Hasegawa A, Okada S (1990) Induction of DNA damage by dimethylarsine, a metabolite of inorganic arsenics, is for the major part likely due to its peroxyl radical. Biochem Biophys Res Commun 168:58–64
Yoshinaga T, Sano S (1980) Coproporphyrinogen oxidase. II. Reaction mechanism and role of tyrosine residues on the activity. J Biol Chem 255:4727–4731
Zwennis WCM, Franssen AC, Wijnans MJ (1990) Use of zinc protoporphyrin in screening individuals for exposure to lead. Clin Chem 36:1456–1459

CHAPTER 3

Membrane Transporters as Sites of Action and Routes of Entry for Toxic Metals

D.C. DAWSON and N. BALLATORI

A. Introduction: Metals and Membranes

There is a long history of interest in the interaction of toxic metals with membranes for two reasons. First, it was recognized early on that plasma membranes of cells were important potential targets for metals and that the effects of metals on membrane proteins might provide a basis for understanding the diverse toxic effects of these substances on the nervous system, the kidney, and the GI tract (CLARKSON 1972; FOULKES 1986; KINTER and PRITCHARD 1977; OEHME 1978; ROTHSTEIN 1970; TEMPLETON and CHERIAN 1983). Equally important historically, however, has been the interest of membrane physiologists in metals, particularly mercurials, as probes of membrane proteins that could be used experimentally to investigate mechanisms of transport and cellular homeostasis (CURRAN 1972; FERREIRA 1978; FRENKEL et al. 1975; HILLYARD et al. 1979; ROTHSTEIN 1970; SCHAEFFER et al. 1973; SCHOLTZ and ZEISKE 1988; SCHWARTZ and FLAMENBAUM 1976; STIRLING 1975). A major problem in both areas, however, has been the multiplicity of actions of metals on cell membranes; so that specific sites of action and cause-and-effect relationships have been difficult to sort out. As a result there appears to be a relatively large literature implicating biological membranes as sites of action for metals, but the mechanistic details of the cellular effects that lead to modulation of transport and ultimately to organ dysfunction have not been clearly defined. Imagine, for example, a cell that is exposed to a metal, such as mercury, cadmium, or lead, in a physiological salt solution. The initial interaction of the metal with the membrane could occur in one of several ways that are not mutually exclusive. Depending on the chemical forms of the metal and their abundance in the solution, the metal or some metal complex could associate with a membrane protein and bind covalently or noncovalently. This membrane binding event could have several different consequences. The properties of the protein could be altered or impaired in some way that could lead, for example, to an alteration in solute or water transport or a disturbance in cell volume (BALLATORI et al. 1988; LAMBERT et al. 1984; ROTHSTEIN and MACK 1991). The binding event could trigger the activation or inactivation of an intracellular second messenger system (i.e., intracellular calcium) that could secondarily modulate membrane transport events (NEMETH and SCARPA 1987; SMITH et al.

1989). The binding event could result in the translocation of the metal or some metal complex into the cell interior where interaction with intracellular membranes, enzymes, or transporters (ANNER and MOOSMAYER 1992; IMESCH et al. 1992; STOCKAND et al. 1993) could ensue (BRUNDER et al. 1988; CHANG and DAWSON 1988; SKROCH et al. 1993). In view of these possibilities, it will clearly be difficult to distinguish "primary" and "secondary" effects of metals on cellular transport processes.

In surveying the literature documenting the effects of metals on transport, one area stands out as offering some insight as to specific membrane proteins that could be sites of interaction with metals, namely those transport proteins that serve as routes for metal entry into cells. Here the use of specific blockers and the direct determination of metal transport has permitted metal-protein interactions to be identified with some certainty. In addition, the identity of the specific transporter responsible for metal permeation provides insights into the variety of chemical forms of the metal that may exist in the environment of the membrane. Metals can enter cells (Fig. 1) as a divalent (or monovalent) cation, as a neutral, lipid-soluble complex, as an anionic complex with Cl^-, OH^-, HCO_3^-, CO_3^{2-}, or as a complex with an organic moiety such as cysteine. This review will focus on several well-documented examples of permeation paths for metals, which serve to identify the proteins interacting with specific metal forms. The physical state of metals in physiological salt solutions will be briefly reviewed and then examples of permeation paths for the neutral, cationic, anionic, and organic forms of the metals will be presented. Although these considerations do not by themselves lead to an understanding of the molecular basis for the effects of metals on cells, an appreciation of these processes is a prerequisite for the design of any rational experimental investigation of the molecular mechanisms of metal toxicity.

B. Chemical Properties of Metals in Solutions

The major determinant of the chemical interaction that will ensue following exposure of a cell to a given metal is the speciation of the metal in the extracellular media. Metal binding to the membrane and penetration into the cell is regulated by the affinity and accessibility of a specific chemical form to various membrane ligands, as will be discussed in detail in subsequent sections. Unfortunately, identification of the chemical forms of metals in biological fluids is usually a daunting task due to the abundance of potential ligands and the kinetic lability of most metal complexes. Metal ligands include nearly all chemical groups capable of associating with hydrogen ions, such as sulfhydryl, carboxyl, amino, imidazole, phosphoryl, and endiols (Table 1). As illustrated in Table 1, metals also display a relatively low ligand specificity and will bind a variety of chemical groups. The affinity of a metal for a specific chemical group can be modulated by chelation to

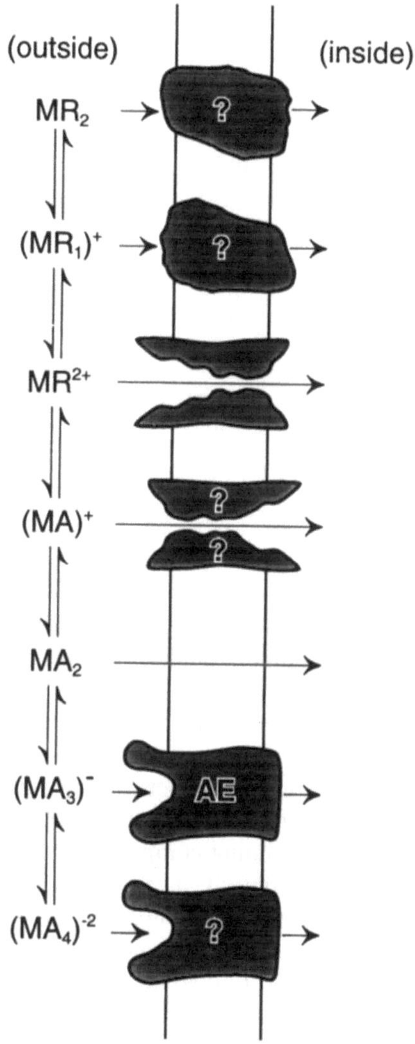

Fig. 1. Diagrammatic representation of metal permeation pathways that might be available in a particular cell. Shown are pathways for cations, neutral complexes, anions, and organic complexes. The availability of such pathways is expected to depend on the metal and on the particular cell of interest

additional chemical groups, or by binding to specialized ligands (e.g., porphyrin ring), or by the formation of polymetallic clusters (e.g., iron-sulfur cluster, 4Fe4S) (MARTELL 1981). Ligand accessibility also limits reactivity. For example, a cysteine residue (sulfhydryl group) deeply embedded within a protein may be unable to react with a metal such as mercury, despite the high affinity of mercury for sulfhydryl groups (Table 1). More-

Table 1. First association constants ($\log K_1$) of various metal complexes with small molecules

Cation	Sulfide[a]	Cysteine	Glycine	NH_3	Imidazole	Acetate
Hg^{2+}	53.5	14.2	10.3	8.8	3.6	4.0
Ag^{+}	50.0	15.0	3.7	3.2	3.8	0.7
Cu^{2+}	41.5		8.6	4.2	4.4	2.2
Pb^{2+}	27.5	12.2	5.1			2.0
Cd^{2+}	27.2	10.5	6.0	2.7	2.8	1.3
Ni^{2+}	27.0	9.6	6.2	2.8	3.3	0.7
Co^{2+}	26.7	9.3	5.2	2.1	2.4	1.5
Zn^{2+}	25.2	9.8	5.9	2.8	2.6	1.0

Values for the association constants were obtained from MADSEN (1963), SMITH and MARTELL (1976) and SILLEN and MARTELL (1971).
[a] pK of the solubility product.

over, competition for a given ligand by other ions in solution may limit the ability of that ligand to interact with a metal. Because of these competing reactions, ligands with the highest stability constants for a given metal may not necessarily be the predominant binding species in solution (MARTELL 1981).

The multiplicity of possible reactions that can occur under physiological conditions often makes it impossible to predict the site of action for a metal. One possible exception is that silver, copper, and in particular mercury have a relatively high affinity for reduced sulfhydryl groups (Table 1), and are bound largely to sulfhydryl-containing compounds in biological tissues. The low molecular weight organomercurial, methylmercury, is almost invariably associated with sulfhydryl groups in biological tissues (RABENSTEIN and FAIRHURST 1975). Despite this relative selectivity of binding for methylmercury, it has nevertheless been impossible to identify which of the many sulfhydryl-containing cellular constituents are targets for methylmercury. The sulfhydryl group is found in most proteins, and is present in low-molecular-weight compounds such as glutathione, cysteine, coenzyme A (CoA), lipoate, and thioglycolate.

A further complication in the identification of target sites and chemical forms of metals is the kinetic lability of coordinate covalent bonds. Metal ligands exchange rapidly in and out of the coordination sphere, in particular for first-row transition metals. This kinetic lability varies between metals, and, as indicated above, is influenced by the nature of the ligand, whether mono- or multidentate, and by the pH and ionic strength of its immediate environment. Copper, for example, forms relatively low affinity complexes with albumin or amino acids, but is tightly bound to ceruloplasmin. Similarly, mercury and cadmium form kinetically labile complexes with amino acids, glutathione, or albumin, but more stable chelates with metallothionein.

Although mercury has an extremely high affinity for sulfhydryl groups (Table 1), the resulting complexes are quite labile. Methylmercury, for example, can exchange between sulfhydryl groups of glutathione at rates that approach those of diffusion-controlled bimolecular reactions ($5.8 \times 10^8 M^{-1} \cdot s^{-1}$; RABENSTEIN and FAIRHURST 1975). These fast rates of exchange are due to the facile nucleophilic displacement of complexed glutathione by glutathione thiolate on the linear two coordinate mercury, and is the reason for the rapid mobility of methylmercury in biological systems (CLARKSON 1972, 1993).

Despite the difficulties inherent in unraveling the solution chemistry of metals, significant progress has been made in identifying membrane target sites under experimental conditions where the extracellular environment is controlled to limit the number of metal species. When metal salts such as $HgCl_2$, $CdCl_2$, or $Pb(NO_3)_2$ are dissolved in protein-free physiological salt solution (e.g., Ringer's) the metals exist predominantly in the form of mononuclear complexes in which a single central metal ion is associated with either chloride ion, hydroxyl ion, or bicarbonate ion. These polyanionic complexes can exist as neutral moieties, or may carry a net positive or negative charge. For example, in Cl-containing solutions mercury can exist in at least five forms, divalent mercury ion (Hg^{2+}), the monochloride complex ($HgCl^+$) bearing a net charge of +1, the neutral dichloride complex ($HgCl_2$), the trichloride complex ($HgCl_3^-$), a monovalent anion, and the tetrachloride complex ($HgCl_4^{2-}$) bearing a net charge of −2. The relative abundance of these forms is determined by the individual stability constants and the concentration of Cl^- in the solution. Table 2, taken from the comprehensive review by WEBB (1966), illustrates the relative abundance of the different forms in the presence of different Cl^- concentrations, and shows the dramatic effect of the chloride concentration on this equilibrium. The salient features that emerge from this comparison are the following: (a) Over a 500-fold range of chloride concentration, divalent Hg^{2+} is by far the least abundant form, becoming appreciable only at the lowest Cl^- concentration. (b) In a physiological salt solution appropriate for terrestrial vertebrates, mercury is predominantly in three forms in roughly equal concentrations, $HgCl_2$, $HgCl_3^-$, and $HgCl_4^{2-}$. (c) Increasing the Cl con-

Table 2. Distribution of mercuric chloride complexes as fractions of the total mercury in media of different Cl^- concentrations

Fraction	$(Cl^-) = 1\,mM$	Krebs-Ringer medium $(Cl^-) = 126\,mM$	Sea water $(Cl^-) = 515\,mM$
$(Hg^{2+})/(Hg_t)$	6.03×10^{-3}	1.26×10^{-12}	9.8×10^{-15}
$(HgCl^+)/(Hg_t)$	3.31×10^{-4}	8.68×10^{-7}	2.8×10^{-8}
$(HgCl_2)/(Hg_t)$	0.9925	0.331	0.0428
$(HgCl_3^-)/(Hg_t)$	7.09×10^{-3}	0.296	0.156
$(HgCl_4^-)/(Hg_t)$	7.09×10^{-5}	0.373	0.801

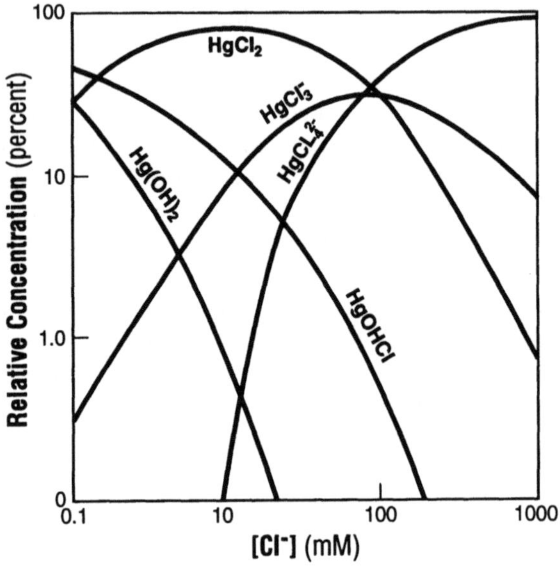

Fig. 2. Plot of the relative distribution of anionic complexes of mercury. (Redrawn from GUTKNECHT 1981)

centration increases the abundance of the charged polyanionic complexes, $HgCl_3^-$ and $HgCl_4^{2-}$. The net charge carried by the metal-anion complex is expected to exert a dominant influence on the interaction of the complex with the external surface of biological membranes.

Generally speaking the consideration of metal-chloride complexes alone is incomplete due to the existence of complexes with hydroxyl ions, carbonate ions, and bicarbonate in physiological solutions. For example, the affinity of Hg^{2+} for OH^- is actually greater than that for Cl^-, but the enormous relative abundance of Cl^- over OH^- means that, at a pH of 7 or below, at Cl concentrations of 10–150 m*M*, Cl^- complexes are likely to dominate. Figure 2 contains a plot redrawn from GUTKNECHT (1981) that shows the relative abundance of Hg complexes at chloride concentrations from 1.0 to 1000 m*M* and includes the hydroxyl complexes $Hg(OH)_2$ and Hg(OH)Cl. Complexation with HCO_3^- is another potentially important event, particularly for cadmium for which the affinity for Cl^- is lower than that for mercury (BUTLER 1964). Formation of polyanionic complexes is also a major consideration for organic mercurial compounds such as methylmercury, in the absence of organic ligands. Although the affinities for the respective ligands are somewhat less than that for Hg^{2+}, the dominant form of methyl mercury in solution is expected to be complexes with Cl^- and OH^-. It is important to remember, however, that even though the chemical forms of a metal in solution may be known, the form that interacts with or traverses the membrane may be difficult to identify. As the metal comes in contact with the cell surface, it may undergo ligand exchange, or a change in net charge or oxidation state, prior to interacting with its target sites. Moreover, cells normally release amino acids, glutathione, and other constituents into the extracellular space, and these may bind metals and alter their disposition.

The general tendency of metals to form complexes with organic and inorganic moieties in solutions creates a genuine dilemma when it comes to designing experiments for the purpose of testing mechanisms of metal action on cells. How is one to predict the abundance of the various forms of the metal in the solutions bathing the cells or tissues? How does one determine if experimental maneuvers that may involve, for example, the application of blocking ligands or other compounds, alter the concentrations of the various forms of a metal? The stability constants describing the equilibria for metals and various organic and inorganic molecules are available (BUTLER 1964; WEBB 1966; SMITH and MARTELL 1976; MARTELL 1981), but, in general, are not sufficiently complete to account for every constituent in an experimental bathing solution. To make matters worse relatively little is known about the interaction of metals with the wide variety of organic compounds such as buffers, hormones, and other modulators that might be used in such an experiment. For cadmium and copper, there exist reasonably selective electrodes that can be used to monitor directly the apparent activity of the divalent form of the ion. LOU et al. (1991) used a cadmium-selective electrode to determine the effect of anions on cadmium activity. ALDA TORRUBIA and GARAY (1990) similarly calibrated copper activity with a copper-selective electrode and used it to show that Hepes, Tris, and Bis-Tris, but not MOPS or MES, were able to complex copper ion. In the case of mercury, an electrode specifically designed to monitor the activity of the divalent species is not available, but a technical bulletin from Orion Research Inc. (1970) suggests that an iodide electrode exhibits a response to mercury, possibly due to the formation of HgI and the generation of monovalent Ag^+. Also suggested for this purpose was a cyanide electrode (personal communication, Orion Inc.). In the absence of a zinc-selective electrode, KALFAKAKOU and SIMONS (1990) used a zinc-selective dye, "Zincon," that yielded an optical signal that was linear over the range of $1-50\,\mu M$ total zinc. This approach may also be useful for copper (KALFAKAKOU and SIMON 1990).

C. Model Systems

The multiplicity of forms that heavy metals can take in solutions combined with the large number of potential targets for binding or mercaptide bond formation in cells provides ample impetus for investigating the effects of heavy metals in model systems in which the number of variables can be reduced. An excellent example that is particularly germane to a consideration of the various forms that heavy metals may adopt in aqueous solution is the study by GUTKNECHT (1981) of the movement of inorganic mercury through planar lipid bilayer membranes. Protein-free planar bilayers were formed using a mixture of egg lecithin and cholesterol in tetradecane to reduce the retention of solvent. The permeability of the membrane was determined directly using ^{203}Hg, with careful attention to the potential effect of unstirred layers. In the presence of physiological salt concentrations (10–100 m*M* NaCl) permeability was high ($4–10 \times 10^{-4}$ cm/s, comparable to that of water) and was unaffected by voltage, as if mercury crossed the bilayer in a nonionic form. The permeability exhibited a strong dependence on the concentration of Cl^- that was predictable on the basis of the calculated abundance of the electrically neutral form, $HgCl_2$, in solution. In as much as the lecithin-cholesterol bilayer is a model for the lipid portion of cell membranes, these results suggest that uptake of neutral chloride complexes of Hg and other heavy metals could be an important consideration in the evaluation of the results of experiments in which cells are exposed to metal salts in solution, although variability in membrane composition is likely to dictate that the permeability of neutral metal complexes, like that of water, can vary widely from one cell to another.

Mercury's ability to permeate lipid bilayers as a chloride complex may allow it to function as an ionophore, mediating electroneutral Cl^-/OH^- exchange (KARNISKI 1992). Using artificial liposomes, KARNISKI (1992) demonstrated that chloride can be driven above its electrochemical equilibrium by an OH^- gradient and that an inwardly directed chloride gradient generates a pH gradient in liposomes treated with either Hg^{2+} or Cu^+. Exchange of Cl^- for OH^- was suggested to be due to the movement of $HgCl_2$ from the chloride-containing side of the lipid bilayer to the chloride-free side, where chloride exchanges for hydroxyl, and returns as HgOHCl. Similarly, permeation of CuCl and exchange of Cl^- for OH^- could affect electroneutral Cl^-/OH^- exchange after exposure to Cu^+. This property may contribute to cellular toxicity of these metals by collapsing pH gradients across cell membranes and intracellular organelles.

The study of DELMONDEDIEU et al. (1989) suggests that the binding of anionic forms of mercury to bilayer constituents could contribute to the action of a metal. They used fluorescence polarization to evaluate the effects of mercury on temperature-induced phase transition in membranes composed of different phospholipids. They concluded that a negatively charged mercury complex might alter the properties of the bilayer by association with positively charged amino groups of the phospholipids.

A very useful model system that is becoming increasingly available for the study of membrane proteins is that created by expressing the protein in some heterologous cell type by means of the cDNA clone (GAMBA et al. 1993; PRESTON et al. 1993). Ion channels and other transporters, for example, can be expressed at high levels in *Xenopus* oocytes as well as a variety of cell lines. A number of advantages accrue from this approach. First, the high level of expression that is often obtainable makes it relatively easy to separate the transport process of interest from background transport activities. This means that in a single oocyte, for example, ionic currents due to an expressed channel protein can reach the level of microamperes, whereas the background currents are a thousandfold less, in the nanoampere range. For electrically silent transporters single oocytes can sometimes be used for determination of isotope uptake or efflux (GAMBA et al. 1993). A related advantage is that cells or oocytes not expressing the proteins can be used to systematically document the background transport activity and to identify actions of the heavy metals that do not directly involve the protein of interest.

Expression systems will be particularly useful for investigating the nature of the interaction of a toxic metal with membrane transport proteins because site-directed mutagenesis can be used to define the amino acid residues that are critical for the metal-protein interaction. Although this is not a trivial task, in the long term such experiments could yield important insights into the particular structural features that render some membrane proteins vulnerable to metals. An example is the manipulation of cysteine residues in the acetylcholine receptor channel that was used as a technique for mapping the region of the pore accessible to sulfhydryl reactive compounds (AKABAS et al. 1992). Particularly pertinent in the present context are the recent experiments of PRESTON et al. (1993) in which site-directed mutagenesis was used to identify a cysteine residue that is critical for the blockade of a water channel, CHIP28, by mercury. CHIP28 is a 128-kDa membrane protein that is thought to function as a constituitively active water channel in red blood cells and renal proximal tubule. The channel can be expressed in oocytes and the osmotic permeability of the oocytes assayed by optical determination of cell swelling in the presence of an osmotic gradient. The role of each of the four cysteines in the molecule was determined by replacing them individually with a serine. The serine substitution did not reduce the expression of water permeability, but one of the cysteines, C189, proved to be crucial for inhibition by mercury. The mutant, C189S, gave rise to water permeability that was insensitive to mercury. Similar results were obtained by ZHANG et al. (1993). The structure of the CHIP28 water pore is not known, but these experiments provide a tantalizing glimpse of the use of this approach to provide a high-resolution assay for sites of metal action within specific membrane proteins. For example, a similar approach may be used to identify the target sites on other membrane transporters that are known to be inhibited by mercury, including amino acid transport system A (CHILES et al. 1988), the Na^+-Pi cotransporter (LOGHMAN-ADHAM 1992), the Na-K pump (ANNER et al. 1992), and a number of other amino acid and sugar transporters (see ROTHSTEIN 1970 and KINTER and PRITCHARD 1977 for reviews). A Na^+-dependent neutral amino acid transporter with properties characteristic of system A has recently been cloned (KONG et al. 1993), which should facilitate identification of the reactive cysteine residue on this amino acid carrier.

D. Mercury Inhibition of NaCl Cotransport: An Example Problem with a Model System

WILKINSON et al. (1993) used the flounder urinary bladder to investigate targets for inorganic mercury, and the results provide a useful example of some of the problems encountered in studies of metal effects in biological models. The flounder bladder is unique among flat-sheet epithelia in the expression, in the apical membrane, of a thiazide-sensitive NaCl cotransporter that was recently cloned by expression (GAMBA et al. 1993) and has homologues in the mammalian kidney. The bladder, as indicated in Fig. 3, normally absorbs NaCl and secretes K, and the latter process can be monitored as a short-circuit current. Preliminary experiments (CHANG et al. 1985) showed that micromolar concentrations of $HgCl_2$ in the mucosal bath blocked K secretion. The effect was readily reversible, simply by replacing the mucosal bath, suggesting that Hg^{2+} might be, like Ba^{2+}, a reversible blocker of apical K channels. Subsequent experiments, however, revealed that: (a) inhibition of K secretion was likely to be an *indirect* effect of $HgCl_2$ and (b) the inhibition was not likely to be mediated by Hg^{2+}. It was found that inhibition of K secretion by $HgCl_2$ was actually accompanied by an *increase* in apical K conductance, i.e., K channels, rather than being blocked, were activated. A key to sorting out these disparate results was the observation that the diuretic hydrochlorothiazide, a specific blocker of the apical NaCl cotransporter, produced effects that were identical to those seen with $HgCl_2$.

Thiazides block the NaCl entry step and, by an indirect mechanism that is not well understood, also attenuate K secretion, despite the fact that

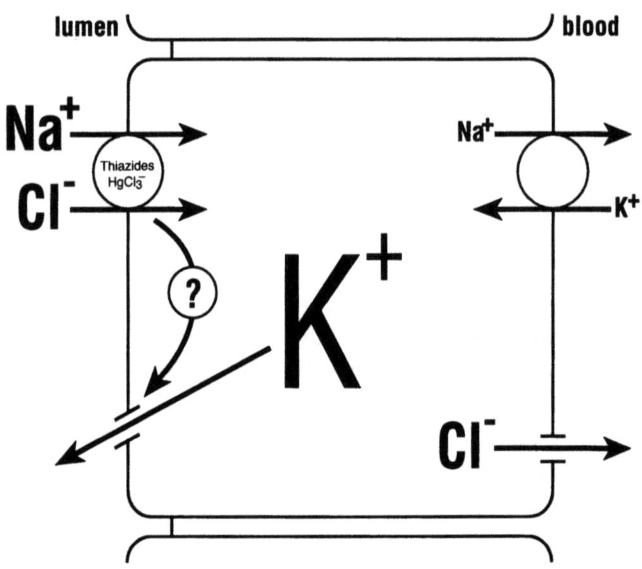

Fig. 3. Working model for NaCl absorption and K^+ secretion by flounder urinary bladder

apical K conductance is actually *elevated* (WILKINSON et al. 1993). Concentrations of $HgCl_2$ of the order of $0.5\,\mu M$ in the mucosal bath reduced NaCl absorption, abolished K secretion, and increased apical K conductance in a manner very suggestive of the thiazides. Similar effects could be obtained by depriving the obligatory cotransporter of either Na or Cl. These observations, although indirect, were consistent with the notion that inorganic mercury blocked the apical thiazide-sensitive cotransporter. A signal feature of the inhibition, however, was its rapid onset and equally rapid reversibility by simply washing the mucosal surface of the tissue. This behavior suggested that the action of mercury was at the outer surface of the membrane and did not involve the formation of a mercaptide bond. The experiments were conducted in physiological salt solutions in which Cl was the dominant anion ($[Cl] \simeq 150\,mM$) and the concentration of organic ligands is expected to be low, so that most of the mercury would be in one of three polyanionic forms in roughly equal amounts, $HgCl_2$, $HgCl_3^-$, and $HgCl_3^{2-}$. The authors speculated that the monovalent anionic form might bind to the chloride site on the cotransporter and reduce or abolish cotransport. This is an example in which possible *transport* of anionic forms of a metal could provide an important clue as to its mechanism of action. If thiazide-sensitive uptake of ^{203}Hg by bladder cells can be demonstrated, this would provide strong evidence that a mercury complex interacts with the transporter. The availability of the cloned transporter that exhibits robust expression in oocytes should facilitate such studies (GAMBA et al. 1993).

E. Metal Entry into Cells

The behavior of metal ions in solution suggests that, in general, we may expect to find cationic, neutral, and anionic forms, the relative abundance of which will depend on the nature of the metal and the ionic environment. If proteins are present, then some fraction of the metal will be bound due to reactions with SH and other groups. The net charge on the metal or metal complex is expected to determine the types of pathways that are available to enter the cell, and it turns out that there are examples of each for metal ions.

I. Permeation in a Lipid-Soluble Form

Despite the existence of neutral metal complexes in solution that might be expected to permeate the lipid portion of cell membranes, there seem to be few systematic studies of this phenomenon in cells. In a study of the interaction of mercury with erythrocytes, WEED, EBER, and ROTHSTEIN (1962) describe the rapid uptake of ^{203}Hg and in a later review ROTHSTEIN (1970) compared this to the much slower uptake of organic mercurials, *P*-chloromercuribenzoic acid (PCMB), chlormerodrin, and *P*-chloromercuribenzene sulfonic acid (PCMBS). In a study of zinc uptake by human red blood cells, KALFAKAKOU and SIMON (1990) suggested that in the presence of thiocyanate or salicylate a component of zinc uptake appeared that could represent permeation of a neutral lipid-soluble complex.

GREGUS et al. (1992) recently demonstrated that acute administration of high doses of dihydrolipoic acid, a lipid-soluble coenzyme containing two sulfhydryl groups, stimulates hepatic uptake and biliary excretion of mercury. The enhanced uptake presumably results from diffusion of the hydrophobic metal complex. However, because lipoic acid is normally localized to mitochondria, it may not play a major role in metal uptake under physiological conditions. Mercury and cadmium can also form lipid-soluble complexes with selenium (MAGOS and WEBB 1980), but the mechanisms involved have not been elucidated.

II. Permeation as a Cation

Divalent metals such as cadmium and mercury are well known as blockers of voltage-dependent (NACHSHEN 1984; HURWITZ 1986; BENNDORF and NILIUS 1988; SHEETS and HANCK 1992) and receptor-activated (HUGHES and BARRITT 1989) calcium-selective channels. Less widely appreciated, perhaps, is the fact that toxic metals can permeate cell membranes via these channels (FUKUDA and KAWA 1977; HINKLE et al. 1987, 1992; CROFTS and BARRITT 1990; BLAZKA and SHAIKH 1991; FLANAGAN and FRIEDMAN 1991). Using larval muscle fibers from a beetle, FUKUDA and KAWA (1977) demonstrated that when Mn^{2+}, Cd^{2+}, Zn^{2+}, or Be^{2+} were substituted for Ca^{2+} in the bathing media, these metals were able to maintain action potentials. Because this tissue normally generates relatively pure calcium spikes, these findings indicate that these metals permeate calcium channels as cations. In contrast, Co^{2+} and Ni^{2+} were unable to maintain membrane potentials, although they blocked normal action potential generation, indicating a different affinity or interaction with the channels (FUKUDA and KAWA 1977).

HINKLE et al. (1987) carried out a detailed study of cadmium permeation in the pituitary cell line GH_4Cl. They found that cadmium entry ($^{109}Cd^{2+}$) was enhanced by the calcium channel agonist, BAY·K8644, and was inhibited by the organic calcium channel blocker nimodipine. In a later study from the same laboratory (HINKLE et al. 1992), these observations were confirmed using the intracellularly trapped flourescent dyes, fura2 and quin2. Accordingly, it was found that cadmium toxicity, measured as the percentage cell death after 24 h exposure to $30\,\mu M$ $CdCl_2$, was markedly reduced by organic calcium channel blockers. The same compounds had no effect on the cytosolic concentration of metallothionein. In cells that did not express voltage-dependent calcium channels, cadmium toxicity was not affected by calcium channel blockers. The authors concluded that, despite the fact that the cadmium influx rate is less than 1% that of calcium, voltage-dependent calcium channels represent an important potential route of cadmium transport into cells and that cadmium entry by this route is primarily responsible for the generalized toxic effects of the metal.

CROFTS and BARRITT (1990) assessed the uptake of Mn^{2+}, Zn^{2+}, Co^{2+}, Ni^{2+}, and Cd^{2+} in isolated rat hepatocytes by measuring the quenching of intracellular quin2 fluorescence. Although hepatocytes lack voltage-dependent calcium channels, they have both receptor-activated channels and a less well defined "basal" calcium inflow mechanism (CROFTS and BARRITT 1990). These investigators demonstrated that Mn^{2+}, Zn^{2+}, Co^{2+}, Ni^{2+}, and Cd^{2+} enter hepatocytes and decrease quin2 fluorescence, and that the rate of fluorescence quenching is stimulated by vasopressin and angiotensin II. The effect of vasopressin was inhibited by verapamil, indicating the presence of an agonist-stimulated calcium inflow system that admits a number of divalent metals. Additional evidence that Cd^{2+} can enter hepatocytes through receptor-activated Ca^{2+} channels was presented by BLAZKA and SHAIKH (1991). They demonstrated that cadmium uptake by hepatocytes in primary culture was inhibited by diltiazam, verapamil, nifedipine, and nitrendipine, whereas vasopressin had a mild stimulatory effect. The stimulatory effect of vasopressin was blocked by verapamil. In contrast to Cd^{2+}, Hg^{2+} uptake into hepatocytes was unaffected by these agents, indicating a different mechanism of transport (BLAZKA and SHAIKH 1991).

Cadmium accumulation by MDCK cells also appears to be mediated in part by agonist-stimulated calcium channels (FLANAGAN and FRIEDMAN 1991). Parathyroid hormone, a stimulator of calcium uptake in MDCK cells, increased cadmium uptake. The dihydropyridine agonist BAY K8644 augmented the effect of parathyroid hormone, an effect that was competitively inhibited by the calcium channel antagonist nifedipine (FLANAGAN and FRIEDMAN 1991).

The selectivity of calcium channels would suggest that cadmium permeates as a cation, perhaps Cd^{2+} or $CdCl^{+}$. The stability constants provided by BUTLER (1964) suggest that, at a chloride concentration of 100 mM, of the order of 60% of the cadmium would be in a positively charged form, although in the experiments considered here other competing ligands, such as HCO_3^- and serum proteins would be expected to reduce the abundance of the charged form of the metal.

SIMONS and POCOCK (1987) obtained similar results in a study of lead uptake by adrenal medullary cells. They found that metal influx was stimulated by depolarizing the cells in high external K, but that the stimulated influx was blocked by methoxyverapamil (D600) with an affinity that was higher than that for blockade of Ca^{2+} influx (0.4 vs. 4.7 μM). Lead uptake was stimulated by the calcium channel agonist BAY·K8644. As was seen with cadmium, these results are consistent with the notion that voltage-dependent calcium channels could be an important route for the entry of lead into cells. Pb permeation was not affected if Cl^- was replaced by NO_3^-, an observation consistent with the relatively low affinity of Pb^{2+} for

Cl^- complexation (BUTLER 1964). TOMSIG and SUSZKIW (1990) used fura2 as a probe for lead entry into chromaffin cells and found that, once inside the cell, the lead was effective in releasing norepinephrine at concentrations three orders of magnitude lower than that required for calcium-induced exocytosis.

Cd^{2+} and Mn^{2+} may also be substrates for ion efflux mechanisms such as the Na-Ca exchanger (FRAME and MILANICK 1991). It has long been appreciated that a number of metals, and in particular Cd^{2+} and Mn^{2+}, inhibit this exchanger (PHILIPSON 1985). Recent studies by FRAME and MILANICK (1991) demonstrate that this inhibition is competitive, and that these metals are in fact transported as substitutes for Ca^{2+}. The K_m for Ca^{2+}, Cd^{2+}, and Mn^{2+} uptake by ferret red blood cells is roughly similar, $\sim 10\,\mu M$, as is the V_{max} (FRAME and MILANICK 1991), indicating that the transport properties of cadmium and manganese are similar to those of calcium in this system.

In contrast to the Na-Ca exchanger, Cd^{2+} does not appear to be a substrate for the Ca ATPase, even though it is a powerful inhibitor of pump activity (SCHATZMANN 1982). Indeed, the K_i for cadmium inhibition of the calcium pump is two orders of magnitude lower than the K_m for Ca^{2+} transport, indicating the presence of a high-affinity cadmium-binding site on the protein, most likely a thiol group (VERBOST et al. 1989).

Zinc efflux from human red blood cells may also involve a Ca^{2+} exchange mechanism (SIMONS 1991), but the transport system involved is not well characterized. ^{65}Zn release from human erythrocytes is stimulated by extracellular Ca^{2+}, but is unaffected by ouabain, 4,4′-diisothiocyano-2,2′-stilbene disulfonate (DIDS), or cytochalasin B, suggesting that a Zn-Ca exchanger mediates zinc efflux in these cells.

Bacteria have specialized transport systems for exporting toxic metals (SILVER et al. 1989; KAUR and ROSEN 1992). Metal ion resistance in bacteria is commonly associated with the induction of membrane ATPases that function to export toxic metals as either anions or cations, including Hg^{2+}, Ag^+, AsO_2^-, Cd^{2+}, CrO_2^{2-}. Recent evidence suggest that in humans Menkes disease is caused by a mutation in a gene that encodes a copper-transporting ATPase (VULPE et al. 1993).

The ability of toxic metals to increase membrane permeability to cations, and in particular to K^+, is well established (ROTHSTEIN 1970); however, it is not known whether these cation-selective pathways also function to transport the toxic metals across the membrane. Changes in cation permeability appear to be the result of an interaction of toxic metals with sulfhydryl groups of membrane proteins that modulate cation permeability (BALLATORI et al. 1988; JUNGWIRTH et al. 1991a,b; KONE et al. 1988, 1990). These proteins have not yet been identified.

III. Permeation as an Anion

The classic example of permeation as an anion is the ability of oxyanions of toxic metals to traverse cell membranes on either the phosphate or sulfate carriers (reviewed by WETTERHAHN-JENNETTE 1981; CLARKSON 1993). Vanadate and arsenate are structurally similar to phosphate and compete with phosphate for transport, as well as intracellular binding sites. Indeed, their toxicity is thought to be directly related to this competition (CLARKSON 1993). Similarly, chromate, selenate, and molybdate are structurally similar to sulfate, and are substrates for sulfate transporters.

The anion exchanger has been implicated as a potential route of entry for cadmium, copper, lead, and zinc into cells. LOU et al. (1991) studied the initial rate of uptake of cadmium by human red blood cells using atomic absorption spectrophotometry. They used a cadmium electrode to monitor the activity of Cd^{2+} in the experimental solutions so that the extent of anionic complexation of Cd^{2+} could be ascertained. It could be shown, for example, that, using an NO_3^- solution as a reference, the addition of 20 m*M* bicarbonate resulted in a 47% decrease in the activity of Cd^{2+}. When bathed by a medium containing 130 m*M* Cl^- and 15 m*M* HCO_3^-, red blood cells took up cadmium and the uptake was virtually abolished by DIDS, implicating the anion exchanger as the route of entry. Efflux of cadmium from the cells was undetectable, suggesting high-affinity binding to hemoglobin or other intracellular proteins.

The uptake of cadmium by red blood cells was found to be enhanced by a variety of experimental maneuvers that would be expected to increase the

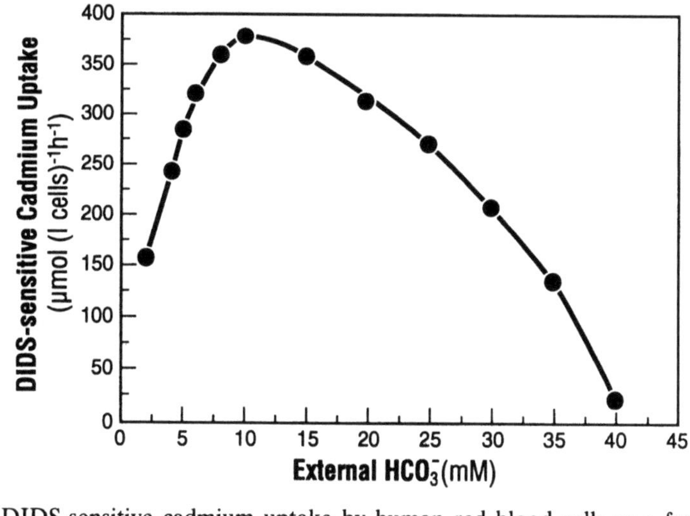

Fig. 4. DIDS-sensitive cadmium uptake by human red blood cells as a function of HCO_3^- concentration. (Redrawn from LOU et al. 1991)

abundance of polyanionic cadmium complexes. As shown in Fig. 4, redrawn from LOU et al. (1991), DIDS-sensitive cadmium uptake exhibited a concentration-dependent stimulation by HCO_3^- in the range of 0–15 m*M* but declined thereafter, as might be expected if the apparent valence of the dominant cadmium complex was altered at higher HCO_3^- concentrations. The stimulation of cadmium uptake by HCO_3^- was enhanced in Cl^--containing solutions although, in the absence of HCO_3^-, Cl^- alone produced little stimulation. In the presence of both Cl^- (130 m*M*) and HCO_3^- (15 m*M*), cadmium uptake was markedly enhanced by raising the pH from 6.2 to 7.8. These observations led the authors to conclude that the transported species was a univalent, anionic complex comprising one cadmium and HCO_3^-, OH^-, and Cl^-, i.e., $[Cd(OH)(HCO_3^-)Cl]^-$ or $[Cd(OH)(HCO_3)_2]^-$. The reduction in Cd uptake at higher concentrations of HCO_3 was attributed to increased formation of the divalent complexes such as $[Cd(OH)(HCO_3)_3]^{2-}$ that would be expected to be transported less readily than the monovalent forms. In a nitrate medium a DIDS-insensitive component of cadmium uptake was also stimulated by HCO_3 that may represent, in part, uptake of a neutral, more lipid soluble complex.

ALDA TORRUBIA and GARAY (1990) obtained evidence that copper uptake by human red blood cells was largely due to the influx of an anionic complex via the anion exchanger. Copper uptake was inhibited 80% by DIDS (5–10 μM), whereas compounds like amiloride or ouabain were without effect. A copper-selective electrode was used to monitor Cu^{2+} activity and to determine the extent of copper complexation under various conditions. Complex formation was negligible in NO_3^- solutions but was extensive in Cl-containing solutions. Accordingly, copper uptake by the cells was stimulated by replacing NO_3^- with Cl^-. Copper uptake was also stimulated by replacing NO_3^- with HCO_3. Copper activity was highly dependent on pH due to the formation of $Cu(OH)_2$, the most abundant species at pH 7.0–7.4. These authors speculated that the primary transported species are $Cu(OH)_2Cl^-$ and $Cu(OH)_2HCO_3^-$, although other monovalent and divalent forms could not be excluded. The use of a Cu^{2+}-sensitive electrode in this study made it possible to ascertain that several buffers (HEPES, TRIS, BIS-TRIS) bound or complexed copper and, although it was not directly tested, the experiments raised the possibility that the inhibitor used to operationally define the exchanger fluxes, DIDS, might also bind copper. The authors used histidine and cysteine to test the effect of binding by organic constituents, and micromolar concentrations of either compound produced a dramatic reduction in DIDS-sensitive copper uptake as expected due to the high affinity of Cu^{2+} for these organic ligands.

Lead (SIMONS 1986) and zinc (ALDA TORRUBIA and GARAY 1989; KALFAKAKOU and SIMONS 1990) also enter red blood cells as anionic complexes. In the case of lead, a modest sensitivity of the DIDS-sensitive uptake to membrane potential prompted the speculation that the neutral species, $PbCO_3$, might be exchanged for a monovalent anion but the transport of the ternary complex, $PbCO_3$-anion, could not be excluded. In the case of zinc, the anion exchanger mediated influx was thought to be due to the movement of the complexes $ZnCO_3Cl^-$ or $Zn(HCO_3)ClOH^-$. In the presence of thiocyananate or salicylate, however, there was a large, DIDS-insensitive influx of Zn that may represent uptake of a neutral, lipid-soluble complex with these anions.

In a study of the effects of organic mercurials on the cation permeability of human red blood cells, KNAUF and ROTHSTEIN (1971) found that the rate of uptake of PCMBS was diminished by 50% in cells exposed to SITS (4-acetamido-4′-isothiocyano-stilbene 2,2′-disulfonic acid), so that the total uptake could be divided into roughly equal SITS-sensitive and SITS-insensitive components. If the chloride in the bathing solution was replaced with either sulfate or phosphate, the SITS-sensitive component of PCMBS uptake was virtually unchanged whereas the SITS-insensitive component was substantially reduced. One interpretation of these results is that PCMBS enters by two routes, one the anion transporter that serves as a route for anionic complexes, and the other the lipid layer that permits the permeation of a neutral Cl complex of PCMBS, the latter perhaps being less abundant in the anion substituted media. Alternatively, a cationic form of the mercurial could permeate via some other membrane protein.

Some additional insight into the role of charge on the permeation of organic mercurials was derived from a comparison of the properties of two organic mercurials PCMBS and PCMB, the former having the more strongly ionizable sulfonic acid groups in place of the carboxyl. At pH 9, the uptake of the two compounds was comparable, but, at an acidic pH of 6, uptake of PCMB exceeded that for PCMBS by ten fold, an effect attributed to the titration of the carboxyl group on the PCMB.

IV. Transport of Organic Complexes

Recent studies demonstrate that toxic metals can enter cells complexed with endogenous organic ligands whose overall structures mimic those of native substrates for membrane transporters. In particular, membrane carriers for amino acids and glutathione may be responsible for the transport of some metals as mercaptides (KERPER et al. 1992; DUTCZAK and BALLATORI 1994; BALLATORI 1994). Although this mechanism had long been considered a theoretical possibility, direct experimental support at the cellular and subcellular level was lacking.

Methylmercury transport across the blood-brain barrier appears to be mediated by the large neutral amino acid transport system (system L) on the luminal surface of brain capillary endothelial cells (KERPER et al. 1992). Previous in vivo studies had shown that the amino acid, L-cysteine, accelerates methylmercury uptake into brain in vivo, but the mechanism was not identified. Because the methylmercury-L-cysteine complex has close structural similarity to L-methionine, a substrate for the L system, it was possible that the metal complex could also enter brain capillary endothelial cells via this amino acid carrier. The L system has a broad substrate specificity and is the major route of entry of neutral amino acids into brain endothelial cells from blood plasma. KERPER et al. (1992) provide the first kinetic data in support of this hypothesis. Using the rapid carotid infusion technique, these investigators demonstrated that uptake of the methylmercury-L-cysteine complex is saturable, stereoselective, and inhibited by large neutral amino acids. Furthermore, brain uptake of L-[^{14}C] methionine was inhibited by the methylmercury-L-cysteine complex, but not by methylmercury chloride, providing further strong evidence for transport of the methylmercury-L-cysteine complex on amino acid transport system L. There was minimal brain uptake of methylmercury when administered as a complex with either albumin or the D enantiomorph of cysteine (KERPER et al. 1992), indicating that other, less selective, transport mechanisms (e.g., ligand exchange, or simple diffusion as a lipid-soluble complex) do not play significant roles under these experimental conditions.

Methylmercury is also transported on membrane carriers for glutathione, as a glutathione mercaptide (CH_3Hg-SG; BALLATORI and CLARKSON 1983; DUTCZAK and BALLATORI 1994). Using isolated rat liver canalicular plasma membrane vesicles, DUTCZAK and BALLATORI (1994) examined the mechanism for the transport of methylmercury out of hepatocytes into bile, a major pathway for elimination of this metal. The results indicate that CH_3Hg-SG is not a substrate for the ATP-dependent canalicular GSSG or glutathione *S*-conjugate carriers, but appears to be a substrate for canalicular carriers that also transport GSH (DUTCZAK and BALLATORI 1994). CH_3Hg-SG transport was *cis*-inhibited and *trans*-stimulated by GSH indicating shared transport mechanisms. Because efflux systems for GSH are found in all mammalian cells, transport of glutathione-metal complexes by such carriers may be a ubiquitous mechanism for the removal of specific metals from cells.

Zinc uptake into red blood cells appears to be mediated in part by an amino acid carrier (AIKEN et al. 1992). A significant fraction of plasma zinc exists complexed with amino acids, in particular, L-histidine and L-cysteine. AIKEN et al. (1992) examined the role of histidine in zinc uptake by erythrocytes, and reported that transport is stimulated by L-histidine, whereas it is inhibited by D-histidine. Uptake of the L-histidine-zinc complex is Na^+-dependent and temperature sensitive, and is unaffected by DIDS, an inhibitor of the anion exchanger. Thus, red bloods can take up zinc as an anion (AIDA TORRUBIA and GARAY 1989), or as an amino acid complex.

A significant fraction of plasma copper(II) is present as the L-histidine complex ($CuHis_2$); however, this complex does not appear to be transported intact across cell membranes. DARWISH et al. (1984) demonstrated that, although L-histidine enhances copper uptake by isolated rat hepatocytes, the amino acid is not cotransported with copper. The actual chemical form of transported copper is not known. Similarly, copper uptake from ^{67}Cu-labeled ceruloplasmin by K-562 cells, a human erythroleukemic cell line, occurs after dissociation of the metal from the protein (PERCIVAL and HARRIS 1990). ^{67}Cu was taken up by these cells, whereas the protein moiety of ceruloplasmin was not. The mechanism by which the copper is extracted from ceruloplasmin and transported across the cell membrane is also undefined. Cu^{2+} appears to be reduced to Cu^{+} during the extraction/translocation process (MCARDLE 1992).

An additional mechanism for transport of metal complexes is by endocytosis/exocytosis (for review see BALLATORI 1991). Fluid-phase, adsorptive, and receptor-mediated endocytosis make a major contribution to the transport of metals that are bound to high molecular weight ligands, and in particular to ligands such as ferritin, transferrin, and other proteins that are selectively cleared by receptor-mediated endocytosis. Because these proteins also have some affinity for toxic metals, they may play an important role in their transport across cell membranes (BALLATORI 1991). The mechanism by which metallothionein and its associated metals are removed from the circulation is not known, but the kidney appears to be the principal site of removal (TANAKA et al. 1975). When rats are injected intravenously with ^{109}Cd-labeled metallothionein, the radioactivity is rapidly and nearly completely accumulated in the kidney (TANAKA et al. 1975).

F. Physiological Significance of Metal Permeation Pathways

The focus of this review has been the mechanism by which metals cross cell membranes, and a survey of the literature suggests that metals and metal complexes may take a variety of routes into cells. The identification of these pathways, which may vary from cell to cell and with different experimental conditions, would seem to be essential for any comprehensive understanding of the mechanisms by which metals can impact cellular activities. An assessment of the physiological or pathophysiological significance of these pathways in relation to metal toxicity must begin with an evaluation of the forms of the particular metals that are likely to be presented to the cells due to the ingestion or uptake of metals by the body. A primary consideration is the strong association of metals to proteins and amino acid in plasma due to binding to SH and other groups that is expected to markedly reduce the free concentration of the divalent species and the various charged or uncharged complexes (BALLATORI 1991; HUGHES 1957). The availability of forms of cadmium, for example, that could enter cells via calcium channels or the anion exchanger could be vanishingly small, so that these mechanisms might be less important as mechanisms for acute metal effects. On the other hand chronic metal exposure over long periods could result in some of these mechanisms assuming greater importance as regards the long-term accumulation of metals in cells.

Acknowledgements. The authors thank Marie Samida for her help in preparing the manuscript. The work of the authors was supported by the NIEHS Center for Membrane Toxicity Studies at The Mount Desert Island Biological Lab (ES03828) and by grant ES06484 and by The National Institute of Diabetes, Digestive and Kidney Diseases (DK29786, DK45880).

References

Aiken SP, Horn NM Saunders NR (1992) Effects of amino acids on zinc transport in rat erythrocytes. J Physiol (Lond) 445:69–80

Akabas MH, Stauffer DA, Xu M, Karlin A (1992) Acetylcholine receptor channel structure probed in cysteine-substitution mutants. Science 258:307–310

Alda Torrubia JO, Garay R (1989) Evidence for a major route for zinc uptake in human red blood cells: $[Zn(HCO_3)_2\,Cl]^-$ influx through the $[Cl^-/HCO_3^-]$ anion exchanger. J Cell Physiol 138:316–322

Alda Torrubia JO, Garay R (1990) Chloride (or bicarbonate)-dependent copper uptake through the anion exchanger in human red blood cells. Am J Physiol 259:C570–C576

Anner BM, Moosmayer M (1992) Mercury inhibits Na-K-ATPase primarily at the cytoplasmic side. Am J Physiol 262:F843–F848

Anner BM, Moosmayer M, Imesch E (1992) Mercury blocks Na-K-ATPase by a ligand-dependent and reversible mechanism. Am J Physiol 262:F830–F836

Ballatori N (1991) Mechanisms of metal transport across liver cell plasma membranes. Drug Metabol Rev 23(1,2):83–132

Ballatori N (1994) Glutathione mercaptides as transport forms of metals. In: Anders MW, Dekant W (eds) Conjugation-dependent carcinogenicity and toxicity of foreign compounds. Academic, Florida, pp 271–298

Ballatori N, Clarkson TW (1983) Biliary transport of glutathione and methylmercury. Am J Physiol 244:G435–G441

Ballatori N, Shi C, Boyer JL (1988) Altered plasma membrane ion permeability in mercury-induced cell injury: studies in hepatocytes of elasmobranch *Raja erinacea*. Toxicol Appl Pharmacol 95:279–291

Benndorf K, Nilius B (1988) Different blocking effects of Cd^{++} and Hg^{++} on the early outward current in myocardial mouse cells. Gen Physiol Biophys 7:345–352

Blazka ME, Shaikh ZA (1991) Differences in cadmium and mercury uptakes by hepatocytes: role of calcium channels. Toxicol Appl Pharmacol 110:355–363

Brunder DG, Dettbarn C, Palade P (1988) heavy metal-induced Ca^{2+} release from sarcoplasmic reticulum. J Biol Chem 263:18785–18792

Butler JN (1964) Introduction to complex formation equilibria. Ionic equilibrium, a mathematical approach. Addison-Wesley, Reading, Massachusetts Palo Alto London, p 261

Chang D, Dawson DC (1988) Digitonin-permeabilized colonic cell layers. J Gen Physiol 92:281–306

Chang D, Betz L, Dawson DC (1985) Mercury reversibly blocks apical K channels in the urinary bladder of the winter flounder, *Pseudopleuronectes americanus*. Bull Mt Des Isl Biol Lab 25:44–45

Chiles TC, Dudeck-Collart KL, Kilberg MS (1988) Inactivation of amino acid transport in rat hepatocytes and hepatoma cells by PCMBS. Am J Physiol 255:C340–C345

Clarkson TW (1972) The pharmacology of mercury compounds. Annu Rev Pharmacol 22:375–406

Clarkson TW (1993) Molecular and ionic mimicry of toxic metals. Annu Rev Pharmacol Toxicol 32:545–571

Crofts JN, Barritt GJ (1990) The liver cell plasma membrane Ca^{2+} inflow systems exhibit a broad specificity for divalent metal ions. Biochem J 269:579–587

Curran PF (1972) Effect of silver ion on permeability properties of frog skin. Biochim Biophys Acta 288:97

Darwish HM, Cheney JC, Schmitt RC, Ettinger MJ (1984) Mobilization of copper (II) from plasma components and mechanism of hepatic copper transport. Am J Physiol 246:G72–G79

Delmondedieu M, Boudou A, Desmazes J-P, Georgescauld D (1989) Interaction of mercury chloride with the primary amine group of model membranes containing phosphatidylserine and phosphatidylethanolamine. Biochim Biophys Acta 986:191–199

Dutczak WJ, Ballatori N (1994) Transport of the glutathione-methylmercury complex across liver canalicular membranes on GSH carriers. J Biol Chem 269: 9746–9751

Ferreira KTG (1978) The effect of copper on frog skin. The role of sulfhydryl groups. Biochim Biophys Acta 510:298–304

Flanagan JL, Friedman PA (1991) Pathathyroid hormone-stimulated cadmium accumulation in Madin-Darby canine kidney cells. Toxicol Appl Pharmacol 109:241–250

Foulkes EC (1986) Cadmium. In: Born GVR, Farah A, Herken H, Welch AD (eds) Handbook of experimental pharmacology. Springer, Berlin Heidelberg New York, p 400

Frame MDS, Milanick MA (1991) Mn and Cd transport by the Na-Ca exchanger of ferret red blood cells. Am J Physiol 261:C467–C475

Frenkel A, Ekblad EBM, Edelman IS (1975) Effects of sulfhydryl reagents on basal and vasopressin-stimulated Na^+ transport in the toad bladder. In: Eisenberg H, Katchalski-Katair E, Manson LA (eds) Biomembranes, vol 7. Plenum, New York, pp 61–80

Fukuda J, Kawa K (1977) Permeation of manganese, cadmium, zinc, and beryllium through calcium channels of an insect muscle membrane. Science 196:309–311

Gamba G, Saltzberg SN, Lombardi M, Miyanoshita A, Lytton J, Hediger MA, Brenner BM, Hebert SC (1993) Primary structure and functional expression of a cDNA encoding the thiazide-sensitive, electroneutral sodium-chloride cotransporter. Proc Natl Acad Sci USA 90:2749–2753

Gregus Z, Stein AF, Varga F, Klaassen CD (1992) Effect of lipoic acid on biliary excretion of glutathione and metals. Toxicol Appl Pharmacol 114:88–96

Gutknecht J (1981) Inorganic mercury (Hg^{2+}) transport through lipid bilayer membranes. J Membr Biol 61:61–66

Hillyard SD, Sera R, Gonick HC (1979) Effects of Cd^{++} on short-circuit current across epithelial membranes. II. Studies with the isolated frog skin epithelium, urinary bladder, and large intestine. J Membr Biol 46:283–294

Hinkle PM, Kinsella PA, Osterhoudt KC (1987) Cadmium uptake and toxicity via voltage-sensitive calcium channels. J Biol Chem 262:16333–16337

Hinkle PM, Shanshala ED II, Nelson EJ (1992) Measurement of intracellular cadmium with fluorescent dyes. J Biol Chem 267:25553–25559

Hughes BP, Barritt GJ (1989) Inhibition of the liver cell receptor-activated Ca^{2+} inflow system by metal ion inhibitors of voltage-operated Ca^{2+} channels but not by other inhibitors of Ca^{2+} inflow. Biochim Biophys Acta 1013:197–205

Hughes WH (1957) A physiochemical rationale for the biological activity of mercury and its compounds. Ann NY Acad Sci 65:454–460

Hurwitz L (1986) Pharmacology of calcium channels and smooth muscle. Annu Rev Pharmacol Toxicol 26:225–258

Imesch E, Moosmayer M, Anner BM (1992) Mercury weakens membrane anchoring of Na-K-ATPase. Am J Physiol 262:F837–F842

Jungwirth A, Ritter M, Lang F (1991a) Influence of mercury ions on electrical properties of rat proximal and distal renal tubules. Nephron 58:229–232

Jungwirth A, Ritter M, Paulmichl M, Lang F (1991b) Activation of cell membrane potassium conductance by mercury in cultured renal epitheloid (MDCK) cells. J Cell Physiol 146:25–33

Kalfakakou V, Simons TJB (1990) Anionic mechanisms of zinc uptake across the human red cell membrane. J Physiol (Lond) 421:485–497
Karniski LP (1992) Hg^{2+} and Cu^{+} are ionophores, mediating Cl^{-}/OH^{-} exchange in liposomes and rabbit renal brush border membranes. J Biol Chem 267: 19218–19225
Kaur P, Rosen BP (1992) Plasmid-encoded resistance to arsenic and antimony. Plasmid 27:29–40
Kerper LE, Ballatori N, Clarkson TW (1992) Methylmercury transport across the blood-brain barrier by an amino acid carrier. Am J Physiol 262:R761–R765
Kinter WB, Pritchard JB (1977) Altered permeability of cell membranes. In: DHK See (ed) Handbook of physiology—reactions to environmental agents. American Physiol Society, Baltimore, MD, pp 563–576
Knauf PA, Rothstein A (1971) Effects of sulfhydryl and amino reactive reagents on anion and cation permeability of the human red blood cell. J Gen Physiol 58:190–210
Kone BC, Kaleta M, Gullens SR (1988) Silver ion (Ag^{+})-induced increases in cell membrane K^{+} and Na^{+} permeability in the renal proximal tubule: reversal by thiol reagents. J Membr Biol 102:11–19
Kone BC, Brenner RM, Gullens SR (1990) Sulfhydryl-reactive heavy metals increase cell membrane K^{+} and Ca^{2+} transport in renal proximal tubule. J Membr Biol 113:1–12
Kong C-T, Yet S-F, Lever JE (1993) Cloning and expression of a mammalian Na^{+}/amino acid cotransporter with sequence similarities to Na^{+}/glucose cotransporters. J Biol Chem 268:1509–1512
Lambert IH, Kramhoft B, Hoffmann EK (1984) Effect of copper on volume regulation in Ehrlich ascites tumour cells. Mol Physiol 6:83–98
Loghman-Adham M (1992) Inhibition of renal Na^{+}-Pi cotransporter by mercuric chloride: role of sulfhydryl groups. J Cell Biochem 49:199–207
Lou M, Garay R, Alda Torrubia JO (1991) Cadmium uptake through the anion exchanger in human red blood cells. J Physiol (Lond) 443:123–136
Madsen NB (1963) Mercaptide-forming agents. In: Hoschester RM, Quastell JH (eds) Metabolic inhibitors. Academic, New York
Magos L, Webb M (1980) The interactions of selenium with cadmium and mercury. CRC Critical Rev Toxicol 1–38
Martell AE (1981) Chemistry of carcinogenic metals. Environ Health Perspect 40:207–226
McArdle HF (1992) The transport of iron and copper across the cell membrane: different mechanisms for different metals? Proc Nutr Soc 51:199–209
Nachshen DA (1984) Selectivity of the Ca binding site in synaptosome Ca channels: inhibition of Ca influx by multivalent metal cations. J Gen Physiol 83:941–967
Nemeth EF, Scarpa A (1987) Rapid mobilization of cellular Ca^{2+} in bovine parathyroid cells evoked by extracellular divalent cations. J Biol Chem 262:5188–5196
Oehme FW (1978) Mechanisms of heavy metal inorganic toxicities. In: Oehme (ed) Toxicity of heavy metals in the environment. Dekker, New York, pp 69–85
Orion Research Inc (1970) Mercury by electrode. Orion Newslett 11:41–42
Ortiz DF, Kreppel L, Speiser DM, Scheel G, McDonald G, Ow DW (1992) Heavy metal tolerance in the fission yeast requires an ATP-binding cassette-type vacuolar membrane transporter. EMBO J 11:3491–3499
Percival SS, Harris ED (1990) Copper transport ceruloplasmin: characterization of the cellular uptake mechanism. Am J Physiol 258:C140–C146
Philipson KD (1985) Sodium-calcium exchange in plasma membrane vesicles. Annu Rev Physiol 47:561–571
Preston GM, Jung JS, Guggino WB, Agre P (1993) The mercury-sensitive residue at cysteine 189 in the CHIP28 water channel. J Biol Chem 268:17–20

Rabenstein DL, Fairhurst MT (1975) Nuclear magnetic resonance studies of the solution chemistry of metal complexes. XI. The binding of methylmercury by sulfhydryl-containing amino acids and by glutathione. J Am Chem Soc 97: 2086–2092
Rothstein A (1970) Sulfhydryl groups in membrane structure and function. In: Bronner F, Kleinzeller A (eds) Current topics in membranes and transport. Academic, New York, pp 135–176
Rothstein A, Mack E (1991) Actions of mercurials on cell volume regulation of dissociated MDCK cells. Am J Physiol 260:C113–C121
Schaeffer JF, Preston RL, Curran PF (1973) Inhibition of amino acid transport in rabbit intestine by *P*-chloromercuriphenyl sulfonic acid. J Gen Physiol 672: 131–146
Schatzmann HJ (1982) The plasma membrane calcium pump of erythrocytes and other animal cells. In: Carafoli E (ed) Membrane transport of calcium. Academic, New York, pp 41–108
Scholtz E, Zeiske W (1988) A novel synergistic stimulation of Na^+-transport across frog skin (Xenopus laevis) by external Cd^{2+}- and Ca^{2+}-ions. Pflugers Arch 413:174–180
Schwartz JH, Flamenbaum W (1976) Heavy metal-induced alterations in ion transport by turtle urinary bladder. Am J Physiol 230:1582–1589
Sheets MF, Hanck DA (1992) Mechanisms of extracellular divalent and trivalent cation block of the sodium current in canine cardiac purkinje cells. J Physiol (Lond) 454:299–320
Sillen LG, Martell AE (1971) Stability constants of metal ion complexes. Chemical Society, Special publication no 25
Silver S, Nucifora G, Chu L, Misra TK (1989) Bacterial resistance ATPases: primary pumps for exporting toxic cations and anions. Trends Biochem Sci 14:76–80
Simons TJB (1986) The role of anion transport in the passive movement of lead across the human red cell membrane. J Physiol (Lond) 378:287–312
Simons, TJB (1991) Calcium-dependent zinc efflux in human red blood cells. J Membr Biol 123:73–82
Simons TJB, Pocock G (1987) Lead enters bovine adrenal medullary cells through calcium channels. J Neurochem 48:383–389
Skroch P, Buchman C, Karin M (1993) Regulation of human and yeast metallothionein gene transcription by heavy metal ions. Prog Clin Biol Res 380:113–128
Smith JB, Dwyer SD, Smith L (1989) Cadmium evokes inositol polyphosphate formation and calcium mobilization. J Biol Chem 264:7115–7118
Smith MW, Phelps PC, Trump BF (1991) Cytosolic Ca^{2+} deregulation and blebbing after $HgCl_2$ injury to cultured rabbit proximal tubule cells as determined by digital imaging microscopy. Proc Natl Acad Sci USA 88:4926–4930
Smith RM, Martell AE (1976) Critical stability constants. Plenum, New York
Stirling CE (1975) Mercurial perturbation of brush border membrane permeability in rabbit ileum. J Membr Biol 23:33–56
Stockand J, Sultan A, Molony D, DuBose T Jr, Sansom S (1993) Interactions of cadmium and nickel with K channels of vascular smooth muscle. Toxicol Appl Pharmacol 121:30–35
Tanaka K, Sueda K, Onosaka S, Okahara K (1975) Fate of ^{109}Cd-labeled metallothionein in rats. Toxicol Appl Pharmacol 33:258–266
Templeton DM, Cherian MG (1983) Cadmium and hypertension. TIPS Rev 4: 501–503
Tomsig JL, Suszkiw JB (1990) Pb^{2+}-induced secretion from bovine chromaffin cells: fura-2 as a probe for Pb^{2+}. Am J Physiol 259:C762–C768
Verbost PM, Flik G, Pang PKT, Lock RAC, Bonga SEW (1989) Cadmium inhibition of the erythrocyte Ca^{2+} pump. J Biol Chem 264:5613–5615

Vulpe C, Levinson B, Whitney S, Packman S, Gitschier J (1993) Isolation of a candidate gene for Menkes disease and evidence that it encodes a copper-transporting ATPase. Nature Genetics 3:7–13
Webb JL (1966) Mercurials. Enzyme and metabolic inhibitors, vol 2, chap 7. Academic, New York, p 729
Weed R, Eber J, Rothstein A (1962) Interaction of mercury with human erythrocytes. J Gen Physiol 45:395–410
Wetterhahn-Jennette K (1981) The role of metals in carcinogenesis: biochemistry and metabolism. Environ Health Perspect 40:233–252
Wilkinson DJ, Post MA, Venglarik C, Chang D, Dawson D (1993) Mercury blockade of thiazide-sensitive NaCl cotransport in flounder urinary bladder. Toxicol Appl Pharmacol 122:170–176
Zhang R, vanHoek AN, Biwersi J, Verkman AS (1993) A point mutation at cysteine 189 blocks the water permeability of rat kidney water channel CHIP28k. Biochemistry 32:2938–2941

CHAPTER 4

Immunotoxicology of Metals

L. PELLETIER and P. DRUET

A. Introduction

Numerous metals are responsible for immunologically mediated disorders in humans (Table 1). One of the most common is contact dermatitis, which is probably due, to delayed hypersensitivity reactions. Thus, chromium salts, nickel, cobalt, beryllium, mercury, gold salts, and platinum salts may induce contact dermatitis. For example, a study of patients suffering from contact dermatitis has found that 13% showed a positive patch test to nickel salts, 8% to potassium dichromate, and 5% to mercurials (BUEHLER 1983). Other manifestations may be observed. Thus, in fluorescent lamp workers beryllium may induce a chronic pulmonary granulomatosis, so-called berylliosis (REEVES 1983). It has been suggested that this affection is autoimmune due to an immune response to an autoantigen modified by beryllium. It is of interest to note that hypersensitivity reactions to beryllium are found in patients with contact dermatitis but are often absent in patients with severe berylliosis. Besides toxic effects on the kidney and the central nervous system, mercurials may induce immunologically mediated glomerulopathies (FILLASTRE et al. 1988). Histologically, this type of glomerulopathy can be membranous glomerulopathy (MG), showing granular IgG deposits with immunofluorescence, or a minimal glomerular change disease thought to be mediated by T lymphocytes. In some cases, immunoglobulins were found linearly deposited along the glomerular basement membrane (GBM) and were either associated or not associated with granular IgG deposits (LINDQVIST et al. 1974; FILLASTRE et al. 1988). This suggests that, in some patients, anti-GBM antibodies are produced. An increased prevalence of anti-laminin antibodies (laminin is a component of GBM) was reported in workers exposed to mercury vapor (LAUWERYS et al. 1983), but this observation was not confirmed in a later study (BERNARD et al. 1987). Besides skin manifestations, gold salts may induce cytopenia, hepatitis, and glomerulopathies. Thus, abnormal proteinuria is found in 6%–17% of rheumatoid arthritis patients treated with gold salts and, in most cases, the glomerulopathy is MG (DRUET et al. 1987; HALL 1982; KATZ et al. 1984). There is a causal relationship between the occurrence of MG since such glomerular lesions have been rarely reported among rheumatoid patients who did not receive gold salts or related drugs. The occurrence of MG is not correlated with the

Table 1. Disorders induced by metals in humans

Metal	Disorder
Nickel	Contact dermatitis
Chromium salts	Contact dermatitis
Beryllium	Contact dermatitis, granulomatous pneumonitis
Mercurials	Contact dermatitis, glomerulopathy
Gold salts	Contact dermatitis, cytopenia, hepatitis, glomerulopathy
Platinum salts	Asthma

cumulative dose of gold, with the duration of treatment, or with the gold salts used. Proteinuria is only observed in some patients and is not dose related, which suggests that susceptibility may be genetically determined. This is supported by the fact that patients with HLA-B8 or DRW3 antigens are at higher risk (WOOLEY et al. 1980). Platinum salts are implicated in the occurrence of asthma thought to be mediated by specific IgE (PEPYS 1983).

In most cases, the nature of the mechanisms at play are unknown, and it is not even known if the disease is due to a response against the toxic (hypersensitivity reaction) or to a response against an autoantigen modified or not by the metal. In this chapter, we report first the basis of the immune response that could be useful for understanding how a metal may induce immune adverse reactions. Thereafter, we describe data obtained from T-cell clones derived from patients with nickel or gold hypersensitivity reactions. In the third section, we discuss experimental data demonstrating that metals such as mercury or gold may induce autoimmunity at least in susceptible strains. Finally, we report a model of immunosuppression induced by mercury in Lewis (LEW) rats, resulting from a disregulation of the immune system.

B. Basis of the Immune Response

T cells do not recognize native antigens. Antigens need to be processed by antigen-presenting cells (APCs), which are essentially dendritic or macrophage-monocytes but also Langerhans cells that nonspecifically internalize the antigen (ZIEGLER and UNANUE 1981). B cells may also act as APCs but they only process the antigen they specifically recognize through their surface immunoglobulin (B-cell antigen receptor) (DAVIDSON et al. 1991). Peptides derived from the antigens are taken in charge by MHC class II molecules and conveyed to the cell membrane (BUUS et al. 1986; DAVIDSON et al. 1991). The peptide is then recognized by a specific T-helper CD4+ lymphocyte which is then activated (SCHWARTZ 1985; PAUL 1989; KOURILSKY and CLAVERIE 1989). This activated T cell produces interleukins, proliferates, and allows the differentiation of B or cytotoxic CD8+ T cells (PAUL 1989). B cells activated by a cognate interaction with T cells proliferate and

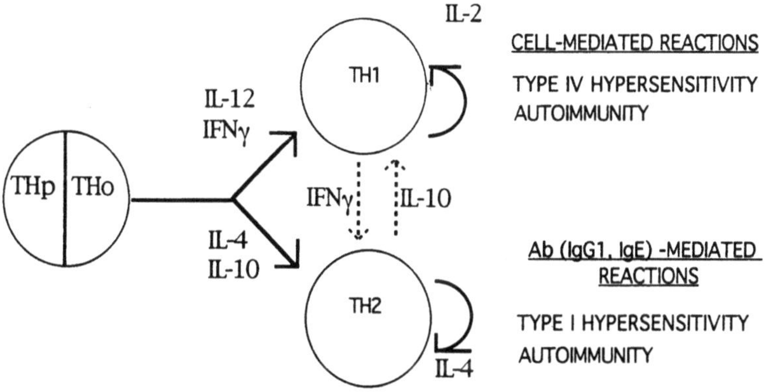

Fig. 1. Differentiation of CD4*.001 T cells. Depending upon the presence of IL-4, IL-10 or IFNg, IL-12, CD4*.001, T cells stimulated by the antigen differentiate either into TH1 or in TH2 cells. The former cells are implicated in DTH and the latter ones mainly in B-cell help

differentiate under the control of various interleukins, in plasma cells secreting antibodies. Depending upon the cytokine production and their functions, CD4+ T cells are divided into TH1 and TH2 cells (MOSSMANN et al. 1986; MOSSMANN and COFFMAN 1989). TH1 cells produce interleukin-2 (IL-2), gamma-interferon (IFNγ), and tumor necrosis factor (TNF)β, are mainly responsible for delayed-type hypersensitivity (DTH) reactions, and can cooperate with B cells for the production of antibodies of the IgG2a isotype in mice. TH2 cells produce IL-4, IL-5, IL-6, and IL-10 and are mainly involved in B-cell help for the production of IgA, IgG1, and IgE isotypes. TH1 and TH2 cells antagonize each other; in addition, IL-4 favors the differentiation of precursor T cells into TH2 while IFNγ favors that of TH1 cells (Fig. 1). TH1 and TH2 cells also exist in humans (ROMAGNANI 1991). The way metals are recognized by the immune system and whether processing of metal-bound proteins is required for T-cell activation are unclear.

There are four main types of hypersensitivity reactions: type I, II, and III hypersensitivity reactions are associated with the production of antibodies while type IV is cell mediated. Immediate hypersensitivity (type I) is due to IgE elicited by the antigen that binds to Fcε receptor of mast cells or eosinophils, leading to their activation. Asthma due to platinum salts might be due to specific IgEs that have been found in sensitized patients. In addition, it seems that this agent triggers IgE production in rats (PEPYS 1983). However, to our knowledge, this model has not been extensively studied. Type IV or delayed type hypersensitivity (DTH) is probably due to TH1 cells and is probably frequently involved in metal-mediated adverse side reactions.

C. Hypersensitivity Reactions

Useful tools have been provided by deriving T-cell clones from patients suffering from nickel- or gold-contact dermatitis that is probably the consequence of DTH reactions (SINIGAGLIA et al. 1985; KAPSENBERG et al. 1988; ROMAGNOLI et al. 1991; ROMAGNOLI et al. 1992). The clones have been obtained either from lesional tissue biopsies or from peripheral blood of patients. Proliferation of some clones is restricted by MHC class II encoded molecules mainly by HLA-DR (DR-5, DR-W11, etc.), but also by the HLA-DQ element for some lines (SINIGAGLIA et al. 1985; KAPSENBERG et al. 1988). All the T-cell lines are CD3+CD4+CD8−. All the clones produce high levels of IL-2 and IFNγ, which could explain their role in the hypersensitivity lesions (SINIGAGLIA et al. 1985). Surprisingly, the clones are able to polyclonally activate autologous B cells for the production of IgM, IgG, and IgA. The production of IL-4 and the other TH2-dependent cytokines has not been looked for and we do not know if the lines are TH1 cells or if they belong to a subset able to secrete both TH1 and TH2 cytokines. The T-cell repertoire of antinickel lines is broad even in one patient (KAPSENBERG et al. 1988). In some clones, the pattern of HLA-class II restriction is unusual since the proliferation of the line seems to require an interaction with nonpolymorphic elements of the MHC class II molecule. Finally, nickel interacts directly with some T cells to mediate their proliferation in the absence of APCs (SINIGAGLIA et al. 1985). In that respect, it would mimick the effect of lectins or of some anti-CD3 antibodies. Since nickel has a high affinity for histidine residues, it appeared that this metal could bind some T-cell receptors, MHC class II molecules, or MHC-bound peptides through histidine residue, which would explain the diversity of the T-cell activation pathways. It has been elegantly shown that nickel interacts with some peptides presented by DRw11 molecules, leading to an absence of stimulation of a T-cell clone specific for the peptide (ROMAGNOLI et al. 1991). In addition, the authors showed that, if they use a variant peptide in which the histidine residue was replaced by a lysine, the new peptide was always recognized by the T-cell clone but nickel became unable to block the T-cell stimulation. Using a similar approach, ROMAGNOLI et al. (1992) analyzed gold-specific T-lymphocyte clones obtained from a patient with rheumatoid arthritis who developed DTH reactions to gold. The clones recognized gold in the context of DR1 molecules and the proliferation did not require antigen processing since fixation of APCs by glutaraldehyde does not affect their ability to "present gold." Au(I) is as efficient as Au(III) in stimulating the T-cell lines. As for nickel, gold may inhibit the recognition of some peptides to the specific hybridomas. It is possible that the beneficial effect of gold on rheumatoid arthritis, a typical autoimmune disease, was due to an interference of gold with the presentation of a self peptide to the autoaggressive T-cell line. In turn, the complex gold-peptide-MHC class II molecule might trigger specific T cells responsible for DTH reactions. It would be interesting to correlate the efficacy of the treatment with the occurrence of DTH reaction. In any case, the molecular basis of the gold-peptide or gold-MHC class II interaction remains to be determined. Another point to be underlined is that anti-gold T cells have not been elicited from patients with gold-induced cytopenia or glomerulopathies (VERWILGHEN et al. 1992). Thus, it is possible that pathogenesis of these diseases is different from those implied in DTH reactions.

D. Experimental Models of Metal-Induced Autoimmunity

It is not known whether metals may trigger autoimmune manifestations in humans, although the development of experimental models suggests that they have the potential to do so with certain genetic backgrounds.

I. Description of the Model

1. $HgCl_2$-Induced Autoimmunity in Rats

Brown Norway (BN) rats injected with nontoxic amounts of $HgCl_2$ (100 μg/10 g body wt. three times a week) develop T-dependent B-cell polyclonal activation which is responsible for the considerable increase in serum IgE concentration (from less than 10 μg/ml to over 10 mg/ml) and for the production of many antibodies directed against exogenous antigens [trinitrophenol (TNP), red blood cells, etc.] and autoantibodies (DNA, laminin, and other elements of glomerular basement membrane antibodies such as collagen IV and entactin, thyroglobulin, etc.) (SAPIN et al. 1977; PROUVOST-DANON et al. 1981; HIRSCH et al. 1982; BELLON et al. 1982; PELLETIER et al. 1988a; ATEN et al. 1988; PUSEY et al. 1990). Thymic atrophy is found as early as day 7. The disease is characterized by a glomerulopathy evolving in two phases: first linear anti-GBM antibody deposition along the glomerular capillary pattern and, second, a change in the pattern of immunofluorescence with the appearance of granular glomerular IgG deposits (SAPIN et al. 1977; DRUET et al. 1978; BELLON et al. 1982). This second phase corresponds to what is found in patients with mercury-induced MG. Clinically, rats exhibit an abnormal proteinuria frequently associated with the nephrotic syndrome but there is no renal failure (DRUET et al. 1978; BELLON et al. 1982). Rats also develop mucositis, Sjögren's syndrome (ATEN et al. 1988), and vasculitis of the gut (MATHIESON et al. 1992), associated with antimyeloperoxidase antibodies (ESNAULT et al. 1992). The B-cell polyclonal activation does not affect all B-cell clones, because, for example, antimyelin basic protein antibodies are not found during the mercury disease (not shown). The role of T cells in targeting autoimmunity has been well demonstrated; thus, BN rats deprived of T cells either genetically or after adult thymectomy, lethal irradiation, and reconstitution with fetal liver cells are completely protected from mercury-induced autoimmunity (Pelletier et al. 1987a). In addition, under some experimental conditions, T cells from diseased BN rats may be sufficient to transfer the disease in naive syngeneic BN rats (Pelletier et al. 1988b).

Susceptibility to mercury-induced autoimmunity is genetically controlled (Druet et al. 1977; Sapin et al. 1982; Aten et al. 1991), and some strains such as LEW rats are completely resistant even when injected with high doses (400 instead of 100 μg/100 g body wt.). Some strains such as BN (Druet et al. 1977; Sapin et al. 1982; Aten et al. 1991) or MAXX (Henry et al. 1992) rats develop linear IgG deposits along the glomerular capillary wall while others such as DZB rats develop only membranous glomerulopathy (Aten et al. 1992). It is interesting to note that $HgCl_2$ induces a B-cell polyclonal activation in both BN (Hirsch et al. 1982) and DZB rats (Aten et al. 1992), associated with the occurrence of glomerulopathy. In addition, in the two strains, there is a tremendous increase in serum IgE concentration. Susceptibility depends upon three to four genes, depending upon the parameter looked for, one of which is major histocompatibility complex (MHC)-linked. The development of histological lesions is inherited as a dominant trait while it is suggested that the development of proteinuria in rats which possess genes for mercury-induced susceptibility requires homozygosity for one gene of the RT1 complex and at least one gene of non-RT1 background (Aten et al. 1991). The development of $HgCl_2$-induced immune complex type glomerulonephritis in the absence of linear deposition of anti-GBM antibodies is also genetically determined and one gene is linked to MHC class II molecules (Aten et al. 1991). Thus, AO strain (RT1-u) is susceptible while AO.1P congenic rats differing from the former at the RT1B and RT1D loci(l) are resistant. The fact that $HgCl_2$-injected DZB rats (RT1-u) develop significant proteinuria while AO rats do not could be related to the presence of non-RT1 genes of BN origin in the former strain.

$HgCl_2$ given intravenously, orally, or intratracheally induces the disease in a similar way (Bernaudin et al. 1981). Exposure to mercury vapor also induces the autoimmune disease in BN rats (Hua et al. 1993). Methylmercury or pharmaceutical ointments and solutions containing organic mercury are effective even when these products are applied on wounds or even on normal skin (Druet et al. 1981).

In the susceptible strain, the effects of $HgCl_2$ depend upon dosage (P. Druet et al. 1978, unpublished observations). Thus, the increase in serum IgE concentration is positively correlated to the dosage and this parameter is less well regulated for low doses. Serum IgE concentration of rats injected with low doses (5 or 10 μg/100 g body wt. three times a week) exhibit kinetics with multiple waves.

2. $HgCl_2$-Induced Autoimmunity in Other Species

Mercury-induced immune disorders have also been described in mice and rabbits (ENESTRÖM and HULTMAN 1984; HULTMAN and ENESTRÖM 1987, 1988; ROMAN-FRANCO et al. 1978). Thus, ROMAN-FRANCO et al. (1978) described in rabbits anti-GBM-mediated glomerulopathy resembling that described in BN rats. Some strains of mice given $HgCl_2$ produce antinucleolear antibodies (ROBINSON et al. 1984, 1986; MIRTCHEVA et al. 1989); MHC-linked genes play an important role in this response and H-2^s mice are particularly susceptible. H-2^s mice also develop increased serum IgG and IgE concentrations (PIETSCH et al. 1989). The increase in serum IgE concentration is dependent upon IL-4 production since treatment with an anti-IL-4 monoclonal antibody prevents modification of IgE concentration; however, the titer of antinuclear antibodies is not modified (OCHEL et al. 1991). Some strains develop glomerulopathy; BALB/c mice develop glomerular IgG deposits (ENESTRÖM and HULTMAN 1984; HULTMAN and ENESTRÖM 1987, 1988), and antinuclear antibodies have been recovered from the kidneys (HULTMAN and ENESTRÖM 1988).

3. Gold-Induced Autoimmunity

Brown Norway rats and not LEW rats injected with gold salts also develop a disease very similar to that induced by $HgCl_2$ (TOURNADE et al. 1991); a shared idiotype is found in both models (GUÉRY et al. 1990), suggesting that the same B-cell clones are activated. However, preliminary data suggest that genetic control of susceptibility to gold-induced autoimmunity is different from that of mercury-induced autoimmune disease. In the former case, in contrast to the latter, MHC class II genes are not implicated or are permissive in both susceptible and resistant strains (KERMARREC et al., in preparation). As for mercury, gold salts induce a rise in serum IgE concentration and antinuclear antibody production in H-2^s-bearing mice (PIETSCH et al. 1989; ROBINSON et al. 1986).

II. Mechanisms of Induction

Three sets of experiments have been carried out to elucidate the mechanisms of induction of the disease:

1. Incubation of normal BN T cells with noncytotoxic amounts of $HgCl_2$ for 2 h was sufficient to trigger mRNA for IL-4. In addition, the protein was probably synthesized since the culture supernatant induced an increase in MHC class II expression on B cells (PRIGENT et al., in preparation). This phenomenon was not found when using LEW T cells. Therefore it seems that $HgCl_2$ acts on a factor controlling the transcription of IL-4 in BN T cells. So far, it has not been clear whether $HgCl_2$ allows transcription of the other genes implicated in T-cell activation such as IL-2 or RIL-2.

2. Autoreactive anti-class II T cells have been detected in BN rats injected with $HgCl_2$ or gold salts. In addition, in the latter, T-cell lines have been obtained. We have shown that they are CD4+CD8−, that they proliferate in the presence of syngeneic and not allogeneic APCs, and that their proliferation is completely abolished in the presence of an anti-class II monoclonal antibody. In addition, these cells are able to cooperate in vitro with normal BN B cells to trigger a rise in Ig and IgE production as well as the synthesis of anti-laminin, anti-DNA, and anti-TNP antibodies, which are all markers of the disease (CASTEDO et al. 1992). These data suggest that these T cells are at play in the B-cell polyclonal activation in vivo.

3. Neonatal administration of $HgCl_2$ to BN rats does not trigger the disease, protects adult rats from rechallenge to mercury, but does not modify the course of the disease induced by gold salts in adult rats (FILLION et al., in preparation). This suggests that $HgCl_2$ is able to specifically render animals tolerant to Hg-induced immune disorders.

Several points merit discussion: mRNA for IL-4 is probably induced as a consequence of a direct effect of $HgCl_2$ on IL-4 expression, while specific

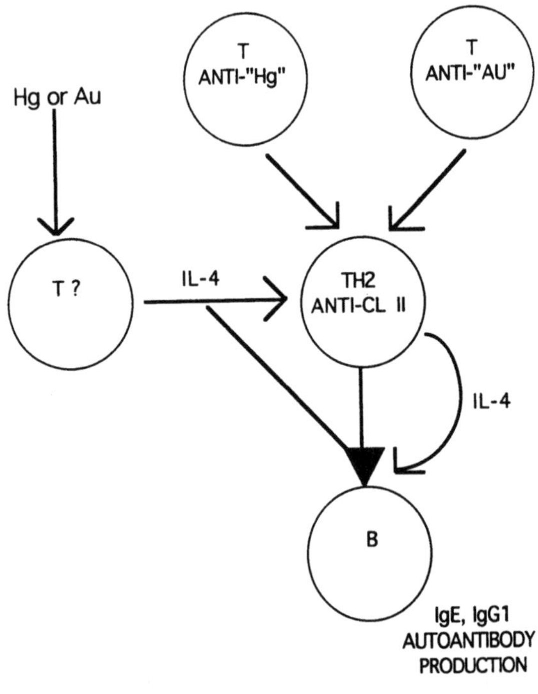

Fig. 2. Model of induction of autoreactive anti-class II T cells in BN rats treated by $HgCl_2$ or gold salts. On one hand, these metals trigger IL-4 synthesis by T cells and, On the other hand, these metals would activate specific T cells that would probably recognize the complex MHC class II*.001 metal and, as a consequence, autoreactive T cells specific for native MHC class II molecules would be activated. Due to the presence of IL-4 these T cells would differentiate into TH2 cells

elicitation of tolerance to mercury implies that the immune system recognizes mercury or a structure modified by mercury. Anti-class II T cells are present in both gold- and mercury-induced immunity and seem to play a role in the B-cell polyclonal activation. However, elicitation of tolerance to mercury does not give protection from gold-induced disease, which suggests that their mode of induction is different in the two models. These data lead us to propose the following model (Fig. 2).

1. Mercury or gold triggers the transcription of the gene for IL-4 by a method that remains to be elucidated. In this respect it is of note that organic and inorganic mercury may induce a rise in the intracellular concentration of calcium, a step that is implicated in cellular activation (TAN and TANG 1993). It is also of note that metals such as zinc, gold, copper, or platinium may interact with DNA-binding factors and thus modify gene expression (HANDEL et al. 1991; FÜRST et al. 1988; PIL and LIPPARD 1992). In the future, it will be important to study the ways by which metals may activate lymphocytes (second messengers, ion channels, enzymes at play, etc.).

2. Mercury or gold triggers anti-mercury and anti-gold T cells, respectively. As described in the hypersensitivity reactions, these cells could recognize either class II modified by mercury or gold or a self-peptide presented by a given MHC class II molecule. At variance with what is found in DTH reactions, in the presence of IL-4, these cells would differentiate into TH2 cells specialized in B-cell help. Autoreactive anti-class II T cells able to interact with native class II molecules expressed by normal B cells would be induced in turn as has been described in other situations (LIN et al. 1991). These latter cells would be responsible for the B-cell polyclonal activation induced by the transfer of T cells from diseased rats into normal BN recipients.

Several other questions still need to be answered:

How are autoreactive anti-native class II T cells induced and do they recognize empty class II molecules or a self peptide presented by class II molecules? In this respect, it is of note that peptides derived from class II molecules themselves (RUDENSKY et al. 1991; LIU et al. 1992) can be presented by MHC class II molecules. Thus, an autoreactive T-cell clone has been shown to recognize a peptide from DR1 expressed by DR1 molecules (LIU et al. 1992); it cannot be excluded that $HgCl_2$-induced anti-class II T cells recognize a peptide from MHC class II molecules presented by the intact class II molecule.

Do "anti-mercury or anti-gold" T cells exist in BN rats as has been suggested in mice (GLEICHMANN et al. 1989) and what do they recognize exactly? For example, it would be of interest to determine whether the profile of self-peptides expressed by MHC class II monecules is modified by mercury or gold.

III. Autoregulation

$HgCl_2$- and gold salt-induced autoimmunity are spontaneously regulated in surviving BN rats even if toxic injections are continued. Multiple factors are probably implied in spontaneous regulation and it is possible that the presence of only one of them is sufficient for spontaneous regulation. CD8+ suppressor cells able to inhibit autoreactive cells have been described as being concomitant with the improvement of rats given $HgCl_2$ (ROSSERT et al. 1988; BOWMAN et al. 1984), but depletion of CD8+ does not prevent regulation (PELLETIER et al. 1990; MATHIESON et al. 1991). However, such treatment renders rats that have improved more susceptible to rechallenge with $HgCl_2$ (MATHIESON et al. 1991), showing that suppressor cells play an important role in the acquired resistance to $HgCl_2$-induced autoimmunity. Anti-idiotypic antibodies have also been demonstrated (CHALOPIN and LOCKWOOD 1984; GUÉRY and DRUET 1990), but their role has not been proven. The decrease in the frequency of autoreactive T cells (ROSSERT et al. 1988) could be due either to death or to functional inactivation of cells. Finally, it has been shown recently that there is a change in the balance TH1/TH2 at the time of recovery with emergence of TH1 cells. Indeed, this phase is associated with the production of IFNγ (ATEN 1992) and a better capacity to produce IL-2 in response to various stimuli than normal BN rats (VAN DER MEIDE et al, unpublished). Moreover, treatment of BN rats with an anti-IL-2 receptor monoclonal antibody, which preferentially impairs TH1 cells, delays the return of serum IgE concentration to normal values (DUBEY et al. 1993).

E. Nonantigen-Specific Immunosuppression Induced by $HgCl_2$

Lewis rats injected with $HgCl_2$ do not exhibit autoimmunity and develop a non-antigen-specific immunosuppression (PELLETIER et al. 1987b). This immunosuppression is mediated by non-antigen-specific suppressor T cells responsible for depression of T-cell functions and protection against organ-specific autoimmune diseases such as Heymann's nephritis (PELLETIER et al. 1987c) or experimental autoimmune encephalomyelitis (EAE) (PELLETIER et al. 1988c). In the latter case, it has been shown that suppressor cells are CD8+ since treatment of $HgCl_2$-injected LEW rats immunized with myelin and treated with an anti-CD8 mAb completely abrogates protection (PELLETIER et al. 1991). As demonstrated by limiting dilution analysis in this model, non-antigen-specific suppressor cells are much more frequent than autoreactive anti-myelin T cells (ROSSERT et al. 1991), showing that, in LEW rats, $HgCl_2$ does not affect activation of myelin-specific autoreactive cells but acts thereafter.

Resistance of LEW rats to mercury-induced autoimmunity is due neither to CD8+ suppressor cells since treatment with an anti-CD8 mAb does not trigger autoimmunity (Pelletier et al. 1990) nor to the absence of autoreactive T cells since $HgCl_2$ triggers autoreactive anti-class II T cells as frequently as in BN rats (Castedo et al. 1993). It could be due to a difference in the nature of the autoreactive T cells in the two strains: they could belong to the TH1 subtype in LEW rats and to the TH2 one in BN rats. Indeed, a CD4+ autoreactive anti-class II T cell line from $HgCl_2$-injected LEW rats has been derived that is probably of the TH1 subtype and that acts as a suppressor/inducer line since it affords protection from active EAE by the bias of suppressor CD8+ cells (Castedo et al. 1993). The fact that LEW autoreactive anti-class II T cells produce IL-2 could contribute to the activation of suppressor cells. It would be useful to develop suppressor lines to study their specificity. It also remains to be determined whether these cells act via a direct interaction with the cell to be suppressed or whether cytokines such as TGFβ (Karpus and Swanborg 1991) are at play.

F. Conclusions

Metals may have the ability to induce dysfunctions of the cells of the immune system. First, they can bind the complex (self)peptide-class II molecule in such a way that they prevent the activation of the specific T cells. They could have a therapeutic effect on autoimmunity by this bias (as for gold in rheumatoid arthritis). Alternatively, they could generate neoantigens leading either to hypersensitive reactions or even to autoimmune reactions. The pattern of cytokines synthesized following T-cell stimulation is probably also important in the type of responses triggered by metals. It could depend upon both the genetic ability to generate TH1 versus TH2 responses and the direct effect of some metals on the transcription of cytokine genes.

References

Aten J (1992) Autoreactivity to renal antigens in models of polyclonal B cell activation. Thesis, University of Leiden, p 22

Aten J, Bosman CB, Rozing J, Stjnen T, Hoedemaeker PJ, Weening JJ (1988) Mercuric chloride-induced autoimmunity in the Brown Norway rat. Cellular kinetics and major histocompatibility complex antigen expression. Am J Pathol 133:127–138

Aten J, Veninga A, de Heer E, Rozing J, Nieuwenhuis P, Hodemaeker P, Weening JJ (1991) Susceptibility to the induction of either autoimmunity or immunosuppression by mercuric chloride is related to the MHC haplotype. Eur J Immunol 21:611–616

Aten J, Veninga A, Bruijn JA, Prins FA, De Heer E, Weening JJ (1992) Antigenic specificities of glomerular-bound autoantibodies in membranous glomerulopathy induced by mercuric chloride. Clin Immunol Immunopathol 63:89–102

Bellon B, Capron M, Druet E, Verroust P, Vial MC, Sapin C, Girard JF, Foidart JM, Mahieu P, Druet P (1982) Mercuric chloride-induced autoimmunity in Brown Norway rats: sequential search for anti-basement membrane antibodies. Eur J Clin Invest 12:127–133

Bernard AM, Roels HR, Foidart JM, Lauwerys RL (1987) Search for anti-laminin antibodies in the serum of workers exposed to cadmium, mercury vapour or lead. Int Arch Occup Environ Health 59:303–309

Bernaudin JF, Druet E, Druet P, Masse P (1981) Inhalation or ingestion of organic or inorganic mercurials produces autoimmune disease in rats. Clin Immunol Immunopathol 20:129–135

Bowman C, Mason DW, Pusey CD, Lockwood CM (1984) Autoregulation of autoantibody synthesis in mercuric chloride in the Brown Norway rat. I. A role for T suppressor cells. Eur J Immunol 14:464–470

Buehler EV (1983) Experimental contact sensitivity. In: Gibson CG, Hubbard R, Parke DV (eds) Immunotoxicology. Academic, London, p 133

Buus S, Sette A, Colon SM, Jenis DM, Grey HM (1986) Isolation and characterization of antigen-Ia complexes involved in T cell recognition. Cell 47:1071–1077

Castedo M, Pelletier L, Druet P (1992) A role for autoreactive anti-class II T cells in gold salt-triggered B cell polyclonal activation in the Brown-Norway (BN) rat. JASN 3:578 (abstract)

Castedo M, Pelletier L, Rossert J, Pasquier R, Villarroya H, Druet P (1993) Mercury-induced autoreactive anti class II T cell line protects from experimental autoimmune encephalomyelitis by the bias of antiergotypic cells in Lewis rats. J Exp Med 177:881–890

Chalopin JM, Lockwood CM (1984) Autoregulation of autoantibody synthesis in mercuric chloride in the Brown Norway rat. II. Presence of antigen augmentable plaque forming cells in the spleen is associated with humoral factors behaving as auto-antiidiotypic antibodies. Eur J Immunol 14:470–475

Davidson HW, Reid PA, Lanzavecchia A, Watts C (1991) Processed antigen binds to newly synthesized MHC class II molecules in antigen-specific B Iymphocytes. Cell 67:105–116

Druet E, Sapin C, Günther E, Feingold N, Druet P (1977) Mercuric chloride induced anti glomerular basement membrane antibodies in the rat. Genetic control. Eur J Immunol 7:348–351

Druet P, Druet E, Potdevin F, Sapin C (1978) Immune type glomerulonephritis induced by $HgCl_2$ in the Brown Norway rat. Ann Immunol (Inst Pasteur) 129C:777–792

Druet P, Jacquot C, Baran D, Kleinknecht D, Fillastre J-P, Mery J-P (1987) Immunologically mediated nephritis induced by toxin and drugs. In: Bach PH, Lock EA (eds) Nephrotoxicity in the experimental and clinical situation. Nijhoff, Amsterdam, p 727

Druet P, Teychenne P, Mandet C, Bascou C, Druet P (1981) Immune type glomerulonephritis induced in the Brown-Norway rat with mercury containing pharmaceutical products. Nephron 28:145–148

Dubey C, Kuhn J, Vial M-C, Druet P, Bellon B (1993) Anti-interleukin 2 receptor monoclonal antibody therapy supports a role for TH1-like cells in $HgCl_2$-induced autoimmunity in rats. Scand J Immunol 37:406–412

Eneström S, Hultman P (1984) Immune-mediated glomerulonephritis induced by mercuric chloride in mice. Experientia 40:1234–1240

Esnault VLM, Mathieson PW, Thiru S, Oliveira DBG, Lockwood CM (1992) Autoantibodies to myeloperoxidase in Brown Norway rats treated with mercuric chloride. Lab Invest 67:114–120

Fillastre J-P, Druet P, Mery J-P (1988) Proteinuria associated with drugs and substances of abuse. In: Cameron JS, Glassock RJ (eds) The nephrotic syndrome. Dekker, New York, p 697

Fillion J, Kuhn J, Druet P, Bellon B (1994) $HgCl_2$ injections at birth induce specific tolerance of mercury disease in Brown-Norway (BN) rats. FASEB Meeting, Anaheim, 24–28 April 1994

Fürst P, Hu S, Hackett R, Hamer D (1988) Copper activates metallothionein gene transcription by altering the conformation of a specific DNA binding protein. Cell 55:705–717

Gleichmann E, Vohr HW, Stringer C, Nuyens J, Gleichmann H (1989) Testing the sensitization of T cells to chemicals. From murine graft-versus-host (GVH) reactions to chemical-induced GVH-like immunological diseases. In: Kammüller ME, Bloksma N, Seinen W (eds) Autoimmunity and toxicology. Elsevier, Amsterdam, p 363

Guéry J-C, Druet P (1990) A spontaneous hybridoma producing autoanti-idiotypic antibodies that recognize a V_k-associated idiotype in mercury-induced autoimmunity. Eur J Immunol 20:1027–1031

Guéry J-C, Tournade H, Pelletier L, Druet E, Druet P (1990) Rat anti-glomerular basement membrane antibodies in toxin-induced autoimmunity and in chronic graft-versus-host reaction share recurrent idiotypes. Eur J Immunol 20:101–105

Hall C (1982) Gold and D-penicillamine induced renal disease. In: Bacon PA, Hadler NM (eds) The kidney and rheumatic disease. Butterworth, London, p 246

Handel ML, de Fazio A, Watts CKW, Day RO, Sutherland R (1991) Inhibition of DNA binding and transcriptional activity of a nuclear receptor transcription factor by aurothiomalate and other metal ions. Mol Pharmacol 40:613–618

Henry GA, Jarnot BM, Steinhoff MM, Bigazzi PE (1992) Mercury-induced renal autoimmunity in the MAXX rat. Clin Immunol Immunopathol 49:187–203

Hirsch F, Couderc J, Sapin C, Fournie G, Druet P (1982) Polyclonal effect of $HgCl_2$ in the rat, its possible role in an experimental autoimmune disease. Eur J Immunol 12:620–625

Hua J, Pelletier L, Berlin M, Druet P (1993) Autoimmune glomerulonephritis induced by mercury vapour exposure in the Brown-Norway rat. Toxicology 79:119–129

Hultman P, Eneström S (1987) The induction of immune complex deposits in mice by peroral and parenteral administration of mercuric chloride: strain dependent susceptibility. Clin Exp Immunol 67:283–292

Hultman P, Eneström S (1988) Mercury-induced antinuclear antibodies in mice: characterization and correlation with renal immune complex deposits. Clin Exp Immunol 71:269–274

Karpus WJ, Swanborg RH (1991) CD4+ suppressor cells inhibit the function of effector cells of experimental autoimmune encephalomyelitis through a mechanism involving transforming growth factor-β. J Immunol 146:1163–1168

Kapsenberg ML, Van der Pouw-Kraan T, Stiekema FE, Schootemeijer A, Bos JD (1988) Direct and indirect nickel-specific stimulation of T lymphocytes from patients with allergic contact dermatitis to nickel. Eur J Immunol 18:977–982

Katz WA, Blodgett RC, Pietrusko RG (1984) Proteinuria in gold-treated rheumatoid arthritis. Ann Intern Med 101:176–178

Kermarrec N, Blanpied C, Pelletier L, Feingold N, Mandet C, Druet P, Hirsch F (1994) Genetis study of gold-salt-induced immune disorders in the rat. Kidney Int

Kourilsky P, Claverie J-M (1989) MHC-antigen interactions: what does the T cell receptor see? Adv Immunol 45:107–193

Lauwerys R, Bernard A, Roels H, Buchet JP, Gennart JP, Mahieu P, Foidart JM (1983) Anti-laminin antibodies in workers exposed to mercury vapour. Toxicol Lett 17:113–116

Lin RH, Mamula MJ, Hardin JA, Janeway CA Jr (1991) Induction of autoreactive B cells allow priming of autoreactive T cells. J Exp Med 173:1433–1439

Lindqvist KY, Makene WJ, Shaba JK, Nantulya V (1974) Immunofluorescence and electron microscopic studies of kidney biopsies from patients with nephrotic syndrome possibly induced by skin lightening creams containing mercury. E Afr Med J 51:168–169

Liu Z, Sun Y-K, Xi Y-P, Harris P, Suciu Foca N (1992) T cell recognition of self-human histocompatibility leukocyte antigen (HLA)-DR molecules. J Exp Med 175:1663–1668

Mathieson PW, Stapleton K, Oliveira DBG, Lockwood CM (1991) Immunoregulation of mercuric chloride-induced autoimmunity in Brown Norway rats: a role for CD8+ T cells revealed by in vivo depletion studies. Eur J Immunol 21: 2105–2109

Mathieson PW, Thiru S, Oliveira DBG (1992) Mercuric chloride-treated Brown-Norway rats develop widespread tissue injury including necrotizing vasculitis. Lab Invest 67:121–129

Mirtcheva J, Pfeiffer C, Bruijn JA, Jaquesmart F, Gleichmann E (1989) Immunological alterations induced by mercury compounds. III. H-2A acts as an immune response and H-2E as an immune "suppression" locus for $HgCl_2$-induced anti-nucleolar autoantibodies. Eur J Immunol 19:2257–2261

Mosmann TR, Coffman RL (1989) Heterogeneity of cytokine suppression patterns and funcitions of helper T cells. Adv Immunol 46:111–147

Mosmann TR, Cherwinski H, Bond MW, Giedlin MA, Coffman RL (1986) Two types of murine helper T cell clones. I. Definition according to profiles of lymphokine activities and secreted proteins. J Immunol 136:2348–2357

Ochel M, Vohr HW, Pfeiffer C, Gleichmann E (1991) IL-4 is required for the IgE and IgG1 increase and IgG1 autoantibody formation in mice treated with mercuric chloride. J Immunol 146:3006–3011

Paul WE (1989) Fundamental immunology. Raven, New York

Pelletier L, Pasquier R, Vial M-C, Mandet C, Moutier R, Salomon JC, Druet P (1987a) Mercury-induced autoimmune glomerulonephritis. Requirement for T cells. Nephrol Dial Transplant 1:211–218

Pelletier L, Pasquier R, Rossert J, Druet P (1987b) $HgCl_2$ induces non specific immunosuppression in LEW rats. Eur J Immnol 17:49–54

Pelletier L, Galceran M, Pasquier R, Ronco P, Verroust P, Bariety J, Druet P (1987c) Down-modulation of Heymann's nephritis by mercuric chloride. Kidney Int 32:227–232

Pelletier L, Pasquier R, Guettier C, Vial M-C, Mandet C, Nochy D, Bazin H, Druet P (1988a) $HgCl_2$ induces T and B cells to proliferate and differentiate in BN rats. Clin Exp Immunol 71:336–342

Pelletier L, Pasquier R, Rossert J, Vial M-C, Mandet C, Druet P (1988b) Autoreactive T cells in mercury-induced autoimmunity. Ability to induce the autoimmune disease. J Immunol 140:750–754

Pelletier L, Rossert J, Pasquier R, Villarroya H, Belair MF, Vial MC, Oriol R, Druet P (1988c) Effect of $HgCl_2$ on experimental allergic encephalomyelitis in Lewis rats. $HgCl_2$-induced down-modulation of the disease. Eur J Immunol 18:243–247

Pelletier L, Rossert J, Pasquier R, Vial M-C, Druet P (1990) Role of CD8+ T cells in mercury-induced autoimmunity or immunosuppression in the rat. Scand J Immunol 30:65–74

Pelletier L, Rossert J, Pasquier R, Villarroya H, Oriol R, Druet P (1991) $HgCl_2$-induced perturbation of the T cell network in experimental allergic encephalomyelitis. II. In vivo demonstration of the role of T suppressor and contra-suppressor cells. Cell Immunol 137:379–388

Pepys J (1983) Allergic reactions of the respiratory tract to low molecular weight chemicals. In: Gibson CG, Hubbard R, Parke DV (eds) Immunotoxicology. Academic, London, p 107

Pietsch P, Vohr HW, Degitz K, Gleichmann E (1989) Immunopathological sings inducible by mercury compounds. II. $HgCl_2$ and gold sodium thiomalate enhance serum IgE and IgG concentrations in susceptible mouse strains. Int Arch Allergy Appl Immun 90:47–53

Pil PM, Lippard SJ (1992) Specific binding of chromosomal protein HMG1 to DNA damaged by the anticancer drug cisplatin. Science 256:234–237

Prigent P, Saoudi A, Druet P, Hirsch F (1993) In vitro effects of mercuric chloride on splenocytes from susceptible Brown-Norway rats and non-susceptible Lewis rats: cytokine involvement. JASN 4:628 (abstract)

Prouvost-Danon A, Abadie A, Sapin C, Bazin H, Druet P (1981) Induction of IgE synthesis and potentiation of anti-ovalbumin IgE response by $HgCl_2$ in the rat. J Immunol 126:699–702

Pusey CD, Bowman C, Morgan A, Weetman AP, Hartley B, Lockwood CM (1990) Kinetics and pathogenicity of autoantibodies induced by mercuric chloride in the Brown Norway rat. Clin Exp Med 81:76–82

Reeves AL (1983) The immunotoxicity of beryllium. In: Gibson CG, Hubbard R, Parke DV (eds) Immunotoxicology. Academic, London, p 261

Robinson CJG, Abraham AA, Balazs T (1984) Induction of anti-nuclear antibodies by mercuric chloride in mice. Clin Exp Immunol 58:300–306

Robinson CJG, Balazs T, Egorov IK (1986) Mercuric chloride, gold sodium thiomalate, and D-penicillamine-induced anti-nuclear antibodies in mice. Toxcol Appl Pharmacol 86:159–169

Romagnan S (1991) Human T_H1 and T_H2 subsets: doubt no more. Immunol. Today 12:256–259

Romagnoli P, Labhardt AM, Sinigaglia F (1991) Selective interaction of Ni with a MHC-bound peptide. EMBO J 10:1303–1306

Romagnoli P, Spinas GA, Sinigaglia F (1992) Gold-specific T cells in rheumatoid arthritis treated with gold. J Clin Invest 89:254–258

Roman-Franco AA, Turiello M, Albini B, Ossi E, Milgrom F, Andres GA (1978) Anti-basement membrane antibodies and antigen-antibody complex in rabbits injected with mercuric chloride. Clin Immunol Immunopathol 9:464–481

Rossert J, Pelletier L, Pasquier R, Druet P (1988) Autoreactive T cells in mercury-induced autoimmune disease. Demonstration by limiting dilution analysis. Eur J Immunol 18:1761–1766

Rossert J, Pelletier L, Pasquier R, Villarroya H, Oriol R, Druet P (1991) $HgCl_2$-induced perturbation of the T cell network in experimental allergic encephalomyelitis. I. Characterization of T cells involved. Cell Immunol 137:367–378

Rudensky AY, Preston-Hurlbut P, Hong S-C, Barlow A, Janeway CA Jr (1991) Sequence analysis of peptides bound to MHC class II molecules. Nature 353: 622–627

Sapin C, Druet E, Druet P (1977) Induction of anti-glomerular basement membrane antibodies in the Brown-Norway rat by mercuric chloride. Clin Exp Immunol 28:173–178

Sapin C, Mandet C, Druet E, Gunther E, Druet P (1982) Immune complex type disease induced by $HgCl_2$ in the Brown-Norway rat. Genetic control of susceptibility. Clin Exp Immunol 48:700–702

Schwartz RH (1985) T-lymphocyte recognition of antigen-Ia complexes in association with gene products of the major histocompatibility complex. Annu Rev Immunol 3:237–261

Sinigaglia F, Scheidegger D, Garotta G, Scheper R, Pletscher M, Lanzavecchia A (1985) Isolation and characterization of Ni-specific T cell clones from patients with Ni-contact dermatitis. J Immunol 135:3929–3932

Tan XX, Tang C (1993) Effects of inorganic and organic mercury on intracellular calcium levels in rat T lymphocytes. J Toxicol Environ Health 38:159–170

Tournade H, Guéry J-C, Pasquier R, Nochy D, Hinglais N, Guilbert B, Druet P, Pelletier L (1991) Experimental gold-induced autoimmunity. Nephrol Dial Transplant 6:621–630

Verwilghen J, Gambling JH, Kingsley GH, Panayi GS (1992) Activation of gold specific lymphocytes in RA patients receiving gold. 8th international congress of immunology. Springer, Berlin Heidelberg New York, p 620 (abstract)

Wooley PH, Griffin J, Panayi GS, Batchelor JR, Welsh KI, Gibson TJ (1980) HLA-DR antigens and toxic reaction to socium aurothiomalate and D-penicillamine in patients with rheumatoid arthritis. N Engl J Med 303:300–302
Ziegler K, Unanue ER (1981) Identification of a macrophage antigen-processing event required for I-region restricted antigen presentation to T-lymphocytes. J Immunol 127:1861–1875

CHAPTER 5

Effects of Metals on Gene Expression

J. KOROPATNICK and M.E.I. LEIBBRANDT

A. Introduction

Heavy metals are both ubiquitous and long-lived in the environment (reviewed by GOYER 1991) and have enormously varied effects on cells. They are an absolute requirement for the function of both prokaryotic and eukaryotic cells (17 out of the 30 elements essential for life are metals; COTTON and WILKINSON 1980), but are toxic to cells and organs through different pathways and to different degrees (reviewed in several chapters of this volume). Some metals have no known function in cells, but have toxic effects: cadmium and arsenic are examples. Cells have developed mechanisms to keep toxic metal species away from critical targets, and some of those mechanisms will be covered in this review. Others are essential for normal cellular function, but are toxic under certain circumstances and at particular concentrations: metals that fall into this category include copper (an essential cofactor for many oxidative enzymes, including catalase, peroxidase, cytochrome oxides, and others – but also a dangerous cellular toxin; HORN 1984), cobalt (an essential cofactor for vitamin B_{12}), manganese (a cofactor in many enzymatic reactions involving phosphorylation, cholesterol, and fatty acid synthesis), iron (required for haemoglobin), selenium (essential for glutathione peroxidase; HOGBERG and ALEXANDER 1986), and molybdenum (an essential cofactor for xanthine oxidase and aldehyde oxidase, and required in plants for fixing atmospheric nitrogen by bacteria). The ability of metals to both nurture and destroy has resulted in the development of multiple, complex cellular mechanisms to regulate intracellular availability and location. Cellular strategies control the availability of metals to cellular components requiring them for function, but minimize exposure to other cellular molecules and structures that are sensitive to their harmful effects. The fact that a large number of metals falls into this group (and so increases the potential for "cross-talk" between cellular components associating with inappropriate metal cofactors) has only increased the variety of these strategies. Some of them will be explored in this chapter. Finally, there are metals that are essential and almost entirely non-toxic. Zinc is a good example – it is required for more than 70 metalloenzymes, and is an essential component of many transcription factors (reviewed by KÄGI, Chap. 15, this volume) and cytoplasmic receptors which transduce external stimuli and regulate the transcription of specific genes by interaction with gene regulatory sequences (reviewed by O'HALLORAN 1989). Some of the transcription factors are metalloregulatory proteins which mediate cellular responses to heavy metal ions specifically; they will be addressed in this review.

A very large number of biologically important molecules require metals as cofactors. These molecules include proteins that have a direct role in signal transduction (including zinc fingers, leucine zippers, helix-turn-helix motifs, among others, that exist in hormone receptors and other transcription factors, and other zinc-dependent proteins; reviewed by PABO and SAUER 1992). Whether metals modulate the level of transcription in vivo by addition or removal from these factors is unknown. Furthermore, metals act as cofactors in metalloenzymes that function in processes that are important in mRNA translation (see the discussion on transferrin and ferritin below) and protein function, but are not directly involved in signal transduction or gene transcription. This is critical in executing the cellular response to increased intracellular signalling and transcription. Since both post-transcriptional and post-translational events are necessary for genes to be expressed, metals that function in this way can be considered to have a role in gene expression.

Metals, therefore, are employed in at least four ways to affect the expression of eukaryotic genes:

1. As the environmental signal that elicits a signal transduction response to alter gene transcription in cells.
2. As the cofactor that mediates signalling protein function in transducing the information from the environment to the genome.
3. As a modifier of mRNA activity or stability.
4. As a cofactor for protein activity.

Although zinc appears to predominate in processes (1) and (2) (CHESTER 1992), many metals act as environmental signals.

Since metals are involved in so many cellular processes, metal bioavailability becomes paramount in understanding how metal ions might affect gene expression at many levels, ranging from signal transduction, transcription factor action, mRNA function, and protein activity. Recent advances in understanding how cells transport metals will be addressed in this review.

The multiple roles of metals within cells makes a detailed review of all of them a daunting task. As a consequence, the area of metal-regulated gene expression in prokaryotes and plants is regretfully omitted; the reader is referred to the recent excellent review of that field by RALSTON and O'HALLORAN (1990). I will focus on the effect of metals on the expression of mammalian genes, especially those important in regulating intracellular metal concentration and availability. Metallothionein gene expression as a model metal-regulated system is of central importance, and special attention will be paid to its regulation and potential function in mammals. For detailed analysis of its regulation by metals in other eukaryotes, the reader will find recent reviews by ZHOU and coworkers (1992) (yeast metallothionein) and ANDREWS (1990) (avian and other metallothioneins) of great value. The recent concise reviews by CHESTERS (1992) and O'HALLORAN (1993) are also valuable in understanding the general area of trace element-gene interactions. The omission of a description of much excellent and illuminating work by my colleagues is, however, unavoidable, and I apologize in advance for these lacunae.

B. Molecular Control of Gene Expression

Biochemists, toxicologists, cell biologists, and geneticists of many backgrounds and interests find that metal-regulated gene expression has an effect, direct or indirect, on the cellular phenomena which they study. A few basic molecular biology terms to facilitate reading by a wide audience are defined here. Transcription is the process by which RNA polymerase synthesizes RNA using DNA sequences as an information-carrying template. Transcription is initiated by the intimate association of RNA polymerase with specific DNA sequences called promoters. These regions lie just prior to ("upstream of", or "5′ to") DNA sequences that carry RNA-encoding information. Eukaryotes have complex promoter regions that often require interaction with multiple other proteins (transcription factors) before transcription can take place. Because gene transcription must be a stringently controlled process to generate a specific cellular phenotype, with repression of transcription in some genes but enhanced transcription in others at specific stages of development or in response to extracellular signals, these multiple interactions of "*trans*"-acting transcription factors with "*cis*"-acting gene promoter regions provide enormous potential for controlled expression of cellular DNA.

C. Eukaryotic Strategies of Signal Transfer

Cells sense external stimuli and respond by changing the expression of a specific set of genes. To do so, transfer of information from the cell surface to the nucleus is required. That process is termed signal transduction. The first event in that process is specific, tight association between the extracellular signalling molecule or atom (hormone or peptide growth factor, osmotic pressure, and metal ions, among others) and a cellular receptor molecule. That association causes an allosteric change in the receptor that causes a single or multistep movement of signal from the cell surface, into the cytoplasm, and finally to the nucleus. The result is modification of specific genes and/or the activity of specific enzymes. The existence of multiple signal transduction pathways is required to get an appropriate response in situations where several different signals from extracellular ligands are received concurrently.

I. Multiple Factor Signal Transduction Systems

Multiple factor signal transduction systems employ separate proteins to receive signals, transduce signals, and transfer the signal to the nucleus by a second messenger. These multiple factors often act in a cascade array, with a change in one inducing changes in others. In eukaryotes, two of the best-described multiple factor signal transduction pathways are those employing guanine nucleotide-binding proteins (G proteins) and tyrosine-specific protein kinases.

G proteins are localized in cell membranes, and are intermediates in the signalling pathways which link receptors receiving extracellular signals with intracellular proteins that, in turn, induce changes in second messengers (cAMP, for example) (for reviews, see LAMB and PUGH 1992, PFISTER et al. 1993, and LAD et al. 1992). Cyclic AMP (among other second messengers) continues to induce changes in other intermediates in a cascade of signals that ultimately alters expression of a specific subset of genes [for example, through binding of transcription factors to *c*AMP *r*esponsive *e*lements (CREs) in the promoters of those genes].

Tyrosine kinases (TKs) are a family of cell surface enzymes that each contain an extracellular N-terminal receptor domain and an intracellular C-terminal protein kinase activity (for review, see FANTL et al. 1993, and KARIN 1992). Interaction of receptor with signalling ligand (for example, epidermal growth factor, insulin, and many others) stimulates the cytoplasmic protein tyrosine kinase to phosphorylate target second messenger proteins. This, in turn, induces a cascade of events that lead to increased expression of genes (such as the cellular oncogenes fos and myc) important in cell cycle progression and DNA synthesis (hence the importance of TKs in malignant transformation) and stimulation of ion and glucose transport.

II. Single Factor Signal Transduction Systems

Single component signal transduction systems differ from multiple component systems in that the functions of separate members of the signal cascade are consolidated into a single protein. In general, the protein receptor of a signalling ligand doubles as a transcription factor; among the best-described systems of this type are the steroid and thyroid hormone receptors (glucocorticoid or oestrogen receptors) (for review, see MULLER and RENKAWITZ 1991). It is of interest to note that the glucocorticoid receptor (GR) is dependent upon binding to zinc for its activity. Other receptors (including the retinoic acid receptor and viral oncogene *erb*A) are related.

In general, the cellular receptor interacts with hormonal messenger to produce a multimeric protein capable, when translocated to the nucleus, of binding DNA at specific promoter regions and altering gene transcription (JENSEN and DESOMBRE 1972; OPPENHEIMER et al. 1972; IVARIE and O'FARRELL 1978). Hormone/receptor complexes interact with *h*ormone-*r*esponsive *e*lements (HREs) in promoters. In the case of the GR, binding of hormone by receptor may unmask domains capable of activating transcription; deleting C-terminal sequences of the receptor produced a constitutively active transcription factor (HOLLENBERG et al. 1987; GODOWSKI et al. 1987; BECKER et al. 1986). Thus, binding of hormone to receptor appears to induce allosteric changes that promote the ability of the receptor to enhance gene transcription. Hormone binding to a single receptor is necessary, but not sufficient, to mediate gene expression; the 90-kDa *h*eat *s*hock *p*rotein (HSP90) is also involved (SANCHEZ et al. 1985; CATELLI et al. 1985; DENIS and GUSTAFSSON 1989). The GR does not bind steroid unless it is associated with two HSP90 molecules and a 59-kDa protein (p59) (GEHRING 1993). Once steroid binding occurs, HSP90 dissociates from the receptor to facilitate translocation into the nucleus and allow GR association with DNA (GEHRING 1993; HUTCHISON et al. 1993). Because purified steroid and thyroid hormone receptors bind HREs in vitro in the absence of hormone (WILLMAN and BEATO 1986; SCHAUER et al. 1989; GRAUPNER et al. 1989), it appears that steroid binding to receptor is not a requirement for transcriptional activation by GR. Rather, steroid binding mediates removal of HSP90 to allow translocation to the nucleus (MULLER and RENKAWITZ 1991). In summary, signal transduction in eukaryotes requires:

1. Sensor molecules capable of undergoing a conformational change detectable by sensitive cellular proteins.
2. Transducer molecules that receive the signal and effect changes in cells, either by participating directly (single factor signal transduction), or transferring the signal to a cascade of secondary transducers (multiple factor signal transduction) that ultimately affect gene transcription. In single factor systems, multiple protein interactions with the transducing protein may be required for translocation to the nucleus, and transcriptional activity when it gets there.
3. Protein phosphorylation in some cases.
4. Direct interaction of protein factors whose activity is modified by the signal transduction process, with specific DNA sequences. These DNA sequences (termed "enhancer" elements) are often upstream of genes, and proteins bound to them affect the initiation of transcription.

As described below, and elsewhere in this volume, signal transduction molecules and transcription factors are, in many cases, dependent upon zinc or copper for structural integrity and activity. A key question for the future is whether the bioavailability of these metal ions is capable of regulating either the level or specificity of gene transcription, and whether the transport and storage of metals into and within cells are mechanisms by which gene expression through signal transduction is regulated.

D. Transduction of Metal Signals in Eukaryotes

Because metal ions have the capacity to both damage cells and mediate necessary physiological processes, cells must be very sensitive to their presence and level. Many cellular systems are responsive to heavy metal signals, and the specific proteins and their nucleic acid binding sites have been partly or completely defined. It appears that those proteins that are the receptors of metals and the modifiers of DNA (or RNA, as described below) activity are one and the same; thus, all eukaryotic metal signal transduction mechanisms so far defined are single factor systems. Those metal-binding regulatory factors have been termed metalloregulatory proteins (O'HALLORAN 1989, 1993; ZHOU and THIELE 1993). Examples of such metal-responsive factors are described below.

I. Entry, Binding, and Storage of Essential Metals

1. Iron

In mammals, Fe is both necessary for normal cell function and toxic to cells under certain circumstances (reviewed by Templeton, Chap. 14, this volume; GOYER 1991). Consequently, two proteins control the uptake and availability of iron in cells. The serum protein transferrin (Tf) binds two Fe(III) ions in a pH- and carbonate-dependent manner. Fe_2Tf is a ligand for the Tf receptor (TfR) when TfR is on the cell surface. The entire complex is transferred to the interior of the cell and iron released from Tf after acidification of involuted vesicles. In a way still poorly understood, iron is then diffused or transported into cytosolic and mitochondrial proteins. Ultimately, iron is localized into heme cofactors, ferritin (which binds iron to protect cells from the toxicity of free ionic iron; HALLIWELL and GUTTERIDGE 1985) and other iron proteins, and iron-sulphur centres (O'HALLORAN 1993). The major regulation of TfR and ferritin activity in cells is not by altered transcription rate of transferrin and TfR genes, but by altered translational availability of their mRNAs (KLAUSNER and HARFORD 1989). This control is accomplished by an RNA-binding metalloregulatory protein termed IRE-BP (*i*ron-*r*esponsive *e*lement *b*inding *p*rotein). The IRE-BP associates with IREs, which are stable stem-loop structures present in both TfR and ferritin mRNAs, and the mRNAs of δ-aminolevulinate synthases (ALASs) (HENTZE et al. 1987; WALDEN et al. 1988; ROUAULT et al. 1988; BROWN et al. 1989); ALASs catalyse the synthesis of the first intermediate in heme biosynthetic pathways, and are involved in Fe movement between cytosol and mitochondria (O'HALLORAN 1993). In the presence of high iron concentrations, Fe binds IRE-BP to form a complex with at least three different functions:

1. Iron association inhibits BP binding to IREs in the 5′ untranslated region of the ferritin mRNA and the mRNAs of ALASs (WALDEN et al. 1988; ROUALT et al. 1988; BROWN et al. 1989). The coding regions of these genes are rendered available for protein production in the presence of high iron. Thus, iron increases ferritin protein production and decreases the amount of available, unbound iron within cells, and concomitantly increases ALAS production to promote movement of iron from the cytosol into the mitochondria in the process of porphyrin Fe (heme) production – an elegant feedback control mechanism (LEIBOLD et al. 1990). The IREs, when not associated with IRE-BP, are not neutral elements. They appear to function as positive enhancers of translation of the mRNAs in which they reside (DIX et al. 1992). Interestingly, a heme-responsive element (Cys-Pro-X-Asp-His, where X is any amino acid) is also present in the precursors of ALAS proteins, but not in the mature enzymes (LATHROP and TIMKO 1993). This HRM (*h*eme *r*egulatory *m*otif) might bind heme when its cytosolic concentration is high and prevent transport of ALAS into the mitochondria – its final site of action. Multiple copies of an HRM are also present in the heme-sensing domain of the HAP1 heme receptor of yeast (PFEIFER et al. 1989). In spite of sequence similarities between HAP1 and ALAS motifs, there is no direct biochemical evidence of heme binding at these sites. Thus, Fe in different forms (Fe/IRE-BP complex, and Fe porphyrin) may control the activity of ALASs at the level of mRNA translation and intracellular transport, respectively.

2. For the TfR mRNA, multiple copies of the IRE are located in the untranslated downstream 3′ region of the mRNA. In the absence of Fe, BP binds to IREs and helps stabilize the mRNA and prevent its degradation. High iron levels favour formation of the Fe-BP complex to inhibit IRE-BP association, with resulting increased mRNA degradation and decreased transferrin receptor protein production. Thus, high iron levels are capable of interacting, through association with an IRE-binding protein, with sequences embedded in ferritin and TfR mRNA. This regulates, at the translational level, production of proteins important in maintaining available iron molecules within cells. It is interesting to note that the association of mRNA stem-loop structures with regulatory proteins is not a unique strategy for control of gene expression, since a similar interaction between thymidylate synthase and its own mRNA has rcently been reported (CHU et al. 1993).

3. Fe-IRE-BP also acts as an iron sulphide-containing enzyme cytosolic aconitase (c-aconitase), which converts citrate to isocitrate via a *cis*-aconitrate intermediate in a reversible reaction. Although m-aconitase is a mitochondrial enzyme mediating entry of glycolysis products into the tricarboxylic acid (TCA) cycle, there is no obvious physiological role for c-aconitase (reviewed by O'HALLORAN 1993).

Eukaryotic cells have developed exquisitely sensitive methods to enhance the physiological functions of iron while minimizing its toxic effects. Those methods are aimed mainly at control of mRNA translation and intracellular transport, and involve interactions between cytosolic Fe-binding proteins and proteins (and, in some cases, their mRNAs) involved in iron metabolism. However, transcriptional control may function in some circumstances. In rat, the light (but not the heavy) ferritin subunit mRNA is transcriptionally induced by iron (WHITE and MUNRO 1988), and iron induces ferritin gene transcription (as measured by run-on transcription in isolated nuclei) 45-fold in response to iron (LESCURE et al. 1991). Thus, eukaryotic cells have utilized a wide array of strategies to regulate iron metabolism at the level of gene transcription, mRNA translation, and protein and metal transport.

2. Copper

Copper is a trace element essential for all life forms. Its importance is due to its ability to act as a cofactor for electron donation and acceptance for many enzymatic reactions, including oxygenation, hydroxylation, and dismutation (HAMER 1993). Like iron, copper has the capacity to damage due to its ability to oxidize proteins and lipids and to enhance formation of toxic intra- and extracellular free radicals. Although eukaryotes protect themselves against toxic copper by synthesizing metallothionein (a small cysteine-rich protein; see below), virtually nothing is known about the processes important in importing and exporting Cu, with the exception of the Menkes'-associated gene defect (see below). Nor is the biochemistry of incorporation of Cu into apoenzymes and other proteins under normal conditions well understood (HAMER 1986).

Copper is associated with two main plasma proteins in eukaryotes: albumin and caeruloplasmin. Both of these Cu-binding proteins have been suggested as the major copper transport proteins (HARRIS and STEVENS 1985). Caeruloplasmin and the copper-dependent enzyme copper-zinc superoxide dismutase (SOD) (which catalyses the dismutation of superoxide anions to hydrogen peroxide) have been used as indicators of copper status in animals. Tissue concentrations of both proteins respond to variation in dietary copper (SUTTLE 1986). Caeruloplasmin mRNA concentration may also be responsive to dietary copper, suggesting control at the transcriptional level (DANKS and MERCER 1988), although caeruloplasmin mRNA is not affected by extracellular copper concentration in isolated hepatocytes (MCARDLE et al. 1990). In fact, copper deficiency in rats in vivo sufficient to eliminate caeruloplasmin-oxidase activity did not affect caeruloplasmin mRNA and apoprotein concentration in liver (GITLIN et al. 1992); therefore, lack of copper prevents only the final activation of enzymatic activity of this protein, and appears not to influence transcription. In this scenario, the control of intracellular copper levels may be the critical point at which regulation of gene expression is achieved. A candidate gene for copper transport in humans (*MnC1*) has recently been identified (VULPE et al. 1993; CHELLY et al. 1993; MERCER et al. 1993; reviewed by HAMER 1993) by cloning mutant forms from cells of patients suffering from Menkes' disease. Menkes' disease is an inherited disorder of Cu metabolism characterized by neurologic degeneration and mental retardation, unusual hair, connective tissue and vascular defects, and death in early childhood (DANKS 1989). Hepatocytes of affected individuals are deficient in copper (DARWISH et al. 1983). The expression of the *MnC1* transcript is altered in 23 out of 32 patients (VULPE et al. 1993; MERCER et al. 1993), and non-overlapping portions of the gene are deleted in 16 out of 100 unrelated patients (CHELLY et al. 1993). The *MnC1* protein appears to be a P-type (that is, utilizing an aspartyl phosphate intermediate to transport cations across a membrane) cation-transporting ATPase, with similarity to a bacterial copper-transporting ATPase. In addition, it has six putative metal-binding motifs at the N-terminus. Because those protein regions have more similarity with other copper-binding motifs than Hg and Cd-binding prokaryotic motifs, the protein may be specific for copper transport. In fact, cadmium and zinc transport in Menkes' cells (and in mottled, a mutant mouse model for Menkes' disease; DANKS 1989) is normal (PACKMAN et al. 1983; PACKMAN and O'TOOLE 1984). Loss of *MnC1* activity in cultured Menkes' fibroblasts leads to loss of copper efflux, but with no apparent defect in Cu uptake. VULPE and coworkers (1993) suggest that, based on their results and detailed copper efflux and cell fractionation studies (PACKMAN and O'TOOLE 1984; HERD et al. 1987; PACKMAN et al. 1987), the *MnC1* copper-transporting ATPase is inserted in the membrane of a subcellular compartment (VULPE et al. 1993).

These data represent the first description of a eukaryotic gene directly involved in heavy metal transport and, as such, open the question of aberrant metal transport in other diseases that result in unusual copper accumulation. Both Wilson's disease and X-linked cutis laxa cause abnormal copper accumulation in humans. Cutis laxa results in connective tissue and copper transport defects similar to that seen in Menkes' patients, without severe neurologic abnormalities (KUIVANIEMI et al. 1981), and may represent an allele of Menkes' resulting in a milder form of the disease (VULPE et al. 1993). Wilson's disease is a genetic disorder involving excretion of copper; unlike Menkes' disease, however, a chronic toxic buildup of the metal occurs in the liver, which undergoes a slow necrosis with the eventual development of cirrhosis. The genetic defect responsible for the disease has been localized to chromosome 13 (FRYDMAN et al. 1985), suggesting that the X-linked *MnC1* gene is not directly responsible (KAPUR et al. 1987). Although Vulpe and coworkers have suggested that a defect in a different, and perhaps related, transport protein may cause the accumulation of copper in Wilson's disease, the genetic basis is not known. Alternatively, the major intracellular protein with which copper associates (metallothionein) may have aberrant expression or accumulation that leads to unusually high copper levels. Metallothioneins (MTs) are ubiquitous, low molecular weight, cysteine-rich proteins with high metal-binding capacity whose expression has been reviewed elsewhere (Chap. 6, this volume). An autosomal recessive mutation [toxic milk (*tx*)] which alters Cu homeostasis in mice (RAUCH 1983; BIEMPICA et al. 1988) has been described which is characterized by reduced hepatic Cu concentration, hypopigmentation, reduced growth, abnormal locomotor behaviour, tremors, and death at 2 weeks of age. Tx mice allowed to suckle from normal lactating dams survive and accumulate as much as 30 times as much hepatic Cu in adulthood as normal mice, and have resulting liver degeneration (BIEMPICA et al. 1988). The concomitant increase in hepatic copper loading and liver degeneration suggests that tx may model Wilson's disease. Little is known of changes in hepatic Cu-binding proteins in tx mice, although serum caeruloplasmin is low (RAUCH et al. 1986). The major portion of hepatic Cu in tx mice is bound to MT (KOROPATNICK and CHERIAN 1993). MT accumulation is greater than 100 times that in wild-type animals, but MT mRNA accumulation is not elevated (KOROPATNICK and CHERIAN 1993). Importantly, the stability of half-life of MT is also 80% higher, raising the intriguing possibility that altered MT metabolism rather than control of transcription may play a role in Cu accumulation. Little is known of control of MT degradation to manipulate intracellular availability of copper and other metals, but several avenues of exploration are suggested by recent reports in the literature. Lysosomal cysteine protease activity (especially cathepsin B) has been implicated in breakdown of MT (MIN et al. 1992; MCKIM et al. 1992; CHOUDHURI et al. 1992). Although cysteine proteases do not appear to require metals for enzymatic activity (SLOANE et al. 1990), zinc-dependent metalloproteinases have been reported to activate cathepsin B in vivo (HARA et al. 1988). Biological availability of zinc has also been reported to affect the rate of MT synthesis and degradation in a chicken macrophage cell line (LAURIN et al. 1990). The question of metal-dependent degradation of MT (which, in turn, may play a role in intracellular metal homeostasis) deserves further scrutiny.

II. Essential Metals as Regulators of Metabolism

1. Iron

As described above, cytosolic iron in several forms affects post-transcriptional regulation of iron metabolism. Iron in the form of heme can also affect the transcription of genes in the heme catabolic and respiratory pathways. Hemin, for example induces heme oxygenase (HO) (essential for heme catabolism) in human macrophages (YOSHIDA et al. 1988). Cadmium or heme also increase HO mRNA levels in mouse hepatoma cells (ALAM et al. 1989), and heme induces rat cytochrome P-450 genes and the yeast genes encoding iso-1-cytochrome c (CYC1) (GUARENTE and MASON 1983), iso-2-cytochrome c (CYC7), and catalase T (CYT1) (ZITOMER et al. 1987). In yeast, heme controls transcription by inducing the metalloregulatory protein HAP1 to bind to cis-acting promoter sequences termed UASs (*u*pstream *a*ctivator *s*ites) (GUARENTE and MASON 1983; WINKLER et al. 1988; PFEIFER et al. 1987a,b). Deletion analysis of HAP1 suggests that, like the glucocorticoid receptor described above, DNA-binding activity may be masked in the absence of heme and revealed by allosteric changes induced by heme binding (GODOWSKI et al. 1987; DENIS and GUSTAFSSON 1989). Heme can also inhibit transcription; for example, transcription in yeast of the V_b subunit of the cytochrome c oxidase gene (*COX5b*) (TRUEBLOOD et al. 1988) and the 5-aminolevinulate synthase gene (the first enzyme in the heme biosynthesis pathway) (SRIVASTAVA et al. 1988) are inhibited by heme, suggesting that heme-responsive regulators other than HAP1 may be at work.

UASs do not regulate the heme oxygenase gene in rats. Rather, the rat HO gene has two copies of the MT metal responsive element (see below) and heavy metal regulation of transcription may be analogous to that seen for MT (MILLER et al. 1985).

2. Copper

Copper(II) affects gene expression in a wide variety of organisms including yeast, mammals, and bacteria. The best-described copper-regulated genes encode metallothioneins (MTs); the biochemistry and biology of this protein have been extensively reviewed (Chap. 6, this volume; HAMER 1986; ANDREWS 1990). Its role as a model and potential mediator of transcription are discussed here.

III. Metallothionein and Other Genes as Models for Metal Regulation

Metallothioneins are low molecular weight, cysteine-rich metal-binding proteins that are transcriptionally regulated by a variety of metals, hormones, and developmental signals. They certainly play a role in resistance to heavy metal toxicity, and have been postulated to mediate several roles important in metal homeostasis in eukaryotes (KÄGI and SCHAFFER 1988). They are encoded by a family of genes in mammals. A single functional *MT-2* gene, a brain-specific *MT-3* gene, and at least five functional *MT-1* genes direct production of different MT isoforms in primates. Different isoforms may be differentially inducible under different conditions. In crab, for example, four MT genes with differential inducibility or inhibition by copper and zinc have been reported by BROUWER et al. (1992). The authors suggest that metal detoxification functions are exclusive to some isoforms, and zinc and copper homeostasis to others. KERSHAW and KLAASSEN (1992) have reported differential degradation of MT-1 and MT-2 isoforms independent of metal composition in zinc-treated rats, and WAN and coworkers (1993) have reported differential accumulation of MT-1a, MT-2a, MT-2d, and MT-2e in cultured rabbit kidney cells in response to induction with cadmium or zinc; they point out that differential transcription rates among gene isoforms alone cannot readily explain differences in accumulation, and suggest that a variety of regulatory processes (including differences in isoform degradation rate) may be operating. Differences in function and expression between MT isoforms in eukaryotes, however, remain a largely unknown and potentially exciting area of exploration.

1. Metal Regulation in Yeast

The yeast gene *CUP1* (copper resistance gene locus) encodes a copper-binding metallothionein that is transcriptionally controlled by the *m*etal-loregulatory *t*ranscription *f*actors (MRTFs) ACE1 (*a*ctivator of *C*UP1 *e*xpression) in Saccharomyces cerevisiae) and AMT1 (*a*ctivator of *m*etal-lothionein *t*ranscription) in Candida glabrata. These metalloregulatory factors regulate transcription through interaction with cis-acting DNA receptors in the regulatory region of the metal-responsive genes (including *CUP1*). Regulation by these factors in yeast has been recently reviewed by ZHOU and coworkers (1992) and the reader is referred to it for a detailed description. The yeast copper-zinc superoxide dismutase (*SOD1*) gene has also been shown to contain an ACE1-binding site in its promoter (CARRI et al. 1991; GRALLA et al. 1991), and its transcription is regulated by ACE1 (O'HALLORAN 1993) and is inducible by copper. Its expression in yeast and higher eukaryotes in relation to metals and oxidative stress is discussed at greater length below.

2. Metal Regulation in Mammals

Metal-responsive transcription of mammalian MT genes is mediated through multiple cis-acting DNA elements termed MREs (*m*etal-*r*esponsive *e*lements) in the promoters of MT genes. These are imperfect 13–15 base pair repeats found in both orientations in the promoters of all described MT genes, as well as other genes (osteopontin, for example; SAUK et al. 1991, as described below). They can act at a distance, like transcriptional enhancer sequences (WESTIN and SCHAFFNER 1988), and are not the only cis-acting regulatory element in MT genes; there are multiple other regulatory elements inter-digitated among MREs (reviewed by ANDREWS 1990). The core sequence "TGCRCNC", where R is any purine, is required for efficient metal-inducible transcription (SEARLE et al. 1987; CULOTTA and HAMER et al. 1989), and is different from the binding sites for copper-activated ACE1 and AMT1 in yeast . There are 6 MREs (MREa-f) in the mouse MT-1 promoter with differential ability to mediate induction of gene expression; MREd is the most effective.

a) Metal-Responsive Transcription Factors

Similar to metal-induced transcription in yeast, MREs mediate transcription through positively acting, metal-responsive transcription factors (MRTFs) (SÉGUIN et al. 1984). Because de novo protein synthesis is not required for gene activation by metals, MRTFs must pre-exist (KARIN et al. 1981). The characteristics of MRTFs in higher eukaryotes have been recently reviewed by THIELE (1992). Six mammalian MRTF activities in different species have been reported in the literature – whether they perform distinct metalloregulatory functions is not known. They are:

α) *MEP-1*. *M*etal *e*lement *b*inding *p*rotein-1 (MEP-1) is a zinc-dependent, 108- to 110-kDa unglycosylated protein factor found in mouse cells. Its specific binding to MRE sequences is abolished by a point mutation in the core consensus sequence, and it appears to be one of the major proteins interacting with MREs (LABBÉ et al. 1991; SÉGUIN 1991).

β) *ZAP*. Zinc-*a*ctivated *p*rotein (ZAP) is a zinc-dependent MRE-binding protein found in rat liver nuclei (SEARLE 1990). Both MEP-1 and ZAP, however, are not stimulated to bind an MRE by Cd and Cu. Since these metals induce MT gene transcription, this suggests that MEP-1 and ZAP are not the only MRTFs operating in rat cells.

γ) *MTF-1*. Human HeLa cells have a zinc-inducible mMT-1 MRE-binding activity termed MTF-1 (WESTIN and SCHAFFNER 1988).

δ) *MRE-BP*. A zinc-independent MRE-binding protein (MRE-BP) has also been described by KOIZUMI and coworkers (1992a). It is 112000 Da in size, and recognizes different MREs of the human *MT-2a* gene with varying affinity: MREc and MREg interact strongly with MRE-BP, but MREd was only partially protected, and MREe was unprotected, in nuclease footprinting assays. Interestingly, both Cd and Zn at high concentrations inhibited MRE-BP binding to MREs, while low metal levels stimulated binding. MRE-BP did not require zinc for binding to MREs. Thus, there may be a role for MRE-BP in negative regulation of MT. In this scenario, transcriptional regulatory proteins could exert both positive and negative regulation through interaction with heavy metals (including cadmium and zinc), possibly by competing for overlapping target sequences around the MRE core. At higher metal concentrations, MRE-BP could lose affinity for MREs to make them fully accessible to positive regulatory factors (ZAP, MTF-1, and/or MEP-1, for example), At lower metal levels, metal-dependent positive factors would decrease in activity and MRE-BP would bind to prevent unnecessary expression of MT genes. In fact, deinduction of zinc-induced MT protein expression has been reported in HeLa cells (KARIN et al. 1981) and constitutively high MT mRNA and protein levels are decreased in mouse melanoma cells (KOROPATNICK and PEARSON 1990) in the presence of prolonged exposure to cadmium or zinc. The possible role of negative regulatory elements in these processes is intriguing, but unexplored.

ε) *MBF-1*. A zinc-dependent 74-kDa *m*etal response element *b*inding *f*actor 1 (MBF-1) has been isolated from mouse L cells, and binds specifically to some MREs. It can induce MT gene transcription in vitro (IMBERT et al. 1989).

ζ) *ZRF*. A zinc-dependent DNA-binding *z*inc *r*egulatory *f*actor (ZRF) has been described by KOIZUMI et al. (1992b) in HeLa cell nuclear extracts. It appears to confer zinc specificity (in contrast to inducibility with a broad range of heavy metals) on the MREa of the human *MT-2a* gene.

Whether these MRTFs mediate separate activities with respect to positive and negative transcriptional regulation, and metal ion specificity and/or specificity for different MREs (and exactly how they mediate), remains unresolved. However, it is clear that there are multiple MRE-binding factors with metal-dependent (and, in the case of MRE-BP, metal-independent) DNA-binding activity. Since a single MRE is responsive to several different metal ions, it could be that:

1. Multiple MRTFs (perhaps with distinct enhancing and repressing capability) exist to respond to different metals – a possibility that is puzzling in view of the inability of MEP-1, MTF-1, or ZAP to respond to Cd, or
2. A single MRTF exists in higher eukaryotes with the ability to respond to several different metals, but which could assume a DNA domain-binding structure irrespective of the metal ligand. This "flexible protein" model (THIELE 1992) would require the existence of a single higher eukaryotic DNA-binding factor responsive to Cd, Zn, Cu and other metals – a species that has not yet been described.
3. DNA-binding MRTF (s) specifically responsive to zinc exist; however, their zinc status depends upon a metalloregulatory protein or proteins that have no DNA-binding activity. Such a metalloregulatory protein (MP) would respond to many different metal ions by removing or adding zinc, and zinc alone, from MRTFs. Metallothioneins may be good candidates for such a role, especially since they exhibit a hierarchy of non-covalent metal binding strengths; stability constants for CuMT (10^{19}–10^{17}) and CdMT (10^{17}–10^{15}) are at least 10- to 100-fold greater than for ZnMT (10^{14}–10^{11}) (reviewed by HAMER 1986). MTs associated with zinc (as they are in cells and animal tissues uninduced with heavy metals) release zinc by displacement upon exposure to either copper or cadmium (SUZUKI et al. 1990); released zinc would then be available to mediate MRTF activity, especially if the newly released zinc were present in the nucleus. MT has, indeed, been localized to the nucleus during development (PANEMANGALORE et al. 1983; ANDREWS et al. 1987; NARTEY et al. 1987) and in certain human tumours (reviewed in Chap. 6, this volume). Such a metal-responsive zinc carrier model would be part of a multiple factor, rather than a single factor, signal transduction system (see above). Since many metalloenzymes as well as MRTFs require zinc, proteins such as MT could conceivably play a role in supplying them with zinc by a similar mechanism. Although MT has been

shown in vitro to exchange Zn atoms with SP1 or TFIIIa (ZENG et al. 1991a,b), proof of such a model will require direct biochemical observation of interactions between MT (or some other zinc carrier) and MRTFs, and/or perturbation of MRTF metal status, DNA-binding activity, and transcriptional activity in response to genetic up- or downregulation of MT in cells, Further, it will be critical to determine whether such a scenario would permit transcriptional control through a cellular "decision" to transfer, or not to transfer, zinc from MPs to MRTFs and metal-dependent enzymes (perhaps by regulating the level of MT, its subcellular localization, or the metals with which it is associated), or whether the intermediate transfer of essential metals is indiscriminate and constitutively permissive of transcription in response to signalling.

b) Selenium

In both eukaryotes and bacteria, selenocysteine may be incorporated into selenoproteins through use of the RNA codon U(T)GA (CHAMBERS and HARRISON 1987), and is mediated by an unusual tRNA; this tRNA first incorporates a serine residue and is subsequently converted to selenocysteine while still attached to the tRNA. In bacterial mRNAs with incorporated selenocysteine, putative stem-loop forming sequences have been described (CHESTERS 1992), suggesting a role in translational efficiency or mRNA degradation or accumulation. Although there is no direct evidence for such control in mammals, it is intriguing to note that the major selenoprotein glutathione peroxidase (GP) is linked with dietary selenium content (LI et al. 1990; SAEDI et al. 1988; TOYODA et al. 1989; SUNDE 1990). Altered GP mRNA content, but not transcription rate of the GP gene, is induced by selenium deficiency, suggesting that selenocysteine-tRNA or its translational cofactor may stabilize the mRNA by binding to it in a manner reminiscent of the IRE-BP/IRE interaction described above (CHESTERS 1992).

IV. Metal Bioavailability and Sequestration

Many different mechanisms to mediate heavy metal resistance in prokaryotes exist, and have been carefully reviewed elsewhere (RALSTON and O'HALLORAN 1990; O'HALLORAN 1993). In eukaryotes, this job is done by metallothioneins which sequester heavy metal ions and make them unavailable to exert toxic effects (reviewed by HAMER 1986; ANDREWS 1990). In this regard, heme-hepoxin induces both heme oxygenase and MT expression in human promyelocytic leukaemia cells and mouse hepatoma cells (ALAM et al. 1989). The authors suggest that MT may sequester the non-toxic metal zinc, which would otherwise compete with Fe for occupation of sites on Fe-dependent regulatory proteins such as the IRE-BP. Thus, it is possible that sequestration of metals by MT could mediate, not only the direct toxic effect of such metals (reviewed in Chap. 1, this volume), but also the association of metals with metalloregulatory proteins. The question of whether zinc (and other metal ions) are freely available to mediate structure and function in such proteins, or are tightly controlled intracellularly (possibly as a mechanism to control protein function) becomes important (discussed by O'HALLORAN 1993). A large number of zinc (and copper) requiring proteins (including transcription factors and receptors) could be controlled by; (1) making zinc and/or copper ions more available within cells (possibly by transporting metals into cells or subcellular compartments by the newly described *MnC1* gene product, a P-type ATPase metal transporter, described above) or (2) making them less available by binding them to intracellular metal storage molecules (including MT), or transporting metals out of cells. If, on the other hand, essential metals like zinc are freely available, non-toxic diffusible components, then such control is not required. However, a few points are clear:

1. The availability of unbound, essential metals such as zinc and copper would not be without toxic consequences. As with iron, low molecular weight zinc complexes can catalyse biopolymer hydrolysis (O'HALLORAN 1993); copper is far more toxic, with the ability to participate in Fenton reactions to produce free radicals, lipid peroxidation, and cellular damage (ROWLEY and HALLIWELL 1983). Control of availability is likely to be essential.
2. Other metals such as calcium and iron are tightly controlled with respect to subcellular localization and intracellular metal concentrations at various stages of the cell cycle. In the absence of detailed information on the cell biology of zinc (VALLEE and FALCHUK 1993), it is likely that a similar level of control is required.

It has been suggested that molecules such as MT might actively remove metals from metal-binding sites in metalloregulatory or metal-dependent proteins; both EDTA and MT can remove zinc from SP1 or TFIIIA and inhibit their activity in vitro (MILLER et al. 1985; KADONAGA et al. 1987; ZENG et al. 1991a,b). As suggestive as these experiments are of a role for availability of zinc in controlling metalloprotein activity, they do not establish that such regulation occurs in vivo.

With respect to extracellular availability of essential metals, COUSINS and LEE-AMBROSE (1992) have shown that dietary zinc is rapidly taken up by rat kidney cell nuclei, and that a resulting zinc-binding activity can be isolated and used to bind an oligonucleotide encoding a trout metal regulatory element that is highly conserved in metal-regulated gene promoters. Pretreatment of human mononuclear cells in vitro with zinc at concentrations close to physiological levels (but unassociated with plasma proteins) affects the ability of those cells to respond to *l*ipo*p*oly*s*accharide (LPS) induction of the respiratory burst (LEIBBRANDT and KOROPATNICK 1994). LPS induction of the respiratory burst is accompanied by increased MT mRNA and protein accumulation in these cells, and, intriguingly, downregulation of MT expression by antisense RNA complementary to MT nucleic acids abolishes the LPS-induced respiratory burst without altering cell growth and viability (LEIBBRANDT et al. 1993). These data underscore the importance of extracellular sources of metals, and the poorly understood processes controlling essential metal transport and sequestration within cells and subcellular compartments, in understanding metal-regulated gene regulation in eukaryotes.

E. Other Metal-Regulated Genes

I. Plastocyanin and *cyt c6*

Non-metallothionein genes in algae, fungi, plants, rodents, and humans are responsive to metals. In the algae Chlamydomonas reinhardtii, for example, copper affects expression of both the copper-containing plastocyanin protein and the heme-protein cytochrome C (cyt c6). These proteins act interchangeably as photosynthetic electron carriers: high Cu levels result in increased plastocyanin due to increased stability of CuMT, and low Cu levels result in decreased plastocyanin and increased cyt c6 expression as a result of transcriptional derepression of the cyt c6 gene, probably mediated by a copper-sensing regulatory factor (MERCHANT and BOGORAD 1987; HILL et al. 1991; MERCHANT et al. 1991).

II. Superoxide Dismutase

In yeast, Cu, Zn SOD activity is induced by copper (GREGORY et al. 1974), either by Cu incorporation into pre-existing SOD to activate the protein, or by induction of SOD gene transcription. The MRTF ACE1, activated by binding to copper, binds to the Saccharomyces cerevisiae Cu,Zn SOD (SOD1) gene promoter (GRALLA et al. 1991) to induce elevated SOD1 mRNA (GRECO et al. 1990). Thus, the same metal signalling and MRTF that controls yeast CUP1 protein expression also induces *trans*-activation of SOD. There may be two reasons to explain the coregulation:

1. SOD1 requires Cu for activity; perhaps Cu-ACE1 association indicates that intracellular Cu levels are sufficient for the formation of active holoenzyme. A requirement for production of SOD1 in the presence of high copper may be that Cu stimulates formation of toxic free radicals through redox interactions with $O_2^{\bullet -}$ (HALLIWELL and GUTTERIDGE 1984), and SOD1 is stimulated to detoxify those radicals and prevent cell damage. In this case, metals may play a role in protective mechanisms to oxidative stress (a condition resulting, in part, from the production and accumulation of reactive oxygen species such as superoxide anion ($O_2^{\bullet -}$), hydrogen peroxide (H_2O_2), and hydroxyl radical (HO•). This raises the possibility that metal-

lothioneins may function in that protection, and SATO and BREMNER (1993) have reviewed the role of metallothionein in oxidative stress.

2. Yeast MT and SOD1 may have complementary activities, or separate functions both aimed at mediating response to oxidative stress. Recently, Thiele and coworkers (TAMAI et al. 1993) have demonstrated that both yeast and monkey MTs may functionally substitute for yeast Cu,Zn SOD; they showed that S. cerevisiae lacking the *SOD1* gene are sensitive to oxidative stress [i.e., dioxygen and paraquat (which generates superoxide radical) toxicity]. Cu induction of yeast MT protected the cells from oxidative stress, and monkey MT-1 and MT-2 isoforms expressed under the control of the CUP1 promoter in cells lacking SOD restored growth on lactate. Therefore, mammalian MTs protect cells from oxygen toxicity, which implies that antioxidant function is a feature of higher eukaryotic MT proteins. In this context, it is interesting to note that human cells undergoing oxidative stress concomitantly express MT; lipopolysaccharide treatment of human monocytes induces, among other phenotype changes, a respiratory burst characterized by enhanced production of hydrogen peroxide (PICK and KEISARI 1980). LPS stimulates both H_2O_2 production and MT expression in both freshly isolated human monocytes and the human monocytic leukaemia cell line THP-1 (LEIBBRANDT and KOROPATNICK 1994). Pretreatment of THP-1 cells with zinc induced MT, but subsequent LPS treatment both failed to induce the respiratory burst and decreased both MT mRNA and protein levels. In addition, downregulation of MT in THP-1 cells by transient transfection of an antisense MT RNA expression vector reduced MT expression dramatically and abolished the LPS-induced respiratory burst (LEIBBRANDT et al. 1994). This raises the possibility that MT may not only mediate some cellular responses to oxidant stress (including, perhaps, a protective role against free radical damage), but may also have a role in transducing the LPS signal. MT may protect against oxidative stress in other cells: CHO cells with increased MT (due to zinc induction) or with decreased MT (after stable transfection with an antisense RNA expression vector) showed a good correlation between MT level and resistance to DNA strand scission by hydrogen peroxide (CHUBATSU and MENEGHINI 1993). No SOD, catalase, or glutathione changes were induced by the zinc or antisense RNA expression and most Zn-induced MT was in the nucleus. However, there was no differential resistance to the killing action of H_2O_2, an observation in agreement with the lack of altered sensitivity to ionizing radiation in mammalian cells expressing foreign MT genes (KELLEY et al. 1988; SCHILDER et al. 1990; KOROPATNICK and PEARSON 1993). In human HeLa cells, oxidant stress induced by hydrogen peroxide treatment increased the DNA-binding activity of AP-1 (ANGEL et al. 1987), which activates *MT-2a* gene response to phorbol esters, DNA damaging agents, and other oxidative stress-inducing conditions (DEVARY et al. 1991; ALAM and SMITH 1992; GRALLA and KOSMAN 1992; ANGEL et al. 1987). Several possibilities for MT function in these conditions exist, including:

1. A "scavenging" role for MT in inactivating free radicals to prevent damage.
2. MT-mediated movement of copper to subcellular locations to minimize toxic effects of the metal.
3. A role for MT in providing copper and/or zinc to SOD or other antioxidant proteins.
4. A role for MT in mediating the zinc status of MRTFs important in regulating expression of stress response genes.

Genetic manipulation to specifically alter MT expression (both up and down) in active MT gene transfection, gene knockout, and antisense nucleic acid expression, combined with exposure to physiological events mediating oxidative stress, will provide further insights into the relationship between MT and oxygen radicals in the future.

III. Heat Shock Proteins

The human 70-kDa *h*eat *s*hock *p*rotein (HSP70) is transcriptionally induced by several mechanisms, including heat, arsenic, amino acid analogues, and metals. Transcription is mediated by a *h*eat *s*hock *e*lement (HSE) in the promoter, and not MREs; metals may interact with a *h*eat *s*hock (transcription) *f*actor (HSF) to allow productive interaction of HSF and HSE to induce HSP70 transcription (WILLIAMS and MORIMOTO 1990). Thus, metals may regulate the expression of genes that do not contain MREs. Some evidence exists for these kinds of events in HSF activity (ZIMARINO et al. 1990; LARSON et al. 1988), and suggests that metal interactions with regulatory protein may mediate that regulation.

IV. Acute Phase Proteins, Heme Oxygenase, and Oncogenes

Pb, Cu, Ni, Zn, and Mg induce (in decreasing order of effectiveness) the acute phase protein genes for serum α_1-acid glycoprotein gene (*AGP1*) and C-reactive protein (CRP) in mice (YIANGOU et al. 1991). The AGP1 promoter can confer $HgCl_2$ induction on a *c*hloramphenicol *a*cetyl*t*ransferase (*CAT*) reporter gene in $HEPG_2$ cells, possibly through four regions of homology to the MRE consensus sequence which may be responsive to the same MRTFs as MT genes.

The heme oxygenase (*HO*) gene (which controls a rate-limiting step in heme catabolism) is induced by $CdCl_2$ in both rat and mouse (ALAM et al. 1989); the HO promoter contains two regions with close resemblance to the MREs in MT gene promoters, but their role in induction has not been explored.

The cellular proto-oncogenes myc and jun are induced by cadmium in rat cells (JIN and RINGERTZ 1990), and overexpression of c-ras in mouse fibroblast NIH 3T3 cells results in increased MT expression (ISONISHI et al. 1991). There is a good correlation ($r = 0.95$) between v-ras mRNA accumulation (over a fourfold range) and MT-1 and MT-2 mRNA accumulation (over a threefold range) among seven NIH 3T3 cell clones stably transfected with a c-ras expression vector (J. Koropatnick and A.F. Chambers, unpublished data). In addition, c-fos mRNA levels are enhanced in response to Cd or Zn (without de novo protein synthesis) in Swiss 3T3 cells (EPNER and HERSCHMAN 1991). Although cadmium is a known carcinogen (FRIBERG et al. 1992), the relationship between metal induction, and oncogene and MT expression, is not known.

F. Metal-Induced Changes in Chromatin Structure

Gene transcription in eukaryotes is controlled, not only by DNA sequence information, but by higher order protein-dependent organization of chromosomes into chromatin (CROSTON and KADONAGA 1993; KORNBERG and LORCH 1992). In general, more transcriptionally active DNA resides in "decondensed" chromatin in a more open conformation (due, in part, to removal of H1 histones) (LAYBOURN and KADONAGA 1991). Proteins associated with chromatin, as well as nuclear pores and scaffold, have a high metal content (LEWIS and LAEMMLI 1982) and zinc has been reported to be important in the chromatin folding in human sperm (CASSWALL et al. 1987; KJELLBERG et al. 1992; BIANCHI et al. 1992); zinc deficiency changes chromatin conformation to a more highly condensed form along with alteration in the proportions of histone H1 subforms (CASTRO 1987). Zinc has been suggested to play a role in facilitating H1 histone removal from chromatin to expose specific genes for transcription (CHESTERS 1992).

In addition, metal interactions with DNA and/or chromatin could interfere with transcription by altering cis-acting transcription factor recognition sites, or the transcription factors themselves. The platinum-containing chemotherapeutic agent cisplatin reduces glucocorticoid hormone-induced transcription from the *m*ouse *m*ammary *t*umour *v*irus (MMTV) promoter [as measured by primer extension and activity of a (CAT) reporter gene] without significant cellular toxicity (T. Archer, personal communication). The characteristic increase in MMTV promoter nuclease sensitivity that accompanies transcription (indicative of a more "open" and accessible chromatin conformation) was lost in cisplatin-treated cells, and binding of the transcription factor *n*uclear *f*actor 1 (NF-1) to its site after hormone stimulation was impaired. These data suggest that cisplatin (and, perhaps, other metal-containing ligands) could exert their toxic effect by interacting directly with promoter regions or the transcription-regulating ligands to alter gene transcription. In this regard, nickel (SNOW 1992; KARGACIN et al. 1993) and chromium (KORTENKAMP et al. 1992; SALNIKOW et al. 1992) can interact with DNA to form protein-DNA or DNA-DNA cross-links; the specificity of these interactions (perhaps in decondensed chromatin regions more accessible to metal ions) and their effect on gene transcription rather than toxicity alone are intriguing avenues of future exploration.

G. Summary

Metal ions are both necessary for life and extremely toxic. Cells have maximized their ability to use metals as cofactors for metalloenzymes, metalloregulatory proteins, and oxygen metabolism, and minimized their damaging potential, by expressing genes whose products regulate metal availability within cells. Those genes sense the presence of metals through metal associations with MRTFs, mRNA-binding proteins, metal transporter proteins, metal-dependent enzymes, and, perhaps, interactions between metals and DNA/chromatin itself. Although it is clear that gene transcription is dependent upon essential metals (particularly zinc), dependence may or may not mean that transcription is controlled by metal availability. It is crucial to explore the effect of altered metal transport (both from extracellular to intracellular sites, and between subcellular compartments) and intracellular storage (in metallothioneins and other metal-binding cellular proteins) on transcription of specific genes. To this end, the discovery of new metal transporter genes, and the ability to genetically up- and downregulate their expression (and the expression of metallothionein, the major intracellular metal ligand) will be the experimental techniques most useful in resolving the question.

Acknowledgements. This review is dedicated to my (J.K.) wife Jane Edwards, and my children lan and Reta. I am grateful to my colleagues for communication of their results, both published and unpublished. We thank Janice DeMoor and Jennifer Fraser for critical reading of the manuscript, and the Medical Research Council of Canada for grant support in generating preliminary data reported here.

References

Alam J, Smith A (1992) Heme-hemopexin-mediated induction of metallothionein gene expression. J Biol Chem 267:16379–16384

Alam J, Shibahara S, Smith A (1989) Transcriptional activation of the heme oxygenase gene by heme and cadmium in mouse hepatoma cells. J Biol Chem 264:6371–6375

Andrews GK (1990) Regulation of metallothionein gene expression. Prog Food Nutr Sci 14:193–258

Andrews GK, Gallant KR, Cherian MG (1987) Regulation of the ontogeny of rat liver metallothionein mRNA by zinc. Eur J Biochem 166:527–531

Angel P, Imagawa M, Chiu R, Stein B, Imbra RJ, Rahmsdorf HJ, Jonat C, Herrlich P, Karin M (1987) Phorbol ester-inducible genes contain a common cis element recognized by a TPA-modulated trans-acting factor. Cell 49:729–739

Becker PB, Gloss B, Schmid W, Strahle U, Schutz G (1986) In vivo protein-DNA interactions in a glucocorticoid response element require the presence of the hormone. Nature 324:686–688

Bianchi F, Rousseaux Prevost R, Sautiere P, Rousseaux J (1992) P2 protamines from human sperm are zinc-finger proteins with one CYS2/H1S2 motif. Biochem Biophys Res Commun 182:540–547

Biempica L, Rauch H, Quintana N, Sternlieb I (1988) Morphologic and chemical studies on a murine mutation (toxic milk mice) resulting in hepatic copper toxicosis. Lab Invest 59:500–508

Brouwer M, Schlenk D, Ringwood AH, Brouwer Hoexum T (1992) Metal-specific induction of metallothionein isoforms in the blue crab Callinectes sapidus in response to single- and mixed-metal exposure. Arch Biochem Biophys 294: 461–468

Brown PH, Daniels McQueen S, Walden WE, Patino MM, Gaffield L, Bielser D, Thach RE (1989) Requirements for the translational repression of ferritin transcripts in wheat germ extracts by a 90-kDa protein from rabbit liver. J Biol Chem 264:13383–13386

Carri MT, Galiazzo F, Ciriolo MR, Rotilio G (1991) Evidence for co-regulation of Cu,Zn superoxide dismutase and metallothionein gene expression in yeast through transcriptional control by copper via the ACE 1 factor. FEBS Lett 278:263–266

Castro CE (1987) Nutrient effects on DNA and chromatin structure. Annu Rev Nutr 7:407–421

Casswall TH, Bjorndahl L, Kvist U (1987) Cadmium interacts with the zinc-dependent stability of the human sperm chromatin. J Trace Elem Electrolytes Health Dis 1:85–87

Catelli MG, Binart N, Jung Testas I, Renoir JM, Baulieu EE, Feramisco JR, Welch WJ (1985) The common 90-kd protein component of non-transformed "8S" steroid receptors is a heat-shock protein. EMBO J 4:3131–3135

Chambers I, Harrison PR (1987) A new puzzle in seleno-protein biosynthesis: selenocysteine seems to be encoded by a stop codon UGA. Trends Biochem Sci 12:255–256

Chelly J, Tumer Z, Tonnesen T, Petterson A, Ishikawa Brush Y, Tommerup N, Horn N, Monaco AP (1993) Isolation of a candidate gene for Menkes disease that encodes a potential heavy metal binding protein (see comments). Nat Genet 3:14–19

Chesters JK (1992) Trace element-gene interactions. Nutr Rev 50:217–223

Choudhuri S, McKim JM Jr, Klaassen CD (1992) Role of hepatic lysosomes in the degradation of metallothionein. Toxicol Appl Pharmacol 115:64–71

Chu E, Voeller D, Koeller DM, Drake JC, Takimoto CH, Maley GF, Maley F, Allegra CJ (1993) Identification of an RNA binding site for human thymidylate synthase. Proc Natl Acad Sci USA 90:517–521

Chubatsu LS, Meneghini R (1993) Metallothionein protects DNA from oxidative damage. Biochem J 291:193–198

Cotton FA, Wilkinson G (1980) Advanced inorganic chemistry. A comprehensive text. Wiley, New York

Cousins RJ, Lee-Ambrose LM (1992) Nuclear zinc uptake and interactions and metallothionein gene expression are influenced by dietary zinc in rats. J Nutr 122:56–64

Croston GE, Kadonaga JT (1993) Role of chromatin structure in the regulation of transcription by RNA polymerase II. Curr Opin Cell Biol 5:417–423

Culotta VC, Hamer DH (1989) Fine mapping of a mouse metallothionein gene metal response element. Mol Cell Biol 9:1376–1380

Danks DM (1989) Disorders of copper transport. In: Scrives C, Beaudet A, Sly W, Valle D (eds) The metabolic basis of inherited disease. McGraw-Hill, New York, pp 1411–1432

Danks DM, Mercer JFB (1988) Metallothionein and ceruloplasmin genes. In: Hurley LS, Keen CL, Lunnerdal B, Rucker RB (eds) Trace elements in man and animals, vol 6. Plenum, New York, pp 287–291

Darwish HM, Hoke JE, Ettinger MJ (1983) Kinetics of Cu(II) transport and accumulation by hepatocytes from copper-deficient mice and the brindled mouse model of Menkes disease. J Biol Chem 258:13621–13626

Denis M, Gustafsson JA (1989) The Mr approximately 90 000 heat shock protein: an important modulator of ligand and DNA-binding properties of the glucocorticoid receptor. Cancer Res 49:2275s–2281s

Devary Y, Gottlieb RA, Lau LF, Karin M (1991) Rapid and preferential activation of the c-jun gene during the mammalian UV response. Mol Cell Biol 11: 2804–2811

Dix DJ, Lin PN, Kimata Y, Theil EC (1992) The iron regulatory region of ferritin mRNA is also a positive control element for iron-independent translation. Biochemistry 31:2818–2822

Epner DE, Herschman HR (1991) Heavy metals induce expression of the TPA-inducible sequence (TIS) genes. J Cell Physiol 148:68–74

Fantl WJ, Johnson DE, Williams LT (1993) Signalling by receptor tyrosine kinases. Annu Rev Biochem 62:453–481

Friberg L, Elinder CG, Kjellstrom T (1992) Cadmium. Environmental Health Criteria 134. World Health Organization, Geneva, p 199

Frydman M, Bonne Tamir B, Farrer LA, Conneally PM, Magazanik A, Ashbel S, Goldwitch Z (1985) Assignment of the gene for Wilson's disease to chromosome 13: linkage to the esterase D locus. Proc Natl Acad Sci USA 82:1819–1821

Gehring U (1993) The structure of glucocorticoid receptors. J Steroid Biochem Mol Biol 45:183–190

Gitlin JD, Schroeder JJ, Lee Ambrose LM, Cousins RJ (1992) Mechanisms of caeruloplasmin biosynthesis in normal and copper-deficient rats. Biochem J 282:835–839

Godowski PJ, Rusconi S, Miesfeld R, Yamamoto KR (1987) Glucocorticoid receptor mutants that are constitutive activators of transcriptional enhancement (published erratum appears in Nature (1987) 326:105). Nature 325:365–368

Goyer RA (1991) Toxic effects of metals. In: Amdur MO, Doull J, Klaassen CD (eds) Casarett and Doull's toxicology, 4th edn. Pergamon, New York, pp 652–661

Gralla EB, Kosman DJ (1992) Molecular genetics of superoxide dismutases in yeasts and related fungi. Adv Genet 30:251–319

Gralla EB, Thiele DJ, Silar P, Valentine JS (1991) ACE1, a copper-dependent transcription factor, activates expression of the yeast copper, zinc superoxide dismutase gene. Proc Natl Acad Sci USA 88:8558–8562

Graupner G, Wills KN, Tzukerman M, Zhang XK, Pfahl M (1989) Dual regulatory role for thyroid-hormone receptors allows control of retinoic-acid receptor activity. Nature 340:653–656

Greco MA, Hrab DI, Magner W, Kosman DJ (1990) Cu,Zn superoxide dismutase and copper deprivation and toxicity in Saccharomyces cerevisiae. J Bacteriol 172:317–325

Gregory EM, Goscin SA, Fridovich I (1974) Superoxide dismutase and oxygen toxicity in a eukaryote. J Bacteriol 117:456–460

Guarente L, Mason T (1983) Heme regulates transcription of the CYC1 gene of S. cerevisiae via an upstream activation site. Cell 32:1279–1286

Halliwell B, Gutteridge JM (1984) Oxygen toxicity, oxygen radicals, transition metals and disease. Biochem J 219:1–14

Halliwell B, Gutteridge JM (1985) The importance of free radicals and catalytic metal ions in human diseases. Mol Aspects Med 8:89–193

Hamer DH (1986) Metallothionein. Annu Rev Biochem 55:913–951

Hamer DH (1993) "Kinky hair" disease sheds light on copper metabolism. Nat Genet 3:3–4

Hara K, Kominami E, Katunuma N (1988) Effects of proteinase inhibitors on intracellular processing of cathepsin B, H and L in rat macrophages. FEBS Lett 231:229–231

Harris ED, Stevens MD (1985) Receptors for ceruloplasmin in aortic cell membranes. In: Mills CF, Bremner I, Chester JK (eds) Trace elements in animals and man. Slough, UK, Commonwealth Agricultural Bureau, pp 320–323

Hentze MW, Caughman SW, Rouault TA, Barriocanal JG, Dancis A, Harford JB, Klausner RD (1987) Identification of the iron-responsive element for the translational regulation of human ferritin mRNA. Science 238:1570–1573

Herd SM, Camakaris J, Christofferson R, Wookey P, Danks DM (1987) Uptake and efflux of copper-64 in Menkes' disease and normal continuous lymphoid cell lines. Biochem J 247:341–347

Hill KL, Li HH, Singer J, Merchant S (1991) Isolation and structural characterization of the Chlamydomonas reinhardtii gene for cytochrome c6. Analysis of the kinetics and metal specificity of its copper-responsive expression. J Biol Chem 266:15060–15067

Hogberg J, Alexander J (1986) In: Friberg L, Nordberg GF, Vouk VB (eds) Specific metals, 2nd edn. Elsevier Scientific, Amsterdam, pp 482–512 (Handbook on the toxicology of metals, vol II)

Hollenberg SM, Giguere V, Segui P, Evans RM (1987) Colocalization of DNA-binding and transcriptional activation functions in the human glucocorticoid receptor. Cell 49:39–46

Horn N (1984) Copper metabolism in Menkes' disease. In: Rennert OM, Chan W-Y (eds) Metabolism of trace metals in man, vol II. CRC Press, Boca Raton, pp 26–52

Hutchison KA, Scherrer LC, Czar MJ, Stancato LF, Chow YH, Jove R, Pratt WB (1993) Regulation of glucocorticoid receptor function through assembly of a receptor-heat shock protein complex. Ann N Y Acad Sci 684:35–48

Imbert J, Zafarullah M, Culotta VC, Gedamu L, Hamer D (1989) Transcription factor MBF-1 interacts with metal regulatory elements of higher eucaryotic metallothionein genes. Mol Cell Biol 9:5315–5323

Isonishi S, Horn DK, Thiebaut FB, Mann SC, Andrews PA, Basu A, Lazo JS, Eastman A, Howell SB (1991) Expression of the c-Ha-ras oncogene in mouse NIH 3T3 cells induces resistance to cisplatin. Cancer Res 51:5903–5909

Ivarie RD, O'Farrell PH (1978) The glucocorticoid domain: steroid-mediated changes in the rate of synthesis of rat hepatoma proteins. Cell 13:41–55

Jensen EV, DeSombre ER (1972) Mechanism of action of the female sex hormones. Annu Rev Biochem 41:203–230

Jin P, Ringertz NR (1990) Cadmium induces transcription of proto-oncogenes c-jun and c-myc in rat L6 myoblasts. J Biol Chem 265:14061–14064

Jin P, Sejersen T, Ringertz NR (1991) Recombinant platelet-derived growth factor-BB stimulates growth and inhibits differentiation of rat L6 myoblasts. J Biol Chem 266:1245–1249

Kadonaga JT, Carner KR, Masiarz FR, Tjian R (1987) Isolation of cDNA encoding transcription factor Sp1 and functional analysis of the DNA binding domain. Cell 51:1079–1090

Kagi JH, Schaffer A (1988) Biochemistry of metallothionein. Biochemistry 27: 8509–8515

Kapur S, Higgins JV, Delp K, Rogers B (1987) Menkes syndrome in a girl with X-autosome translocation. Am J Med Genet 26:503–510

Kargacin B, Klein CB, Costa M (1993) Mutagenic responses of nickel oxides and nickel sulfides in Chinese hamster V79 cell lines at the xanthine-guanine phosphoribosyl transferase locus. Mutat Res 300:63–72

Karin M (1992) Signal transduction from cell surface to nucleus in development and disease. FASEB J 6:2581–2590

Karin M, Slater EP, Herschman HR (1981) Regulation of metallothionein synthesis in HeLa cells by heavy metals and glucocorticoids. J Cell Physiol 106:63–74

Kelley SL, Basu A, Teicher BA, Hacker MP, Hamer DH, Lazo JS (1988) Overexpression of metallothionein confers resistance to anticancer drugs. Science 241:1813–1815

Kershaw WC, Klaassen CD (1992) Degradation and metal compositon of hepatic isometallothioneins in rats. Toxicol Appl Pharmacol 112:24–31

Kjellberg S, Bjorndahl L, Kvist U (1992) Sperm chromatin stability and zinc binding properties in semen from men in barren unions. Int J Androl 15:103–113

Klausner RD, Harford JB (1989) cis-trans models for post-transcriptional gene regulation. Science 246:870–872

Koizumi S, Suzuki K, Otsuka F (1992a) A nuclear factor that recognizes the metal-responsive elements of human metallothionein IIA gene. J Biol Chem 267: 18659–18664

Koizumi S, Yamada H, Suzuki K, Otsuka F (1992b) Zinc-specific activation of a HeLa cell nuclear protein which interacts with a metal responsive element of the human metallothionein-IIA gene. Eur J Biochem 210:555–560
Kornberg RD, Lorch Y (1992) Chromatin structure and transcription. Annu Rev Cell Biol 8:563–587
Koropatnick J, Cherian MG (1993) A mutant mouse (tx) with increased hepatic metallothionein stability and accumulation. Biochem J 296:443–449
Koropatnick J, Pearson J (1990) Zinc treatment, metallothionein expression, and resistance to cisplatin in mouse melanoma cells. Somat Cell Mol Genet 16: 529–537
Koropatnick J, Pearson J (1993) Altered cisplatin and cadmium resistance and cell survival in Chinese hamster ovary cells expressing mouse metallothionein. Mol Pharmacol 44:44–50
Kortenkamp A, Curran B, O'Brien P (1992) Defining conditions for the efficient in vitro cross-linking of proteins to DNA by chromium(III) compounds. Carcinogenesis 13:307–308
Kuivaniemi H, Peltonen L, Palotie A, Kaitila I, Kivirikko K (1981) Abnormal copper metabolism and deficient lysyl oxidase activity in a heritable connective tissue disorder. J Clin Invest 69:730–733
Labbe S, Prevost J, Remondelli P, Leone A, Seguin C (1991) A nuclear factor binds to the metal regulatory elements of the mouse gene encoding metallothionein-1. Nucleic Acids Res 19:4225–4231
Lad PM, Kaptein JS, Lin CK, Kalunta CI, Scott SJ, Gu DG (1992) G-proteins and the role of second messengers in the regulation of the human neutrophil. Immunol Ser 57:107–136
Lamb TD, Pugh ENJ (1992) G-protein cascades: gain and kinetics. Trends Neurosci 15:291–298
Larson JS, Schuetz TJ, Kingston RE (1988) Activation in vitro of sequence-specific DNA binding by a human regulatory factor. Nature 335:372–375
Lathrop JT, Timko MP (1993) Regulation by heme of mitochondrial protein transport through a conserved amino acid motif. Science 259:522–525
Laurin DE, Barnes DM, Klasing KC (1990) Rates of metallothionein synthesis, degradation and accretion in a chicken macrophage cell line. Proc Soc Exp Biol Med 194:157–164
Laybourn PJ, Kadonaga JT (1991) Role of nucleosomal cores and histone H1 in regulation of transcription by RNA polymerase II. Science 254:238–245
Leibbrandt MEI, Koropatnick J (1994) Activation of human monocytes with lipopolysaccharide induces metallothionein expression and is diminished by zinc. Toxicol Appl Pharmacol 124:72–81
Leibold EA, Laudano A, Yu Y (1990) Structural requirements of iron-responsive elements for binding of the protein involved in both transferrin receptor and ferritin mRNA post-transcriptional regulation. Nucleic Acids Res 18:1819–1824
Leibbrandt MEI,Khokha R, Koropatnick J (1994) Antisense down-regulation of metallothionein in a human monocytic cell line alters adherence, invasion and the respiratory burst. Cell Growth Diff 5:17–25
Lescure AM, Proudhon D, Pesey H, Ragland M, Theil EC, Briat JF (1991) Ferritin gene transcription is regulated by iron in soybean cell cultures. Proc Natl Acad Sci USA 88:8222–8226
Lewis CD, Laemmli UK (1982) Higher order metaphase chromosome structure: evidence for metalloprotein interactions. Cell 29:171–181
Li NQ, Reddy PS, Thyagaraju K, Reddy AP, Hsu BL, Scholz RW, Tu CP, Reddy CC (1990) Elevation of rat liver mRNA for selenium-dependent glutathione peroxidase by selenium deficiency. J Biol Chem 265:108–113
McArdle HJ, Mercer JF, Sargeson AM, Danks DM (1990) Effects of cellular copper content on copper uptake and metallothionein and ceruloplasmin mRNA levels in mouse hepatocytes. J Nutr 120:1370–1375

McKim JM Jr, Choudhuri S, Klaassen CD (1992) In vitro degradation of apo-, zinc-, and cadmium-metallothionein by cathepsins B, C, and D. Toxicol Appl Pharmacol 116:117–124
Mercer JF, Livingston J, Hall B, Paynter JA, Begy C, Chandrasekharappa S, Lockhart P, Grimes A, Bhave M, Siemieniak D et al (1993) Isolation of a partial candidate gene for Menkes disease by positional cloning (see comments). Nat Genet 3:20–25
Merchant S, Bogorad L (1987) The Cu(II)-repressible plastidic cytochrome c. Cloning and sequence of a complementary DNA for the pre-apoprotein. J Biol Chem 262:9062–9067
Merchant S, Hill K, Howe G (1991) Dynamic interplay between two copper-titrating components in the transcriptional regulation of cyt c6 (published erratum appears in EMBO J (1991) 10:23201). EMBO J 10:1383–1389
Miller J, McLachlan AD, Klug A (1985) Repetitive zinc-binding domains in the protein transcription factor IIIA from Xenopus oocytes. EMBO J 4:1609–1614
Min KS, Nakatsubo T, Fujita Y, Onosaka S, Tanaka K (1992) Degradation of cadmium metallothionein in vitro by lysosomal proteases. Toxicol Appl Pharmacol 113:299–305
Muller M, Renkawitz R (1991) The glucocorticoid receptor. Biochim Biophys Acta 1088:171–182
Nartey NO, Banerjee D, Cherian MG (1987) Immunohistochemical localization of metallothionein in cell nucleus and cytoplasm of fetal human liver and kidney and its changes during development. Pathology 19:233–238
O'Halloran TV (1989) In: Sigel H (ed) Metal ions in biological systems, vol 25. Dekker, New York, p 105
O'Halloran TV (1993) Transition metals in control of gene expression (see comments). Science 261:715–725
Oppenheimer JH, Koerner D, Schwartz HL, Surks MI (1972) Specific nuclear triiodothyronine binding sites in rat liver and kidney. J Clin Endocrinol Metab 35:330–333
Pabo CO, Sauer RT (1992) Transcription factors: structural families and principles of DNA recognition. Annu Rev Biochem 61:1053–1095
Packman S, O'Toole T (1984) Trace metal metabolism in cultured skin fibroblasts of the mottled mouse: response to metallothionein inducers. Pediatr Res 18: 1282–1286
Packman S, O'Toole C, Price DC, Thaler MM (1983) Cadmium, zinc, and copper metabolism in the mottled mouse, an animal model for Menkes' kinky hair syndrome. J Inorg Biochem 19:203–211
Packman S, Sample S, Whitney W (1987) Defective intracellular copper translocation in Menkes' kinky hair syndrome. Pediatr Res [Suppl]21:293s
Panemangalore M, Banerjee D, Onosaka S, Cherian MG (1983) Changes in the intracellular accumulation and distribution of metallothionein in rat liver and kidney during postnatal development. Dev Biol 97:95–102
Pfeifer K, Arcangioli B, Guarente L (1987a) Yeast HAP1 activator competes with the factor RC2 for binding to the upstream activator site UAS1 of the CYC1 gene. Cell 49:9–18
Pfeifer K, Prezant T, Guarente L (1987b) Yeast HAP1 activator binds to two upstream activation sites of different sequence. Cell 49:19–27
Pfeifer K, Kim KS, Kogan S, Guarente L (1989) Functional dissection and sequence of yeast HAP1 activator. Cell 56:291–301
Pfister C, Bennett N, Bruckert F, Catty P, Clerc A, Pages F, Deterre P (1993) Interactions of a G-protein with its effector: transducin and cGMP phosphodiesterase in retinal rods. Cell Signal 5:235–241
Pick E, Keisari Y (1980) A simple colorimetric method for the measurement of hydrogen peroxide produced by cells in culture. J Immunol Methods 38:161–170
Ralston DM, O'Halloran TV (1990) Metalloregulatory proteins and molecular mechanisms of heavy metal signal transduction. Adv Inorg Biochem 8:1–31

Rauch H (1983) Toxic milk, a new mutation affecting copper metabolism in the mouse. J Hered 74:141–144
Rauch H, Dupuy D, Stockert RJ, Sternlieb (1986) Hepatic copper and superoxide dismutase activity in toxic milk mutant mice. In: Rotilio G (ed) Superoxide and superoxide dismutase in chemistry, biology and medicine. Elsevier, New York, pp 304–306
Rouault TA, Hentze MW, Caughman SW, Harford JB, Klausner RD (1988) Binding of a cytosolic protein to the iron-responsive element of human ferritin messenger RNA. Science 241:1207–1210
Rowley DA, Halliwell B (1983) Superoxide-dependent and ascorbate-dependent formation of hydroxyl radicals in the presence of copper salts: a physiologically significant reaction? Arch Biochem Biophys 225:279–284
Saedi MS, Smith CG, Frampton J, Chambers I, Harrison PR, Sunde RA (1988) Effect of selenium status on mRNA levels for glutathione peroxidase in rat liver. Biochem Biophys Res Commun 153:855–861
Salnikow K, Zhitkovich A, Costa M (1992) Analysis of the binding sites of chromium to DNA and protein in vitro and in intact cells. Carcinogenesis 13:2341–2346
Sanchez ER, Toft DO, Schlesinger MJ, Pratt WB (1985) Evidence that the 90-kDa phosphoprotein associated with the untransformed L-cell glucocorticoid receptor is a murine heat shock protein. J Biol Chem 260:12398–12401
Sato M, Bremner I (1993) Oxygen free radicals and metallothionein. Free Radic Biol Med 14:325–337
Sauk JJ, Norris K, Kerr JM, Somerman MJ, Young MF (1991) Diverse forms of stress result in changes in cellular levels of osteonectin/SPARC without altering mRNA levels in osteoligament cells. Calcif Tissue Int 49:58–62
Schauer M, Chalepakis G, Willmann T, Beato M (1989) Binding of hormone accelerates the kinetics of glucocorticoid and progesterone receptor binding to DNA. Proc Natl Acad Sci USA 86:1123–1127
Schilder RJ, Hall L, Monks A, Handel LM, Fornace AJ Jr, Ozols RF, Fojo AT, Hamilton TC (1990) Metallothionein gene expression and resistance to cisplatin in human ovarian cancer. Int J Cancer 45:416–422
Searle PF (1990) Zinc dependent binding of a liver nuclear factor to metal response element MRE-a of the mouse metallothionein-1 gene and variant sequences. Nucleic Acids Res 18:4683–4690
Searle PF, Stuart GW, Palmiter RD (1987) Metal regulatory elements of the mouse metallothionein-1 gene. In: Metallothionein II. Birkhauser, Basel, pp 407–414 (Experientia supplementum 52)
Seguin C (1991) A nuclear factor requires Zn^{2+} to bind a regulatory MRE element of the mouse gene encoding metallothionein-1. Gene 97:295–300
Séguin C, Felber BK, Carter AD, Hamer DH (1984) Competition for cellular factors that activate metallothionein gene transcription. Nature 312:781–785
Sloane BF, Moin K, Krepela E, Rozhin J (1990) Cathepsin B and its endogenous inhibitors: the role in tumor malignancy. Cancer Metastasis Rev 9:333–352
Snow ET (1992) Metal carcinogenesis: mechanistic implications. Pharmacol Ther 53:31–65
Srivastava G, Borthwick IA, Maguire DJ, Elferink CJ, Bawden MJ, Mercer JF, May BK (1988) Regulation of 5-aminolevulinate synthase mRNA in different rat tissues. J Biol Chem 263:5202–5209
Sunde RA (1990) Molecular biology of selenoproteins. Annu Rev Nutr 10:451–474
Suttle NF (1986) Copper deficiency in ruminants; recent developments. Vet Rec 119:519–522
Suzuki CA, Ohta H, Albores A, Koropatnick J, Cherian MG (1990) Induction of metallothionein synthesis by zinc in cadmium pretreated rats. Toxicology 63: 273–284
Tamai KT, Gralla EB, Ellerby LM, Valentine JS, Thiele DJ (1993) Yeast and mammalian metallothioneins functionally substitute for yeast copper-zinc superoxide dismutase. Proc Natl Acad Sci USA 90:8013–8017

Thiele DJ (1992) Metal-regulated transcription in eukaryotes. Nucleic Acids Res 20:1183–1191
Toyoda H, Himeno S, Imura N (1989) The regulation of glutathione peroxidase gene expression; implications for species differences and the effect of dietary selenium manipulation. In: Wendel A (ed) Selenium in biology and medicine. Springer, Berlin Heidelberg New York, pp 3–7
Trueblood CE, Wright RM, Poyton RO (1988) Differential regulation of the two genes encoding Saccharomyces cerevisiae cytochrome c oxidase subunit V by heme and the HAP2 and REO1 genes. Mol Cell Biol 8:4537–4540
Vallee BL, Falchuk FH (1993) The biochemical basis of zinc physiology. Physiol Rev 73:79–118
Vulpe C, Levinson B, Whitney S, Packman S, Gitschier J (1993) Isolation of a candidate gene for Menkes disease and evidence that it encodes a copper-transporting ATPase. Nat Genet 3:7–13
Walden WE, Daniels McQueen S, Brown PH, Gaffield L, Russell DA, Bielser D, Bailey LD, Thach RE (1988) Translational repression in eukaryotes: partial purification and characterization of a repressor of ferritin mRNA translation. Proc Natl Acad Sci USA 85:9503–9507
Wan M, Hunziker PE, Kagi JH (1993) Induction of metallothionein synthesis by cadmium and zinc in cultured rabbit kidney cells (RK-13). Biochem J 292: 609–615
Westin G, Schaffner W (1988) Heavy metal ions in transcription factors from HeLa cells: Sp1, but not octamer transcription factor requires zinc for DNA binding and for activator function. Nucleic Acids Res 16:5771–5781
White K, Munro HN (1988) Induction of ferritin subunit synthesis by iron is regulated at both the transcriptional and translational levels. J Biol Chem 263:8938–8942
Williams GT, Morimoto RI (1990) Maximal stress-induced transcription from the human HSP70 promoter requires interactions with the basal promoter elements independent of rotational alignment. Mol Cell Biol 10:3125–3136
Willmann T, Beato M (1986) Steroid-free glucocorticoid receptor binds specifically to mouse mammary tumour virus DNA. Nature 324:688–691
Winkler H, Adam G, Mattes E, Schanz M, Hartig A, Ruis H (1988) Co-ordinate control of synthesis of mitochondrial and non-mitochondrial hemoproteins: a binding site for the HAP-1 (CYP1) protein in the UAS region of the yeast catalase T gene. EMBO J 7:1799–1804
Yiangou M, Ge X, Carter KD, Papaconstantinou J (1991) Induction of several acute-phase protein genes by heavy metals: a new class of metal-responsive genes. Biochemistry 30:3798–3806
Yoshida T, Biro P, Cohen T, Muller RM, Shibahara S (1988) Human heme oxygenase cDNA and induction of its mRNA by hemin. Eur J Biochem 171: 457–461
Zeng J, Heuchel R, Schaffner W, Kagi JH (1991a) Thionein (apometallothionein) can modulate DNA binding and transcription activation by zinc finger containing factor Sp1. FEBS Lett 279:310–312
Zeng J, Vallee BL, Kagi JH (1991b) Zinc transfer from transcription factor IIIA fingers to thionein clusters. Proc Natl Acad Sci USA 88:9984–9988
Zhou P, Thiele DJ (1993) Copper and gene regulation in yeast. Biofactors 4:105–115
Zhou P, Szczypka MS, Sosinowski T, Thiele DJ (1992) Expression of a yeast metallothionein gene family is activated by a single metalloregulatory transcription factor. Mol Cell Biol 12:3766–3775
Zimarino V, Wilson S, Wu C (1990) Antibody-mediated activation of Drosophila heat shock factor in vitro. Science 249:546–549
Zitomer RS, Seller JW, McCarter DW, Hastings GA, Wick P, Lowry CV (1987) Elements involved in oxygen regulation of the Saccharomyces cerevisiae CYC7 gene. Mol Cell Biol 7:2212–2220

CHAPTER 6

Metallothionein and Its Interaction with Metals

M.G. CHERIAN

A. Introduction

Although metals are not metabolized in a similar manner to organic compounds, they are constantly converted to different ionic forms and also transferred to various ligands intracellularly. Metals are involved in various metabolic reactions as constituents of metalloenzymes or as cofactors and are important cellular factors for synthesis of nucleic acids, proteins, and carbohydrates. Nutritional studies have shown that these essential metals themselves can interact with each other and can cause deficiency symptoms of one in the presence of an excess of certain others (ERSHOFF 1948; PORTER et al. 1977). In addition to essential metals, we are constantly exposed to various nonessential metals in our diet, air, and water. In certain cases, exposure to toxic metals occurs at the workplace or in the treatment of diseases along with environmental exposure. Several toxic metals are closely related to essential metals in their chemistry and they can interact either directly or through their binding to various cellular components. These interactions play a major role in the nutrition and toxicology of metals. These interactions can result in changes in absorption, disposition, retention, excretion, and toxicity of metals (FRIEDEN 1976). This review will deal with one such interaction, namely the role of metallothionein (MT) in the nutrition and toxicology of metals.

About 3 decades ago, MT was isolated as a cadmium-binding protein from horse kidney, but this protein can also bind with essential metals such as zinc and copper (MARGOSHES and VALLEE 1957). Its structure and induced synthesis have attracted scientists from various disciplines. The solution structure as determined by ^{113}Cd NMR and crystal structure by X-ray crystallography have demonstrated that MT has a unique structure, with two distinct adamantine-like metal-binding clusters which are of a dynamic nature (OTVOS and ARMITAGE 1980; KÄGI 1991). The natural occurrence of a large amount of MT-bound zinc and copper in certain tissues during mammalian development and its induced synthesis by metals and various other organic compounds suggests an important role for this protein in the metabolism of essential metals and also in the detoxification of certain toxic metals and free radicals (CHERIAN and CHAN 1993). Increased expression of MT has been reported in metabolic disorders of essential metals and also in certain human tumors (Koropatnick and Cherian 1993; Cherian et al. 1993).

The physicochemical properties of mammalian MT can be summarized as:

1. Low molecular weight (<9 kDa) protein with a conserved structure and high cysteine content (30%) but no disulfide bonds.
2. High metal content and binding with both essential and toxic metals through thiolate bonds and oligonuclear complexes (clusters).
3. Low levels found in adult mammalian tissues but its synthesis is inducible by metals, hormones, cytokines, growth factors, and stress conditions.
4. High levels found in fetal mammalian livers, certain human tumors, and proliferating epithelial cells, bound to zinc and copper.
5. Division of the multiple isoforms into four structurally distinct subgroups (MT-1, MT-2, MT-3, MT-4), but MT-1 and MT-2 are the two major isoforms found in most mammalian tissues.
6. Found mainly intracellularly in cytoplasm of cells in adult mammalian tissues while it can be detected in cell nucleus of fetal tissue and in epithelial cells of certain tumors.

B. Metal Binding and Dynamic Aspects of Metallothionein Structure

The high content of cysteine (30%) and its special occurrence in the polypeptide as Cys-Cys, Cys-X-Cys, and Cys-X-Y-Cys sequences (where X and Y are amino acids other than cysteine) are the major reasons for the high affinity of metals in MT. All the thiol groups in MT participate in metal binding, and the coordination properties of the metal-binding sites have been elucidated. Studies have shown a tetrahedral tetrathiolate coordination of metals such as cadmium and zinc and a trigonal trithiolate coordination for copper and silver (Winge 1991). These structures are possible by a sharing of the thiolate ligands between adjacent metal ions as bridging thiols. In addition, in mammalian and invertebrate MTs, the metals are arranged in two distinct clusters and they can also accommodate mixed metal ions (Kägi 1991). These structural features of MT are completely elucidated by a two-dimensional NMR solution structure and X-ray crystallography of isolated crystals of Cd,Zn-MT-2 (Messerle et al. 1992; Robbins et al. 1991). Although metals are bound to MT with very high affinity, there are differences in their dissociation constants, for example the apparent average dssociation constants $K'\text{Cd} = 5 \times 10^{-17} M$ and $K'\text{Zn} = 5 \times 10^{-13} M$ have been calculated for Cd and Zn respectively in MT (Vasak and Kägi 1983). It has also been shown that metals in the three-metal cluster of MT can be released more easily than those in the four-metal cluster. These differences in binding will allow displacement and rearrangement of various different metals between two-cluster domains within the MT molecule and also between MT molecules. However, the metal bound to one domain does not influence the reactivity of the metal in the other cluster domain. Most of the information on the dynamic aspects of MT structure is derived from metal NMR studies and has been reviewed in recent articles by OTVOS et al. (1993) and KÄGI (1993).

In most of the mammalian tissues studied, there are two major isoforms of MT (termed MT-1 and MT-2), which differ slightly in amino acid sequence and net charge. These MT isoforms bind metals identically but are encoded by two distinct genes. Recently two other mammalian MT genes (termed MT-3 and MT-4) have been identified in certain mammal tissues (PALMITER et al. 1993), but their significance is not yet understood. In contrast, only one MT gene has been isolated from avian species, chicken, and turkey (FERNANDO and ANDREWS 1989).

The role of amino acids other than cysteine in the binding of metals in MT is unclear. Recent studies by site-directed mutagenesis and recombinant DNA techniques have replaced the interdomain lysines in MT-2 from chinese hamster ovary (CHO) cells with glutamic acid and/or glutamine and demonstrated that these lysine replacements did not affect the metal-binding ability of MT (CODY and HUANG 1993). However, these changes reduced the expression of MT by these cells and thus increased the cytotoxicity of cadmium in vitro in these cells. These studies confirm previous reports on chemical modification of lysine residues which did not affect metal binding to MT (TEMPLETON and CHERIAN 1984b; PANDE et al. 1985). Thus all the studies so far suggest that only cysteine residues are involved in the binding of metals to MT. Thus a protective role of MT against cytotoxicity of metals can be expected, if those metals can bind with MT.

C. Induction of Metallothionein and Excretion of Metals

One of the major effects of induction of MT is the decreased excretion of metals which can avidly bind with MT. This is not surprising because MT has been considered as an intracellular metal storage protein which is shown to be synthesized on free polysomes in rat liver, the site of synthesis of storage proteins such as ferritin (CHERIAN et al. 1981). Induction of hepatic MT by pretreatment of rats with cadmium salts increased the zinc content of liver and decreased significantly the biliary excretion of cadmium, suggesting that once cadmium is bound to hepatic MT it is no longer available for biliary excretion (CHERIAN 1977; KLAASSEN 1978). The biliary excretion is a major excretory route for cadmium, especially after a single high-dose exposure. Although the major intracellular form of cadmium is protein bound, the form of cadmium in the bile is a low molecular weight complex, partially characterized as a glutathione conjugate (CHERIAN and VOSTAL 1977). It is also known that the induction of MT synthesis by injection of metals can reduce the diffusible form of cadmium in the body and thereby decrease the biliary excretion of cadmium (CHERIAN and VOSTAL 1974). Biliary excretion of copper, mercury, zinc, and silver is similarly decreased by cadmium pretreatment in rats (KLAASSEN 1978). Preinjection of experimental animals with zinc or copper also decreased the biliary excretion of cadmium (STOWE 1976; CHERIAN 1980). It has also been shown that the biliary excretion of a bolus of zinc, cadmium, or mercury was significantly decreased when MT was chronically induced by feeding a high-zinc diet containing 1150 μg Zn/g in rats (JAW and JEFFERY 1988). Thus there is significant experimental evidence which suggests that the induced synthesis of MT within a few hours after injection of cadmium or chronic feeding of zinc has a marked effect on the toxicokinetics of certain metals. The long biological half-life of cadmium in mammals may be due to its strong binding to an intracellular protein like MT. When cadmium-pretreated rats were fed a zinc-deficient diet, the levels of MT and zinc in the pancreas markedly decreased without any change in the cadmium content (TEMPLETON and CHERIAN 1984a). Further analysis showed that cadmium was transferred from MT to high molecular weight proteins in the cytosol of pancreas. Nutritional studies have also shown that newborn rat pups, when allowed to suckle from a zinc-deficient mother, resulted in a rapid degradation of hepatic MT and release of zinc, suggesting the role of MT as a zinc storage protein in newborn rats (GALLANT and CHERIAN 1987).

The effects of zinc pretreatment on renal accumulation, intrarenal distribution, and urinary excretion of inorganic mercury were studied in control and unilateral nephrectomized rats (ZALUPS and CHERIAN 1992). In control rats, most of the injected inorganic mercury is accumulated in kidney, especially in the outer medulla, with subsequent urinary excretion, which is the major route of excretion for inorganic mercury. Injection of zinc salts resulted in induced synthesis of MT in the kidney, and subsequent injection of inorganic mercury showed increased accumulation of mercury in kidney. This increase in renal mercury levels was primarily due to an increase in accumulation of mercury in the cortex rather than in the outer medulla as in control rats. These results indicate that zinc pretreatment causes a shift in the pattern of intrarenal accumulation of inorganic mercury. This may be related to the synthesis of MT in epithelial cells of renal cortex and may show a positive relationship between zinc-induced synthesis of MT and the accumulation of inorganic mercury in the kidney. There is also a significant decrease in the urinary excretion of inorganic mercury in rats following pretreatment with zinc salts.

After induction of MT synthesis, the hepatic accumulation of inorganic mercury is also increased in rats. While both hepatic and renal accumulation of inorganic mercury and hepatic zinc content increased in association with zinc pretreatment, the urinary and fecal excretions of mercury were decreased (ZALUPS and CHERIAN 1992). These results support an intracellular metal-binding role for preinduced MT in the liver and kidney. The decreased fecal excretion of inorganic mercury observed in this study was consistent with the decreased biliary excretion of metals after induction of hepatic MT synthesis (CHERIAN 1977; KLAASSEN 1978). Thus the presence of MT in tissues has a marked effect on accumulation of certaim metals and their excretion. The intracellular binding of cadmium to MT is somewhat specific, while metals such as mercury, copper, and zinc bind to other proteins along with MT. After induced synthesis of MT, cadmium bound to other proteins is transferred exclusively to MT in most of the tissues, especially in liver.

In Wilson's disease, where there is a defect in copper excretion pathways, high levels of MT saturated with copper are detected in the liver (NARTEY et al. 1987b; DANKS 1989).

D. Detoxification of Metals

A reduction in the toxicity of certain metal ions such as cadmium and mercury following induction of MT in animals has long been recognized (PISCATOR 1964). Resistance to metals has also been demonstrated in cultured cells with elevated levels of MT (RUGSTARD and NORSETH 1975). In an earlier review, WEBB (1987) discussed these effects, which were demonstrated repeatedly in various systems under different experimental conditions. However, it should be pointed out that some of these observed effects are applicable only to certain experimental conditions and may not be relevant to normal physiological conditions.

Studies have shown that, after pretreatment with a low dose of cadmium, the development of tolerance to an otherwise lethal challenge dose of cadmium salts occurs in experimental animals within 24 h (GOERING and KLAASSEN 1984b). The tissue levels of MT in adult animals may be lower than that of glutathione. But induction of MT with metals can increase tissue levels severalfold and thus MT can become one of the major sulfur-containing proteins within the cell (ONOSAKA et al. 1984). Induction of MT synthesis by zinc or cadmium can protect rats from the acute hepatotoxicity of injected cadmium salts. This tolerance appears to be due to increased cadmium binding to cytosolic MT after its induced synthesis and also to a decreased binding of cadmium to critical organelles such as nuclei, mitochondria, and endoplasmic reticulum in the liver (GOERING and KLAASSEN 1984a). Thus the induced synthesis of MT can alter the binding and subcellular distribution of cadmium and also decrease its acute toxicity. Most of the reported protective effects of MT in animals and cell culture show a short-term effect and it is doubtful whether MT can provide long-term protection against toxicity of metals. However, with long-term low-level exposure to cadmium, hepatic toxicity is rarely reported and kidney is considered to be the critical organ. This could be due to continued synthesis of MT in the liver and movement of cadmium from liver to kidney, under these conditions.

Elevations of both glutathione and MT have been reported in mesangial cells in culture, exposed to cadmium salts (CHIN and TEMPLETON 1993). In previous comparative studies using rat liver slices, it was shown that both glutathione and MT can play a role in the protection against the cytotoxicity of cadmium and menadione (CHAN and CHERIAN 1992; CHAN et al. 1992). It was also shown that MT alone can provide an effective protective role in the acute cytotoxicity of cadmium even when the intracellular glutathione level has been significantly depleted by buthionine sulfoximine injection.

The type of metals, chemicals, or conditions used to induce MT synthesis may also influence the tolerance to toxicity of metals. Studies have shown that unilateral nephrectomy in rats increase the ability of the remnant kidney to synthesize MT and accumulate more of the injected mercury. In additon, pretreatment with zinc can protect kidney from the toxicity of mercury in both control rats and those after unilateral nephrectomy (ZALUPS and CHERIAN 1992). This is in contrast to a previous report where MT induced by cadmium pretreatment did not provide complete protection against nephrotoxicity of inorganic mercury (WEBB and MAGOS 1978). Only a small fraction of renal mercury binds to MT under these experimental conditions (WEBB and MAGOS 1976). It is possible that mercury displaces zinc from MT more readily than cadmium because of the weak binding of zinc and sulfhydryl groups in the MT molecule compared to cadmium binding. Recent studies on transgenic mice where both MT-1 and MT-2 genes were inactivated show exclusively that these mice were more sensitive to cadmium toxicity than control mice, providing the proof that MT is definitely involved in the protection against metal toxicity (MICHAISKA and CHOO 1993; MASTERS et al. 1994).

In low-level chronic exposure, MT sequesters cadmium intracellularly as the cadmium-MT complex and thereby decreases the toxic effects of the metal. By contrast, extracellular cadmium-MT has been shown to be nephrotoxic to experimental animals. Parenteral injection of cadmium in the form of cadmium-MT can cause acute renal damage in rats and mice (NORDBERG et al. 1975; CHERIAN et al. 1976; SQUIBB et al. 1984; MAITANI et al. 1988). These toxic effects were similar to those observed after repeated exposure to cadmium salts; but the critical renal concentration of cadmium is much lower after injection of cadmium-MT (10 μg/g) than after repeated injections of cadmium salts (200 μg/g). The low molecular weight cadmium-MT is freely filtered by the glomerulus and reabsorbed by the proximal convoluted tubules and can cause acute damage to the renal tubular epithelial cells (CHERIAN et al. 1976; GOYER et al. 1984; DORIAN et al. 1992). It has been proposed that the hepatic cadmium-Mt is released and transported to kidney in blood plasma and that the nephrotoxicity occurs at a certain renal concentration of cadmium with chronic exposure (GOYER et al. 1978, 1984; DUDLEY et al. 1985). In a recent liver transplant experiment, the movement of cadmium-MT form liver to kidney was demonstrated in rats where the liver with cadmium-MT was transplanted to a control rat (CHAN et al. 1993).

Thus there is good evidence for the movement of cadmium from liver to kidney in the form of cadmium-MT and a direct role for cadmium-MT in the in vivo nephrotoxicity of cadmium.

E. Regulation of Zinc and Copper Metabolism

Although the protective effects of MT in metal toxicity have been studied extensively both in animal experiments and in cell culture systems, detoxification is by no means universally accepted as the primary function of MT (WEBB 1987; TEMPLETON and CHERIAN 1991). The high endogenous hepatic levels of zinc- and copper-bound MT in fetal and early neonatal life in several mammals suggest that MT may serve as an intracellular storage protein for these essential metals during perinatal development. However, the proportion of these essential metals associated with MT may vary with the species as well as fetal and neonatal age. In rat, rabbit, mouse, Syrian, and Chinese hamsters, the hepatic MT concentration is maximal at or soon after birth (WONG and KLAASSEN 1979; WEBB and CAIN 1982; PANEMANGALORE et al. 1983), while in humans (RIORDAN and RICHARDS 1980; BAKKA and WEBB 1981; NARTEY et al. 1987a), sheep (BREMNER et al. 1977), and guinea pigs (LUI 1987) the maximum amount of MT is found during mid or late gestation. The amount of zinc or copper associated with MT may also vary depending on the species. For example, in neonatal rat liver zinc is the major MT-bound metal, while in guinea pig it is mainly copper, and in human fetal liver both zinc and copper are found in high proportions with MT (RYDEN and DEUTSCH 1978). Differences in the two major isoforms, MT^{-1} and MT^{-2}, have also been reported in rat and mice liver (SUZUKI et al. 1983; LEHMAN-MCKEEMAN et al. 1991). During development rat liver contained more MT^{-2} than MT^{-1} while mouse liver contained more MT^{-1} than MT^{-2}. Recent studies also report changes in distribution of these two isoforms in livers of various mammalian species, and these differences may be related to changes in metabolism of different metals in these species (CHERIAN and CHAN 1993).

Although MT is mainly localized in the cytoplasm in adult hepatocytes, its presence in nucleus and cytoplasm is observed in both fetal and neonatal livers (PANEMANGALORE et al. 1983; NARTEY et al. 1987a). The significance of the nuclear-cytoplasmic localization of MT in hepatocytes and the functions of specific isoforms of MT are not yet clearly understood. In general, studies suggest that MT may serve as a major zinc and copper storage protein, reducing the ionic concentrations of these metals in liver during the fetal and neonatal period in several mammalian species. This "buffering" effect of MT is similar to the role of ferritin in binding of iron during development. Thus the marked changes in MT levels and the cellular localization in mammalian livers during development and also its presence in certain undifferentiated tumors (CHERIAN et al. 1993) suggest a biological function for this protein in cellular differentiation and maturation.

Studies on analyses of MT gene expression in the mouse reproductive tract and embryo have established that, from the time of implantation to late in gestation, the mouse embryo is surrounded by cells, interposed between the maternal and embryonic environments, that actively express MT genes (DE et al. 1989; OUELLETTE 1982; ANDREWS et al. 1993). Developmentally regulated high level expression of MT genes occurs within the reproductive tract (deciduum, placental spongiotrophoblasts) and in extraembryonic membranes (visceral endoderm) and also in male germ cells (DE et al. 1991). The precise roles and mechanisms of regulation of MT genes during mammalian development are not well defined. It has been suggested that both endogenous zinc and glucocorticoids may cotribute to high expression of MT. Both MT and its mRNA levels can be decreased in rat fetal liver when the dams were fed a zinc-deficient diet during late gestation (GALLANT and CHERIAN 1986; ANDREWS et al. 1987). However, depletion of glutathione in the dams with a BSO injection or with a sulfur-deficient diet did not affect the hepatic MT synthesis in rat fetus. But inhibition of the cystathionase pathway by repeated injection of propargylglycine decreased the hepatic levels of MT in newborn rats (GALLANT and CHERIAN 1989). In human fetus, the cystathionase or trans-sulfuration pathway is absent and thus MT may play a role as a cysteine storage protein in early human development (ZLOTKIN and CHERIAN 1988). Analysis of metal regulation of MT genes in mid-gestation mouse embryos in vivo and in vitro during the teratogenic period in mice (days 9–10 of gestation) suggested that, following maternal injection, cadmium but not zinc was prevented from reaching the embryo (ANDREWS et al. 1993). High-affinity binding of cadmium by MT in deciduum and placenta may provide a barrier for cadmium from reaching the postimplantation embryo. Moreover, MT has been localized in various cell types in full-term human placenta (WAALKES et al. 1984; GOYER et al. 1992) and thus MT may play a role in preventing the teratogenic effects of cadmium (Chap. 1, this volume).

It has been shown that incubation of the zinc-finger transcription factor *Sp1* with apo-MT can remove zinc from the transcription factor to form zinc-MT (ZENG et al. 1991). The displacement of zinc from zinc-finger transcription factors can affect its binding to DNA with decreased ability to induce transcription of functional genes. Thus it is possible that regulation of cellular levels of MT and apo-MT is an important mechanism in the alteration of cellular functions (Chap. 15, this volume). However, intracellular apo-MT is known to have a very short half-life and therefore it is important to demonstrate changes in apo-MT levels in association with zinc-finger transcription factor dependent gene expression in in vivo systems. Activation of certain apometalloenzymes has also been shown after in vitro addition of zinc-MT (UDOM and BRADY 1980).

The evidence for any specific roles for MT in the gastrointestinal absorption of zinc and copper and also in the hepatotoxicity of copper in certain genetic diseases is critically discussed by BREMNER (1993). At very high oral doses of zinc the intestinal MT content is increased, and this in concert with another cysteine-rich intestinal protein (CRIP) can regulate zinc absorption (HEMPE and COUSINS 1991). CRIP is a saturable intracellular zinc-binding protein that can facilitate the uptake and transport of zinc ions across the enterocyte and its transfer at the basolateral membrane. The low intestinal MT concentrations will allow high binding of zinc to CRIP and its absorption. But at high intestinal MT levels, both the binding of zinc to CRIP and its absorption are decreased because of the active sequestration of zinc by intestinal MT. However, the mechanism of this control depends on high mucosal MT concentrations with very high zinc intakes which occurs after the breakdown of normal homeostatic control. The induction of MT in intestinal mucosal cells can result in sequestration of dietary copper which is subsequently excreted as Cu-MT when the intestinal cells are desquamated. Therefore zinc supplementation has been used for an effective treatment for Wilson's disease patients (BREWER et al. 1991). Although these effects in zinc and copper absorption are found at high levels of intestinal MT levels, the role of MT in the normal absorption of zinc and copper is not yet understood.

Direct evidence for a causative role of MT in metal resistance has come from studies of copper sensitivity. A series of copper-resistant hepatoma cell lines have been isolated and they demonstrate a level of resistance proportional to the cellular concentration of Cu-MT (FREEDMAN and PEISACH 1989). Yeast requires *CUP1* gene, encoding an MT-like protein, to utilize Cu and avoid Cu-induced damage. Deletion of this gene renders yeast sensitive to Cu, and introduction of a plasmid construct containing a mammalian MT gene confers both copper and cadmium resistance in a copy-dependent manner. Thus the resistant phenotype is consistently associated with MT, can be modified by removal of MT, and can be restored by reintroduction of the MT gene (THIELE 1992). However, in mammals, increased accumulation of hepatic copper results in liver damage, although there is a concurrent increase in Cu-MT. These effects are found in genetic diseases such as Wilson's disease in humans (NARTEY et al. 1987b), LEC rats (SUGAWARA et al. 1993), and toxic milk mutant mice (MERCER et al. 1992). There is evidence that Cu-MT can increase lipid peroxidation initiated by an iron-chelate and cause oxidative cell damage in neonatal guinea pig liver (SUNTRES and LUI 1991) and also in an in vitro liver microsomal system. These studies imply that excessive Cu-MT may act as a pro-oxidant under certain conditions, but there is no evidence that Cu-MT itself can initiate lipid peroxidation (STEPHENSON et al. 1994).

The recent studies on cloning of the Menkes' gene (VULPE et al. 1993; MERCER et al. 1993) and Wilson's disease gene (BULL et al. 1993) show that these gene products have strong homology to P-type ATPase proteins, similar to those involved in inorganic ion transport in metal-resistant bacteria (SILVER and WALDERHAUG 1992). The N-terminal domain of this protein contains six repeats of a CysXXCys motif which is similar to the bacterial metal detoxification proteins and shows a weak resemblance to the metal-binding sites in MT. The Wilson's disease gene has been mapped to chromosome 13, and in two patients with Wilson's disease a seven-base deletion with the coding region of P-type ATPase has been detected (Bull et al. 1993). The implications of these results are discussed in detail in another chapter (Chap. 19, this volume).

F. Lipid Peroxidation and Oxidative Stress

Although both metals and glucocorticoids can directly induce MT synthesis, divalent metals like cadmium and zinc are the best-known inducers of MT synthesis. Metal-responsive elements (MREs) and glucocorticoid-responsive elements (GREs) have been identified in MT genes (Palmiter 1987). In addition to these inducers, several other compounds (which do not even bind with MT), oxidative stress, and inflammatory conditions have been shown to induce MT synthesis in various systems. Several mechanisms are proposed for such induction of MT synthesis and they may involve lipid peroxidation with free radical formation, release of cytokines during inflammation, and changes in tissue distribution of zinc.

Several studies (Min et al. 1993; Sato et al. 1993; Sato and Bremner 1993) have described the mechanisms involved in the induction of MT synthesis by various organic compounds, oxidative stress and inflammatory conditions, and also the physiological role of MT as a free radical scavenger. Most of the organic compounds induced MT synthesis indirectly either by release of certain cytokines from macrophages, monocytes, and other cell types or by an increase in cellular zinc concentration. A role for glucocorticoid in the induction of MT synthesis in inflammation has also been suggested. For example, injection of turpentine into mice can cause both induction of hepatic MT synthesis and acute inflammatory responses, such as an increase in plasma fibrinogen and ceruloplasmin; but all these effects were prevented by pretreatment with dexamethasone (Min et al. 1993). It should be pointed out that the hepatic MT levels induced by glucocorticoids alone are usually lower than those induced by inflammation or nonmetallic compounds. Therefore changes in glucocorticoids alone may not explain induction of MT in inflammation. Moreover, since induction of hepatic MT synthesis by these compounds has been demonstrated in adrenalectomized mice, glucocorticoids may not be the primary inducer of MT (Min et al. 1992). It was also shown in mice injected with menadione or carbon tetrachloride and in rats injected with paraquat that the induction of hepatic MT synthesis was independent of lipid peroxidation (Sato et al. 1993). Depletion of glutathione also did not affect the induced synthesis of MT, suggesting that the cysteine pools for these two thiol compounds may be different. Both MT mRNA and protein were induced in livers of mice following whole body X-irradiation but no such changes could be demonstrated in cultured cells, suggesting an indirect effect such as in the inflammatory response (Koropatnick et al. 1989; Koropatnick and Pearson 1990, 1993). The cytokines released during oxidative stress and inflammation may affect various other cellular functions.

The potential role of cytokines in induction of MT synthesis has been studied extensively, which demonstrates that interleukin (IL)-1, IL-6, tumor necrosis factor, and interferon can induce MT synthesis in the liver (Friedman and Stark 1985; Cousins and Leinart 1988; De et al. 1989). It has been established that IL-6 is a major mediator of acute phase protein synthesis in hepatocytes in response to infection and tissue injury. The induction of MT synthesis in hepatocytes by IL-6 is enhanced by both zinc and glucocorticoids (Schroeder and Cousins 1990). In addition, differential regulation of the isoforms of MT by IL-6 has been reported in various tissues in rats, suggesting a direct role for IL-6 in the induction of MT in inflammation. Induction of several acute-phase protein genes, such as those for acid glycoprotein and C-reactive proteins by metals shows significant sequence homology (Yiangou et al. 1991) to the metal-responsive elements in the MT gene. However, it is unclear whether there is any relation between this homology and induced synthesis of MT in inflammation.

Several studies suggest that induction of MT synthesis can play a role as a free radical scavenger in X-irradiation (Matsubara et al. 1986; Koropatnick et al. 1989) and after treatment with Adriamycin (Nagamuma et al. 1988), carbon tetrachloride, tumor necrosis factor (Sato et al. 1993), and other anticancer drugs (Satoh et al. 1993). Injection of zinc-MT to rats can suppress the formation of gastric ulcers induced by various stress conditions and exposure to chemicals (Mimura et al. 1988). Although induction of zinc-MT in certain conditions can scavenge hydroxyl and other oxygen radicals and can provide some protection against lipid peroxidation and DNA damage, the relative importance of MT in comparison with other antioxidant defense systems (such as GSH and SOD) in normal physiological conditions is not well understood. The co-expression of MT and Cu,Zn-superoxide dismutase (SOD) genes has been reported in yeast where MT may act both as donor of metals to the enzymes and as an antioxidant (Carri et al. 1991). In vitro studies have shown that in GSH-depleted cells induction of MT can protect against the cytotoxic effects of quinone (Chan et al. 1992) and hydroperoxide (Ochi 1988). It is important to consider the subcellular distribution of MT and the ability of cells to synthesize MT in evaluating its role as an antioxidant. However, it is unlikely that MT will provide long-term protection in continuous exposure to oxidant stress.

G. Summary

The elevated cellular content of MT in the cytoplasm and/or nucleus can alter the intracellular distribution and binding patterns of several metals which can avidly bind with MT. Since the intracellular content of MT is much higher than its extracellular levels under physiological conditions, induced synthesis of MT can increase the intracellular concentrations of certain metals and decrease their excretion rate. A major biological function of MT may be in the regulation of the metabolic pathways of zinc and copper, which are essential for various biological functions of cells, including gene transcription and cell proliferation. The chemical structure of MT-containing polythiols can effectively sequester different metals with very high affinity in two distinct clusters. The changing cellular content of thionein may be a useful method to control the cellular requirement, metabolism, and nutritional status of essential metals. Overexpression of MT under various experimental conditions can be considered as a cellular adaptive mechanism to modify certain cellular responses, including cell injury. In some cases, MT may provide short-term protection against acute toxicity of metals and free radicals. Induced synthesis of MT may be one of several protective cellular mechanisms against the toxic effects of metals and drugs. However, the mechanisms involved in the selection of these responses or signals to protect against toxic insults are unclear. Further studies on the mechanisms of celluar uptake, intracellular binding, and excretory pathways of environmental chemicals such as metals and drugs are essential to understand the modulation of metal toxicity and development of drug resistance at a cellular and molecular level.

References

Andrews GK, Gallant KR, Cherian MG (1987) Regulation of the ontogeny of rat liver metallothionein mRNA by zinc. Eur J Biochem 166:527–531

Andrews GK, McMaster MT, De SK, Paria BC, Dey SK (1993) In: Suzuki KT, Imura N, Kimura M (eds) Cell-specific expression and regulation of the mouse metallothionein I and II genes in the reproductive tract and preimplantation embryo. Metallothionein III. Birkhauser, Basel, pp 363–380

Bakka A, Webb M (1981) Metabolism of zinc and copper in the neonate: changes in the concentration and contents of the thionein bound Zn and Cu with age in the livers of the newborn of various species. Biochem Pharmacol 30:721–725

Bremner I (1993) Involvement of metallothionein in the regulation of mineral metabolism. In: Suzuki KT, Imura N, Kimura M (eds) Metallothionein III. Birkhauser, Basel, pp 111–124

Bremner I, Williams RB, Young BW (1977) Distribution of copper and zinc in the liver of the developing fetus. Br J Nutr 38:87–92

Brewer GJ, Yuzbasiyan-Gurkan V, Johnson V (1991) Treatment of Wilson's disease with zinc IX: response of serum lipids. J Lab Clin Med 118:446–470

Bull PC, Thomas GR, Rommens JM, Forbes JR, Cox DW (1993) The Wilson disease gene is a putative copper transporting P-type ATPase similar to the Menkes gene. Nature Genet 5:327–337

Carri MT, Galiazzo F, Ciriolo MR, Rotilio G (1991) Evidence for co-regulation of Cu, Zn superoxide dismutase and metallothionein gene expression in yeast through transcriptional control by copper via the ACE 1 factor. FEBS Lett 278:263–266

Chan HM, Cherian MG (1992) Protective roles of metallothionein and glutathione in hepatotoxicity of cadmium. Toxicology 72:281–290

Chan HM, Tabarrok R, Tamura Y, Cherian MG (1992) The relative importance of glutathione and metallothionein on protection of hepatotoxicity of menadione in rats. Chem Biol Inter 84:113–124

Chan HM, Zhu LF, Zhong R, Grant D, Goyer RA, Cherian MG (1993) Nephrotoxicity in rats following liver transplantation from cadmium-exposed rats. Toxicol Appl Pharmacol 123:89–96

Cherian MG (1977) Biliary excretion of cadmium in rat. II. The role of metallothionein in the hepatobiliary transport of cadmium. J Toxicol Environ Health 2:955–961

Cherian MG (1980) Biliary excretion of cadmium in rat. III. Effects of chelation agents and change in intracellular thiol content on biliary transport and tissue distribution of cadmium. J Toxicol Environ Health 6:379–391

Cherian MG, Chan HM (1993) Biological functions of metallothionein. In: Suzuki KT, Imura N, Kimura K (eds) Metallothionein III. Birkhäuser, Basel, pp 87–109

Cherian MG, Vostal JJ (1974) Biliary excretion of cadmium in rat. Toxicol Appl Pharmacol 29:141

Cherian MG, Vostal JJ (1977) Biliary excretion of cadmium in rat. 1. Dose-related biliary excretion and the form of cadmium in the bile. J Toxicol Environ Health 2:945–954

Cherian MG, Goyer RA, Delaquierriere-Richardson L (1976) Cadmium-metallothionein induced nephropathy. Toxicol Appl Pharmacol 38:399–408

Cherian MG, Yu S, Redman CM (1981) Site of synthesis of metallothionein in rat liver. Can J Biochem 59:301–306

Cherian MG, Huang PC, Klaassen CD, Liu YP, Longfellow DG, Waalkes MP (1993) National Cancer Instiute Workshop on the possible roles of metallothionein in carcinogenesis. Cancer Res 53:922–925

Chin TA, Templeton DM (1993) Protective elevations of glutathione and metallothionein in cadmium-exposed mesangial cells. Toxicology 77:145–156

Cody CW, Huang PC (1993) Metallothionein detoxification function is impaired by replacement of both conserved lysines with glutamines in the Hinge between the two domains. Biochemistry 32:5127–5131

Cousins RJ, Leinart AS (1988) Tissue specific regulation of zinc metabolism and metallothionein genes by interleukin-1. FASEB J 2:2884–2890

Danks DM (1989) Disorders of copper transport. In: Scriver CR, Beaudet AL, Sly WS, Valle D (eds) Metabolic basis of inherited disease, 6th edn. McGraw-Hill, New York, pp 1411–1431

De SK, McMaster MT, Dey SK, Andrews GK (1989) Cell-specific metallothionein gene expression mouse decidua and placentae. Development 107:611–621

De SK, Enders GC, Andrews GK (1991) High levels of metallothionein mRNAs in male germ cells of the adult mouse. Mol Endocrinol 5:628–636

Dorian C, Gattone II VH, Klaassen CD (1992) Renal cadmium deposition and injury as a result of accumulation of cadmium-metallothionein (Cd-MT) by the proximal convoluter tubules. Toxicol Appl Pharmacol 114:173–181

Dudley RE, Gammal LM, Klaassen CD (1985) Cadmium-induced hepatic and renal injury in chronically exposed rats: likely role of hepatic cadmium-metallothionein in nephrotoxicity. Toxicol Appl Pharmacol 77:414–426

Ershoff BH (1948) Conditioning factors in nutritional disease. Physiol Rev 28:107–137

Fernando LP, Andrews GK (1989) Cloning and expression of an avian metallothionein-encoding gene. Gene 81:177–183

Freedman JH, Peisach J (1989) Resistance of cultured hepatoma cells to copper toxicity. Purification and characterization of the hepatoma metallothionein. Biochim Biophys Acta 992:145–154

Frieden E (1976) Copper and iron metalloprotein. Trend Biochem Sci 1:273–274

Friedman RL, Stark GR (1985) γ-Interferon-induced transcription of HLA and metallothionein genes containing homologous upstream sequences. Nature 314: 637–639

Gallant KR, Cherian MG (1986) Influence of maternal mineral deficiency on the hepatic metallothionein and zinc in newborn rats. Biochem Cell Biol 64:8–12

Gallant KR, Cherian MG (1987) Changes in dietary zinc result in specific alterations of metallothionein concentrations in newborn rat liver. J Nutr 117:706–716

Gallant KR, Cherian MG (1989) Metabolic changes in glutathione and metallothionein in newborn rat liver. J Pharmacol Exp Ther 249:631–637

Goering PL, Klaassen CD (1984a) Tolerance to cadmium-induced toxicity depends on presynthesized metallothionein in liver. J Toxicol Environ Health 14:803–812

Goering PL, Klaassen CD (1984b) Tolerance to cadmium-induced hepatotoxicity following cadmium pretreatment. Toxicol Appl Pharmacol 74:308–313

Goyer RA, Cherian MG, Richardson LD (1978) Renal effects of cadmium. In: Cadmium 77; proceedings of the 1st international cadmium conference, San Francisco, CA Metal Bulletin, pp 183–185

Goyer RA, Cherian MG, Delaquierriere-Richardson LD (1984) Correlation of parameters of cadmium exposure with onset of cadmium-induced nephropathy in rats. J Environ Pathol Toxicol Oncol 5:89–100

Goyer RA, Haust MD, Cherian MG (1992) Cellular localization of metallothionein in human term placenta. Placenta 13:349–355

Hempe JM, Cousins RJ (1991) Cysteine-rich intestinal protein binds zinc during transmucosal zinc transport. Proc Natl Acad Sci USA 88:9671–9674

Jaw S, Jeffery EH (1988) The effect of dietary zinc status on biliary metal excretion. J Nutr 118:1385–1390

Kägi JHR (1991) Overview of metallothionein. Methods Enzymol 205:613–626

Kägi JHR (1993) Evolution, structure and chemical activity of class I metallothioneins: an overview. In: Suzuki KT, Imura N, Kimura K (eds) Metallothionein III. Birkhäuser, Basel, pp 29–55

Klaassen CD (1978) Effect of metallothionein on the hepatic disposition of metals. Am J Physiol 234:E47–E53

Koropatnick J, Cherian MG (1993) A mutant mouse (tx) with increased hepatic metallothionein stability and accumulatin. Biochem J 296:443–449

Koropatnick J, Pearson J (1990) Zinc pretreatment, metallothionein expression and resistance to cisplatin in mouse melanoma cells. Somat Cell Mol Genet 16:529–537

Koropatnick J, Pearson J (1993) Altered cisplatin and cadmium resistance and cell survival in chinese hamster ovary cells expressing mouse metallothionein. Mol Pharmacol 44:44–50

Koropatnick J, Leibbrandt M, Cherian MG (1989) Organ-specific metallothionein induction in mice by x-irradiation. Radiat Res 119:356–365

Lehman-McKeeman LD, Kershaw WC, Klaassen CD (1991) Species differences in metallothionein regulation: a comparison of the induction of isometallothioneins in rats and mice. In: Klaassen CD, Suzuki KT (eds) Metallothionein in biology and medicine, CRC, Boca Raton, pp 121–132

Lui EMK (1987) Metabolism of copper and zinc in the liver and bone of perinatal guinea pig. Comp Biochem Physiol 86:173–183

Maitani T, Cuppage FE, Klaassen CD (1988) Nephrotoxicity of intravenously injected cadmium-metallothionein: critical concentration and tolerance. Fund Appl Toxicol 10:98–108

Margoshes M, Vallee BL (1957) A cadmium protein from equine kidney cortex. J Am Chem Soc 79:4813–4814

Masters BA, Kelly EJ, Quaife CJ, Brinster RL, Palmiter RD (1994) Targeted disruption of metallothionein I and II genes increases sensitivity to cadmium. Proc Natl Acad Sci USA 91:584–588

Matsubara J, Shida T, Ishioka K, Dgawa S, Inada T, Machida K (1986) Protective effect of zinc against lethality in irradiated mice. Environ Res 41:558–567

Mercer JFB, Grimes A, Rauch H (1992) Hepatic metallothionein gene expression in toxic milk mice. J Nutr 122:1254–1259

Mercer JFB, Livingston J, Hall B, Paynter JA, Begy C, Chandrasekarappa S, Lockhart P, Grimes A, Bhave M, Siemieniak D, Glover TW (1993) Isolation of a partial candidate gene for Menkes by positional cloning. Nature Genet 3:20–25

Messerle BA, Schäffer A, Väsak M, Kägi JHR, Wüthrich K (1992) Comparison of the solution conformations of human [Zn_7]-metallothionein-2 and [Cd_7]-metallothionein-2 using nuclear magnetic resonance spectroscopy. J Mol Biol 225: 433–443

Michalska AE, Choo KHA (1993) Targeting and germ-line transmission of a null mutation at the metallothionein I and II loci in mouse. Proc Natl Acad Sci USA 90:8088–8092

Mimura T, Tsujikawa K, Yasuda N, Nakajima H, Haruyama M, Ohmura T, Okage M (1988) Suppression of gastric ulcer induced by stress and HCL-ethanol by intravenously administered metallothionein. Biochem Biophys Res Commun 151:725–729

Min KS, Terano Y, Onosaka S, Tanaka K (1992) Induction of metallothionein synthesis by menadione or carbon tetrachloride is independent of free radical production. Toxicol Appl Pharmacol 113:74–79

Min KS, Itoh N, Okamoto H, Tanaka K (1993) Indirect induction of metallothionein by organic compounds. In: Suzuki KT, Imura N, Kimura M (eds) Metallothionein III. Birkhauser, Basel, pp 159–174

Naganuma A, Satoh M, Imura N (1988) Specific reduction of toxic side effects of adriamycin by induction of metallothionein in mice. Jpn J Cancer Res 79:406–411

Nartey NO, Banerjee D, Cherian MG (1987a) Immunohistochemical localization of metallothionein in cell nucleus and cytoplasm of fetal human liver and kidney and its changes during development. Pathology 19:233–238

Nartey NO, Frei JV, Cherian MG (1987b) Hepatic copper and metallothionein in Wilson's disease (hepatolenticular degeneration). Lab Invest 57:397–401

Nordberg GF, Goyer RA, Nordberg M (1975) Comparative toxicity of cadmium-metallothionein and cadmium chloride on mouse kidney. Arch Pathol 99:192–197

Ochi T (1988) Effects of glutathione depletion and induction of metallothioneins on the cytotoxicity of an orgainc hydroperoxide in cultured mammalian cells. Toxicology 50:257–268

Onosaka S, Tanaka K, Cherian MG (1984) Effects of cadmium and zinc on tissue levels of metallothionein. Environ Health Perspect 54:67–72

Otvos JD, Armitage IM (1980) Structure of the metal clusters in rabbit liver metallothionein. Proc Natl Acad Sci USA 77:7094–7098

Otvos JD, Liu X, Li H, Shen G, Basti M (1993) Dynamic aspects of metallothionein structure. In: Suzuki KT, Imura N, Kimura K (eds) Metallothionein III. Birkhauser, Basel, pp 57–74

Ouellette AJ (1982) Metallothionein mRNA expression in fetal mouse organs. Dev Biol 92:240–246

Palmiter RD (1987) Molecular biology of metallothionein gene expression. In: Kägi JHR, Kojima Y (eds) Metallothionein II. Birkhauser, Basel, pp 63–80

Palmiter RD, Sandgren EP, Koeller DM, Findley SD, Brinster RL (1993) In: Suzuki KT, Imura N, Kimura K (eds) Metallothionein III. Birkhauser, Basel, pp 399–406

Pande J, Vasak M, Kägi JHR (1985) Interaction of lysine residues with the metal thiolate clusters in metallothionein. Biochemistry 24:6717–6722

Panemangalore M, Banerjee D, Onosaka S, Cherian MG (1983) Changes in intracellular accumulation and distribution of metallothionein in rat liver and kidney during postnatal development. Dev Biol 97:95–102
Piscator M (1964) On cadmium in normal human kidneys together with a report on the isolation of metallothionein from livers of cadmium-exposed rabbits. Nord Hyg Tidskv 48:76–82
Porter KG, McMaster D, Elmes ME, Love AH (1977) Anaemia and low serum-copper during zinc therapy. Lancet 2:774
Riordan JR, Richards V (1980) Human fetal liver contains both copper-rich forms of metallothionein. J Biol Chem 255:5380–5383
Robbins AH, McRae DE, Williamson M, Collett SA, Xuong NH, Furey WF, Wang BC, Stout CD (1991) Refined crystal structure of Cd.Zn metallothionein at 2.0 Å resolution. J Mol Biol 221:1269–1293
Rugstad HE, Norseth T (1975) Cadmium resistance and content of cadmium-binding protein in cultured human cells. Nature 257:136
Ryden L, Deutsch HF (1978) Preparation and properties of the major copper-binding component in human fetal liver. J Biol Chem 253:519–524
Sato M, Bremner I (1993) Oxygen free radicals and metallothionein. Free Radic Biol Med 14:325–337
Sato M, Sasaki M, Hojo H (1993) Induction of metallothionein synthesis by oxidative stress and possible role in acute phase response. In: Suzuki KT, Imura N, Kimura M (eds) Metallothionein III. Birkhauser, Basel, pp 125–140
Satoh M, Tsuchiya T, Kumada Y, Naganuma A, Imura N (1993) Protection against lethal toxicity of various anticancer drugs by preinduction of metallothionein synthesis in mice. J Trace Element Exp Med 6:41–44
Schroeder JJ, Cousins RJ (1990) Interleukin 6 regulates metallothionein gene expression and zinc metabolism in hepatocyte monolayer cultures. Proc Natl Acad Sci USA 87:3137–3141
Silver S, Walderhaug M (1992) Gene regulation of plasmid- and chromosome-determined inorganic ion transport in bacteria. Microbiol Rev 56:195–228
Squibb KS, Pritchard JB, Fowler BA (1984) Cadmium-metallothionein nephropathy: relationships between ultrastructural/biochemical alterations and intracellular cadmium binding. J Pharmacol Exp Ther 229:311–321
Stephenson GF, Chan HM, Cherian MG (1994) Copper-metallothionein from the Toxic Milk Mutant mouse enhances lipid peroxidation initiated by an organic hydroperoxide. Toxicol Appl Pharmacol 125:90–96
Stowe HD (1976) Biliary excretion of cadmium by rats. Effects of zinc, cadmium and selenium pretreatments. J Toxicol Environ Health 2:45–53
Sugawara N, Sugawara C, Sato M, Mori M (1993) Copper metabolism at two stages in the onset of spontaneous hepatitis in new mutant Long-E vans cinnamon (LEC) rats. J Trace Element Exp Med 6:15–21
Sunters ZE, Lui EMK (1991) Age-related differences in iron-nitrilotriacetate hepatotoxicity in the guinea pig: role of copper metallothionein. J Pharmacol Exp Ther 258:797–806
Suzuki KT, Ebihara Y, Akitomi H, Nishikawa M, Kawamura R (1983) Change in ratio of the two hepatic isometallothioneins with development from prenatal to neonatal rats. Comp Biochem Physiol 76:33–38
Templeton DM, Cherian MG (1984a) Effects of zinc deficiency on pre-existing cadmium-metallothionein in the pancreas. Toxicology 29:251–260
Templeton DM, Cherian MG (1984b) Chemical modification of metallothionein. Biochem J 221:569–575
Templeton DM, Cherian MG (1991) Toxicological significance of metallothionein. Methods Enzymol 205:11–24
Thiele DJ (1992) Metal-regulated transcription in eukaryotes. Nucleic Acids Res 20:1183–1191
Udom AO, Brady FO (1980) Reactivation in vitro of zinc requiring apo-enzymes by rat liver zinc thionein. Biochem J 187:329–335

Vasak M, Kägi JHR (1983) Spectroscopic properties of metallothionein. Met Ions Biol Syst 15:213–273
Vulpe C, Levinson B, Whitney S, Packman S, Gitschier J (1993) Isolation of a candidate gene for Menkes disease and evidence that it encodes a copper-transporting ATPase. Nature Genet 3:7–13
Waalkes MP, Poisner AM, Wood GM, Klaassen CD (1984) Metallothionein like proteins in human placenta and fetal membranes. Toxicol Appl Pharmacol 74:179–184
Webb M (1987) Toxicological significance of metallothionein. In: Kägi JHR, Kojima Y (eds) Metallothionein II. Birkhauser, Basel, pp 109–134
Webb M, Cain K (1982) Functions of metallothionein. Biochem Pharmacol 31:137–142
Webb M, Magos L (1976) Cadmium-thionein and the protection by cadmium against the nephrotoxicity of mercury. Chem Biol Interact 14:357–369
Webb M, Magos L (1978) Maleate induced changes in the kidney binding of mercury in rats pretreated with cadmium. Chem Biol Interact 21:215–226
Winge DR (1991) Limited proteolysis of metallothioneins. Methods Enzymol 205: 438–447
Wong KL, Klaassen CD (1979) Isolation and characterization of metallothionein which is highly concentrated in newborn rat liver. J Biol Chem 254:12399–12403
Yiangou M, Ge X, Carter KC, Papaconstantinou J (1991) Induction of several acute-phase protein genes by heavy metals: a new class of metal-responsive genes. Biochemistry 30:3798–3806
Zalups RK, Cherian MG (1992) Renal metallothionein metabolism after a reduction of renal mass. II. Effect of zinc pretreatment on the renal toxicity and intrarenal accumulation of inorganic mercury. Toxicology 71:103–117
Zeng J, Heuchel R, Schaffner W, Kägi JHR (1991) Thionein (apometallothionein) can modulate DNA binding and transcription activation by zinc finger containing Sp1. FEBS Lett 279:310–312
Zlotkin SH, Cherian MG (1988) Hepatic metallothionein as a source of zinc and cysteine during the first year of life. Pediatr Res 24:326–329

CHAPTER 7
Biochemical Mechanisms of Aluminum Toxicity

E.H. JEFFERY

A. Introduction

In considering Al[1] toxicity, one is struck by the normally successful exclusion of this ubiquitous element from animal tissues. Al is the third most prevalent element in the world, naturally present in air, soil, and water. In addition, Al enjoys a GRAS (Generally Regarded As Safe) rating by the FDA, allowing unregulated addition not only to most municipal water supplies, but to many processed foods and medications. Even with such wide exposure, tissue Al levels are normally very low. Under the unusual conditions where tissue Al levels start to rise, several distinct lesions, including an encephalopathy, osteodystrophy, and anemia are seen to occur. Some tissues accumulate Al more readily than others. Whether Al has one dominant effect, with organ selectivity of lesions accounted for by the uneven Al accumulation, or several organ-specific effects on cellular biochemistry, is yet to be determined. Several reviews on Al toxicity focus on bioavailability (e.g., DE VOTO and YOKEL 1994; VAN DER VOET 1992b), but few focus on biochemical mechanisms of toxicity (EXLEY and BIRCHALL 1992; ABREO and GLASS 1993).

In 1972, Alfrey and coworkers alerted nephrologists to a neurological disease in patients on hemodialysis. The encephalopathy was characterized by speech disturbances progressing through a personality change to seizures and dementia. In their seminal paper (ALFREY et al. 1976) they suggested that this dialysis dementia was possibly an Al intoxication. Also in 1976, a similar encephalopathy was reported in patients receiving Al-contaminated dialysis fluid (FLENDRIG et al. 1976). While this report did much to confirm a role for Al in the syndrome, it also strengthened the belief that enteral binding gels were not responsible. Only later was it found that frequently, during renal insufficiency, Al bioaccumulates from oral phosphate binding gels. As a consequence, Ca-based binding gels are now being developed. During normal kidney function, Al is cleared even when doses as large as 3–5 g/day are ingested. In contrast, uremics often accumulate Al to toxic levels. In individuals with normal kidney function, aluminum can accumulate

[1] The symbols Al, Fe, etc., will be used to denote total metal, regardless of ionic form.

if Al-contaminated total parental nutrition (TPN) fluids are administered. Even an oral dose may become bioavailable under certain circumstances, for example in the presence of citrate. Rapid onset of dementia, leading to death, has been reported in patients who were not on dialysis, but were given oral sodium citrate in combination with phosphate binding gels made up of aluminum hydroxide (KIRSCHBAUM and SCHOOLWERTH 1989).

However, some dialysis patients do not show Al accumulation, dementia, or bone disease, while undergoing prolonged binding gel treatment that successfully controls their phosphate levels (KERR and WARD 1988). Furthermore, variability in Al disposition has not been limited to the clinic. Lack of reproducible data on tissue accumulation, even within a single laboratory and under strict analytical conditions, has nurtured the skepticism of the scientific community. Three factors play a prominent role in producing this uncertainty. One is the difficulty in measuring Al accurately, due to the generally low levels in tissues compared to the high levels ubiquitous in the environment, making contamination of samples a serious problem. KERR and WARD (1988) state that Alfrey's success in identifying Al as causative in dialysis dementia was based on his improved methodology for Al estimation. Lack of further advances in measurement are again holding back the field. Experts in the field are unable to agree as to whether or not Al accumulates in the brains of patients with Alzheimer's disease (AD). Yet the roles that environment and genetic makeup play in the etiology of AD cannot be readily evaluated without this knowledge. A second factor is the lack of information on speciation and equilibration between Al species in biological systems (MARTIN 1988, 1992), a particularly serious problem in attempting to use in vitro systems to learn about in vivo effects of Al. The third factor, responsible for much of the lack of reproducibility, is our lack of knowledge of controlling factors in Al disposition. Distribution and cellular uptake studies now clearly identify a role for transferrin (ABREO and GLASS 1993). However, absorption mechanisms are still not clearly defined. The bioavailability of Al in the presence of dietary ligands such as citrate and F requires considerably more study. While patients have grown delirious and died within a week of combined gel and citrate treatment (KIRSCHBAUM and SCHOOLWERTH 1989), antacids containing these two components are available over the counter with no warning of a daily limit.

Aluminum exposure is by inhalation, ingestion, and contamination of intravenous fluids, and Al has been reported to accumulate in several organs, including bone, liver, and speen (DOMINGO et al. 1991). While Al is not reported to be essential to life, once it has penetrated the body, it is biologically active. In its ionic form, any metal will interact with the functioning of a cell. However, few metals can penetrate the cell without a specialized uptake system, and none has been identified for Al. Thus bioavailability becomes an important issue when discussing Al toxicity. Under any condition, the amount of free Al ion is likely to be exceedingly low, i.e., in the femtomolar range (MARTIN 1988). However, Al must bind to a ligand at the site of toxicity to effect toxicity, and this action alone shifts the equilibrium between bound and free Al. Thus interaction at the site of toxicity is dependent upon the total Al load, the concentration of ligands and their relative affinities for Al, and the concentration and affinities of ions that compete for those ligands. Once inside the cell, Al appears to have a number of effects that might be explained variously by interaction with Fe-, Mg-, or Ca-dependent systems.

B. Aluminum Species in Biological Systems

In aqueous solution below pH 5, free Al exists predominantly as $Al(H_2O)_6^{3+}$ (most frequently abbreviated to Al^{3+}). As the pH rises to neutral, Al becomes mostly the extremely insoluble $Al(OH)_3$ which polymerizes and forms a precipitate that redissolves as the pH rises above 8 to form, predominantly, the more soluble $Al(OH)_4^-$. Thus at physiological pH and in the absence of ligands, Al is mostly undissolved and unavailable for interaction with the body. In considering likely biological ligands for interaction with Al, Al binds strongly to oxygen (e.g., OH^-, PO_4^{3-}, SO_4^{2-}, and CH_3COO^-) and fluorine, less strongly to nitrogen, and has essentially no affinity for sulfur ligands.

Ions with which Al might compete for ligands in biological systems are Fe, Ca, and Mg. MARTIN (1988) has pointed out that one characteristic of Al that sets it aside from most essential metals is the rate of ligand exchange in and out of the metal ion coordination sphere. This is very slow (10^4 slower than Mg^{2+}, 10^7 slower than Ca^{2+}, but only ten times slower than Fe^{3+}). In ionic size, Al is closest to Fe, and is found interacting with transferrin (Tf) and ferritin (TRAPP 1983; FLEMING and JOSHI 1987). Mg is somewhat larger, and Ca very much larger: in the eightfold coordination in which Ca is often found bound to proteins, it is seven times larger than the sixfold coordination of Al^{3+}. Therefore, if Ca is to compete with Al, it would most likely be for small ligands such as phosphate. Since at physiological pH Al forms less soluble phosphates than Ca, this has been proposed to account for its accumulation in bone (EXLEY and BIRCHALL 1992). However, if Al does replace Ca, the 10^7 slower exchange rate will assure inhibition. To reiterate, similarity of charge, size, and ligand exchange rates between Al and Fe may make sharing of ligands likely but Al inhibition of Fe-dependent reactions of little biological significance. On the other hand, Al has greater binding affinity and slower reaction kinetics than do Mg and Ca, so that Al should readily inhibit many reactions involving these two ions.

Theoretically predicted interactions, or even reports of experimental interactions in an in vitro system, are not necessarily an indication of in vivo effects on a particular system. This depends upon the ability of Al to reach that system, and to be in a form available for interaction. The difficulty in extrapolation from in vitro to in vivo is, to a great extent, because of the altered speciation and ligand affinities seen when the milieu is changed. Thus Farrar reported unusual Fe and Ga binding for Tf from patients suffering from AD, implicating altered Al- (because of the similarity between Al and Ga) and Fe-Tf kinetics in the etiology of AD (FARRER et al. 1990). However others, unable to repeat his work, suggested that Tf binding characteristics are unchanged in AD (McGREGOR et al. 1991). McGregor attributed the change that FARRER and coworkers saw to altered electrolyte balance following column chromatography, which could have altered metal ion competition.

In another in vitro study, Leterrier evaluated the effect of Al on neurofilament conformation (LETERRIER et al. 1992). The study sought to identify a role for Al in formation of neurofibrillary tangles. The Al was presented to the protein in association with a plant product, maltol, as ligand. Lactate and ascorbate also enhanced availability, while citrate, presumably because it had greater affinity for Al than did the neurofilament, decreased, rather than increased, the binding of Al to the neurofilament subunit. Such reliance on the form of Al for availability and subsequent (toxic) effect, and our ignorance as to competing ligands and ions inside and outside the neuron during in vivo exposure to Al, greatly hinders our ability to draw meaningful conclusions from such studies.

Similar to the problem of extrapolating from in vitro to in vivo, is the difficulty in extrapolating Al bioavailability data from aquatic to terrestrial animals. Both citrate and F have been used effectively to chelate Al in the aquatic medium and reverse Al toxicity in brook trout (DRISCOLL et al. 1980). Citrate has frequently been shown to enhance gastrointestinal absorption of Al in man (WEBERG and BERSTAD 1986) and animals (e.g., FULTON and JEFFERY 1990). On the other hand, once in plasma, not only does Martin propose that little or no Al would normally be citrate-bound (MARTIN 1988 and YOKEL 1994), but Abreo has shown little or no Al uptake from Al citrate, and proposes that only Al-Tf is available to cells (ABREO and GLASS 1993). On the other hand, while Al has been used to decrease F absorption in animals, F increased Al accumulation when F and Al were coadministered intravenously (STEVENS et al. 1987), and we find that even oral F causes the coaccumulation of Al in bones of rats and rabbits (AHN et al. 1994).

C. Bioavailability of Aluminum

I. Exposure

Inhalation is mostly an occupational exposure route (DE VOTO and YOKEL 1994). Approximately 3% is absorbed by this route. Welders exposed to a maximum of 10 mg/m^3 had raised plasma and urinary Al levels, although they showed no overt signs of toxicity (DE VOTO and YOKEL 1994). Outside the work place, exposure via inhalation is minimal.

Ingestion of Al is unavoidable because of the addition of Al to many processed foods (e.g., leavening and antiadherance agents in baked goods), to tablet formulations (e.g., the buffering component of buffered aspirins) and through municipal water-processing plants. It has been estimated that man ingests, on average, 30–50 mg Al/day, and that of this approximately 1% is absorbed (GANROT 1986). However, individuals ingesting buffered aspirin or Al-containing antacids may take in several grams a day, similar in magnitude to the 1–2 g daily dose typical for those ingesting aluminum hydroxide phosphate-binding gels. Oral absorption is greatly enhanced by the presence of citrate (WEBERG and BERSTAD 1986) and most cases of Al accumulation to toxic levels from oral exposure of humans can be linked to combined Al and citrate administration. The special case of aluminum accumulation in uremics given oral Al is not associated with decreased urinary excretion, which actually increases. Accumulation may be associated with altered distribution, increased absorption (ITTEL et al. 1987), or disturbed renal vitamin D and/or PTH regulation of calcium homeostasis (VAN DER VOET 1992a). Parenteral exposure is essentially limited to two groups of patients; uremics receiving Al-contaminated hemodialysis fluids, and children receiving Al-contaminated parenteral nutrition. The latter is due to contamination of protein components during preparation and both exposures can be largely overcome by alternative preparation techniques and careful monitoring programs.

II. Gastrointestinal Absorption

Experimental animals given oral Al exhibit increased plasma and urinary Al, showing unequivocally that absorption does occur from the gastrointestinal tract (DE VOTO and YOKEL 1994). Furthermore, normal, healthy adult humans given a single oral dose of Al show raised urinary Al levels for up to 4 days (WEBERG and BERSTAD 1986). Because of its greater solubility at low pH, Al is considered to be readily absorbable by simple diffusion from the stomach. Some intestinal uptake studies have been limited to acidic conditions in an attempt to maintain Al in solution (ADLER and BERLYNE 1985; VAN DER VOET and DE WOLFF 1987). Other studies have shown that uptake is not limited to the low pH that would only be found in the stomach (FROMENT et al. 1989; COCHRAN et al. 1990). In the presence of citrate, Al crosses the mucosa at the upper intestine, with the plasma Al peak coinciding with the peak for coadministered D[1-^{3}H]-glucose (FROMENT et al. 1989).

It is now clear that bioavailability varies with the form of Al. The chemical form of Al at the mucosal wall depends not only on the form ingested, but also on the pH, concentration of the dose, and the presence of dietary complexing factors such as citrate and F. Many experiments have been performed with soluble $AlCl_3$, which produces a low pH solution. If the pH is then raised, artificially or physiologically during entrance into the intestine, insoluble aluminum hydroxide will form unless alternative ligands are available. Enhanced bioavailability in the presence of citrate has frequently been explained on the basis of improved solubility of Al at physiological pH. However, at physiological pH, the neutral citrate complex, most abundant at pH 3, is almost entirely replaced by an anionic complex, the most abundant form above pH 4. Partridge and coworkers found that citrate-enhanced absorption is proportional to the concentration of Al in solution in the lumen of the intestine over the entire physiological range of pH 2–7.4, suggesting that not just the neutral complex, but the cationic and anionic complexes are also bioavailable (PARTRIDGE et al. 1992). The authors also showed enhanced ethyl acetate: water partition over the same broad pH range. They concluded that citrate complexation with Al both solubilizes the Al and permits simple diffusion across the mucosal wall regardless of pH or the charge of the citrate complex.

While much has been written on variability of absorption with pH of the lumen or form of the Al administered (see reviews by VAN DER VOET 1992; DE VOTO and YOKEL 1994), variability between laboratories is still greater than the variability caused by any of these factors! For example, one laboratory reported 27% absorption of a dose of $AlCl_3$ given rats (GUPTA et al. 1986), while another reported less than 1% for a similar dose of the same compound under similar conditions of acidity to the same species (ITTEL et al. 1987). Clearly, there are unrecognized varibles here which, when identified, may help us to understand why one patient can tolerate binding gel treatment, while another rapidly accumulates Al to neurotoxic levels.

III. Transcellular Uptake

The mechanism of Al absorption is proposed to be a carrier-mediated, energy-dependent process (FEINROTH et al. 1982). In studies where the upper intestine of rats was perfused in situ, uptake into the mucosal cell was considerably more rapid than release into blood, suggesting that uptake is a two-step transcellular process (VAN DER VOET and DE WOLFF 1987). Uptake may utilize the Ca uptake system (VAN DER VOET 1992). Working with perfused rat duodenum at pH 2, ADLER and BERLYNE (1985) identified a vitamin D-dependent pathway in competition with Ca. Studies using a jejunal everted gut sac system in rat at pH 7.4 showed an energy-dependent mucosal uptake of Al which appeared independent of Ca, although overall passage of Al into the portal system was negatively affected by calcium (FEINROTH et al. 1982). This suggests that the second step of transcellular uptake might be in competition with Ca. Energy-dependent uptake of $AlCl_3$, significantly decreased by Ca-channel blockers, has been identified in both the duodenum (COCHRAN et al. 1990) and jejunum (PROVAN and YOKEL 1988a). However, in later work using an everted rat gut sac, Provan and Yokel (1988b) could not inhibit uptake with the classic Ca channel blockers verapamil or 4-aminopyridine. Interestingly, in this second study, Provan and YOKEL found that uptake was inhibited by kinetin, a known blocker of the paracellular system. Since paracellular passage of compounds is inhibited by Ca, which serves to tighten gap juctions, information gathered to date and interpreted to support the interaction of Al with the Ca uptake system may need to be revised. Also, with the widely different pHs used for these studies, varying from 2.0 to 8.5, the different laboratories may have been studying the uptake of very different Al species.

It has been suggested that, since Al binds to both transferrin (TRAPP 1983) and ferritin (FLEMING and JOSHI 1987), the Fe uptake system might be involved in Al absorption, and some evidence exists that Fe competes with Al for uptake (FERNANDEZ MENDEZ et al. 1991; VAN DER VOET and DE WOLFF 1987; VAN DER VOET 1992). Whether or not Al interacts with the Fe uptake system at the intestinal mucosa, there is strong evidence that Al interacts with Fe transport and metabolism (see Sects. C.V, C.VII, D). These possibilities do not preclude uptake via other, as yet unidentified, systems also. For example, there are no studies on the interaction between Al and Mg uptake. This may be because out knowledge of Mg uptake at the intestinal mucosal membrane is far from complete (SHILS 1988).

IV. Paracellular Uptake

Three studies have addressed the possibility of paracellular uptake, where luminal components leak between cells of the intestinal mucosa rather than passing through them during normal transcellular uptake (PROVAN and YOKEL 1988b; FROMENT et al. 1989; VAN DER VOET 1992b). Ruthenium red, a dye that does not enter cells, was able to pass between mucosal cells in the presence of Al citrate, but not $AlCl_3$, suggesting that the citrate supported opening of the tight junctions between cells (FROMENT et al. 1989). Sodium citrate had earlier been shown to exert this effect. In electrophysiological experiments on transcellular resistance, $AlCl_3$ had no effect on the tight junctions while Al citrate decreased resistance, indicative of disruption of the junctions (FROMENT et al. 1989). Al citrate would be in solution mostly as an anionic species at this pH, which may be able to pass the mucosal barrier via the paracellular uptake system regardless of charge, since there are no membranes to traverse. Froment proposed that the mechanism of paracellular opening is most probably by chelation of the Ca necessary for tight junction maintenance. If this were so, then enhanced Al uptake might also be effected by a Ca-deficient state, or the addition of any Ca-chelator to the diet. However, Al uptake appears to be decreased in Ca deficiency (VAN DER VOET 1992a). In recent experiments, citrate-enhanced Al uptake did not show energy dependence (which would have indicated transcellular uptake) but did show almost complete inhibition by the paracellular blocker 2,4,6-triaminopyrimidine (VAN DER VOET 1992b). Unfortunately, controls to show that known paracellular transport was inhibited while the mucosa was not sufficiently damaged to disturb known transcellular transport pathways, were not reported. General poisoning of the mucosal membrane by triaminopyrimidine cannot be dismissed.

V. Systemic Transport

Once in plasma, the most likely ligands for transporting Al are citrate and transferrin. The former has about 100 μM unbound sites available, the latter 50 μM (37 μM Tf, two sites per molecule and 30% Fe-saturated). MARTIN has calculated (1988) that at the pH of plasma, the affinity of trivalent Al for Tf is sufficiently greater than for citrate, that Al would be essentially all Tf-bound. However, this does not preclude a role for citrate in presenting Al to Tf in a soluble, available form (MOSHTAGHIE and BAZRAFSHAN 1992). Although Fe binds to Tf with far greater affinity than does Al, Tf is normally only 30% saturated with Fe. Thus even if all plasma Al is Tf-bound (as high as 85% is generally reported; KHALIL-MANESH et al. 1989) Tf would only be 30% Al-saturated at a plasma Al of 25 μM, a concentration that is rarely exceeded, even in patients with overt aluminum toxicosis (BUYS and KUSHNER 1989). While unaffected humans have plasma Al values of 0.1 or 0.2 μM, typical plasma Al levels associated with clinical signs of Al toxicosis are 1.5–4 μM. Thus, theoretically, Al should not be expected to compete with Fe for transport. In addition to binding to Tf, Al has been found bound to a small (8000 Da) protein that, after desferrioxamine treatment, increased and became the major Al-binding protein (KHALIL-MANESH et al. 1989). A major problem still in the evaluation of plasma binding for Al is that chromatography may alter binding by altering both ligands and competing ions.

VI. Accumulation in Erythrocytes

Early studies suggested equal distribution of Al between plasma and erythrocytes. However, studies over a broader range of blood Al levels (GUPTA et al. 1986; FULTON and JEFFERY 1990) show that the proportion in erythrocytes varies with the Al load. At low blood Al, a majority is present inside the erythrocyte (FULTON and JEFFERY 1990). As the blood Al is raised, so erythrocyte binding appears to saturate (MOXON and JEFFERY 1991), causing the proportion of Al in the erythrocyte to fall as higher doses of Al remain in the plasma. A number of studies have suggested that, as the plasma Al rises, so the percentage that is plasma protein-bound also rises, and the percentage excreted in urine decreases (RAHMAN et al. 1985). This may be because clearance is through filtration (YOKEL and MCNAMARA 1985), a system not available to protein-bound moieties.

Early evaluation of plasma Al measurement methods revealed that heparin was contaminated with Al and warned against the use of heparin in sample preparation (KOSTYNIAK 1983). Recent studies not only show that citrate also is contaminated (MAY et al. 1992) but that citrate and EDTA are able to draw Al out of the erythrocyte and into the plasma (MOXON and JEFFERY 1991). Because Al accumulates preferentially in the erythrocyte at low blood Al levels, use of citrate or EDTA for sample preparation may significantly raise free plasma Al, while at very high blood Al levels contamination of plasma with erythrocyte Al would not be expected to be as significant and would increase the percentage free Al very little. Binding inside the erythrocyte has not been studied, although MARTIN (1992) has proposed that ATP and 2,3-diphosphoglycerate would be likely candidate ligands, present at 1 and 3 m*M*, respectively.

VII. Cellular Uptake

Comparing cellular uptake of Al from Tf and citrate, Abreo has proposed that Al-Tf is essentially the only form of systemic Al available to cells (ABREO and GLASS 1993). Others have confirmed the superiority of Al-Tf availability using an osteoblast-like cell line (KESAI et al. 1991) and human erythroleukemia cells (MCGREGOR et al. 1990) although uptake of $AlCl_3$ into cultured hepatocytes has also been reported (MULLER and WILLHELM 1987). Abreo found virtually no Tf-independent uptake from Friend erythroleukemia (FE) cells (ABREO et al. 1990), although there was some uptake from Al citrate in hepatocytes (ABREO et al. 1991). He suggested that this seeming Tf-independent uptake might be due to either trace amounts of Tf present in the serum added to cell cultures, or to secretion of Tf by hepatocytes. Trace contamination by Tf from serum in the culture medium appears an unlikely cause because (a) serum is typically heat inactivated prior to use in media and (b) the FE cells were also exposed to 10% fetal calf serum in the culture medium, yet showed no Tf-independent uptake of Al. One alternative possibility is that hepatocytes have less specific uptake mechanisms than other cells, in their role as detoxifiers of foreign chemicals.

VIII. Aluminum Interactions with Desferrioxamine

Presently desferrioxamine (DFO) is the only clinical treatment for removal of Al, and has the advantage that the DFO-Al complex is removed by dialysis. Interestingly, while Fe has a greater affinity for DFO in a simple in vitro system, in plasma only a small fraction (10%) of Fe is found bound as ferrioxamine, while 80% of plasma Al has been reported as aluminoxamine (VASILIKAKIS et al. 1992), confirming that Al, unlike Fe, can be mobilized from Tf by DFO. DFO enters cells and, in experiments where rats were given aluminum either in the diet (aluminum hydroxide plus citrate) or intraperitoneally (as aluminum lactate) for 29 days before DFO treatment, the DFO-dependent increase in plasma Al correlated with both the Al dose and bone Al load (GREGER and POWERS 1992). However, in an extensive clinical study, the DFO-dependent increase in plasma Al was only found to correlate with bone Al in patients who had received Al during the preceding 6 months (PEI et al. 1992). Patients having terminated Al-gel treatment prior to 6 months before the test still showed high bone Al levels, but DFO treatment did not mobilize any of the Al into plasma. These data suggest that Al is slowly deposited into a bone store unavailable to DFO.

D. Aluminum-Related Anemia

Patients suffering from renal failure and receiving dialysis treatment frequently suffer from a normocytic anemia successfully treated with human recombinant erythropoietin, a regulatory protein synthesized by the kidney. However, patients often also exhibit a microcytic, hypochromic anemia that can be separated from the normocytic anemia of renal insufficiency, and that is refractory to erythropoietin (ROSENLOF et al. 1990). Al administration to animals produces a similar microcytic anemia (TOUAM et al. 1983; FULTON and JEFFERY 1993). In addition, not only can the anemia be reversed by desferrioxamine treatment (TIELMANS et al. 1985), but a prospective clinical study correlated plasma Al levels with severity of the anemia (SHORT et al. 1980). Patients undergoing hemodialysis with Al-contaminated fluids showed an increased plasma Al and decreased hematocrit; both parameters were then reversed when the patients were switched to an Al-free dialysate.

The etiology of microcytic hypochromic anemias as a class can be ascribed to decreased hemoglobin synthesis. Al has been shown to inhibit hemoglobin synthesis in Friend erythroleukemia cells (ABREO et al. 1990) and in bone marrow cells (ZAMEN et al. 1992), where it also accumulates. In vitro studies evaluating incorporation of $^{59}Fe^{2+}$ into heme have identified heme, rather than globin, synthesis as the inhibited pathway in uremia (MORIYAMA et al. 1975). The most common cause of a fault in heme synthesis leading to microcytic anemia is iron deficiency or lack of availability. Although Al-related anemia is refractory to Fe, the anemia could be caused by an interaction between Al and Fe metabolism.

Uptake of aluminum into erythroid cells in culture is dependent upon Al being presented bound to Tf, and does not occur in the presence of fully iron-saturated Tf (MLADENOVIC 1988). Furthermore, while one study reports that Al-Tf inhibited iron uptake (McGREGOR et al. 1990), another study reports exposure of erythroid cells to Al-Tf actually increased, rather than decreased, the rate of Fe-Tf uptake (ABREO et al. 1990). This finding suggests that the inhibitory effect of Al is not on uptake of Fe, but does not preclude the possibility that Al affects a later step in Fe transfer into protoporphyrin. Tf receptors are reported to be upregulated (WARD et al. 1984). A block in Fe utilization might well cause an upregulation in Tf receptors that could result in the increased uptake reported. Hemodialysis patients exhibit protoporphyrin accumulation (SWARTZ et al. 1985) and animals treated with Al also show porphyrin accumulation (BERLYNE and YAGIL 1973), suggesting a block in ferrochelatase-dependent Fe insertion into protoporphyrin. While decreased ferrochelatase activity has been reported in dialysis patients (MOREB et al. 1983), no chemical rationale has been offered to explain Al inhibition of ferrochelatase. Only divalent cations have been reported to act as alternative substrates for this enzyme (KAPPAS et al. 1983). Availability of reduced Fe for ferrochelatase might prove to be a fruitful area of study. Fe incorporated into ferritin may not be released in the presence of Al. Al has been reported to inhibit ceruloplasmin, which can catalyze the release of Fe from ferritin (HUBER and FRIEDEN 1970). A detailed mechanism of Fe reduction is still unknown.

A number of enzymes in the heme synthetic pathway are sensitive to foreign metals, most particularly to lead. Al has been reported to activate erythrocyte delta aminolevulinic acid dehydratase (ALA-D), the second step in heme synthesis, both in vitro and in vivo (MEREDITH et al. 1977). However, these researchers found inhibition as the dose of Al was increased and other groups have also reported that ALA-D is inhibited by Al (ZAMEN et al. 1993). The interaction of ALA-D with Al is not easily explained at the molecular level, since the vulnerability of ALA-D to lead depends upon interaction of lead with essential sulfhydryl groups of ALA-D, for which Al would not bear any great affinity. Another enzyme of heme synthesis, uroporphyrinogen decarboxylase, has been reported to be inhibited in uremics (MCGONIGLE and PARSONS 1985), although later studies have not repeated this finding (UMEDA et al. 1990). However, BUYS and KUSHNER (1989) have pointed out that since both ALA-D and uroporphyrin decarboxylase enzymes are in great excess in developing erythrocytes, a change in enzyme activity for either of these would be expected to have little or no clinical outcome.

Heme oxygenase, the enzyme responsible for the breakdown of heme, is stimulated by a wide variety of metals (MAINES and KAPPAS 1976). (FULTON and JEFFERY 1993) Al stimulates heme oxygenase in liver and spleen (unpublished communication), organs which both accumulate Al preferentially (DOMINGO et al. 1991). An increase in heme oxygenase activity could cause both a decreased availability of heme for hemoglobin synthesis and increased turnover of hemoglobin. While the increase in hepatic heme oxygenase appears sufficient to cause loss of the hepatic heme-containing enzyme cytochrome P450 (FULTON and JEFFERY 1993), there is not yet any evidence that Al action on heme oxygenase plays any role in heme availability for hemoglobin and subsequent anemia.

E. Aluminum-Related Bone Disease

Experimental Al administration to animals produces osteomalacia (GOODMAN 1986). Essentially all patients suffering from chronic renal failure exhibit some degree of renal osteodystrophy. In kidney disease there is a decreased production of 1,25-dihydroxyvitamin D_3, a loss of Ca (hypercalcuria), and hyperplasia of the parathyroid leading to high serum PTH, probably secondary to Ca loss. In some patients, PTH levels are low or normal, and Al accumulation at the osteoid surface, or growing front, of the bone is typically seen in these patients. This Al accumulation appears to be secondary to kidney disease, in that it is corrected by kidney transplant (NORDAL et al. 1992). The role of Al in renal osteodystrophy is not yet clear.

Aluminum-related renal osteodystrophy includes both osteitis fibrosa (OF) and osteomalasia (OM). In addition, aplastic bone disease frequently accompanies OM. All are refractory to vitamin D treatment (PIERIDES 1980). Osteitis fibrosa is characterized by increased bone remodeling, high osteoblast and osteoclast activity, but normal mineralization, whereas osteomalacia is characterized by decreased bone remodeling, increased osteoid surface, or unmineralized bone matrix, and impaired mineralization. In aplastic bone disease remodeling is almost completely absent, described as an adynamic state. Patients with OF have little accumulation of Al in bone, but show an increased serum Al following DFO treatment, suggesting a body burden of Al. Bone samples from patients with OM exhibit strong staining for Al at the osteoid, or mineralization front. Because of the localization of Al at the mineralization front (COURNOT-WITMER et al. 1981), it was early proposed that Al might act physically to interfere with mineralization. Only later studies showing Al interaction with biochemical processes within osteoblasts put this idea to rest. Administration of Al has been shown to decrease bone mineralization (RODRIGUEZ et al. 1990) and inhibit osteoblast proliferation (GOODMAN 1986) in the whole animal. In cell cultures, Al had been reported to both inhibit osteoblast proliferation (BLAIR et al. 1989) and stimulate osteoclast activity (SPRAGUE and BUSHINSKY 1990). These studies implicate Al as the inhibitor of calcification in Al-related renal osteodystrophy (GOODMAN 1986).

In studies using lower doses, Al has been reported to be mitogenic, suggesting a possible biphasic dose response. Looking with hindsight at Al loads in patients with OF and OM, this picture could be indicative of an activation of osteoblasts at low Al doses (OF) and an inhibition at higher doses (OM). In vitro and in vivo experiments support the possibility that Al may have a biphasic effect on bone growth. Low levels of Al exhibit a PTH-dependent osteogenic effect in dogs (QUARLES et al. 1989). A biphasic effect of Al has been identified on in vitro bone growth (LIEBERHERR et al. 1987; LAU et al. 1991). Lieberherr found stimulation of ^{45}Ca uptake in mouse fetal limb bones in culture at very low Al doses, and a reversal of this effect at doses above 1.5 μM. Lau, working with chick primary calvarial cell culture and human osteoblast cell culture, showed that Al was mitogenic in these systems, increasing ^{3}H-thymidine incorporation into DNA. Furthermore, correlating well with the in vivo studies of QUARLES et al. (1989), they found the effect was potentiated by PTH. Above 50 μM, the effect was reversed and Al became inhibitory. Lau therefore proposed that the effect of Al is biphasic, although he found that even in his own laboratory the dose giving maximal mitogenic action varied from experiment to experiment. Lau suggested environmental contamination as the cause of this variability. Lack of regulation of Al speciation in the cultures, giving variable free Al concentration, is another possible cause. Use of Al-Tf might overcome some of this variability.

Interestingly, Quarles found osteoblast cells in culture required serum for complete mitogenic action (QUARLES et al. 1991). While he considered that this might be to trigger the signal transduction pathway (e.g., PTH contamination) an alternative possibility is that Tf inclusion in serum improved Al delivery to its intracellular site of action. Al modulation of G protein function is an attractive theory for the biphasic mechanism(s) of mitogenic action and inhibition (see Sect. G), particularly since Al could act alone or combined with F to have differing effects.

F. Aluminum Neurotoxicity

Aluminum administration to animals causes a neurotoxicity that is accompanied by neurofibrillary tangle formation (MUNOZ-GARCIA et al. 1986). In neuroblastoma cultures, Al can also induce tangle formation (SINGER et al. 1990). Alzheimer's disease is associated with the formation of tangles, and, while tangles do not appear in patients suffering from dialysis dementia (EDWARDSON et al. 1992), AD is associated with some of the same behavioral changes as dialysis dementia. These similarities have suggested a possible common etiology for the two diseases. However, after considerable effort in the field, a role for Al in AD remains an unproven and controversial hypothesis. While both experimental Al and AD cause formation of neurofibrillary tangles, the morphology differs (MUNOZ-GARCIA et al. 1986). Furthermore, while it is agreed that Al does not accumulate in AD to the extent seen in dialysis dementia, whether or not Al accumulates at all in the brains of AD patients is still a question (CHAFI et al. 1991).

Studies have evaluated the effect of Al on the physiology and biochemistry of many neuronal systems. Crapper McLaughlin cites over 70 reports of biochemical effects of Al that might play a role in neurotoxicity, identifying potential interactive sites that are nuclear, cytoplasmic, cytoskeletal, membranous, and external to the cell (CRAPPER MCLAUGHLIN 1989). However, most of the work has been carried out using purified systems and has not been related to the whole organism, nor even to a unified hypothesis of action. Crapper McLaughlin proposes that DNA interactions may have most clinical significance, citing nuclear Al accumulation in neurons of AD patients (PEARL and BRODY 1980) as support. Altered expression of cytoskeletal genes also supports this hypothesis (MUMA et al. 1988). However, Pearl's group, using improved technology, no longer support nuclear accumulation of Al, claiming tangles are the only intracellular site of Al deposition (GOOD et al. 1992). Furthermore, considering affinities, Al is unlikely to compete successfully with other metal ions for DNA binding (MARTIN 1988).

Human and animal studies have implicated disturbances of cholinergic function in memory loss. Specifically, a deficiency in cholinergic function and decreased acetyl choline levels have been identified in AD (MARQUIS 1989). Al-dependent loss of choline acetyltransferase and acetylcholinesterase activities in rabbits injected intracisternally with $AlCl_3$ (YATES et al. 1980) cannot be consistently repeated, although Al binds to pruified acetylcholiesterase (MARQUIS 1989). Marquis reports a biphasic effect of Al on acetylcholinesterase (activation at low doses, inhibition at higher doses) and suggests that the activation seen at low doses may play a role in Al neurotoxicity. She points out that the inhibition of acetylcholinetransferase seen at higher doses would increase, rather than decrease, cholinergic function. $AlCl_3$ is reported to inhibit choline uptake by rat brain synaptosomes (LAI et al. 1982). However, in this study the uptake of a number of other neurotransmitters was inhibited, suggesting that there could have been a generalize toxic effect on the membrane, leading to swelling and leaking of the synaptosome (DELEERS et al. 1986). No unified theory for in vivo Al action on the cholinergic system is identified from these data.

Pathological evaluation of neurofibrillary tangles has shown that the hyperphosphorylation of proteins seen in AD is also seen following Al treatment (JOHNSON and JOPE 1988), implicating either increased phosphorylation or decreased dephosphorylation. Experiments with purified phosphatase show inhibition by $AlCl_3$ (YAMAMOTO et al. 1990). Phosphorylation is partly regulated by a G-protein-stimulated cAMP-dependent system. Cyclic AMP has been shown to increase in hippocampus and cortex of rats receiving Al (JOHNSON and JOPE 1987) and in neuroblastoma cells treated with Al (SINGER et al. 1990). Al has no direct effect on adenyl cyclase, but may stimulate adenyl cyclase via the G protein G_s (JOHNSON et al. 1989). Alternatively, incubation of Al with radiolabeled triphosphopyridine nucleotides and purified protein is reported to cause the transfer of the triphosphate moiety from ATP, GTP, or CTP onto tau, the major microtubule-associated protein of AD (ABDEL-GHENY et al. 1993). Hyperphosphorylation of proteins resulting from Al-catalyzed triphosphorylation is an intriguing idea that requires confirmation.

Patients with dialysis dementia respond positively to an antibody against the hyperphosphorylated microtubule-associated protein, tau, that accumulates in AD tangles (GUY et al. 1991). The neurofilament protein tubulin contains an acidic tau binding site to which Al might bind to cause tau accumulation. However, not only do dialysis dementia patients characteristically show no tangle formation, but the tangles seen in animals treated with Al are predominantly made up of neurofilament subunits such as tubulin, rather than microtubule-associated protein. A potential role for Al in G-protein-regulated neurofilament polymerization is discussed in Sect. G.

In addition to a role in hyperphosphorylation, Al has been proposed to play a role in tangle formation by interacting with the hyperphosphorylated proteins. Leterrier has proposed, from studies on purified neurofilaments, that Al can interact directly with neurofilament subunits to cause intramolecular conformation changes at low Al concentrations and intermolecular cross-linking at higher concentrations (LETERRIER et al. 1992). The implication is that the two tangle morphologies seen in Al toxicity and AD could be produced from a single agent. However, prolonged dephosphorylation of the filament subunits did not block Al binding, suggesting that, if such an interaction occurs in vivo, it is not related to hyperphosphorylation of proteins. Incubation of Al with isolated neurofilament subunits has been shown to result in subunit cross-linking β-sheet formation, similar to that seen in neurofibrillary tangles of AD (HOLLOSI et al. 1992). Hollosi and coworkers propose that Al binds phosphate or glutamate linkages to induce subunit cross-linking β-sheet formation. They proposed that, while abnormal phosphorylation and high Al will rapidly form β-sheet aggregation, normal phosphorylation may support β-sheet formation in the presence of excess Al, or conversely very little Al may be necessary in the presence of abnormally high phosphorylation. However, they did not test these possibilities experimentally.

G. Aluminum and Second Messenger Systems

EXLEY and BIRCHALL (1992) have suggested that disruption of the second-messenger system, most probably at the G-protein cycle, may account for all Al toxicity. However, Al probably interacts with second messenger systems in at least three or four different ways.

I. Fluoroaluminate Stimulation of G-Protein Systems

In purified systems, fluoroaluminates (AlF_x) activate the guanine nucleotide binding regulatory component of adenylate cyclase (STERNWEIS and GILMAN 1982). Bigay has proposed that this is by AlF_4^- mimicking the terminal phosphate of GTP in its binding to the α-subunit, permitting the α-subunit-GDP-AlF_4^- complex to have the activity of the normal GTP complex (BIGAY et al. 1987). Persistent stimulation by the G protein is proposed to result because the GDP-AlF_4^- complex lacks the possibility of hydrolysis, the usual inactivation step. Many different G-protein systems support a role for AlF_x in G-protein stimulation (see MESTERS et al. 1993). However, the exact form of the AlF_x complex is debated. Antonny and Charbre have proposed that the AlF_x complex that interacts with GDP may be $AlF_3(OH)^-$ (ANTONNY and CHABRE 1992), and may contain Mg as well as Al and F (ANTONNY et al. 1993). Recently, work on bacterial peptide chain-elongating factor EF-Tu suggests that in this system AlF_x stimulation may be somewhat more complex still, requiring inclusion of ribosomes for AlF_x to inhibit GTP hydrolysis (MESTERS et al. 1993).

It would appear that in in vitro systems AlF_x interferes with the functioning of many, but not all, G-protein systems. Adenylate cyclase stimulation was first reported in an in vitro system (STERNWEIS and GILMAN 1982). In hepatocytes, AlF_x causes a rise in intracellular Ca, suggesting that the G protein activated is coupled to phospholipase C-dependent phosphatidyl inositide hydrolysis (BLACKMORE et al. 1985), which appears to be the case in a number of cell types, e.g., parotid cells (MERTZ et al. 1990). AlF_x is reported to activate many other isolated G proteins, such as the G protein that activates K^+ channels in guinea pig atrial cells (YATANI and BROWN 1991), that do not interact with either adenyl cyclase or the phosphatidyl inositide pathway.

II. Fluoride Stimulation of Second Messenger Systems

Fluoride alone also affects cell signal systems. F inhibits phosphatases and is a tissue-specific mitogen, causing growth and differentiation of osteoblasts, in the absence of Al. Recently Lau compared the mitogenic action of F and Al (LAU et al. 1991) and showed that, unlike F, the Al effect is not limited to a single cell type. Because F mitogenic activity is dependent upon growth factors such as insulin and IGF-1 that stimulate osteoblast growth via tyrosyl protein kinase systems, it has been proposed that F acts by inhibiting phosphotyrosyl protein phosphatase, increasing cellular levels of tyrosyl phosphorylation and maintaining stimulation of osteoblast proliferation (LAU et al. 1989). Again unlike F, Al stimulation does not require endogenous growth factors, being unaffected by medium change which removes growth factors and deletes F mitogenicity (LAU et al. 1991).

Unless specific controls are included to differentiate effects of F alone from those of AlF_x, it is not always possible to identify which may be the activator. For example, a study on the effect of F addition to a G-protein-gated muscarinic atrial K^+ channel system in guinea pig atrial cells showed that the stimulating effect of F could be reversed by DFO chelation of Al (YATANI and BROWN 1991). Interestingly, F and Al have been found to coaccumulate in bone following the administration of both, or even of fluoride alone (AHN et al. 1994). While no studies have related this directly to G-protein activation, F treatment results in increased bone mass and Al greatly magnifies F stimulation of DNA synthesis in an osteoblast-like cell line (KAWASE et al. 1991). With the recent finding that Al stimulates insulin-like growth factor production (LAU et al. 1993), a further interaction between Al and F should be considered. Through phosphatase inhibition, F could synergize the action of insulin-like growth factors, the production of which had been stimulated by Al.

III. Aluminum Stimulation of Second Messenger Systems

Aluminum is reported to stimulate second messenger systems in the absence of F. McDONALD and MAMRACK (1988) identified an effect of Al alone on phosphatidylinositol hydrolysis. While the phospholipase C that normally carries out this function is regulated by a G protein, their work involved the purified enzyme with no G-protein component. They found inhibition of hydrolysis of phosphatidylinositol-4,5-bisphosphate, but stimulation of the hydrolysis of phosphatidylinositol. The stimulation should therefore lead to diacylglycerol-dependent increases in intracellular Ca. Their work clearly showed that Al and not AlF was the active factor, by using F to chelate Al and reverse the stimulation.

Aluminum has been found to be mitogenic in an osteoblast cell line (QUARLES et al. 1991). Like McDonald and Mamrock, these researchers suggested that Al is acting neither via AlF_x, nor at a G protein. They dismiss any action of F because none was added to the system. They proposed no interaction with a G-protein system because pertussis vaccine, a known G-protein inhibitor, did not affect Al stimulation of mitogenesis. They proposed that Al may stimulate production of growth factors. While they acknowledged that pertussis vaccine may not inhibit all G proteins, independence from a G-protein system is supported by work of LAU et al. (1991), showing no stimulation of cAMP production by a mitogenic dose of Al in cultured osteoblasts, although F-mitogenesis is associated with raised cAMP levels. Rather, Lau saw an enhanced 1,25-dihydroxyvitamin D_3-dependent stimulation of osteocalcin, an effect normally inhibited following adenyl cyclase stimulation. Lau also concluded against G-protein involvement, although he pointed out that Al might activate a G protein not associated with cAMP production. More recently, Lau and collaborators have shown that Al stimulates production of mitogenic insulin-like growth factors, which could account for its mitogenic action (LAU et al. 1993). In contrast to bone, in brain Al causes an accumulation of cAMP (see Sect. E), implicating G-protein stimulation of adenyl cyclase.

IV. GTP Interaction with Aluminum

Martin has studied the effect of Al on the interaction of GTP with tubulin, a major tangle component in Al toxicity. He reports a slowing of GTP hydrolysis, which in this system causes promotion of polymerization (MARTIN 1988). Martin proposes that this effect is due to formation of an Al-GTP chelate in place of the normal Mg-GTP chelate. The G protein transducin, which couples the light receptor protein rhodopsin to cGMP phosphodiesterase during visual transduction, is activated by F plus Al, but inhibited by Al alone (MILLER et al. 1989). Using chelators, Miller manipulated free Al^{3+} to allow Al to compete with Mg for GTP and saw inhibition of GTP hydrolysis in a cell-free system. This occurred even at a 10^8-fold excess of

Mg, important since normally intracellular ionic Mg is maintained at approximately $5 \times 10^{-4} M$, while free Al is only in the femtomolar range (Martin 1988). More recently, Huang and Bittar (1991) found that GTP chelation of Al stopped GTP stimulating the G-protein system that activates Na efflux in barnacle fibers.

Miller et al. (1989) proposed that even if Al and AlF both inhibit GTP hydrolysis, they could have opposite effects on the G-protein system, as he had seen in his experiments. In an attempt to reconcile his own data with those of Martin, Miller argued that, while Al was also seen to inhibit GTP hydrolysis in the GTP-transducin system that he used, the primary effect must be to slow GTP replacement of GDP on the G protein and thus inhibit the action of the G protein. Secondarily, Al replacement of Mg could inhibit GTP hydrolysis on the G protein. Miller did not consider an alternative possibility, that the GTP-Al interaction varies with Al concentration and/or free Mg so that the Al-GTP effect is biphasic, activating at low levels of Al that inhibit hydrolysis and inhibiting at levels that inhibit GTP/GDP exchange. Alternatively, the interaction of Al-GTP with one G protein may have a predominant effect on GTP/GDP exchange, while interaction with another could result predominantly in slowing hydrolysis.

References

Abdel-Ghany M, El-Sebae AK, Shalloway D (1993) Aluminum-induced nonenzymatic phosphoincorporation into human tau and other proteins. J Biol Chem 11976–11981

Abreo K, Glass J (1993) Cellular, biochemical, and molecular mechanisms of aluminium toxicity. Nephrol Dial Transplant [Suppl] 1:5–11

Abreo K, Glass J, Sella M (1990) Aluminum inhibits hemoglobin synthesis but enhances iron uptake in Friend erythroleukemia cells. Kidney Int 37:677–681

Abreo K, Jangula J, Jain SK, Sella M, Glass J (1991) Aluminum uptake and toxicity in cultured mouse hepatocytes. J Am Soc Nephrol 1:1299–1304

Adler AJ, Berlyne GM (1985) Duodenal aluminum absorption in the rat: effect of vitamin E. Gastrointest Liver Physiol 12:G209–G213

Ahn HW, Futton B, Moxon D, Jeffery EH (1994) The interactive effects of Fluoride and Aluminum Uptake and accumulation in bones of rabbits administered both agents in their drinking water. Toxicol Environ Health (in press)

Alfrey AC, LeGendre GR, Kaehny WD (1976) The dialysis encephalopathy syndrome. Possible aluminium intoxication. N Engl J Med 294:184–188

Antonny B, Chabre M (1992) Characterization of the aluminum and beryllium fluoride species which activate transducin. J Biol Chem 267:6710–6718

Antonny B, Sukumar M, Bigay J, Chabre M, Higashijima T (1993) The mechanism of aluminum-independent G-protein activation by fluoride and magnesium. J Biol Chem 268:2393–2402

Berlyne GM, Yagil R (1973) Aluminum induced porphyria in rats. Lancet ii:1502–1503

Bigay J, Deterre P, Pfister C, Chabre M (1987) Fluoride complexes of aluminium or beryllium act on G-proteins as reversibly bound analogues of the γ phosphate of GTP. EMBO J 6:2907–2913

Blackmore PF, Bocckino SB, Waynick LE, Exton JH (1985) Role of guanine nucleotide-binding regulatory protein in the hydrolysis of hepatocyte phosphatidylinositol 4,5-biphosphate by calcium-mobilizing hormones and the control of cell calcium. J Biol Chem 260:14477–14483

Blair HC, Finch JL, Avioli R, Crouch EC, Slatopolsky E, Teitelbaum SL (1989) Micromolar aluminum levels reduce ^{3}H-thymidine incorporation by cell line UMR 106-01. Kidney Int 35:1119–1125
Buys SS, Kushner JP (1989) Hematological effects of aluminum toxicity. In: Gitelman HJ (ed) Aluminum and health: a critical review. Dekker, New York, pp 235–256
Chafi AH, Hauw J-J, Rancurel G, Berry J-P, Galle C (1991) Absence of aluminium in Alzheimer's disease brain tissue: electron microprobe and ion microprobe studies. Neurosci Lett 123:61–64
Cochran M, Goddard G, Ludwigson N (1990) Aluminium absorption by rat duodenum: further evidence of energy-dependent uptake. Toxicol Lett 51:287–294
Cournot-Witmer G, Zingraff J, Plachott JJ, Escaig F, Lefevre R, Boumati P, Bourdeau A, Garabedian M, Galle P, Bourdon R, Drueke T, Balsan S (1981) Aluminum iocalization in bone from hemodialyzed patients: relationship to matrix mineralization. Kindey Int 20:375–385
Crapper McLaughlin DR (1989) Aluminum neurotoxicity: criteria for assigning a role in Alzheimer's disease. In: Lewis TE (ed) Environmental chemistry and toxicology of aluminum. Lewis, Chelsea, MI, pp 299–315
De Voto E, Yokel RA (1994) Aluminum speciation and toxicokinetics. Environ Health Perspect (in press)
Deleers M, Servais JP, Wulfert E (1986) Neurotoxic cations induce membrane rigidification and membrane fusion at micromolar concentrations. Biochim Biophys Acta 855:271–276
Domingo JL, Gomez M, Llobet JM, Corbella J (1991) Influence of some dietary constituents on aluminum absorption and retention in rats. Kidney Int 39:598–601
Driscoll CT Jr, Baker JP, Bisogni JJ Jr, Schofield CL (1980) Effect of aluminium speciation on fish in dilute acidified waters. Nature 284:161–164
Edwardson JA, Candy JM, Ince PG, McArthu FK, Morris CM, Oakley AE, Taylor GA, Bjertness E (1992) Aluminum accumulation, β-amyloid deposition and neurofibrillary changes in the central nervous system. In: Chadwick DJ, Whelan J (eds) Aluminum in biology and medicine. Wiley, New York, pp 165–185 (Ciba foundation symposium 169)
Exley C, Birchall JD (1992) The cellular toxicity of aluminum. J Theor Biol 159:83–98
Farrer G, Altman P, Welch S, Wychrij O, Ghose B, Lejeune J, Cornett J, Prasher V, Blair J (1990) Defective gallium-transferrin binding in Alzheimer disease and Down syndrome: possible mechanism for accumulation of aluminum in brain. Lancet 335:747–750
Feinroth M, Feinroth MV, Berlyne GM (1982) Aluminum absorption in the rat everted gut sac. Miner Electrolyte Metabol 8:29–35
Fernandez Mendez MJ, Fell GS, Brock JH (1991) Aluminium uptake by intestinal cells; effect of iron status and precomplexation. Nephrol Dial Transplant 6:672–674
Fleming JT, Joshi JG (1987) Ferritin: the role of aluminum in ferritin function. Neurobiol Aging 12:413–418
Flendrig JA, Kruis H, Das HA (1976) Aluminium intoxication: the cause of dialysis dementia? Proc Eur Dial Transplant Assoc 13:355–361
Froment DPH, Molitoris BA, Buddington B, Miller N, Alfrey AC (1989) Site and mechanism of enhanced gastrointestinal absorption of aluminium by citrate. Kidney Int 36:978–984
Fulton B, Jeffery EH (1990) Absorption and retention of aluminum from drinking water. Fund Appl Toxicol 14:788–796
Fulton B, Jeffery EH (1994) Heme oxygenase induction: a possible factor in alumin um-associated anemia. Biol Trace Elem Res 40:9–19
Ganrot PO (1986) Metabolism and possible health effects of aluminum. Environ Health Perspect 65:363–441

Good PF, Pearl DP, Bierer LM, Schmeidler J (1992) Selective accumulation of aluminum and iron in the neurofibrillary tangles of Alzheimer's disease: a laser microprobe (LAMMA) study. Ann Neurol 31:286–292
Goodman WG (1986) Experimental aluminum-induced bone disease: studies in vivo. Kidney Int 29:S32–S36
Greger JL, Powers CF (1992) Assessment of exposure to parenteral and oral Al with and without citrate using a desferrioxamine test in rats. Toxicology 76:119–132
Gupta SK, Waters DH, Gwilt PR (1986) Absorption and disposition of aluminum in the rat. J Pharm Sci 75:586–589
Guy SP, Jones D, Mann DMA, Itzhaki RF (1991) Human neuroblastoma cells treated with aluminium express, an epitope associated with Alzheimer's disease neurofibrillary tangles. Neurosci Lett 121:166–168
Hollosi M, Urge L, Perczel A, Kajtar J, Teplan I, Otvos L Jr, Fasman GD (1992) Metal ion-induced conformational changes of phosphorylated fragments of human neurofilament (NF-M) protein. J Mol Biol 223:673–682
Huang Y-P, Bittar EE (1991) Protection by GTP from the effects of aluminum on the sodium efflux in barnacle muscle fibers. Biochim Biophys Acta 1062:255–263
Huber CT, Frieden E (1970) The inhibition of ferroxidase by trivalent and other metal ions. J Biol Chem 245:3979–3984
Ittel TH, Buddington B, Miller NL, Alfrey AC (1987) Enhanced gastrointestinal absorption of aluminum in uremic rats. Kidney Int 32:821–826
Johnson GVW, Jope RS (1987) Aluminum alters cyclic AMP and cyclic GMP levels but not presynaptic cholinergic markers in rat brain in vivo. Brain Res 403:1–6
Johnson GVW, Jope RS (1988) Phosphorylation of rat brain cytoskeletal proteins is increased after orally administered aluminum. Brain Res 456:95–103
Johnson GVW, Li X, Jope RS (1989) Aluminum increases agonist-stimulated cyclic AMP production in rat cerebral cortical slices. J Neurochem 53:258–263
Kappas A, Sassa S, Anderson KE (1983) The porphyrias. In: Stanbury JB, Wyngaarden JB, Fredrickson DS, Goldstein JL, Brown MS (eds) The metabolic basis of inherited disease, 5th edn. McGraw-Hill, New York, pp 1301–1384
Kasai K, Hori MT, Goodman WG (1991) Transferrin enhances the antiproliferative effect of aluminium on osteoblast-like cells. Am J Physiol 260:E537–E543
Kawase T, Orikasa M, Suzuki A (1991) Aluminofluoride- and epidermal growth factor-stimulated DNA synthesis in MOB 3-4-F2 cells. Pharmacol Toxicol 69:330–337
Kerr DSN, Ward MK (1988) Aluminum intoxication: history of its clinical recognition and management. In: Siegel H (ed) Aluminum and its role in biology. Dekker, New York, pp 217–258
Khalil-Manesh F, Agness C, Gonick HC (1989) Aluminum-binding protein in dialysis dementia. Nephron 52:323–328
Kirschbaum BB, Schoolwerth AC (1989) Acute aluminum toxicity associated with oral citrate and aluminum containing antacids. Am J Med Sci 297:9–11
Kostyniak PJ (1983) An electrothermal atomic absorption method for aluminum analysis in plasma: identification of sources of contamination in blood sampling procedures. J Anal Toxicol 7:20–23
Lai JCK, Lim L, Davison AM (1982) Effect of Cd^{2+}, Mn^{2+}, and Al^{3+} on rat brain synaptosomal uptake of noradrenaline and serotonin. J Inorg Biochem 17:215–225
Lau K, Farley JR, Freeman TK, Baylink DJ (1989) A proposed mechanism of the mitogenic action of fluoride on bone cells: inhibition of the activity of an osteoblastic acid phosphatase. Metabolism 38:858–868
Lau KHW, Yoo A, Wang SP (1991) Aluminum stimulates the proliferation and differentiation of osteoblasts in vitro by a mechanism that is different from fluoride. Mol Cell Biochem 105:93–105
Lau KHW, Utrapiromsuk S, Yoo A, Mohan S, Strong DD, Baylink DJ (1993) Mechanism of mitogenic action of aluminum ion on human bone cells: potential

involvement of the insulin-like growth factor regulatory system. Arch Biochem Biophys 303:267–273
Leterrier JF, Langui D, Probst A, Ulrich J (1992) A molecular mechanism for the induction of neurofilament bundling by aluminum ions. J Neurochem 58:2060–2070
Lieberherr M, Grosse B, Cournot-Witmer G, Hermann-Erlee MPM, Balsan S (1987) Aluminum action on mouse bone cell metabolism and response to PTH and 1,25$(OH)_2D_3$. Kidney Int 31:736–743
Maines MD, Kappas A (1976) Studies on the mechanism of induction of haem oxygenase by colbalt and other metal ions. Biochem J 154:125–131
Marquis J (1989) Neurotoxicity of aluminum. In: Lewis TE (ed) Environmental chemistry and toxicology of aluminum. Lewis, Chelsea, MI, pp 289–298
Martin RB (1988) Bioinorganic chemistry of aluminum. In: Sigel H (ed) Metal ions in biological systems, vol 24. Dekker, New York, pp 1–58
Martin RB (1992) Aluminum speciation in biology. In: Chadwick DJ, Whelan J (eds) Aluminum in biology and medicine. Wiley, New York, pp 5–25 (Ciba foundation symposium 169)
May JC, Rains TC, Yu LJ, Etz NM (1992) Aluminum content of source plasma and sodium citrate anticoagulant. Vox Sang 62:65–69
McDonald LJ, Mamrack MD (1988) Aluminum affects phosphoinositide hydrolysis by phosphoinositidase C. Biochem Biophys Res Commun 155:203–208
McGonigle R, Parsons V (1985) Aluminium-induced anaemia in haemodialysis patients. Nephron 39:1–9
McGregor SJ, Naves ML, Oria R, Vass K, Brock JH (1990) Effect of aluminium on iron uptake and transferrin-receptor expression by human erythroleukaemia K562 cells. Biochem J 272:377–382
McGregor SJ, Brown D, Brock JH (1991) Transferrin-gallium binding in Alzheimer's disease. Lancet 338:1394–1395
Meredith PA, Moore MA, Goldberg A (1977) Effects of aluminium, lead and zinc on delta-aminolaevulinic acid dehydratase. Enzyme 22:22–27
Mertz LM, Horn VJ, Baum BJ, Ambudkar IS (1990) Calcium entry in rat parotid acini: activation by carbachol and aluminum fluroride. Am J Physiol 258:C654–C661
Mesters JR, deGraaf JM, Kraal B (1993) Divergent effects of fluoroaluminates on the peptide chain elongation factors EF-Tu and EF-G as members of the GTPase superfamily. FEBS 321:149–152
Miller JL, Hubbard CM, Litman BJ, Macdonald TL (1989) Inhibition of transducin activation and guanosine triphosphatase activity by aluminum ion. J Biol Chem 264:243–250
Mladenovic J (1988) Aluminum inhibits erythropoiesis in vitro. J Clin Invest 81: 1661–1665
Moreb J, Popovtzer MM, Friedlaenderm MM, Konijn AM, Hershko C (1983) Evaluation of iron status in patients on chronic dialysis: usefulness of bone marrow hemosiderin, serum ferritin, transferrin saturation, MCV and red cell protoporphyrin. Nephron 35:196–200
Moriyama Y, Rege A, Fisher JW (1975) Studies on an inhibitor of erythropoesis ii. Inhibitory effects of serum from uremic rabbits on heme synthesis in rabbit bone marrow culture. Proc Soc Exp Biol Med 148:94–97
Moshtaghie AA, Bazrafshan MR (1992) Comparative binding study of aluminum and chromium to human transferrin: effect of iron. Biol Trance Elem Res 32:39–45
Moxon DR, Jeffery EH (1991) Aluminum distribution between plasma and erythrocytes varies with aluminum load and the use of anticoagulants. FASEB J 5:A876
Muller L, Wilhelm M (1987) Effects of cadmium in rat hepatocytes: interaction with aluminum. Toxicology 44:193–201

Muma NA, Troncoso JC, Hoffman PN, Koo EH, Price DL (1988) Aluminium neurotoxicity: altered expression of cytsokeletal genes. Mol Brain Res 3:115–122
Munoz-Garcia D, Pendlebury WW, Kessler JB, Pearl D (1986) An immunocytochemical comparison of cytoskeletal proteins in aluminum-induced and Alzheimer type neurofibrillary tangles. Acta Neuropathol (Berl) 70:243–248
Nordal KP, Dahl E, Halse J, Aksnes L, Thomassen Y, Flatmark A (1992) Aluminum metabolism and bone histology after kidney transplantation: a one-year follow-up study. J Clin Endrocrinol Metab 74:1140–1145
Partridge NA, Regnier FE, Reed WM, White JL, Hem SL (1992) Contribution of soluble aluminium species to absorption of aluminium from the rat gut in situ. Clin Sci 83:425–430
Pearl DP, Brody AR (1980) Alzheimer's disease: x-ray spectrographic evidence of aluminum accumulation in neurofibrillary tangle-bearing neurons. Science 208: 297–299
Pei Y, Hercz G, Greenwood C, Sharrard D, Segre G, Manuel A, Saiphoo C, Fenton S (1992) Noninvasive prediction of aluminum bone disease in hemo- and peritoneal dialysis patients. Kidney Int 41:1374–1382
Pierides AM (1980) Haemodialysis osteodystrophy with osteomalacia as the main finding – relation to aluminium intoxication. Calcif Tissue Res 31:65–453
Provan SD, Yokel RA (1988a) Influence of calcium on aluminum accumulation by the rat jejunal slice. Res Commun Chem Pathrol Pharmacol 59:79–92
Provan SD, Yokel RA (1988b) Aluminium uptake by the in situ rat gut preparation. J Pharmacol Exp Ther 245:928–931
Quarles LD, Gitelman HJ, Drezner MK (1989) Aluminum-induced neo-osteogenesis: attenuation by parathyroid hormone deficiency. J Clin Invest 83:1644–1650
Quarles LD, Wenstrup RJ, Catillo SA, Drezner MK (1991) Aluminum-induced mitogenesis in MC3T3-E1 osteoblasts: potential mechanism underlying neo-osteogenesis. Endrocrinology 128:3144–3151
Rahman H, Skillen AW, Channon SM, Ward MK, Kerr DNS (1985) Methods for studying the binding of aluminum by serum protein. Clin Chem 31:1969–1973
Rodrigues M, Felsenfeld AJ, Llach F (1990) Aluminum administration in the rat separately affects the osteoblast and bone mineralization. J Bone Miner Res 5:59–67
Rosenlof K, Fyhrquist F, Tenhunen R (1990) Erythropoietin, aluminium, and anaemia in patients on haemodialysis. Lancet 335:247–249
Shils ME (1988) Magnesium in health and disease. Annu Rev Nutr 1988:429–460
Short AIK, Winey RJ, Robson JS (1980) Reversible microcytic hypochromic anemia in dialysis patients due to aluminium intoxication. Proc EDTA 17:226–233
Singer HS, Serles CD, Hahn IH, March JL, Troncoso JC (1990) The effect of aluminum on markers for synaptic neurotransmission, cyclic AMP, and neurofilaments in a neuroblastoma × glioma hybridoma (NG108-15). Brain Res 528:73–79
Sprague SM, Bushinsky DA (1990) Mechanism of aluminum-induced calcium efflux from cultured neonatal mouse calvariae. Am J Physiol 258:F583–F588
Sternweis PC, Gilman AG (1982) Aluminum: a requirement for activation of the regulatory component of adenylate cyclase by fluoride. Proc Natl Acad Aci USA 1982 79:4888–4891
Stevens BJ, Willis GL, Humphrey TJ, Atkins RC (1987) Combined toxicity of aluminum and fluoride in the rat: possible implications in hemodialysis. Tr Elem Med 4:61–66
Swartz RD, Dombrouski J, Burnatowska-Hledin M, Mayor GH (1985) Microcytic anemia: a marker for the diagnosis of aluminum toxicity. Proc EDTA 22:101–105
Tielmans C, Collart F, Wens R, Smeyers-Verbeeke J, Van Hooff I, Dratwa M, Verbeelen D (1985) Improvement of anemia with deferroxiamine in hemodialysis in patients with aluminum-induced bone disease. Clin Nephrol 24:237–241

Touam M, Martinex F, Lacour B, Bourdon R, Zingraff J, DiGuilio S, Drueke T (1983) Aluminum-induced, reversible microcytic anemia in chronic renal failure: clinical and experimental studies. Clin Nephrol 19:295–298

Trapp GA (1983) Plasma aluminium is bound to transferrin. Life Sci 33:311–316

Umeda M, Tsurusaki K, Kamikawa S, Izumi N, Tasumoto R, Kishimoto T, Maekawa M (1990) Red blood cell aluminum in patients with renal failure and effect of desferrioxamine infusion. Blood Purif 8:295–300

Van der Voet GB (1992a) Intestinal absorption of aluminum – relation to neurotoxicity. In: Isaacson RL, Jensen KF (eds) The vulnerable brain and environmental risks, vol 12: toxins in food. Plenum, New York, pp 35–47

Van der Voet GB (1992b) Intestinal absorption of aluminum. In: Chadwick DJ, Whelan J (eds) Aluminum in biology and medicine. Wiley, New York, pp 117–122 (Ciba foundation symposium 169)

Van der Voet GB, De Wolff FA (1987) The effect of di- and trivalent iron on the intestinal absorption of aluminium in rats. Toxicol Appl Pharmacol 90:190–197

Vasilakakis DM, O'Haese PC, Lamberts LV, Lemoniatou E, Dignensis PN, De Broe ME (1992) Removal of aluminoxamine and ferrioxamine by charcoal hemoperfusion and hemodialysis. Kidney Int 41:1400–1407

Ward JH, Jordan I, Kushner JP, Kaplan J (1984) Heme regulation of HeLa cell transferrin receptor number. J Biol Chem 259:13235–13240

Weberg R, Berstad A (1986) Gastrointestinal absorption of Al from single doses of aluminum-containing antacids in man. Eur J Clin Invest 16:428–432

Yamamoto H, Saitoh Y, Yasugawa S, Miyamoto E (1990) Dephosphorylation of tau factor by protein phosphatase 2A synaptosomal cytosol fractions and inhibition by aluminum. J Neurochem 55:683–690

Yatani A, Brown AM (1991) Mechanism of fluoride activation of G protein-gated muscarinic atrial K^+ channels. J Biol Chem 266:22872–222877

Yates CM, Simpson J, Russel D, Gordon A (1980) Cholinergic enzymes in neurofibrillary degeneration produced by aluminium. Brain Res 197:269–274

Yokel RA, McNamara PJ (1985) Aluminum bioavailability and disposition in adult and immature rabbits. Toxicol Appl Pharmacol 77:344–352

Zaman K, Mukhtar M, Siddique H, Miszta H (1992) The effect of aluminum on the stromal cells in vitro on bone marrow in rats. Toxicol Ind Health 8:103–109

Zaman K, Zaman W, Dabrowski Z, Miszta H (1993) Inhibition of delta aminoevulinic acid dehydratase activity by aluminum. Comp Biochem Physiol 104C: 269–273

CHAPTER 8

Mercury Toxicity

K. Miura, A. Naganuma, S. Himeno, and N. Imura

A. Introduction

Methyl, mercuric, and elemental mercury are the major chemical forms of mercury present in the environment and widely recognized as highly toxic substances for mammals including humans, the degree of toxicity being dependent on the ingestion route. Although the toxicity of mercury has been recognized since 1500 B.C., many people are still exposed to the risk of poisoning by mercury compounds. Even quite recently, heavy environmental mercury pollution due to goldmining activities in the Amazon attracted public attention. Mercury is widely used in dental treatment for tooth amalgam fillings, and the exposure of dentists, dental technicians, and treated patients to mercury vapor is regarded as a serious health problem. Furthermore, Who (1990) has recently emphasized the risk of prenatal exposure to methylmercury from maternal consumption of fish, which contains this metal alkyl naturally in small quantities.

This chapter primarily describes the most recent findings on the mechanisms of toxicity and metabolism of methyl, mercuric, and elemental mercury. Since methylmercury is known as one of the most important mercurials to which the general population is frequently exposed in the environment by food consumption, the mechanisms of its toxicity and disposition are discussed in detail.

B. Organic Mercury

Generally organic mercury compounds are defined as those having a carbon-mercury bond, such as methyl, ethyl, phenyl, and alkoxyalkylmercury. Of these organic mercurials, methylmercury (MeHg) is generally believed to be the most resistant to biodegradation in the environment and the general population is primarily exposed to this metal alkyl through diet, especially fish, in which mercury is predominantly found in the form of MeHg. Therefore, this section deals mainly with MeHg as the most abundant form of organic mercury in the environment.

I. Methylmercury

1. Mechanism of Uptake and Excretion

a) Blood-Brain Barrier

Methylmercury is primarily neurotoxic and the damage is almost exclusively limited to the central nervous system in the adult human. This means that MeHg accumulates in the brain up to a level high enough to cause neuronal damage. Before entry into the brain, exogenous chemicals in the blood first come into contact with the blood-brain barrier, which is formed by a continuous layer characterized by tight junctions between the endothelial cells, with astrocyte foot processes surrounding each capillary (GOLDSTEIN and BETZ 1986). Therefore, MeHg has to pass through the plasma membranes before entering the brain. It has been assumed that the ease of passage of MeHg compounds across this barrier largely depends on its lipophilicity (LAKOWICZ and ANDERSON 1980; HAMMOND and BELILES 1980). LAKOWICZ and ANDERSON (1980) showed that lipid bilayers in vesicles of dimyristoyl phosphatidylcholine and of dioleoyl phosphatidylcholine are highly permeable to MeHgCl and that partitioning of MeHgCl into bilayers is slight, which means it rapidly diffuses across membranes. The octanol/water distribution ratio, an indicator of lipophilicity, of MeHg is the highest among the mercurials examined by HALBACH (1990), i.e., 2.5, 1.0, 0.5, and 0.02 in MeHg, $HgCl_2$, bromomercurihydroxypropane (BMHP), and chlormerodrin, respectively. These results also showed that the inhibitory effects of mercurials on the force of contraction of isolated guinea pig papillary muscles increased with their lipid solubility (HALBACH 1990).

On the other hand, the fact that water-soluble complexes of MeHg with cysteine and glutathione have been identified in blood (RABENSTEIN and FAIRHURST 1975; NAGANUMA and IMURA 1979) and brain (THOMAS and SMITH 1982) suggests the entry of MeHg through a specific carrier-mediated transport system of these neutral amino acids, which are essential for neurotransmitter synthesis (OLDENDORF 1970) in the brain. HIRAYAMA (1980, 1985) showed that the initial entry rate of MeHg injected intravenously into brains was enhanced by coadministration of L-cysteine. Using a single capillary pass method (OLDENDORF 1970), ASCHNER and CLARKSON (1988) found that the transport of cysteine complexes of MeHg (MeHg-L-cysteine) was completely blocked by coadministration of methionine. They then demonstrated that MeHg-L-cysteine complexes pass through the L system in isolated bovine endothelial capillary suspensions (ASCHNER and CLARKSON 1989). Since the MeHg-L-cysteine complex has a close structural similarity to L-methionine, a substrate for the L system, they assumed that the metal complex may enter the brain capillary endothelial cells via this amino acid carrier. Recently, KERPER et al. (1992) confirmed this hypothesis by in vivo study in rat using the rapid carotid infusion technique. Their results indicated that: (a) $Me^{203}Hg$-L-cysteine transport was inhibited by methionine and an amino acid analogue (2-aminobicyclo-[2,2,1]heptane-2-carboxylic acid, BCH) of L system substrate, but not by α-methylaminoisobutyric acid, a substrate for the alanine-preferring system. (b) L-[^{14}C]Methionine transport was inhibited by MeHg-L-cysteine but not by MeHgCl. (c) Uptake of $Me^{203}Hg$-glutathione was inhibited by L-methionine and BCH but not D-methionine. They then concluded that MeHg might enter the brain as a cysteine complex via the L system and that plasma MeHg-glutathione serves as a source of MeHg-cysteine.

b) Renal Uptake and Excretion

Kidney is one of the major organs in which MeHg accumulates. The mechanism of renal uptake of MeHg has been well studied. Richardson and Murphy (1975) demonstrated that pretreatment of rats with GSH-depleting agents significantly reduced MeHg uptake by the kidneys. Naganuma et al. (1988) examined the role of extrarenal GSH on the renal uptake of MeHg using mice pretreated with 1,2-dichloro-4-nitrobenzene (DCNB), which depletes GSH levels in the liver and plasma without affecting the renal GSH level. The specific depletion of hepatic and plasma GSH levels significantly reduced the renal uptake of MeHg. A substantial increase in the renal uptake of mercury was observed when methylmercury was administered to mice or rats as a GSH complex (Alexander and Aaseth 1982; Hirayama 1985; Naganuma et al. 1988). Thus, MeHg may be translocated to the kidney as a complex with GSH. After glomerular filtration in the kidneys, the GSH moiety of MeHg-SG may be hydrolyzed by the action of γ-glutamyltranspeptidase (γ-GTP) and dipeptidase located in the brush border membrane, and then incorporated into the kidney cells as MeHg-cysteine. Therefore, administration of γ-GTP inhibitor to animals results in a decrease in MeHg renal uptake and an increase in urinary excretion of MeHg and GSH (Naganuma et al. 1988). Tanaka et al. reported that the strain, sex, and age differences in rate of renal uptake of MeHg in mice can at least partially be explained by the difference in renal γ-GTP activity (Tanaka et al. 1991, 1992b). In female mice, both renal MeHg uptake and γ-GTP activity were increased to the level of males by testosterone treatment (Tanaka et al. 1992b), suggesting that the sex differences in MeHg uptake by kidneys may be due to a difference in renal γ-GTP activity controlled at least in part by testosterone.

Methylmercury may be incorporated into the kidney cells not only through the brush border membrane after glomerular filtration but also through the basolateral plasma membrane from the peritubular capillary. This hypothesis is supported by the experimental results that bilateral ligation of the ureters to obstruct the glomerular filtration reduced the renal accumulation of MeHg to 50% of that of the unobstructed kidneys (Tanaka et al. 1992a). Renal uptake of MeHg in ureter-ligated mouse was significantly decreased by injection of probenecid, an inhibitor of peritubular transport of organic acid, but not by acivicin, an inhibitor of γ-GTP (TANAKA et al. 1992a). These reports suggest that at least two transport systems play major roles in renal MeHg uptake: one is a route from the glomeruli through the brush border membrane which is dependent on the action of γ-GTP, and the other is the route using an organic anion transport system through the basolateral membrane.

c) Biliary Excretion

A considerable amount of MeHg is excreted into rat bile, and more than 90% of the MeHg is reabsorbed from the gut (CLARKSON et al. 1973; NORSETH 1973). Although the species differences in the rates of MeHg biliary excretion have been observed (NAGANUMA and IMURA 1984b; URANO et al. 1988b), bile is one of the main routes for the elimination of MeHg in all species. Biliary excretion of MeHg is significantly decreased by administration of GSH depletors (REFSVIK 1978; BALLATORI and CLARKSON 1983), and the MeHg concentration in the bile is correlated with biliary GSH concentration (BALLATORI and CLARKSON 1983, 1985). Complete inhibition of biliary secretion of GSH, without changing hepatic levels of GSH, also blocked MeHg secretion into bile (BALLATORI and CLARKSON 1983). These experimental results suggest that MeHg is secreted into bile as a complex with GSH, and the secretion of this complex into bile is closely linked to the GSH transport system (BALLATORI and CLARKSON 1983, 1985). In fact, the main form of MeHg existing in rat bile is complexes with GSH or its metabolites, such as cysteinylglycine and cysteine (URANO et al. 1988a,b). It has been reported that part of the GSH secreted from the liver into bile is hydrolyzed into cysteinylglycine and cysteine by γ-GTP and dipeptidase in the biliary ductular epithelium (ABBOTT and MEISTER 1986). MeHg may be secreted into bile as a GSH complex, and then a part of the complex is degraded into MeHg-cysteinylglycine and MeHg-cysteine (URANO et al. 1988a,b).

2. Mechanism of Toxicity

a) Neurotoxicity

The neuronal damage caused by MeHg is almost exclusively limited to the central nervous system in the adult human. The areas of damage to the brain are highly localized (focal), e.g., in the visual cortex and the granular layer on the cerebellum, especially in the infolded regions (sulci). Little is known about the mechanism of selective damage caused by MeHg in the nervous system. In general, the specific toxicity of MeHg to the central nervous system has been ascribed to its relatively easy passage through the blood-brain barrier and the lack of ability in nervous cells to repair the cell damage caused by MeHg. As a model system for studying mechanisms of selective toxicity of MeHg, SAGER and SYVERSEN (1984, 1986) used three types of cultured cells and showed that neuroblastoma cells were more sensitive to MeHg than glioma cells or fibroblasts. In addition, they showed that the neuroblastoma cells accumulated more MeHg than the two less susceptible cell types. SARAFIAN and VERITY (1985, 1986) and VERITY and SARAFIAN (1991) used an in vitro model of cerebellar granule cell suspension to investigate the selective damage caused by MeHg. They found that endogenous GSH level in granule cells was significantly lower than that in glial cells. Furthermore, a decrease in GSH level in granule cells by treatment with buthionine sulfoximine (BSO) caused an increased lipoperoxidation and neuronal cell injury by MeHg, but only a minimal effect was observed in glial cells. The ability to accumulate MeHg and a low level of intracellular GSH appears to make the cells susceptible to MeHg. Using seven sublines with different sensitivities to MeHg established from the PC12 cell line, which has been used as a model for neuronal cells, it was revealed that the sublines were more susceptible to MeHg as their ability to accumulate MeHg increased and as the intracellular GSH level decreased (MIURA and CLARKSON 1993; MIURA et al. 1994a). Furthermore, intracellular GSH levels in neuronal cell lines, such as mouse neuroblastoma and PC12 cell lines, are lower than those in nonneuronal cell lines, such as HeLa, mouse glioma, L cell, and colon 26 (MIURA et al. 1994b).

There has been increasing evidence suggesting that oxidative stresses inducing free radicals play a role in the underlying biochemical mechanisms involved in the neurotoxicity of MeHg (KASUYA 1975; GANTHER 1978; YONAHA et al. 1983; HALLIWELL and GUTTERIDGE 1985; BONDY and MCKEE 1990). LEBEL et al. (1990) demonstrated with the aid of the probe 2′,7′-dichlorofluorescin diacetate (DCFH-DA) that both in vivo and in vitro exposure to MeHg increased the formation rate of reactive oxygen species in mouse brain region, which is believed to be selectively vulnerable to MeHg. Nonfluorescent 2′,7′-dichlorofluorescin (DCFH) is rapidly oxidized by reactive oxygen species to highly fluorescent 2′,7′-dichlorofluorescein (DCF), then providing a direct demonstration of the formation rate of reactive oxygen species. From the observation of the protective effect of desferoxamine, LEBEL and BONDY (1990) suggested the possibility that an iron-peroxide intermediate (ferryl ion) and other reactive oxygen species play a major role in MeHg-induced cell death.

One of the clinical signs associated with MeHg intoxication is a myasthenia gravis-like muscle weakness in adults (RUSTAM et al. 1975), a syndrome which responded well to therapy with neostigmine, a reversible acetylcholinesterase inhibitor. In this syndrome, two effects of MeHg on synaptic transmission at the neuromuscular junction were demonstrated using intracellular microelectrode recording techniques (ATCHISON and NARAHASHI 1982; ATCHISON et al. 1984). First, nerve-evoked, synchronous quantal release of acetylcholine (ACh) is inhibited, as indicated by a decrease in end-plate potential (EPP) amplitude. Second, spontaneous quantal release of ACh is first increased and then decreased, as indicated by an increase and decrease in miniature-end-plate potential (MEPP) frequency. Using the putative Ca antagonist *N*, *N*-bis(3,4-dimethoxyphenylethyl)-*N*-methylamine (YS035), LEVESQUE and ATCHISON (1991) demonstrated that release of Ca from nerve terminal mitochondria contributed to the increased MEPP frequency caused by MeHg. They further showed that the ability of mitochondria, isolated from rat forebrain to sequester Ca, was disrupted by MeHg. MINNEMA et al. (1989) reported MeHg-induced increases in spontaneous release of various transmitters such as dopamine, γ-aminobutyric acid, and ACh in rat brain synaptosomes. From the similarity of the effects, they speculated that MeHg affected the release through processes common to all transmitter systems, suggesting MeHg-induced increase in synaptosomal membrane permeability.

b) Developmental Toxicity

The developing brain appears to be more sensitive to MeHg than the adult brain. A clinical sign of prenatal intoxication by MeHg is unspecific infantile cerebral palsy. Histological features observed are decreased numbers of neurons and distortion of the cytoarchitecture in the cortical areas (MATSUMOTO et al. 1965; CHOI et al. 1978). Recently, PECKHAM and CHOI (1988) demonstrated in mice exposed to MeHg during pregnancy abnormal neuronal migration and anomalous cortical cytoarchitectonic patterning using autoradiographic technique. Using embryonic chick retinal cell aggregation in vitro, KLEINSCHUSTER et al. (1983) examined the effect of MeHg on the ability of embryonic cells to reconstruct tissue-like patterns. REBEL et al. (1983) observed changes in the relative concentrations of specific gangliosides within mouse neuroblastoma cells exposed in vitro to MeHg. They claimed that MeHg alters cell-cell recognition and/or cohesion in reaggregating cultures of dissociated neonatal mouse cerebellar cells in a complex manner. JACOBS et al. (1986) showed that MeHg altered cerebellar cell recognition, suggesting a disruption in the arrangement of specific cell surface recognition molecules. Furthermore, they confirmed that the effect of MeHg on cerebellar cell recognition was mimicked by colchicine, suggesting a mechanism through microtubules as a target molecule.

On the other hand, RODIER et al. (1984) and SAGER et al. (1982) demonstrated that MeHg arrested the cell division during the development of the central nervous system of mice exposed to MeHg in utero. Referring to the evidence for MeHg effects on microtubules as described below, they proposed that the antimitotic action of this metal alkyl is a mechanism which possibly causes degeneration (decreased numbers of neurons) in the developing nervous system.

Early postnatal exposure to MeHg seems to inhibit the development and maturation of the brain. CHOI et al. (1981) observed incomplete arborization of the dendritic tree of Purkinje cells in mice exposed to MeHg in the early postnatal period.

c) Cellular and Molecular Mechanisms

YOSHINO et al. (1966) reported that protein synthesis in nervous tissues from rat injected with MeHg thioacetamide was inhibited before the first clinical signs of intoxication. The effects of MeHg on protein synthesis have since been studied repeatedly (CARMICHAEL et al. 1975; OMATA et al. 1980, 1982; SYVERSEN 1982; FAIR et al. 1987), indicating that MeHg directly interacts with some parts of the protein synthetic machinery.

Studying the steps in the action of MeHg on protein synthesis, CHEUNG and VERITY (1985) showed that the amino acylation of phenylaranyl-tRNA is the most sensitive step in the brains of neonatal rats exposed to MeHg. The effects of MeHg on the activities of six aminoacyl-tRNA synthetase species in the adult rat brain were investigated by HASEGAWA et al. (1988), who demonstrated that three of them were inhibited on MeHg administration, one was elevated, and the other two showed no significant changes. These results appear to be consistent with the reports by KASAMA et al. (1989) in which the effect of MeHg on the protein synthesis in dorsal root ganglia is not uniform for each protein species. Among 200 protein species investigated, 20 showed real stimulation of the synthesis, whereas 7 were moderately inhibited and 16 were inhibited more strongly than the total protein in the tissue.

Disruption of microtubules and inhibition of tubulin polymerization by MeHg has been well documented in cultured cells (MIURA et al. 1978; IMURA et al. 1980; SAGER et al. 1983) and in the in vitro assembly system (IMURA et al. 1980; SAGER et al. 1983; MIURA et al. 1984). The effect of MeHg on microtubules precedes changes in other biochemical functions such as macromolecule biosynthesis (MIURA and IMURA 1987). The selective effect of MeHg on microtubules was confirmed by several researchers using kidney epithelial PtK2 cells and murine embryonal carcinoma (EC) cells. SAGER and MATHESON (1988) and WASTENEYS et al. (1988) showed the relative insensitivities of other cytoskeletal components (vimentin and actin filaments). In addition, the heterogeneous response of neuronal microtubules to MeHg was suggested by using EC cells differentiated with retinoic acid (WASTENEYS et al. 1988).

Since microtubules are associated with many cellular functions, for example, mitosis, cell moving, axonal transport, phagocytosis, and sperm motility, we can reasonably assume that the specific alteration of microtubules by MeHg results in various types of cellular disorders depending on the differential functions of microtubules in each cell type. SAKAI (1972) reported disturbances in rat spermatogenesis. MOHAMED et al. (1986, 1987) observed a decreased percentage of motile spermatozoa and decreased scores for sperm speed at subneurotoxic doses of MeHg with no significant changes in serum testosterone levels. BROWN et al. (1988) observed microtubule damage in lymphocytes isolated from mice receiving a single dose (10 mg/kg) of MeHg. They revealed a good correlation between the degree of microtubule disassembly and the inhibition of mitogen responsiveness. They then proposed that damage to the microtubule system may underlie the effects of the toxicants on lymphocyte functions. The effect of MeHg on proliferating nerve cells in the developing brain in relation to microtubule disruption by MeHg has been reviewed in a Who report (1990).

The effect of microtubule disruption by MeHg on the differentiated nerve cells is uncharacteristic. Abe et al. (1975) described that MeHg locally injected into the ninth dorsal root ganglion of the frog blocked axoplasmic transport. The study on tubulin gene expression indicated tubulin synthesis in animal cells was controlled by an autoregulatory mechanism, in which the increased concentration of unpolymerized tubulin subunits in cytoplasm by the treatment of colchicine inhibited new tubulin synthesis. Since depolymerization of microtubules due to MeHg may increase the size of the pool of unpolymerized tubulin subunits, Miura and Imura (1989) determined tubulin synthetic activity after disruption of microtubules by MeHg in mouse glioma. They demonstrated by two-dimensional gel electrophoresis that β-tubulin synthesis monitored with ^{35}S-methionine was inhibited up to about 50%, while total protein synthesis was only slightly inhibited. Furthermore, they showed that the tubulin mRNA level declined corresponding to the extent of the decreased tubulin protein synthesis, but the transcription rates of tubulin gene were not changed. These results indicated that MeHg causes not only the specific disruption of microtubule structure but also the successive inhibition of tubulin synthesis. The inhibition of protein synthesis by MeHg has been proposed as one of the targets of its toxicity, as described above. Kasama et al. (1989) noticed that tubulin protein was one of the proteins with decreased levels among 200 protein species examined in dorsal root ganglia from MeHg-treated rats.

Of major importance has been the identification of the chemical species of mercury which actually cause cell damage. Suda et al. (1991) and Suda and Takahashi (1992) reported that MeHg and ethylmercury were degraded by hydroxyl radical and hypochlorous acid. Furthermore, they showed that phagocytic cells collected from humans, rats, guinea pigs, and rabbits are able to degrade ethylmercury and MeHg in the intraphagosome (Suda et al. 1992). The microsome fraction in rat liver was also found to be able to degrade MeHg and ethylmercury (Suda and Hirayama 1992). However, the contribution of reactive oxygen-producing systems to the in vivo biotransformation of MeHg remains to be clarified.

d) Modifying Factors of Toxicity

Since the first report by Ganther et al. (1972), selenium has been shown to interfere with the metabolism and the toxic effect of MeHg (Who 1976, 1990). Administration of equivalent amounts or lower doses of selenium to rodents decreased toxicity of MeHg. Selenium deficiency enhanced the fetal toxicity of MeHg (Nishikido et al. 1987). The mechanisms of these protective effects of selenium are still unknown. It has been shown that bismethylmercuric selenide (BMS) is formed from selenium and MeHg in the blood (NAGANUMA and IMURA 1980b). BMS penetrates the blood-brain barrier more easily than MeHg alone (NAGANUMA et al. 1980). However, the contribution of BMS to the protective effects of selenium on MeHg toxicity is still unclear.

CHANG et al. (1978) and WELSH (1979) reported that vitamin E protects MeHg toxicity in rodents. According to GANTHER (1978), vitamin E modifies MeHg metabolism by acting as a radical scavenger.

C. Inorganic Mercury

Mercuric mercury and elemental mercury are known to be toxicologically important delate inorganic mercurials being delate poisonous to humans.

I. Mercuric Mercury

1. Tissue Accumulation and Excretion

a) Renal Uptake and Retention

Of all the organs, the kidney is the primary location for the accumulation of mercuric mercury (Hg^{2+}). It has been shown in the mouse that as much as 50% or more of the administered dose of Hg^{2+} accumulates in the kidneys within 1 h after i.v. injection of $HgCl_2$ (NAGANUMA and IMURA 1984a). In the kidneys of rat (MIURA et al. 1981; ZALUPS 1991), mouse (HULTMAN and ENESTROM 1986; HULTMAN et al. 1985), and rabbit (ZALUPS and BARFUSS 1990), Hg^{2+} preferentially accumulates in proximal tubule cells after administration of $HgCl_2$. The pars recta of the proximal tubules (S3 segment), which is the part of the kidneys most sensitive to Hg^{2+} toxicity, is the primary site for the accumulation of Hg^{2+} in the rat kidney (ZALUPS 1991). Electron microscopic observation of proximal tubular cells of rat kidney after administration of $HgCl_2$ indicated a specific deposition of Hg in the lysosomes (MIURA et al. 1981). ZHAO and WANG (1988) reported that the lysosome was the most sensitive site of the tubular ultramicrostructures to Hg^{2+}. For evaluation of Hg-induced renal injury in humans the determination of urinary intestinal-type alkaline phosphatase (IAP) activity can be useful (NUYTS et al. 1992a), because IAP is a specific and sensitive marker for pathogenic alterations in the S3 segment of the human proximal tubule (NUYTS et al. 1992b).

CHAN et al. (1992) have examined the relative renal toxicity of Hg^{2+} in rats when Hg^{2+} is administered as either $HgCl_2$ or the complex with metallothionein (MT). They observed that the injury caused by Hg-MT localized mainly in the terminal portions of the proximal convoluted tubule and the initial portions of the proximal straight tubule, whereas $HgCl_2$ caused necrosis in pars recta (S3 segments) of the proximal tubule as described above. This experimental result may indicate the difference in renal uptake mechanism of mercury administered as $HgCl_2$ and Hg-MT.

After administration of $HgCl_2$, Hg^{2+} may be transported into the kidneys as a complex with GSH (GSHgSG), which is filtered through the glomerulus and taken up by the kidney cells after degradation of its GSH moiety by γ-GTP as in the case of MeHg (see Sect. B, above) (TANAKA et al. 1990). Indeed, the treatment of rats (BERNDT et al. 1985) or mice (TANAKA et al. 1990) with γ-GTP inhibitor (acivicin) prior to $HgCl_2$ injection resulted in a marked increase in the urinary excretion of both GSH and Hg^{2+}, and a decrease in renal uptake of Hg^{2+} (TANAKA et al. 1990). A specific depletion of hepatic GSH, which is a major source of plasma GSH, prior to $HgCl_2$ injection also substantially reduced renal uptake of Hg^{2+} and consequently reduced the renal damage (TANAKA et al. 1990). Hg^{2+} taken up by the kidney cells induces MT synthesis and binds to this cysteine-rich protein. The Hg^{2+} bound to MT is hardly released into the lumen of tubules and retained in the kidney, whereas non-MT bound Hg^{2+} is easily released.

b) Biliary and Urinary Excretion

The rate of biliary excretion of Hg^{2+} is very low in comparison with methylmercury excretion (BALLATORI and CLARKSON 1984b). The form of Hg^{2+} excreted into the bile has been identified as a complex with GSH (GSHgSG) (BALLATORI and CLARKSON 1984b). Inhibition of biliary secretion of GSH by administration of sulfobromophthalein (BSP) resulted in a parallel inhibition of Hg^{2+} secretion (BALLATORI and CLARKSON 1984a). Thus, the biliary secretion of Hg^{2+} may be in large part dependent on the biliary transport of GSH. The mechanism of urinary excretion of Hg^{2+} has not yet been clarified. As described above, Hg^{2+} is incorporated into the kidneys by a γ-GTP-dependent system, and inhibition of γ-GTP significantly increases urinary excretion of Hg^{2+} (TANAKA et al. 1990). Therefore, GSHgSG which has escaped from the action of γ-GTP may be excreted into the urine.

2. Toxicity

Since mercuric mercury mainly accumulates in the kidney, Hg^{2+} causes severe nephrophathy, characterized by extensive loss of proximal tubule function and viability (GANOTE et al. 1975). The renal toxicity of Hg^{2+} can be estimated by increases in urinary levels of biomarker enzymes, such as lactate dehydrogenase, alkaline phosphatase, aspartate aminotransferase, and γ-glutamyltranspeptidase. Among these marker enzymes, alkaline phosphatase and aspartate aminotransferase are the most sensitive indicators, and lactate dehydrogenase is the most responsive marker to renal mercury toxicity (DIETER et al. 1992).

Glomerular injury is also observed in the kidney of rat after a single high-dose administration of $HgCl_2$. However, the magnitude of the injury observed in the glomerulus is less than that in the tubules (GRITZKA and TRUMP 1968; MIURA et al. 1981). Although exposure to a single- and high-dose of Hg^{2+} may directly injure the kidneys, chronic low-dose exposure to Hg^{2+} induces an immunologic glomerular disease (ROMAN-FRANCO et al. 1978).

3. Mechanism of Renal Toxicity

a) Binding to Thiols

Since Hg^{2+} has a high affinity to thiol groups, it strongly inhibits activities of SH enzymes in vitro (WAKU and NAKAZAWA 1979). Hg^{2+} may bind to thiol groups or to other moieties of essential cellular macromolecules and inhibit their biological activities. Hg^{2+} may also bind to GSH in tissues and form GSHgSG. Treatment of animals with an inhibitor of GSH synthesis or GSH depletors markedly sensitized mice to Hg^{2+} toxicity (BERNDT et al. 1985; NAGANUMA et al. 1990), suggesting that cellular GSH is a major determinant of Hg^{2+} toxicity. A decrease in cellular GSH concentration may increase the population of active mercury, which can easily react with essential cellular macromolecules. An attempt to increase cellular GSH concentrations by preadministration of GSH monoester significantly prevented the renal toxicity of Hg^{2+} (HOUSER et al. 1992; NAGANUMA et al. 1990).

b) Effect on Membrane Function

The membrane of the kidney cells is the first target of Hg^{2+} toxicity. VAN-HOEK et al. (1990) have shown that Hg^{2+} increased NaCl and urea permeability and inhibited water permeability of rat brush border membrane vesicles. Since Hg^{2+} did not affect the permeability of pure phospholipid membrane vesicles, they concluded that the constituents in pathways for water and solutes which were influenced by Hg^{2+} must have been membrane proteins. NAKADA et al. (1978) have shown that the permeability of liposome to glucose was significantly increased by a very low concentration ($10^{-7}\,M$) of Hg^{2+}, suggesting that lipid in biomembrane could be one of the primary targets of Hg^{2+}. KARNISKI (1992) has demonstrated that Hg^{2+} is a potent mediator of Cl^-/OH^- exchange across both liposomes and brush border membrane vesicles. Another study which examined the effect of Hg^{2+} on the thermotropic properties of both liposome and rat red cell membranes has shown that, besides protein thiol groups, Hg^{2+}-lipid-binding sites play an important role in the interaction of Hg^{2+} with red cell membrane (DELNOMDEDIEU and ALLIS 1993).

c) Calcium Homeostasis

In rats given $HgCl_2$, an increase in renal Ca^{2+} concentration was observed (GRITZKA and TRUMP 1968; WEINBERG et al. 1989) and the increase in renal

Ca^{2+} concentrations was significantly associated with the increase in blood urea nitrogen level and the extent of lethal cell injury as determined morphologically (WEINBERG et al. 1989). In a study using renal proximal tubular cells in vitro, the magnitude of Hg^{2+}-induced cytotoxicity was also correlated with the elevation of cellular Ca^{2+} concentrations (AMBUDKAR et al. 1988). Both the elevations of cellular Ca^{2+} and the cytotoxicity induced by Hg^{2+} (10–50 mM) could not be observed in the cells incubated with Ca^{2+} free medium, indicating a possibility that $HgCl_2$-induced renal cell damage involves the entry of Ca^{2+} from the extracellular milieu which potentiates the progression of cellular injury (AMBUDKAR et al. 1988). JUNG and ENDOU (1990) have found that Hg^{2+} at a low concentration (0.1–10 nM) stimulates angiotensin II-induced elevation of cytosolic Ca^{2+} concentration in an isolated early proximal tubule. Since the stimulatory effect of Hg^{2+} was completely inhibited by pretreating with an inhibitor of phospholipase C (PLC), the authors suggested that Hg^{2+} might stimulate the Ca^{2+} increment induced by angiotensin II through an activation of PLC.

A stimulatory effect of Hg^{2+} on release of mitochondrial Ca^{2+} has also been observed in rat kidney mitochondria (CHAVEZ and HOLGUIM 1988). Chavez and his collaborators concluded that the Hg^{2+}-induced Ca^{2+} release may be due to the modification of sulfhydryl groups of mitochondrial membrane proteins (CHAVEZ and HOLGUIN 1988; CHAVEZ et al. 1989, 1991).

d) Lipid Peroxidation

Many investigators have reported that Hg^{2+} administration induces lipid peroxidation in the kidney (BENOV et al. 1990; GSTRAUNTHALER et al. 1983; RUNGBY and ERNST 1992). The Hg^{2+}-induced lipid peroxidation is usually accompanied by a decrease in renal GSH concentration (GSTRAUNTHALER et al. 1983; KEE and SIN 1992; NIELSEN et al. 1991; SIN and TEH 1992; ZALUPS and LASH 1990). LUND et al. (1991) demonstrated that Hg^{2+} depleted mitchondrial GSH and enhanced H_2O_2 formation in the kidney mitochondria under conditions of impaired respiratory chain electron transport. MILLER et al. (1991) also reported that Hg^{2+} enhanced the rate of O_2^--dismutation, leading to increased production of H_2O_2 by renal mitochondria. Recently, Woods and his collaborators have found a stimulative effect of Hg^{2+} on in vitro oxidation of porphyrinogen in the presence of GSH and H_2O_2 (WOODS et al. 1990b), and examined the mechanism of the Hg-induced porphyrinogen oxidation. They suggested that the binding of Hg^{2+} to GSH formed a reactive Hg^{2+}-thiol complex which interacted with H_2O_2 to promote the oxidation of porphyrinogens and possibly other biomolecules (WOODS et al. 1990a,b). In fact, the Hg^{2+} complex with GSH (GSHgSG) catalyzed the decomposition of H_2O_2, and the reaction of GSHgSG with H_2O_2 resulted in high-performance lipid chromatography (HPLC)-detectable products distinct from GSH, GSSG, or GSHgSG (MILLER and WOODS 1993).

e) Autoimmune Reaction

In several species, such as humans, rat, mouse, and rabbit, Hg^{2+} has been known to induce immunologic glomerular disease. The mechanism of induction of the mercury-induced autoimmune disease has been studied extensively in Brown Norway (BN) rats. BN rats treated with Hg^{2+} develop autoantibodies to components of renal basement membranes (such as anti-laminin, anti-collagen IV, anti-nuclear) and an autoimmune glomerulonephritis (as reviewed by DRUET 1991). This autoimmune glomerulonephritis is characterized by IgG deposits along the glomerular basement membrane at an early stage and then by the presence of granular IgG deposits typical of membranous glomerulopathy (DRUET et al. 1978). T cells obtained from Hg^{2+}-injected BN rats recognize normal B cells and are able to transfer the autoimmune disease into normal BN rats (PELLETIER et al. 1988), proving that this disease may be due to a T-cell-dependent polyclonal B cell activation (HIRSCH et al. 1982; PELLETIER et al. 1987), which is probably mediated by the appearance of autoreactive anti-self class II T cells (PELLETIER et al. 1988). Strain specificity in the mercury-induced autoimmune reaction in rats has been reported (DRUET et al. 1977). The genetic background underlying this strain specificity has also been elucidated (DRUET et al. 1982; SAPIN et al. 1982, 1984).

f) Modifying Factors of Toxicity

The lethal toxicity of Hg^{2+} was significantly prevented by preinduction of MT synthesis (YAMANE et al. 1977). Renal MT may bind to Hg^{2+} and suppress its toxic action in the kidney cells. GSH may also be a major determinant of Hg^{2+} toxicity. Depletion of GSH by treatment of mice with an inhibitor of GSH synthesis markedly enhanced (about tenfold) the lethal and renal toxicity of $HgCl_2$ (NAGANUMA et al. 1990).

Selenium (Se), an essential trace element, has a protective effect against Hg^{2+} toxicity. Coadministration of a selenium compound with Hg^{2+} resulted in remarkable depression of acute toxicity of Hg^{2+} through the formation of stable complexes of Hg and Se in various tissues such as liver and kidney (NAGANUMA and IMURA 1980a, 1981, 1984a).

II. Elemental Mercury

1. Exposure to Elemental Mercury

Elemental mercury (Hg^0) exists in the environment as liquid metallic mercury or mercury vapor. In the general population, the major source of exposure to mercury vapor is dental amalgams (WHO 1991). Most conventional silver amalgams contain 50% mercury by weight. Mercury vapor is continuously released from dental amalgam fillings and the release rate is enhanced by chewing or tooth brushing.

Occupational exposure to mercury vapor occurs in a variety of industries such as chloralkali plants, thermometer factories, gold extraction process, and mercury mining. Dentists and dental assistants are also exposed to mercury vapor during insertion, polishing, and removal of amalgam fillings.

2. Metabolism

a) Absorption

Mercury vapor is well absorbed through inhalation (about 80%), while liquid mercury is poorly absorbed by the gastrointestinal tract (about 0.01%). Due to its high lipophilicity and diffusibility, mercury vapor is readily transferred across the cell membranes. When mercury vapor is inhaled, it efficiently enters the blood stream through the alveolar membrane and is quickly distributed between the plasma and red blood cells (RBCs). Little is known about the efficiency of elemental mercury absorption through the skin.

b) Distribution and Disposition

In blood, dissolved mercury vapor enters the RBCs and oxidizes to the divalent ionic form (Hg^{2+}). The catalase-H_2O_2 complex (compound I) is responsible for this oxidation (see below). Oxidized mercury readily reacts with SH groups of protein or glutathione.

Generally similar tissue distribution patterns of mercury after mercury vapor exposure and after intravenous inorganic mercury injection have been reported (WHO 1991). However, two important exceptions are brain and fetus, where a markedly higher amount of mercury was found in mercury vapor-exposed animals than in inorganic mercury-injected animals. These observations indicate that: (a) the oxidation of mercury vapor in RBCs is not sufficient to change all the mercury vapor into ionic mercury, (b) part of the mercury physically dissolved as an elemental form persists in the blood stream for a sufficiently long time to reach the brain and placenta, and (c) this form of mercury readily crosses the blood-brain and placental barriers. Based on the data from human volunteers (HURSH et al. 1976) and from an in vitro study, HURSH et al. (1988) indicated that almost all (97%) of inhaled mercury vapor enters the brain as the unoxidized form. On entering the tissues including the brain, mercury vapor is considered to be oxidized and accumulated.

Until recently, little information has been available concerning detailed distribution of mercury in the brain after mercury vapor exposure. However, a recent improvement in the staining method permitted the detection of mercury deposits in the brains of animals exposed to rather lower concentrations of mercury vapor (MOLLER-MADSEN 1992; WARFVINGE et al. 1992). Mercury was found primarily in the neocortex, thalamus, pons, ependyma, choroid plexuses, cerebellar Purkinje cells, and vessel walls in the brain of rats exposed to mercury vapor.

Distribution of mercury in fetal and neonatal tissues has been extensively investigated in guinea pigs exposed to mercury vapor in utero (YOSHIDA et al. 1987, 1990). Though mercury vapor easily crosses the placental barrier, mercury concentration in the fetal or neonatal brain is lower than that in the maternal brain after in utero exposure. On the contrary, much higher concentrations of mercury were found in the fetal liver than in the maternal liver. Most of the mercury in the fetal liver was found in the fraction of MT in the cytosol. The hepatic concentration of MT is known to increase drastically in the late gestational and early neonatal period. It seems likely that mercury vapor which entered the fetus is readily oxidized in the fetal liver, then trapped by the abundant MT, thus resulting in a lower distribution of mercury to the other fetal tissues.

Acatalasemic mice have very low activity of catalase in the RBCs. The influence of oxidation of mercury in RBCs on the fate of mercury has been well been investigated using these animals. Acatalasemic mice exposed to mercury vapor showed a reduced oxidation of mercury vapor in the RBCs and a higher concentration of mercury in tissues than normal mice (OGATA et al. 1985). The amount of mercury released from the RBCs to plasma in vitro was larger in acatalasemic mice than in normal mice (YAMAMOTO et al. 1992), suggesting that a substantial amount of elemental mercury which escaped from the oxidation in the RBCs is released into the plasma and then distributed to other tissues in acatalasemic mice.

Ethanol can competitively inhibit the oxidation of mercury vapor by catalase. Pretreatment with ethanol reduces mercury deposition in the tissues in mercury vapor-exposed animals. The differences in tissue distribution of mercury after mercury vapor exposure between ethanol-treated animals and acatalasemic mice may be ascribed to the fact that acatalasemic mice lack catalase activity only in the RBCs.

c) Excretion

After exposure to mercury vapor, mercury is eliminated from the body mainly via urine and feces. However, exhalation of a small quantity of mercury is observed even several days after short-term exposure to mercury vapor. Reduction of once deposited inorganic mercury to elemental mercury contributes to this exhalation (see below).

3. Biotransformation

a) Oxidation of Mercury Vapor

An earlier study by CLARKSON et al. (1961) had shown that RBCs have the ability to rapidly convert mercury vapor to mercuric ion. NIELSEN-KUDSK (1969) found that the oxidation of mercury vapor was inhibited by ethanol. Ethanol can be oxidized by peroxidatic reaction by catalase-H_2O_2 complex (compound I). The inhibitory effect of aminotriazole, a known inhibitor on the reaction of compound I and H_2O_2 (Eq. 3, below), on mercury vapor oxidation was also observed (MAGOS et al. 1974). Catalase in its compound I state is characterized by its unique ability to transfer two electrons in a single step, whereby the redox state of Fe in the heme moiety cycles between +III and +V. According to these observations, it is now considered that elemental mercury (Hg^0) is converted to mercuric ion (Hg^{2+}) primarily by the catalase-compound I system, as shown in the following equations. Reactions 1 and 3 represent the decomposition pathway of H_2O_2 by catalase.

$$\text{Catalase} + H_2O_2 \rightarrow \text{Compound I} + H_2O \quad (1)$$
$$\text{Compound I} + Hg^0 \rightarrow \text{Catalase} + Hg^{2+} \quad (2)$$
$$\text{Compound I} + H_2O_2 \rightarrow \text{Catalase} + O_2 + H_2O \quad (3)$$

Participation of catalase in the oxidation of mercury vapor was also confirmed indirectly by the low rate of oxidation in blood from acatalasemic mice and farm ducks, both of which are deficient in catalase (HALBACH 1991). The role of glutathione peroxidase in the oxidation of mercury vapor in RBCs seems to be negligible, although this enzyme may be able to manipulate the rate of oxidation by modifying the supply of H_2O_2 in RBCs.

b) Reduction of Mercury to Elemental Mercury

In a human volunteer experiment, exhalation of mercury vapor was observed up to several days after a single inhalation exposure to mercury vapor. Unexpected detection of volatilized mercury in exhaled air posed the question as to whether this form of mercury is derived from the unoxidized mercury vapor circulating in the blood stream for several days or whether it is derived from the reduction of once-deposited ionic mercury (Hg^{2+}) to elemental mercury (Hg^0).

Possible reduction of inorganic mercury to elemental mercury in mammals was first noted by CLARKSON et al. (1964), who showed evidence of mercury volatilization in rats given mercuric chloride. DUNN et al. (1981a,b) found that an oral dose of ethanol after injection of mercuric mercury dramatically increased the exhalation of mercury in mice. Ethanol elevated the mercury exhalation in a dose-dependent manner. Aminotriazole given to mercuric chloride-injected mice also increased the exhalation of mercury (SUGATA and CLARKSON 1979).

In an in vitro experiment with liver homogenate (DUNN et al. 1981b), mercury volatilization activity was enhanced by the addition of ethanol but disappeared on heating. The highest mercury volatilization activity was found in the cytosol fraction. However, the actual entity responsible for the reduction of mercury to its volatile form is still unclear. OGATA et al. (1987) reported a higher rate of mercury volatilization in vitro in the presence of a superoxide anion-producing system such as xanthine/xanthine oxidase.

These observations indicate that inorganic mercury deposited in the tissue might be more labile than generally recognized, although even enhanced volatilization of mercury by ethanol had minor effects on the tissue distribution of mercury. Elevation of mercury volatilization by inhibitors of catalase such as ethanol and aminotriazole suggests a cycle of mercury reduction and reoxidation. It is conceivable that a small quantity of mercury vapor is generated continuously by an unknown reduction system in the tissue and is rapidly reoxidized by the catalase-compound I system, which can be inhibited by ethanol. A higher rate of mercury exhalation was observed in acatalasemic mice than in normal mice after injection of mercuric chloride (OGATA et al. 1987), supporting the hypothesis that the ability to reoxidize mercury vapor determines the rate of mercury exhalation.

4. Toxicity

Toxicity of elemental mercury has been reviewed (WHO 1991). Symptoms after acute exposure to mercury vapor include chest pains, dyspnea, coughing, hemoptysis, impairment of pulmonary function, and interstitial pneumonitis leading to death. Subacute exposure to mercury vapor causes psychic reactions characterized by delirium, hallucinations, and suicidal tendencies.

Chronic exposure to mercury vapor occurs in the workplace. The classic triad of symptoms involves the oral cavity (gingivitis, salivation, stomatitis), tremor, and psychological changes. Central nervous system symptoms observed in workers exposed to mercury vapor are termed "erethism," which includes insomnia, loss of appetite, pathological shyness, emotional lability, irritability, and memory loss. Renal dysfunction as determined by urinalysis has been observed in workers exposed to mercury vapor. Both glomerular (proteinuria) and tubular (enhanced excretion of tubular enzymes) effects have been reported.

5. Mechanism of Toxicity

Both the central nervous system and kidney are the primary target organs for mercury vapor intoxication.

As mentioned in Sect. C.II.2, above, inhaled mercury vapor persists in the blood stream sufficiently long to reach the blood-brain barrier. Due to its high lipophilicity, mercury vapor rapidly diffuses across the blood-brain barrier, where it is oxidized to ionic mercury by the catalase-compound I system. Oxidized mercury is easily subjected to the reaction with SH-containing ligand in the central nervous system; thus fixation of mercury occurs. Little is known concerning the precise causal relationship between the sites of mercury deposition in the brain and the characteristic neurotoxicity of mercury vapor. It is conceivable that essentially the same mechanism of toxicity as that of ionic mercury is operating at molecular level.

The renal effects of mercury vapor are also considered to result from the conversion of mercury to ionic mercury by the oxidation of elemental mercury.

References

Abbott W, Meister A (1986) Intrahepatic transport and utilization of biliary glutathione and its metabolites. Proc Natl Acad Sci USA 83:1246–1250
Abe T, Haga T, Kurokawa M (1975) Blockage of axoplasmic transport and depolymerization of reassembled microtubules by methylmercury. Brain Res 86:504–508
Alexander J, Aaseth J (1982) Organ distribution and cellular uptake of methylmercury in the rat as influenced by the intra- and extracellular glutathione concentration. Biochem Pharmacol 31:685–690
Ambudkar IS, Smith MW, Phelps PC, Regec AL, Trump BF (1988) Extracellular Ca^{2+}-dependent elevation in cytosolic Ca^{2+} potentiates $HgCl_2$-induced renal proximal tubular cell damage. Toxicol Ind Health 4:107–123
Aschner M, Clarkson TW (1988) Uptake of methylmercury in the rat brain: effects of amino acids. Brain Res 462:32–39
Aschner M, Clarkson TW (1989) Methylmercury uptake across bovine brain capillary endothelial cells in vitro: the role of amino acid. Pharmacol Toxicol 64:293–297
Atchison WD, Narahashi T (1982) Methylmercury-induced depression of neuromuscular transmission in the rat. Neurotoxicology 3:37–50
Atchison WD, Clark AW, Narahashi T (1984) Presynaptic effects of methylmercury at the mammalian neuromuscular junction. In: Narahashi T (ed) Cell and mol neurotoxicol. Raven, New York, pp 23–43
Ballatori N, Clarkson TW (1983) Biliary transport of glutathione and methylmercury. Am J Physiol 244:G435–G441
Ballatori N, Clarkson TM (1984a) Dependence of biliary secretion of inorganic mercury on the biliary transport of glutathione. Biochem Pharmacol 33:1093–1098
Ballatori N, Clarkson TM (1984b) Inorganic mercury secretion into bile as a low molecular weight complex. Biochem Pharmacol 33:1087–1092
Ballatori N, Clarkson TW (1985) Biliary secretion of glutathione and of glutathione-metal complex. Fundam Appl Toxicol 5:816–831
Benov LC, Benchev IC, Monovich OH (1990) Thiol antidotes effect on lipid peroxidation in mercury-poisoned rats. Chem Biol Interact 76:321–332
Berndt WO, Baggett JM, Blacker A, Houser M (1985) Renal glutathione and mercury uptake by kidney. Fundam Appl Toxicol 5:832–839
Bondy SC, McKee M (1990) Prevention of chemically induced synaptosomal changes. J Neurosci Res 25:229–235
Brown D, Ruehl K, Bormann S, Little J (1988) Effects of methylmercury on the microtubule system of mouse lymphocytes. Toxicol Appl Pharmacol 94:66–75
Carmichael N, Cavanagh JB, Rodda RA (1975) Some effects of methylmercury salts on the rabbit nervous system. Acta Neuropathol (Berl) 32:112–125
Chan HM, Satoh M, Zalups RK, Cherian MG (1992) Exogenous metallothionein and renal toxicity of cadmium and mercury in rats. Toxicology 76:15–26
Chang LW, Gilbert M, Sprechler J (1978) Modification of methylmercury neurotoxicity by vitamin E. Environ Res 17:356–366
Chavez E, Holguin JA (1988) Mitochondrial calcium release as induced by Hg^{2+}. J Biol Chem 263:3582–3587
Chavez E, Zazueta C, Diaz E, Holguin JA (1989) Characterization by Hg^{2+} of two different pathways for mitochondrial Ca^{2+} release. Biochim Biophys Acta 986: 27–32
Chavez E, Zazueta C, Osornio A, Holguin JA, Miranda ME (1991) Protective behavior of captopril on $Hg(^{++})$-induced toxicity on kidney mitochondria. In vivo and in vitro experiments. J Pharmacol Exp Ther 256:385–390
Cheung MK, Verity MA (1985) Experimental methylmercury neurotoxicity: locus of mercurial inhibition of brain protein synthesis in vivo and in vitro. J Neurochem 44:1799–1808

Choi BH, Lapham LW, Amin-Zaki L, Saleem T (1978) Abnormal neuronal migration, deranged cerebral cortical organization and diffuse white matter astrocytosis of human fetal brain. A major effect of methylmercury poisoning in utero. J Neuropathol Exp Neurol 37:719–733

Choi BH, Kudo M, Lapham LW (1981) A golgi and electronmicroscopic study of cerebellum in methylmercury poisoned neonatal mice. Acta Neuropathol (Berl) 54:233–237

Clarkson TW, Rothstein A (1964) The excretion of volatile mercury by rats injected with mercuric salts. Health Phys 10:1115–1121

Clarkson TW, Gatzy J, Dalton C (1961) Studies on the equilibration of mercury vapor with blood. University of Rochester, Report UR-582

Clarkson TW, Small H, Norseth T (1973) Excretion and absorption of methyl mercury after polythiol resin treatment. Arch Environ Health 26:173–176

Delnomdedieu M, Allis JW (1993) Interaction of inorganic mercury salts with model and red cell membranes: importance of lipid binding sites. Chem Biol Interact 88:71–87

Dieter MP, Boorman GA, Jameson CW, Eustis SL, Uraih LC (1992) Development of renal toxicity in F344 rats gavaged with mercuric chloride for 2 weeks, or 2,4,6,15, and 24 months. J Toxicol Environ Health 36:319–340

Druet P (1991) Effect of inorganic mercury on the immune system. In: Suzuki T, Imura N, Clarkson TW (eds) Advances in mercury toxicology. Plenum, New York, pp 395–409

Druet E, Sapin C, Gunther E, Feingold N, Druet P (1977) Mercuric chloride induced anti-glomerular basement membrane antibodies in the rat. Eur J Immunol 7:348–351

Druet P, Druet E, Potdevin F, Sapin C (1978) Immune type glomerulonephritis induced by $HgCl_2$ in the Brown Norway rat. Ann Immunol 129C:777–792

Druet E, Sapin C, Fourinie G, Mandet C, Gunther E, Druet P (1982) Genetic control of susceptibility to mercury-induced immune nephritis in various strains of rat. Clin Immunol Immunopathol 25:203–212

Dunn JD, Clarkson TW, Magos L (1981a) Interaction of ethanol and inorganic mercury: generation of mercury vapor in vivo. J Pharmacol Exp Ther 216:19–23

Dunn JD, Clarkson TW, Magos L (1981b) Ethanol reveals novel mercury detoxification step in tissues. Science 213:1123–1125

Fair PH, Balthrop JE, Wade JL, Braddon-Galloway S (1987) In vivo incorporation of ^{14}C-leucine into brain protein of mice treated with methylmercury and thiol complexes of methylmercury. Toxicol Appl Pharmacol 36:213–220

Ganote CE, Reimer KA, Jennings RB (1975) Acute mercuric chloride nephrotoxicity: an electron microscopic and metabolic study. Lab Invest 31:633–647

Ganther HE (1978) Modification of methylmercury toxicity and metabolism by selenium and vitamin E: possible mechanisms. Environ Health Perspect 25:71–76

Ganther HE, Goudie C, Sunde ML, Kopecky MJ, Wager P, Oh SH, Hoekstra WG (1972) Selenium: relation to decreased toxicity of methylmercury added to diets containing tuna. Science 175:1122–1124

Goldstein GW, Betz AL (1986) The blood-brain barrier. Sci Am 255:74–83

Gritzka TL, Trump BF (1968) Renal tubular lesions caused by mercuric chloride. Am J Pathol 52:1225–1277

Gstraunthaler G, Pfaller W, Kotanko P (1983) Glutathione depletion and in vitro lipid peroxidation in mercury or maleate induced acute renal failure. Biochem Pharmacol 32:2969–2972

Halbach S (1990) Mercury compounds: lipophilicity and toxic effects on isolated myocardial tissue, Arch toxicol 64:315–319

Halbach S (1991) Mercury vapor uptake and oxidoreductases in erythrocytes. In: Suzuki T, Imura N, Clarkson TW (eds) Advances in mercury toxicology. Plenum, New York, pp 167–180

Halliwell B, Gutteridge JMC (1985) Oxygen free radicals and the nervous system. Trends Neurosci 8:22–26

Hammond PB, Beliles RP (1980) Metals. In: Casarett LJ, Doull J (eds) Toxicology — the basic science of poisons. Macmillan, New York, pp 409–467

Hasegawa K, Omata S, Sugano H (1988) In vivo and in vitro effects of methylmercury on the activities of aminoacyl-tRNA synthetases in rat brain. Arch Toxicol 62:470–472

Hirsch F, Couderc J, Sapin C, Fournie G, Druet P (1982) Polyclonal effect of $HgCl_2$ in the rat, its possible role in an experimental autoimmune disease. Eur J Immunol 12:620–625

Hirayama K (1980) Effect of amino acids on brain uptake of methylmercury. Toxicol Appl Pharmacol 55:318–323

Hirayama K (1985) Effects of combined administration of thiol compounds and methylmercury chloride on mercury distribution in rats. Biochem Pharmacol 34:2030–2032

Houser MT, Milner LS, Kolbeck PC, Wei SH, Stohs SJ (1992) Glutathione monoethyl ester moderates mercuric chloride-induced acute renal failure. Nephron 61:449–455

Hultman P, Enestrom S (1986) Localization of mercury in the kidney during experimental acute tubular necrosis studied by the cytochemical silver amplification method. Br J Exp Pathol 67:493–503

Hultman P, Enestrom S, Von Schenck H (1985) Renal handling of inorganic mercury in mice: the early excretion phase following single intravenous injection of mercuric chloride studied by silver amplification methods. Virchows Arch [B] 49:209–224

Hursh JB, Clarkson TW, Cherian MG, Vostal JJ, Mallie RV (1976) Clearance of mercury (Hg-197, Hg-203) vapor inhaled by human subjects. Arch Environ Health 31:302–309

Hursh JB, Sichak SP, Clarkson TW (1988) In vitro oxidation of mercury by the blood. Pharmacol Toxicol 63:266–273

Imura N, Miura K, Inokawa M, Nakada S (1980) Mechanism of methylmercury cytotoxicity: by biochemical and morphological experiments using cultured cells. Toxicology 17:241–254

Jacobs AJ, Maniscalco WM, Finkelstein JN (1986) Effects of methylmercuric chloride, cycloheximide, and colchicine on the reaggregation of dissociated mouse cerebellar cells. Toxicol Appl Pharmacol 86:362–3871

Jung KY, Endou H (1990) Mercury chloride as a possible phospholipase C activator: effect on angiotensin II-induced $[Ca^{++}]i$ transient in the rat early proximal tubule. Biochem Biophys Res Commun 173:606–613

Karniski LP (1992) Hg^{2+} and Ca^{2+} are ionophores, mediating Cl^-/OH^- exchange in liposomes and rabbit renal brush border membranes. J Biol Chem 267:19218–19225

Kasama H, Itoh K, Omata S, Sugano H (1989) Differential effects of methylmercury on the synthesis of protein species in dorsal root ganglia of the rat. Arch Toxicol 63:226–230

Kasuya M (1975) The effect of vitamin E on the toxicity of alkyl mercurials on nervous tissue in culture. Toxicol Appl Pharmacol 32:347–354

Kee KN, Sin YM (1992) Effect of mercury on tissue glutathione following intrarenal injection of mercuric chloride. Bull Environ Contam Toxicol 48:509–514

Kerper LE, Ballatori N, Clarkson TW (1992) Methylmercury transport across the blood-brain barrier by an amino acid carrier. Am J Physiol 262 R761–R765

Kleinschuster SJ, Yoneyama M, Sharma RP (1983) A cell aggregation model for the protective effect of selenium and vitamin E on methylmercury toxicity. Toxicology 26:1–9

Lakowicz JR, Anderson CJ (1980) Permeability of lipid bilayers to methylmercuric chloride: quantification by fluorescence quenching of a carbazole-labeled phospholipid. Chem Biol Interact 30:309–323

LeBel CP, Bondy SC (1990) Sensitive and rapid quantitation of reactive oxygen species in rat synaptosomes. Neurochem Int 17:435–440

LeBel CP, Ali SF, McKee M, Bondy SC (1990) Organometal induced increases in oxygen reactive species: the potential of 2′,7′-dichlorofluorescin diacetate as an index of neurotoxic damage. Toxicol Appl Pharmacol 104:17–34

Levesque P, Atchison W (1991) Disruption of brain mitochondrial calcium sequestration by methylmercury. J Pharmacol Exp Ther 256:236–256

Lund BO, Miller DM, Woods JS (1991) Mercury-induced H_2O_2 production and lipid peroxidation in vitro in rat kidney mitochondria. Biochem Pharmacol 11:S181–S187

Magos L, Sugata Y, Clarkson TW (1974) Effects of 3-amino-1,2,3-triazole on mercury uptake by in vitro human blood samples and by whole rats. Toxicol Appl Pharmacol 28:367–373

Matsumoto H, Koya G, Takeuchi T (1965) Fetal Minamata disease. J Neuropathol Exp Neurol 24:563–574

Miller DM, Woods JS (1993) Redox activity of mercury-thiol complexes: implications for mercury-induced porphyria and toxicity. Chem Biol Interact 88:23–35

Miller DM, Lund BO, Woods JS (1991) Reactivity of Hg(II) with superoxide: evidence for the catalytic dismutation of superoxide by Hg(II). J Biochem Toxicol 6:293–298

Minnema DJ, Cooper GP, Greenland RD (1989) Effects of methylmercury on neurotransmitter release from rat brain synaptosomes. Toxicol Appl Pharmacol 99:510–521

Miura K, Clarkson TW (1993) Reduced methylmercury accumulation in a methylmercury-resistant rat pheochromocytoma PC12 cell line. Toxicol Appl Pharmacol 118:39–45

Miura K, Imura N (1987) Mechanism of methylmercury cytotoxicity. CRC Crit Rev Toxicol 18:161–188

Miura K, Imura N (1989) Mechanism of cytotoxicity of methylmercury, with special reference to microtubule disruption. Biol Trace Elem Res 21:313–315

Miura K, Suzuki K, Imura N (1978) Effects of methylmercury on mitotic mouse glioma cells. Environ Res 17:453–471

Miura K, Mori R, Imura N (1981) Effects of selenium on mercury-induced renal lesions and on subcellular mercury distribution. Ecotoxicol Environ Safety 5: 351–367

Miura K, Inokawa M, Imura N (1984) Effects of methylmercury and some metal ions on microtubule networks in mouse glioma cells and in vitro tubulin polymerization. Toxicol Appl Pharmacol 73:218–231

Miura K, Clarkson TW, Ikeda K, Naganuma A, Imura N (1994a) Establishment and characterization of methylmercury-resistant PC12 cell line. Environ Health Perspectives (in press)

Miura K, Clarkson TW, Ikeda K, Naganuma A, Imura N (1994b) Cellular factors modulating sensitivity to methylmercury. Jpn J Toxicol Environ Health 40:19

Mohamed MK, Lee WI, Mottet NK, Burbacher TM (1986) Laser light-scattering study of the toxic effects of methylmercury on sperm motility. J Androl 7:11–15

Mohamed MK, Burbacher TM, Mottet NK (1987) Effects of methyl mercury on testicular functions in Macaca fascicularis monkeys. Pharmacol Toxicol 60:29–36

Moller-Madsen B (1992) Localization of mercury in CNS of the rat. V. Inhalation exposure to metallic mercury. Arch Toxicol 66:79–89

Naganuma A, Imura N (1979) Methylmercury binds to a low molecular weight substance in rabbit and human erythrocytes. Toxicol Appl Pharmacol 47:613–616

Naganuma A, Imura N (1980a) Changes in distribution of mercury and selenium in soluble fractions of rabbit tissues after simultaneous administration. Pharmacol Biochem Behav 13:537–544

Naganuma A, Imura N (1980b) Bis(methylmercuric) selenide as a reaction product from methylmercury and selenite in rabbit blood. Res Commun Chem Pathol Pharmacol 27:163–173

Naganuma A, Imura N (1981) Properties of mercury and selenium in a high-molecular weight substance in rabbit tissues formed by simultaneous administration. Pharmacol Biochem Behav 15:449–454

Naganuma A, Imura N (1984a) Effect of time intervals of selenium administration after injection of mercuric chloride on toxicity and renal concentration of mercury in mice. Ind Health 22:91–96

Naganuma A, Imura N (1984b) Species differences in biliary excretion of methylmercury. Biochem Pharmacol 33:679–682

Nakada S, Inoue K, Nojima S, Imura N (1978) Changes in permeability of liposome caused by methylmercury and inorganic mercury. Chem Biol Interact 22:15–23

Naganuma A, Kojima Y, Imura N (1980) Interaction of methylmercury and selenium in mouse: formation and decomposition of bis (methylmercuric)selenide. Res Commun Chem Pathol Pharmacol 30:301–316

Naganuma A, Oda-Urano N, Tanaka T, Imura N (1988) Possible role of hepatic glutathione in transport of methylmercury into mouse kidney. Biochem Pharmacol 37:291–296

Naganuma A, Anderson ME, Meister A (1990) Cellular glutathione as a determinant of sensitivity to mercuric chloride toxicity. Biochem Pharmacol 40:693–697

Nishikido N, Furuyashiki K, Naganuma A, Suzuki T, Imura N (1978) Maternal selenium deficiency enhances the fetolethal toxicity of methylmercury. Toxicol Appl Pharmacol 88:322–328

Nielsen JB, Andersen HR, Andersen O, Starklint H (1991) Mercuric chloride-induced kidney damage in mice: time course and effect of dose. J Toxicol Environ Health 34:469–483

Nielsen-Kudsk F (1969) Factors influencing in vitro uptake of mercury in blood. Acta Pharmacol Toxicol 27:161–172

Norseth T (1973) Biliary excretion and intestinal reabsorption of mercury in the rat after injection of methyl mercuric chloride. Acta Pharmacol Toxicol 33:280–288

Nuyts GD, Roels HA, Verpooten GF, Bernard AM, Lauwerys RR, De-Broe ME (1992a) Intestinal-type alkaline phosphatase in urine as an indicator of mercury induced effects on the S3 segment of the proximal tubule. Nephrol Dial Transplant 7:225–229

Nuyts GD, Verpooten GF, De-Broe ME (1992b) Intestinal alkaline phosphatase as an indicator of effects on the S3-segment of the human proximal tubule. Eur J Clin Chem Clin Biochem 30:713–715

Ogata M, Kenmotsu K, Hirota N, Meguro T, Aikoh H (1985) Mercury uptake in vivo by normal and acatalasemic mice exposed to metallic mercury vapor and injected with metallic mercury ($^{203}Hg^0$) or mercuric chloride ($^{203}HgCl_2$). Arch Environ Health 40:151–154

Ogata M, Kenmotsu K, Hirota N, Meguro T, Aikoh H (1987) Reduction of mercuric ion and exhalation of mercury in acatalasemic and normal mice. Arch Environ Health 42:26–30

Oldendorf WH (1970) Measurement of barium uptake of radiolabeled substances using a tritiated water internal standard. Brain Res 24:372–376

Omata S, Horigome T, Momose Y, Kambayashi M, Mochizuki M, Sugano H (1980) Effects of methylmercury chloride on the in vivo rate of protein synthesis in the brain of the rat. Toxicol Appl Pharmacol 56:207–215

Omata S, Momose Y, Ueki H, Sugano H (1982) In vivo effect on methylmercury on protein synthesis in peripheral nervous tissues of the rat. Arch Toxicol 49:203–215

Peckham NH, Choi BH (1988) Abnormal neuronal distribution within the cerebral cortex after prenatal methylmercury intoxication. Acta Neuropathol (Berl) 76: 222–226

Pelletier L, Pasquier R, Vial MC, Mandet C, Moutier R, Salomon JC, Druet P (1987) Mercury-induced autoimmune glomerulonephritis: requirement for T-cells. Nephrol Dial Transplant 1:211–218

Pelletier L, Pasquier R, Rossert J, Vial MC, Mandet C, Druet P (1988) Autoreactive T cells in mercury-induced autoimmunity. Ability to induce the autoimmune disease. J Immunol 140:750–754

Rabenstein DL, Fairhurst MT (1975) Nuclear magnetic resonance studies of the solution chemistry of metal complexes. J Am Chem Soc 97:2086–2092

Rebel G, Guerin P, Prasad KM (1983) Effect of methylmercuric chloride on gangliosides of mouse neuroblastoma cells in culture. Lipids 18:664–667

Refsvik T (1987) Excretion of methylmercury in rat bile: the effect of diethylmaleate, cyclohexene oxide and acrylamide. Acta Pharmacol Toxicol 42:135–141

Richardson RJ, Murphy SD (1975) Effect of glutathione depletion on tissue deposition of methylmercury in rats. Toxicol Appl Pharmacol 31:505–519

Rodier PM, Ashner M, Sager PR (1984) Mitotic arrest in the developing CNS after prenatal exposure to methylmercury. Neurobehav Toxicol Teratol 6:379–385

Roman-Franco AA, Twirello M, Abini B, Ossi E (1978) Anti-basement membrane antibodies with antigen-antibody complex in rabbits injected with mercuric chloride. Clin Immunol Immunopathol 9:404–411

Rungby J, Ernst E (1992) Experimentally induced lipid peroxidation after exposure to chromium, mercury or silver: interactions with carbon tetrachloride. Pharmacol Toxicol 70:205–207

Rustam H, von Burg R, Amin-Zaki L, Elhassani S (1975) Evidence for a neuromuscular disorder in methylmercury poisoning. Arch Environ Health 30:190–195

Sager PR, Matheson D (1988) Mechanisms of neurotoxicity related to selective disruption of microtubules and intermediate filaments. Toxicology 49:479–492

Sager PR, Syversen TLM (1984) Differential responses to methylmercury exposure and recovery in neuroblastoma and glioma cells and fibroblasts. Exp Neurol 85:371–382

Sager PR, Syversen TLM (1986) Disruption of microtubules by methylmercury. In: Clarkson TW, Sager PR, Syversen TLM (eds) The cytoskeleton: a target for toxic agents. Plenum, New York, pp 97–116

Sager PR, Doherty RA, Rodier PM (1982) Morphometric analysis of the effect of methylmercury on developing mouse cerebellar cortex. Toxicologist 2:116

Sager PR, Doherty RA, Olmstead JB (1983) Interaction of methylmercury with microtubules in cultured cells and in vitro. Exp Cell Res 146:127–137

Sakai K (1972) Effects of methyl mercuric chloride on rat spermatogenesis. Kumamoto Med J 25:94–100

Sapin C, Mandet C, Druet E, Gunther E, Druet P (1982) Immune complex type disease induced by $HgCl_2$ in Brown-Norway rats: genetic control of susceptibility. Clin Exp Immunol 48:700–704

Sapin C, Hirsch F, Delaporte JP, Bazin H, Druet P (1984) Polyclonal IgE increase after $HgCl_2$ injections in BN and LEW rats: a genetic analysis. Immunogenetics 20:227–236

Sarafian T, Verity MA (1985) Inhibition of RNA and protein synthesis in isolated cerebellar cells by in vitro and in vivo methylmercury. Neurochem Pathol 3:27–39

Sarafian T, Verity MA (1986) Mechanism of apparent transcription inhibition by methylmercury in cerebellar neurons. J Neurochem 47:625–631

Sin YM, Teh WF (1992) Effect of long-term uptake of mercuric sulphide on thyroid hormones and glutathione in mice. Bull Environ Contam Toxicol 49:847–854

Suda I, Hirayama K (1992) Degradation of methyl and ethyl mercury into inorganic mercury by hydroxyl radical produced from rat liver microsomes. Arch Toxicol 66:398–402

Suda I, Takahashi H (1992) Degradation of methyl and ethyl mercury into inorganic mercury by other reactive oxygen species besides hydroxyl radical. Arch Toxicol 66:34–39

Suda I, Totoki S, Takahashi H (1991) Degradation of methyl and ethyl mercury into inorganic mercury by oxygen free radical-producing systems: involvement of hydroxyl radical. Arch Toxicol 65:129–134

Suda I, Totoki S, Uchida T, Takahashi H (1992) Degradation of methyl and ethyl mercury into inorganic mercury by various phagocytic cells. Arch Toxicol 66:40–44

Sugata Y, Clarkson TW (1979) Exhalation of mercury–further evidence for an oxidation-reduction cycle in mammalian tissues. Biochem Pharmacol 28:3474–3476

Syversen TLM (1982) Changes in protein and RNA synthesis in rat brain neurons after a single dose of methylmercury. Toxicol Lett 10:31–34

Tanaka T, Naganuma A, Imura N (1990) Role of γ-glutamyltranspeptidase in renal uptake and toxicity of inorganic mercury in mice. Toxicology 60:187–198

Tanaka T, Naganuma A, Kobayashi K, Imura N (1991) An explanation for strain and sex differences in renal uptake of methylmercury in mice. Toxicology 69:317–329

Tanaka T, Naganuma A, Imura N (1992a) Routes for renal transport of methylmercury in mice. Eur J Pharmacol 228:9–14

Tanaka T, Naganuma A, Miura N, Imura N (1992b) Role of testosterone in γ-glutamyltranspeptidase dependent renal methylmercury uptake in mice. Toxicol Appl Pharmacol 112:58–63

Thomas DJ, Smith CJ (1982) Effects of coadministered low molecular weight thiol compounds on short term distribution of methylmercury in the rat. Toxicol Appl Pharmacol 62:104–110

Urano T, Naganuma A, Imura N (1988a) Methylmercury-cysteinylglycine constitutes the main form of methylmercury in rat bile. Res Commun Chem Pathol Pharmacol 60:197–210

Urano T, Naganuma A, Imura N (1988b) Species differences in biliary excretion of methylmercury: role of non-protein sulfhydryls in bile. Res Commun Chem Pathol Pharmacol 62:339–351

Van-Hoek AN, de-Jong MD, van-Os CH (1990) Effects of dimethylsulfoxide and mercurial sulfhydryl reagents on water and solute permeability of rat kidney brush border membranes. Biochim Biophys Acta 1030:203–210

Verity MA, Sarafian T (1991) Role of oxidative injury in the pathogenesis of methylmercury neurotoxicity. In: Suzuki T, Imura N, Clarkson TW (eds) Advances in mercury toxicology. Plenum, New York, pp 209–222

Waku K, Nakazawa Y (1979) Toxic effects of several mercury compounds on SH- and non-SH enzymes. Toxicol Lett 4:49–55

Warfvinge K, Hua J, Berlin M (1992) Mercury distribution in the rat brain after mercury vapor exposure. Toxicol Appl Pharmacol 117:46–52

Wasteneys GO, Cardin M, Reuhl KR, Brown DL (1988) The effects of methylmercury on the cytoskeleton of murine embryonal carcinoma cells. Cell Biol Toxicol 4:41–60

Weinberg JM, Johnson KJ, de-la-Iglesia FA, Allen ED (1989) Acute alterations of tissue Ca^{++} and lethal tubular cell injury during $HgCl_2$ nephrotoxicity in the rat. Toxicol Pathol 17:483–493

Welsh SO (1979) The protective effect of vitamin E and N,N′-diphenyl-p-phenylenediamine (DPPD) against methyl mercury toxicity in the rat. J Nutr 109:1673–1681

WHO (1976) Environmental health criteria 1: Mercury. World Health Organization, Geneva, p 132

WHO (1990) Environmental Health Criteria 101: Methylmercury. World Health Organization, Geneva, p 144

WHO (1990) Environmental Health Criteria 118: Inorganic Mercury. World Health Organization, Geneva, p 168

Woods JS, Calas CA, Aicher LD (1990a) Stimulation of porphyrinogen oxidation by mercuric ion. II. Promotion of oxidation from the interaction of mercuric ion,

glutathione, and mitochondria-generated hydrogen peroxide. Mol Pharmacol 38:261–266
Woods JS, Calas CA, Aicher LD, Robinson BH, Mailer C (1990b) Stimulation of porphyrinogen oxidation by mercuric ion. I. Evidence of free radical formation in the presence of thiols and hydrogen peroxide. Mol Pharmacol 38:253–260
Yamamoto H, Ishii K, Meguro T, Taketa K, Ogata M (1992) Impaired in vitro accumulation of mercury in erythrocytes of acatalasemic mice. Acta Med Okayama 46:67–73
Yamane Y, Fukino H, Imagawa M (1977) Suppressive effect of zinc on the toxicity of mercury. Chem Pharm Bull 25:1509–1518
Yonaha M, Saito M, Sagai M (1983) Stimulation of lipid peroxidation by methylmercury in rats. Life Sci 32:1507–1514
Yoshida M, Aoyama H, Satoh H, Yamamura Y (1987) Binding of mercury to metallothionein-like protein in fetal liver of the guinea pig following in utero exposure to mercury vapor. Toxicol Lett 37:1–6
Yoshida M, Satoh H, Kojima S, Yamamura Y (1990) Retention and distribution of mercury in organs of neonatal guinea pigs after in utero exposure to mercury vapor. J Trace Elem Exp Med 3:219–226
Yoshino Y, Mozai T, Nakao K (1966) Biochemical changes in the brain in rats poisoned with an alkyl mercury compound, with special reference to the inhibition of protein synthesis in brain cortex slices. J Neurochem 13:1223–1230
Zalups RK (1991) Autometallographic localization of inorganic mercury in the kidneys of rats: effect of unilateral nephrectomy and compensatory renal growth. Exp Mol Pathol 54:10–21
Zalups RK, Barfuss D (1990) Accumulation of inorganic mercury along the renal proximal tubule of the rabbit. Toxicol Appl Pharmacol 106:245–253
Zalups RK, Lash LH (1990) Effect of uninephrectomy and mercuric chloride on renal glutathione homeostasis. J Pharmacol Exp Ther 254:962–970
Zhao JY, Wang SJ (1988) Experimental study of proteinuria caused by chronic exposure to mercury. Biomed Environ Sci 1:235–246

CHAPTER 9

Toxicology of Cadmium

P.L. GOERING, M.P. WAALKES, and C.D. KLAASSEN

A. Introduction

Cadmium (Cd) is unique among metals because of its diverse toxic effects, extremely protracted biological half-life (approximately 20–30 years in humans), low rate of excretion from the body, and predominant storage in soft tissues (primarily liver and kidney) rather than bone. The health hazards associated with cadmium exposure became known in the 1940s when FRIBERG (1948) reported the occurrence of emphysema and proteinuria in workers exposed to cadmium dust. In the 1960s, cadmium was catapulted into the mainstream of metal toxicology research when cadmium was identified as the major etiological factor in *itai-itai* disease, a condition that afflicted Japanese women exposed to cadmium via their diet which contained cadmium-contaminated rice and water. Cadmium is an extremely toxic element of continuing concern because environmental levels have risen steadily due to continued worldwide anthropogenic mobilization. The mobilization has derived from past and current industrial and agricultural practices.

Cadmium has a diversity of toxic effects including nephrotoxicity, carcinogenicity, teratogenicity, and endocrine and reproductive toxicities. Although cadmium is not essential for growth and development in mammals, it generally followes the metabolic pathways of the essential elements zinc and copper.

I. Production and Uses

Worldwide cadmium production at present is around 17000 metric tons/year and the global emission of cadmium compounds into the atmosphere is estimated to be 7000 metric tons/year, mainly from anthropogenic sources (STOEPPLER 1991).

Cadmium is used very little as the pure metal, but it does have important uses as a constituent of many alloys and in its various salt forms (for review, see WAALKES et al. 1991d; STOEPPLER 1991). The primary uses for the majority of cadmium include batteries, electroplating, stabilizers, and pigments. Cadmium is used as the negative electrode in rechargeable Cd-Ni batteries. Electroplated cadmium provides excellent protective properties for iron and steel against corrosion, particularly for automotive and aircraft applications. Other major uses for cadmium include pigments for paint and as stabilizers in the plastics industry (STOEPPLER 1991).

II. Exposure to Cadmium

Exposure of humans to cadmium occurs both through the environment and occupational settings. Atmospheric emissions from man-made sources exceed those from natural sources by about one order of magnitude (KAZANTZIS 1986).

Food and cigarettes constitute the two major sources of cadmium exposure for the general population. Cadmium is present in most foods, with highest concentrations found in kidney, liver, and shellfish (LAUWERYS 1978). The concentration of cadmium in lungs of smokers is approximately twice that of nonsmokers (LEWIS et al. 1972). The cadmium intake from the atmosphere is negligible compared with that from food for the general population. Much higher concentrations of cadmium are found in industrialized urban areas. In workplaces associated with cadmium use, exposure is mainly due to inhalation of fumes and dusts.

III. Metabolism

In humans, intestinal absorption of cadmium is estimated at 5% (MCLELLAN et al. 1978; FRIBERG et al. 1986). Both human and animal data demonstrate that cadmium absorption from the lungs is 25%–50% (FRIBERG et al. 1986). Others report that up to 90% of cadmium deposited deep in the lung can be absorbed and distributed to other critical organs such as liver, kidney, and testis (OBERDÖRSTER 1986; WAALKES and OBERDÖRSTER 1990).

Once absorbed, clearance of cadmium from the circulation and tissue deposition is rapid. Over 50% of the body burden of cadmium is localized in liver and kidney (BERNARD and LAUWERYS 1986). Concentrations of cadmium in the kidneys are initially much less than those in liver, but increase slowly with time. This is consistent with a slow transfer of cadmium from liver to kidneys. Uptake of cadmium by other tissues includes spleen, pancreas, heart, and testes. In the classical sense of the term, cadmium is not biotransformed by phase I or phase II reactions which would enhance excretion of other xenobiotics. Cadmium is detoxified to a degree by long-term tissue storage. The biological half-life of cadmium in humans is 10–30 years (GOYER 1991). Consequently, with continuous environmental exposure, the concentrations of cadmium in tissues will increase slowly throughout life. It is stored in most tissues bound to the low-molecular protein, metallothionein, which is synthesized in response to cadmium exposure (FRIBERG et al. 1986). Metallothionein will bind cadmium with high affinity, and the intracellular cadmium-metallothionein (Cd-MT) complex is relatively biologically inert. The remarkable capacity of the liver and kidney to synthesize metallothionein is responsible for the high accumulation of cadmium in these organs. Induction of metallothionein prior to exposure to toxic or lethal doses of cadmium will induce tolerance to these effects (GOERING and KLAASSEN 1983; see also Sect. E.I). The relationship of metallothionein to cadmium metabolism is discussed elsewhere in this volume (see Chap. 6).

The major route of excretion of cadmium is via feces and bile. Approximately 95% of oral cadmium dose is excreted in the feces due to the poor absorption of cadmium from the GI tract. Urinary excretion of cadmium may become significant after the onset of renal damage.

B. Molecular and Cellular Effects

The underlying basis for the toxicity of many metals, including cadmium, in biological systems involves the formation of complexes with nucleophilic ligands of target molecules. Cadmium, unlike copper and iron, exists as a stable, bivalent ion and does not generally undergo redox reactions (KÄGI and HAPKE 1984). The affinity of cadmium for numerous ligands under physiological conditions occurs in the following order: thiol (RS^-) > phosphate (RPO^-) > chloride (RCl^-) > carboxyl ($RCOO^-$) (KÄGI and HAPKE 1984). The affinity of cadmium increases for biomolecules containing more than one binding site, e.g., glutathione and metallothionein. Because these ligands are found in many biomolecules, the number of potential targets for cadmium is large. The reader should be cognizant that many of the effects reported below were observed under in vitro conditions, and thus it is important to also evaluate their in vivo significance.

I. Calmodulin-Calcium-Cadmium Interactions

Recent evidence suggests that chemical-induced cell injury and death may result from impaired regulation of calcium (CORCORAN and RAY 1992). Metals may exert their toxicity by interfering with calcium signal transduction proteins, thus interfering with normal cell communication, growth, and differentiation (ROSSI et al. 1991). Cadmium stimulates release of calcium from intracellular storage sites and increases levels of inositol triphosphates (SMITH et al. 1989). This release of calcium may be related to the interaction of cadmium with a plasma membrane sialoprotein receptor (CHEN and SMITH 1992).

Calmodulin is a major intracellular calcium-binding protein which regulates a wide variety of calcium-dependent processes (CHEUNG 1982). The binding of calcium to calmodulin induces a conformational change in the protein to an active form. The calcium-calmodulin complex is then capable of regulating a host of other molecules such as phosphodiesterase, adenylate kinase, protein kinases, and Ca-ATPase. Cadmium has been shown to interact with calmodulin (FORSEN et al. 1979; ANDERSSON et al. 1982; CHAO et al. 1984; CHEUNG 1984; MILLS and JOHNSON 1985). In vitro studies have shown that cadmium acts as a calcium agonist by binding to calmodulin and activating calmodulin-dependent reactions (HABERMANN et al. 1983; SUZUKI et al. 1985). If binding of cadmium to calmodulin is sufficient to abnormally prolong the activation of calmodulin via disrupting its regulation by calcium, serious pathologic consequences could result. Depending on the concentration of cadmium and the specific assay conditions employed, other studies have shown that the interaction of cadmium with calmodulin can result in either the activation or the inhibition of calmodulin-sensitive enzymes such as phosphodiesterase (DONNELLY 1978; COX and HARRISON 1983; CHAO et al. 1984; RICHARD et al. 1985) and Ca^{2+}-Mg^{2+} adenosine triphosphatase (AKERMAN et al. 1985). Conversely, calmodulin has been suggested as a mediator for some of the toxic effects of heavy metals, including cadmium (CHAO et al. 1984; CHEUNG 1984).

The importance of these cadmium-calmodulin interactions as a mechanism of toxicity is supported by studies demonstrating that cadmium toxicity is ameliorated by pretreatment with calmodulin inhibitors. Cadmium-induced necrosis of the testes in CF-1 mice, as assessed by extent of hemorrhage, can be prevented by pretreatment with calmodulin inhibitors, such as trifluoperazine, chlorpromazine, or *N*-(6-aminohexyl)-5-chloro-1-naphthalene sulfonamide (W7), all of which inhibit the regulatory actions of calmodulin (NIEWENHUIS and PROZIALECK 1987). Chlorpromazine pretreatment has a less dramatic, though still significant, effect on cadmium-induced testicular toxicity in rats, as pretreatment with the inhibitor reduced cadmium-induced testicular hemorrhage, while modifying cadmium distribution (SHIRAISHI et al. 1994). Cadmium-induced nephropathy and proteinuria may result from perturbations in renal calcium metabolism, which may involve calmodulin (FOWLER et al. 1987; NORDBERG 1989; FOWLER et al. 1991). However, Cd-MT proximal tubule injury is not related to an increased permeability of cell membranes to external calcium, since this change is not evident until after the onset of tubule cell necrosis (GOERING et al. 1985; FOWLER et al. 1987). Recent studies have demonstrated that Cd-MT proximal tubule cell injury is also unrelated to changes in the distribution of intracellular calcium from various storage sites (FOWLER et al. 1991). Thus, the calcuria observed after Cd-MT exposure in laboratory studies, and in cadmium workers, most likely occurs secondary to proteinuria. A mechanism for Cd-MT-induced nephropathy resulting from the activation of calmodulin by cadmium in the absence of, or prior to, changes in calcium distribution (intracellular or extracellular) is plausible, since a protracted activation of calmodulin may lead to cytoskeletal damage and perturb cell-cell communication (TRUMP and BEREZESKY 1987).

II. Other Effects

Because of the physicochemical similarities of cadmium and zinc, cadmium will frequently follow the metabolic pathways of zinc. Cadmium is transported into cells through mechanisms specific for zinc uptake (Stacey and Klaassen 1980; Waalkes and Poirier 1985). Thus, at the molecular level, cadmium toxicity may be due to its substitution for zinc in various biochemical pathways, including metalloenzymes (Vallee and Ulmer 1972; Vallee and Glades 1974). Zinc deficiency can potentially enhance cadmium toxicity (Parzyck et al. 1978; Waalkes 1986), including carcinogenesis (Waalkes et al. 1991b). Cadmium has also been demonstrated to interact with the finger loop domains in *trans*-acting gene-regulating proteins and soluble receptor proteins (Sunderman and Barber 1988; Makowski et al. 1991).

Acute parenteral administration of cadmium decreases hepatic microsomal cytochrome P-450 content and inhibits associated mixed-function oxidase enzyme activities (Hadley et al. 1974; Means et al. 1979; Gregus et al. 1982). Decreases in hepatic mixed function oxidase activities and cytochrome P-450 content have been demonstrated after chronic cadmium dosing for 6 months (Dudley et al. 1985). The reduction in cytochrome P-450 is due to increased heme degradation, which results from cadmium stimulation of heme oxygenase activity (Maines and Kappus 1977; Eaton et al. 1980). Such an effect would alter the capacity of an organism to adequately biotransform endogenous compounds as well as other xenobiotics. In mitochondria, cadmium inhibits the transmembrane transport of calcium (Webb 1979) and oxidative phosphorylation (Sporn et al. 1969).

Cadmium displaces essential metal cofactors, such as zinc and copper, from metalloenzymes, altering their activity, which may lead to toxicity. Such enzymes include alcohol dehydrogenase, carboxypeptidase, carbonic anhydrase, δ-aminolevulinic acid dehydratase, and superoxide dismutase (Vallee and Ulmer 1972; Vallee and Glades 1974; Goering et al. 1987).

Exposure to cadmium may produce oxidative stress, which may result directly in toxicity or may occur secondary to cadmium toxicity. Results from studies using cultured cells have demonstrated cadmium-induced formation of superoxide anion radicals (Amoruso et al. 1982) and implicated superoxide anions in Cd-induced DNA single-strand scissions (Ochi et al. 1983). Cadmium inhibited superoxide dismutase in vivo, resulting in elevated superoxide levels (Shukla et al. 1987). Cadmium has been shown to increase peroxidation of lipids in isolated rat hepatocytes (Stacey et al. 1980) and in other target tissues in vivo and in vitro (Gabor et al. 1978; Wahba and Waalkes 1990); thus, increased levels of lipid peroxides following exposure to cadmium could constitute a source of active oxygen species.

C. Target Organ Toxicity

Cadmium has been shown to be toxic to numerous organ systems in mammals, but the manifestations of toxicity vary considerably depending on the route of administration, dose and time after administration, species, gender, and environmental and nutritional factors.

I. Acute Toxicities

Acute intoxication in humans from high exposures to cadmium can occur from either ingestion of cadmium salts or inhalation of cadmium fumes and dusts. These types of exposures are most toxic to the tissue initially exposed, i.e., lungs or GI tract. Kidney and liver toxicities may also occur following severe acute exposures to high levels of cadmium (WISNIEWSKAKNYPL et al. 1971). Early toxic symptoms are primarily local pulmonary or gastric irritation. Following ingestion, symptoms progress to nausea, vomiting, salivation, diarrhea, abdominal cramps, metabolic acidosis, and eventually death. In rats, high, acute exposure to cadmium by gavage can be more toxic than in humans because of the lack of an emetic response in rats. Thus more cadmium is absorbed with subsequent necrosis of the liver and testes. The same organs are damaged after injection by other routes (FRIBERG et al. 1986). Following inhalation of high cadmium concentrations, signs and symptoms include chest pain, nausea, dyspnea, dizziness, and diarrhea, with progression to or development of pneumonitis and fatal pulmonary edema (DUNPHY 1967; ZAVON and MEADOW 1970; WAALKES et al. 1991d).

II. Chronic Toxicities

The most common and well-understood human toxicities (noncarcinogenic) resulting from chronic cadmium exposures occur in the pulmonary system and kidney. The kidney is affected following either pulmonary or oral exposures; however, lung is affected only after exposure by inhalation.

1. Lung

Cadmium exists in airborne particulate matter ($<2\,\mu m$ diameter) in the oxide form as CdO, generated by various industrial practices, and is also present in cigarettes. The principal long-term effects of low-level exposure are chronic obstructive pulmonary disease and emphysema. Chronic obstructive lung disease results from chronic bronchitis, progressive fibrosis of the lower airways, and alveolar damage leading to emphysema. The associated clinical signs and symptoms include dyspnea, reduced vital capacity, and increased residual volume. The pulmonary toxicity of cadmium may be related to the inhibition of α-1-antitrypsin activity; however, no differences in plasma activity were observed between symptomatic and non-symptomatic Cd-exposed workers (GOYER 1991).

2. Kidney

The renal cortex is considered to be one of the major target tissues for cadmium poisoning. The effects on renal function and morphology have been extensively studied in both humans and experimental animals. Renal dysfunction has been shown in cadmium workers (MASON et al. 1988; THUN et al. 1989), and in populations consuming cadmium-contaminated diets and living in urban areas with cadmium pollution (NOGAWA et al. 1989; KIDO et al. 1988, 1990; LAUWERYS et al. 1991). The chronic effects of cadmium on kidney are characterized by proximal tubular dysfunction. Manifestations of cadmium-induced nephropathy include elevated urinary cadmium levels, low molecular weight proteinuria, aminoaciduria, glucosuria, and phosphaturia related to decreased renal absorption of phosphate (KLAASSEN 1990; GOYER 1991). These findings are consistent with the tubular dysfunction characteristic of Fanconi's syndrome. The mechanism of kidney injury by cadmium is related to the reabsorption of circulating Cd-MT in the proximal tubule (see Sect. E.I).

The proteinuria observed after cadmium exposure is believed to result from a decrease in the normal reabsorption of low molecular weight proteins in the renal proximal tubule. Glomerular dysfunction may occur at later stages of intoxication. Proteinuria has been observed in workers exposed to cadmium oxide dust (FRIBERG 1950). The primary proteins found in urine during cadmium intoxication are β_2-microglobulins, immunoglobulin light chains, retinol-binding protein, lysozyme, and ribonuclease (GOYER 1991). Proteinuria is considered to be the first sign of tubular dysfunction produced by cadmium. Renal dysfunction is believed to occur when the cortical concentration reaches 200 μg/g (FRIBERG et al. 1986; GOYER 1991), but may be less. Evidence from studies on laboratory animals and environmentally exposed and occupationally exposed populations indicates that urinary metallothionein levels are directly related to duration of cadmium exposure and possibly to the degree of renal tubular dysfunction (SHAIKH 1991). Enzymuria has also been observed during cadmium nephropathy; increases in urinary enzymes of lysosomal origin (*N*-acetyl-β-D-glucosaminidase) and of brush border origin (γ-glutamyltransferase and alkaline phosphatase) are observed (MASON et al. 1988).

3. Liver

Reports on chronic effects of cadmium on human liver function are rare. In experimental animals, liver also accumulates substantial amounts of cadmium after both acute and chronic poisoning (KOTSONIS and KLAASSEN 1977, 1978), which results in hepatic injury with both types of exposure (DUDLEY et al. 1982; STOWE et al. 1972; FAEDER et al. 1977). DUDLEY et al. (1982) demonstrated that liver is a target organ after acute exposure and that the liver injury may play a role in the lethality of animals soon after exposure. The prominent morphological changes after an acute high dose of cadmium included parenchymal cell necrosis, deterioration of the rough endoplasmic reticulum, proliferation of the smooth endoplasmic reticulum, and mitochondrial degenerative changes (Dudley et al. 1982). Repeated exposures of rats to low, daily doses of cadmium over 6 months resulted in liver injury prior to the onset of renal damage (Dudley et al. 1985). The effects in liver included elevations in plasma activities of alanine and aspartate aminotransferases, decreased microsomal mixed-function oxidase activities, and decreased structural integrity of hepatocytes. These changes occurred after the liver cadmium concentration reached 60 μg/g. These findings are important because they raise the possibility that, during long-term cadmium exposures, hepatic injury may result in the hepatic release of Cd-MT, with its subsequent translocation to kidney to produce renal injury (Foulkes 1978; Cherian et al. 1976; Squibb et al. 1984; Dudley et al. 1985).

4. Developmental Effects

Cadmium has teratogenic and embryotoxic effects, but these toxicities appear to be species and strain dependent, and also depend on gestational age at time of exposure. Distinguishing between a direct fetal effect of cadmium and effects resulting from maternal toxicity is problematic. Human developmental effects have not been reported. When administered during pregnancy in rodents, cadmium causes necrosis of decidual and/or placental tissues. Maternal/fetal transfer of cadmium is limited by retention of the metal in placenta. Human and rodent placenta contains inducible MT-like proteins (Waalkes et al. 1984; Goyer and Cherian 1992), which may bind cadmium and serve to protect the embryo/fetus from damage by cadmium. Cadmium administered during the period of organogenesis leads to prolonged anorexia, growth retardation, and skeletal, neural tube, limb bud, and craniofacial abnormalities (Ishizu et al. 1973; Nakashima et al. 1988).

The placenta appears to be a target organ for cadmium in humans. Cadmium accumulates in human placental tissues (Sikorski et al. 1988; Goyer and Cherian 1992), leading to morphological and functional impairment of this tissue. Cigarette smoking during pregnancy is associated with adverse morphological changes in this tissue similar to those observed in experimental animals (van der Velde et al. 1983). In humans, low birth weight and higher relative placental concentration weight (placenta weight to birth weight ratio) have been associated with maternal smoking (Chatterjee et al. 1988).

5. Reproductive Effects

Cadmium can act on the reproductive system by either directly affecting the gonads and accessory organs, or indirectly via interference with the hypothalamus-pituitary-gonadal axis.

Cadmium produces testicular injury in experimental animals after acute exposure but not after chronic exposure (Kotsonis and Klaassen 1977,

1978). Cadmium-induced testicular necrosis has not been reported in humans, although it has been observed in nonhuman primates (GIROD and CHAVINEAU 1964; LOHIYA et al. 1976). Exposure of testis to cadmium causes vascular bed damage and interstitial cell edema, leading to hemorrhage, reduced androgen production by Leydig cells, necrosis of the Sertoli cells, inhibition of spermatogenesis, and eventually testicular atrophy (PARIZEK 1957; MEEK 1959; GUNN et al. 1963a). Cadmium inhibits choline acetyltransferase activity and decreases acetylcholine synthesis in spermatozoa, and impairs sperm motility (DWIVEDI 1983). These changes are observed in rats, susceptible strains of mice, and other scrotal mammalian species. The testes of nonscrotal mammals, e.g., hedgehog and shrew, and the gonads of birds are generally not affected after cadmium exposure (SAMARAWICKRAMA 1979). The testes of newborn rats are also resistant to damage by cadmium (WONG and KLAASSEN 1980).

In animal studies, pre- and postnatal exposure to cadmium has been shown to alter ovarian function and reduce estrogen production, leading to changes in normal pubertal progression of females (PARIZEK et al. 1968; DER et al. 1977). The mechanism for these effects may result from direct inhibition of estrogen-synthesizing enzymes in the ovary or changes in the hypothalamic-pituitary-ovarian axis (VARGA and PAKSY 1991). Cadmium chloride injected subcutaneously in diestrus rats inhibits ovulation and impairs hCG-stimulated progesterone secretion from ovary (PAKSY et al. 1989). In this study, ovulation returned to control levels 50 days after cadmium injection, but the decreased hCG-stimulated progesterone secretions were still apparent. In contrast, chronic exposure to cadmium does not alter ovarian cycling, nor progesterone and estradiol secretion (VARGA et al. 1991). Cadmium exposure also affects ovarian-uterine function by interfering with embryo implantation (GIAVINI et al. 1980). Cadmium accumulates in the pituitary and may affect reproductive function by altering gonadotropin dynamics (DER et al. 1977). As is the case with testes, sufficiently high doses of cadmium can cause acute ovarian necrosis in rodents, which is species and strain dependent (REHM and WAALKES 1988).

6. Bone

Osteomalacia and osteoporosis are reported primarily in patients (95% females) with *itai-itai* disease, and osteomalacia has been observed in cadmium industry workers, the majority of whom are male (GOYER 1991). While the mechanisms for these effects are not well known, cadmium disruption of calcium metabolism may be important. One of the manifestations of renal injury in cadmium-exposed populations is calcuria and negative calcium balance. It is not clear whether these effects on calcium metabolism occur secondary to renal injury, or whether there is a direct effect on calcium pathways (see discussion of calmodulin-calcium-cadmium interactions in Sect. B.I).

The disease *itai-itai* became prevalent in Japan prior to and during the 1940s, and was associated with the ingestion of cadmium-contaminated rice. The rice became contaminated after growing in waste runoff water from a lead-zinc-cadmium mine. Clinical manifestations of the disease included kidney injury, osteomalacia, myalgia, and spontaneous fracture of long bones (TSUCHIYA 1978).

Evidence from experimental animals suggests that cadmium plays a direct role in the pathogenesis of *itai-itai* disease. Because more than 95% of patients with this disease were postmenopausal women who had borne many children (average, six), studies have been conducted to evaluate the effects of dietary cadmium on bone during pregnancy and lactation, and after ovariectomy (BHATTACHARYYA 1991; SACCO-GIBSON et al. 1992). Female animals are susceptible to cadmium-induced bone loss, increased bone resorption, and loss of bone calcium during these experimental conditions. The effects on bone occur prior to kidney dysfunction. Cadmium increases bone resorption in fetal rat limb bone cultures and increases the number of multinucleated osteoclast-like cells in bone marrow cultures (BHATTACHARYYA 1991). These in vitro results demonstrate that cadmium can act directly on bone, which may lead to osteoporosis or osteopenia. Thus, nursing and postmenopausal women exposed to cadmium may be potentially sensitive to bone mineral loss.

While strong evidence exists implicating cadmium as a major causative factor in *itai-itai* disease, other factors, such dietary deficiencies in minerals and vitamins, may have contributed to the disease (TSUCHIYA 1978; KJELLSTRÖM 1986). Serum $1\alpha,25(OH)_2$-vitamin D concentrations were depressed in cadmium-exposed cohorts presenting with clinical nephropathy (NOGAWA et al. 1987), which suggests that cadmium-induced bone effects may result from disruption of vitamin D and parathyroid hormone metabolism. Because kidney injury results from chronic cadmium exposure, a cadmium-related inhibition of the renal conversion of 25(OH)-vitamin D to $1\alpha,25(OH)_2$-vitamin D may lead to decreased calcium reabsorption, demineralization of bone, and eventually osteomalacia (FRIBERG et al. 1986).

7. Immune Effects

The lack of epidemiological studies evaluating the effects of cadmium on the immune system make it difficult to assess its effects in humans. Reports of cadmium immunotoxicity in rodents after long-term exposure would argue for the need for human clinical studies (for reviews, see KOLLER 1980; DESCOTES 1992). Mice chronically exposed to cadmium exhibit an immunosuppressive effect on the humoral immune response, i.e., a decreased antibody response (KOLLER et al. 1975); however, conflicting results demonstrate an elevated antibody response or no response. An immunosuppressive effect of cadmium on cell-mediated immunity, i.e., a decrease in delayed-type hypersensitivity, has also been reported in mice (MÜLLER et al. 1979). Cadmium has been shown to decrease phagocytosis in peritoneal macrophages, natural killer cell activity, and resistance to infections in rodents after chronic exposure (see DESCOTES 1992). Chronic exposure to cadmium produces an immune-mediated glomerulonephritis in rats, characterized by IgG deposition in the glomerular basement membrane as well as antibodies directed against glomerular glycoproteins (DRUET et al. 1982).

D. Carcinogenesis

Cadmium has been recently upgraded to a human carcinogen by the IARC (1993) and is a potent animal carcinogen (IARC 1976; SUNDERMAN 1978; NOMIYAMA 1982; OBERDÖRSTER 1986; OBERDÖRSTER and COX 1990; WAALKES and OBERDÖRSTER 1990; WAALKES et al. 1993a). A study examining cancer risk from occupational cadmium exposure has implicated cadmium as the etiologic factor in lung carcinogenesis in this cohort (STAYNER et al. 1992).

While the mechanism of action of metal carcinogens, including cadmium, is only poorly understood, cadmium carcinogenesis may be related to direct genotoxic effects of cadmium. Cadmium binds to two distinct sites on DNA, including a high-affinity site, where cadmium binding demonstrated positive cooperativity (WAALKES and POIRIER 1984). Cadmium has been shown to intercalate between DNA strands and disrupt transcription by destabilizing the helical structure of DNA (WEBB 1979). Cadmium also inhibits DNA-dependent RNA-polymerase activity (STOLL et al. 1976). Cadmium has a high affinity for nucleic acid bases which results in aberrant nucleic acid metabolism (WACKER and VALLEE 1959; IZATT et al. 1971); it also inhibits DNA polymerases (MIYAKE et al. 1979), induces changes in DNA synthesis indicative of base mispairing (LOEB et al. 1977), inhibits thymidine kinases (CIHAK 1979), and inhibits spindle formation (FRIBERG et al. 1986). Cadmium induces single-strand DNA breaks in mammalian cells indicative of direct genotoxic potential (COOGAN et al. 1992, 1994). Cadmium genotoxicity may be related to the substitution of cadmium for zinc in the finger-loop domains of gene-regulating proteins and other receptors (SUNDERMAN and BARBER 1988; MAKOWSKI et al. 1991). This substitution could potentially interfere with the proposed normal regulation of transcription factors by thioneins (ZENG et al. 1991). Cadmium activates the transcription of protooncogenes c-*jun* and c-*myc* in cultured cells (JIN and RINGERTZ 1990; TANG and ENGER 1992). Cadmium stimulation of tumor growth may involve the binding of cadmium to the external domain of a cell membrane receptor, which is coupled to phospholipase C (SMITH et al. 1989; CHEN and SMITH 1992).

I. Human Studies

Occupational or environmental cadmium exposure has been associated in some studies with development of cancers of the lung, prostate, kidney, liver, hematopoietic system, and stomach (KIPLING and WATERHOUSE 1967;

BERG and BURBANK 1972; KOLONEL 1976; LEMEN et al. 1976; BAKO et al. 1982; THUN et al. 1985; ELINDER et al. 1985; KAZANTZIS 1987; KAZANTZIS et al. 1988; ABD ELGHANY et al. 1990; CAMPBELL et al. 1990; STAYNER et al. 1992). The evidence is presently considered clear enough to establish a linkage between cadmium exposure and human neoplasia, at least within the lung (STAYNER et al. 1992); however, in other tissues the association is controversial (prostate) or preliminary (kidney, liver, stomach, hematopoietic system) until additional confirmatory evidence is available.

The lung is considered to be the best established site of human carcinogenesis by cadmium and several studies have shown an association between occupational exposure and pulmonary carcinogenesis (LEMEN et al. 1976; THUN et al. 1985; ELINDER et al. 1985; LANCET 1986; STAYNER et al. 1992). Cadmium-induced prostatic cancer in humans is much less well defined; however, a relationship between human prostatic cancer and occupational (KIPLING and WATERHOUSE 1967; LEMEN et al. 1976; ABD ELGHANY et al. 1990) and environmental (BAKO et al. 1982) exposure has been suggested. In contrast, other studies do not demonstrate an association between prostatic cancer and cadmium exposure (KAZANTZIS et al. 1988). Kidney (KOLONEL 1976), liver (CAMPBELL et al. 1990), hematopoietic system (BERG and BURBANK 1972), and gastric cancers (KAZANTZIS et al. 1988) have been associated with cadmium exposure, but these effects should be viewed as extremely tentative, because only a single report exists for each. Fully defining the carcinogenic risk of cadmium to humans will require further epidemiological studies.

II. Animal Studies

Cadmium has been recognized as a potent rodent carcinogen at the site of injection for over 30 years (HEATH et al. 1962). Early studies also demonstrated the remarkable ability of zinc to prevent cadmium carcinogenesis at the injection site and in the testes (GUNN et al. 1963b, 1964). This phenomenon may be of potential significance in elucidating mechanism(s) of cadmium carcinogenesis.

1. Lung

The lung has more recently been established as a site of cadmium carcinogenesis in rats (TAKENAKA et al. 1983). The demonstration of lung carcinomas in rats following chronic inhalation of cadmium strengthens the hypothesis that cadmium can induce lung cancer in humans. Rats inhaling cadmium chloride aerosols for as long as 1.5 years had a greater than 70% lung carcinoma incidence at the highest exposure levels (TAKENAKA et al. 1983). Other cadmium compounds, including the oxide, have been shown to be pulmonary carcinogens in rats (OLDIGES et al. 1989; GLASER et al. 1990). Cadmium induces lung cancer after either continuous (TAKENAKA et al. 1983) or discontinuous (OLDIGES et al. 1989) inhalation. The effectiveness of cadmium as a lung carcinogen appears to be species dependent, as the mouse and hamster appear to be resistant (HEINRICH et al. 1989). As is the case with several other sites of cadmium carcinogenesis, zinc will prevent cadmium-induced lung tumors (OLDIGES et al. 1989; OBERDÖRSTER and COX 1990).

2. Prostate

Several recent studies have demonstrated the potential of cadmium to cause prostate cancer in rats. Systemic cadmium exposure, including oral exposure, results in adenomas of the rat prostate (WAALKES et al. 1988a, 1989; WAALKES and REHM 1992). It appears that normal testicular function, i.e., androgen production, is requisite for development of cadmium-induced prostatic tumors.

3. Testes

Testes are a well-established target of cadmium toxicity in rats and mice, and a single injection of cadmium can result in a high incidence of testicular interstitial cell tumors (GUNN et al. 1963b, 1964; ROE et al. 1964; WAALKES et al. 1988a, 1989). Recent work indicates that oral cadmium exposure can also result in induction of interstitial cell tumors in rat testes (WAALKES and REHM 1992). The development of testicular tumors is thought to be related to the chronic degenerative effects of cadmium in this tissue (GUNN et al. 1963b, 1964; WAALKES et al. 1989), though tumors can develop in the absence of degenerative lesions (WAALKES and REHM 1992).

4. Injection Site

Cadmium will induce malignant tumors at subcutaneous or intramuscular injection sites (HEATH et al. 1962; KAZANTZIS 1963; HADDOW et al. 1964; POIRIER et al. 1983; WAALKES et al. 1988a, 1989). This is a common occurrence with metals in general, and may be related to a solid-state phenomenon where malignancies form during encapsulation of a chronically irritating implant, and are not necessarily due to the chemical nature of the implant (WAALKES and OBERDÖRSTER 1990). Cadmium at subcutaneous injection sites will in fact form a distinct calcified area; however, several studies argue against a simple solid-state mechanism. Repeated cadmium injections cause a marked increase in the rate of metastases of injection site tumors, indicating chemical modification of cellular characteristics and enhancement of progression to malignancy from repeated exposure (WAALKES et al. 1988b). Furthermore, zinc prevents the carcinogenic effects of cadmium at the site of injection (GUNN et al. 1963b, 1964), even when given by a totally different route (WAALKES et al. 1989), pointing towards the antagonism of the effects of cadmium by zinc as a mechanism for reduction of carcinogenesis. The strain of rats used has a pronounced effect on the final incidence and latency of cadmium induction of injection site sarcomas

(WAALKES et al. 1991a), indicating a genetic basis of susceptibility. It has been concluded from other mechanistic studies that the susceptibility of some tissues, e.g., testes and prostate, to cadmium carcinogenesis may be related to the relative lack of metallothionein in these tissues (WAALKES et al. 1989, 1992a, 1993a), which would reduce the capacity of these tissues to detoxify cadmium.

5. Hematopoietic

Cadmium will affect tumors of the hematopoietic system in rodents. In mice infected with murine lymphocytic leukemia virus, oral cadmium increases the incidence of lymphocytic leukemia (BLAKLEY 1986). It is suspected that cadmium-induced reduction in immunosurveillance allows emergence of leukemia. Subcutaneous doses of cadmium will also induce lymphoma in mice in a strain-dependent manner (WAALKES et al. 1994; WAALKES and REHM 1994). Dietary cadmium exposure in Wistar rats will cause dose-related increases in large granular lymphocyte (LGL) leukemia (WAALKES et al. 1992b). In contrast, a single high-dose subcutaneous injection of cadmium will markedly decrease the spontaneous incidence of LGL leukemia in Fischer rats (WAALKES et al. 1991a). A reduction in spontaneous lymphoma incidence can also occur in hamsters treated with cadmium (WAALKES et al. 1994).

6. Metal-Metal Interactions

The interaction of other metal ions, e.g., zinc, with cadmium has been shown to influence cadmium carcinogenesis. Zinc ameliorates the carcinogenic effects of cadmium in lung, testes, and at the injection site (GUNN et al. 1963b, 1964; WAALKES et al. 1989; OLDIGES et al. 1989). Other metals, such as calcium and magnesium, are relatively ineffective in this respect (POIRIER et al. 1983; KASPRZAK 1990). The specificity of zinc to antagonize cadmium genotoxicity (COOGAN et al. 1992) and carcinogenesis at a variety of different sites (see above) may be useful data to elucidate the mechanism(s) of cadmium carcinogenesis. Dietary zinc deficiency can enhance the progression of testicular lesions induced by cadmium in rats and increase the incidence of cadmium-induced injection site sarcomas (WAALKES et al. 1991b).

7. Synergism and Antagonism

Chronic experiments using combined exposures to multiple carcinogenic metals, such as would occur most frequently in human exposures (WAALKES and OBERDÖRSTER 1990), have not been carried out with cadmium. However, cadmium can both enhance and antagonize the carcinogenic effects of several organic carcinogens (HARRISON and HEATH 1986; KUROKAWA et al. 1985; WADE et al. 1987; WAALKES et al. 1991c). The incidence of diethylnitrosamine (DEN)-induced hepatic and renal tumors is markedly enhanced in rats by cadmium injections within the 1st week following DEN (WADE et al. 1987). In marked contrast, chronic oral treatment with cadmium starting at least 2 weeks after DEN exposure markedly reduces liver and lung tumor incidence in the mouse (WAALKES et al. 1991c, 1993b).

E. Roles of Metallothionein and Glutathione in Cadmium Toxicity

I. Metallothionein

Discussion of the toxicity of cadmium is not complete without a discussion of the influence of metallothionein (MT), a low molecular weight metal-binding protein, on cadmium toxicity. The protein also influences cadmium metabolism, but this aspect will be considered elsewhere in this volume (Chap. 6). MTs contain numerous thiol ligands, a property which confers a high degree of metal binding. The arrangement of these thiol groups is responsible for the high affinity associated with the binding. Many tissues contain MTs and their synthesis is highly inducible by a diverse group of chemical and physical agents (WAALKES and GOERING 1990).

It is well accepted that MT is responsible for the detoxication of cadmium in many species, and pharmacodynamic tolerance occurs via high-affinity sequestration of the metal. Treatment with low, nontoxic doses of cadmium, zinc, and other xenobiotics that stimulate MT synthesis will prevent the toxicity of subsequent cadmium exposure. The intracellular sequestration of cadmium into an inert Cd-MT complex reduces the interaction of cadmium ion with target molecules (WAALKES and GOERING 1990). Tolerance develops to cadmium-induced lethality, genotoxicity, inhibition of hepatic drug metabolism, testicular necrosis, hepatotoxicity, Cd-MT nephrotoxicity, carcinogenesis, fetotoxicity, and teratogenicity (ITO and SAWAUCHI 1966; YOSHIKAWA 1973; LEBER and MIYA 1976; ROBERTS et al. 1976; FERM and LAYTON 1979; GOERING and KLAASSEN 1983, 1984; SQUIBB et al. 1984; MAITANI et al. 1988; COOGAN et al. 1992, 1994).

The sequestration of cadmium by MT is a double-edged sword, i.e., although Cd-MT is relatively inert when stored as an intracellular complex, it becomes a potent nephrotoxicant after reaching the systemic circulation (CHERIAN et al. 1976; SQUIBB et al. 1984). Human cadmium nephrotoxicity may be related to Cd-MT exposure, because this may be a major form of cadmium in diet (MAITANI et al. 1984). Cadmium salts absorbed from the GI tract or lungs are initially transported to liver, where synthesis of MT is induced. Continual exposure to cadmium results in liver injury with leakage of Cd-MT into the systemic circulation (DUDLEY et al. 1985). The complex is transported to kidney, filtered, and reabsorbed by the proximal tubule, possibly via a mechanism involving receptor mediated endocytosis (FOULKES

1978; SQUIBB et al. 1984; DUDLEY et al. 1985; DORIAN et al. 1992). The reabsorbed complex is degraded by lysosomal enzymes, and the released cadmium ion induces MT in proximal tubule cells. Eventually, the capacity to synthesize MT is exceeded, and cadmium can poison sensitive molecular targets.

Tissues that contain high levels of MT, or are capable of increasing MT concentrations after an inducing stimulus, are protected to a relative degree from cadmium toxicity; however, tissues that are deficient in MT, or incapable of inducing its synthesis, appear to be highly susceptible to cadmium toxicity. For example, rat, mouse, and monkey testes, rat ventral prostate, and hamster ovary are all known to be susceptible to either the acute or chronic effects, including carcinogenesis, of cadmium (REHM and WAALKES 1988; WAALKES et al. 1988a) and appear to be deficient in MT. The proteins which bind cadmium in these tissues appear to be unrelated to MT (for brief discussion, see WAALKES and GOERING 1990). The lack of MT inducibility in these tissues may result in both a decrease in high-affinity cadmium binding and enhanced binding capacity, allowing cadmium greater access to molecular targets.

II. Glutathione

The capacity of cells to synthesize MT after cadmium exposure is an important cellular adaptive mechanism against cadmium toxicity; however, it is not necessarily the only mechanism, and other additional factors may be important (HILDEBRAND et al. 1982). A second important Cd-binding ligand that may influence cadmium toxicity is the ubiquitous tripeptide glutathione (GSH). GSH is the most important nonprotein thiol, and is involved in numerous biochemical pathways. It is important in protecting cells against damage from radiation, oxygen radicals, heat, and sulfhydryl reactive agents, and provides the bulk of sulfhydryl groups for the detoxication of electrophilic xenobiotics. In addition to MT, GSH appears to provide buffering capacity against cadmium toxicity (SINGHAL et al. 1987). GSH status of cells is important in cadmium toxicity; elevations and reductions in tissue GSH concentrations can reduce and exacerbate, respectively, the hepatotoxic and nephrotoxic effects of cadmium (DUDLEY and KLAASSEN 1984; SUZUKI and CHERIAN 1989). Even in the presence of elevated, and otherwise protective, levels of MT, lowered levels of GSH result in increased sensitivity to cadmium toxicity (OCHI et al. 1988; KANG et al. 1989; SUZUKI and CHERIAN 1989). Thus, GSH and MT may represent part of a multitiered cellular defense mechanism against toxic metals like cadmium.

F. Conclusion

The effects of cadmium on experimental animals and humans involve multiple organ systems and many mechanisms have been presented to explain these effects. Due to continued mobilization of cadmium in the environment, further laboratory and epidemiological studies are needed to assess the clinical significance of exposure to cadmium on human health. Mechanistic animal studies should focus on the most suitable choice of species, dose, route, and duration of exposure in order to facilitate the extrapolation of results from laboratory animals to humans. Additional epidemiological studies, including cohorts such as cadmium workers, should address whether cadmium exposure is associated with adverse health effects, particularly carcinogenesis and immune dysfunction.

Note. This article does not represent or contain official policy statements of the Food and Drug Administration.

References

Abd Elghany N, Schumacher MC, Slattery M, West DW, Lee JS (1990) Occupation, cadmium exposure, and prostate cancer. Epidemiol 1:107–115

Akerman KEO, Honkaniemi J, Scott IG, Andersson LC (1985) Interaction of Cd^{2+} with calmodulin activated (Ca^{2+} + Mg^{2+})-ATPase activity of human erythrocyte ghosts. Biochim Biophys Acta 845:48–53

Amoruso MA, Witz G, Goldstein BD (1982) Enhancement of rat and human phagocyte superoxide anion radical production by cadmium in vitro. Toxicol Lett 10:133–138

Andersson T, Drakenberg T, Forsen S, Thulin E (1982) Characterization of the Ca^{2+} binding sites of calmodulin from bovine testis using ^{43}Ca and ^{113}Cd NMR. Eur J Biochem 126:501–505

Bako G, Smith ESO, Hanson J, Dewar R (1982) The geographical distribution of high cadmium concentrations in the environment and prostate cancer in Alberta. Can J Public Health 73:92–94

Berg JW, Burbank F (1972) Correlations between carcinogenic trace metals in the water supplies and cancer mortality. Ann NY Acad Sci 199:249–264

Bernard A, Lauwerys R (1986) Effects of cadmium exposure in man. In: Foulkes EC (ed) Cadmium toxicology. Springer, Berlin Heidelberg New York, pp 135–177 (Handbook of experimental pharmacology, vol 80)

Bhattacharyya MH (1991) Cadmium-induced bone loss: increased susceptibility in females. Water Air Soil Pollution 57–58:665–673

Blakley BR (1986) The effect of cadmium- and viral-induced tumor production in mice. J Appl Toxicol 6:425–429

Campbell TC, Chen J, Liu C, Li J, Parpia B (1990) Nonassociation of aflatoxin with primary liver cancer in a cross-sectional ecological survey in the Peoples Republic of China. Cancer Res 50:6882–6893

Chao SH, Suzuki Y, Zysk JR, Cheung WY (1984) Activation of calmodulin by various metal cations as a function of ionic radius. Mol Pharmacol 26:75–82

Chatterjee MS, Abdel-Rahman M, Klein P, Bogden J (1988) Amniotic fluid cadmium and thiocyanate in pregnant women who smoke. J Reprod Med 33: 417–420

Chen Y-C, Smith JB (1992) A putative lectin-binding receptor mediates cadmium-evoked calcium release. Toxicol Appl Pharmacol 117:249–256

Cherian MG, Goyer RA, Delaquerriere-Richardson L (1976) Cadmium-metallothionein-induced nephrotoxicity. Toxicol Appl Pharmacol 38:399–408

Cheung WY (1982) Calmodulin. Sci Am 246:62–70

Cheung WY (1984) Calmodulin: its potential role in cell proliferation and heavy metal toxicity. Fed Proc 43:2995–2999

Cihak A (1979) Metabolic alterations of liver regeneration. XV. Cadmium-mediated depression of thymidine and thymidylate kinase induction in rats. Chem Biol Interact 25:355–362

Coogan TP, Bare RM, Waalkes MP (1992) Cadmium-induced DNA damage: effects of zinc pretreatment. Toxicol Appl Pharmacol 113:227–233
Coogan TP, Bare RM, Bjornson EJ, Waalkes MP (1994) Enhanced metallothionein gene expression protects against cadmium genotoxicity in cultured rat liver cells. J Toxicol Environ Health (in press)
Corcoran GB, Ray SD (1992) The role of the nucleus and other compartments in toxic cell death produced by alkylating hepatotoxicants. Toxicol Appl Pharmacol 113:167–183
Cox JL, Harrison SD Jr (1983) Correlation of metal toxicity with in vitro calmodulin inhibition. Biochem Biophys Res commun 115:106–111
Der R, Fahim Z, Yousef M, Fahim M (1977) Effects of cadmium on growth, sexual development and metabolism in female rats. Res Commun Chem Pathol Pharmacol 16:485–505
Descotes J (1992) Immunotoxicology of cadmium. In: Nordberg GF, Herber RFM, Alessio L (eds) Cadmium in the human environment: toxicity and carcinogenicity. Intl Agency Res Cancer, Lyon, France, pp 385–390
Donnelly TE (1978) Effects of zinc chloride on the hydrolysis of cyclic GMP and cyclic AMP by the activator-dependent cyclic nucleotide phosphodiesterase from bovine heart. Biochim Biophys Acta 522:151–160
Dorian C, Gattone VH II, Klaassen CD (1992) Renal cadmium disposition and injuries as a result of accumulation of cadmium-metallothionein by the proximal convoluted tubules – a light microscopic autoradiography study with ^{109}CdMT. Toxicol Appl Pharmacol 114:173–181
Druet P, Bernard A, Hirsch F, Weening JJ, Gengoux P, Mahieu P, Birkeland S (1982) Immunologically mediated glomerulonephritis induced by heavy metals. Arch Toxicol 50:187–194
Dudley RE, Klaassen CD (1984) Changes in hepatic glutathione concentration modify cadmium-induced hepatotoxicity. Toxicol Appl Pharmacol 72:530–538
Dudley RE, Svoboda DJ, Klaassen CD (1982) Acute exposure to cadmium causes severe liver injury in rats. Toxicol Appl Pharmacol 65:302–313
Dudley RE, Gammal LM, Klaassen CD (1985) Cadmium-induced hepatic and renal injury in chronically exposed rats: likely role of hepatic cadmium-metallothionein in nephrotoxicity. Toxicol Appl Pharmacol 77:414–426
Dunphy B (1967) Acute occupational cadmium poisoning: a critical review of the literature. J Occup Med 9:22–26
Dwivedi C (1983) Cadmium-induced sterility: possible involvement of the cholinergic system. Arch Environ Contam Toxicol 12:151–156
Eaton DL, Stacey NH, Wong K-L, Klaassen CD (1980) Dose-response effects of various metal ions on rat liver metallothionein, glutathione, heme oxygenase, and cytochrome P-450. Toxicol Appl Pharmacol 55:393–402
Elinder CG, Kjellström T, Hogstedt C, Anderson K, Spang G (1985) Cancer mortality of cadmium workers. Br J Ind Med 42:651–655
Faeder EJ, Chaney SP, King LC, Hinners TA, Bruce R, Fowler BA (1977) Biochemical and ultrastructural changes in livers of cadmium treated rats. Toxicol Appl Pharmacol 39:473–487
Ferm VH, Layton WM Jr (1979) Reduction in cadmium teratogenesis by prior cadmium exposure. Environ Res 18:347–350
Forsen S, Thulin E, Lilja H (1979) ^{113}Cd NMR in the study of calcium binding proteins: troponin C. FEBS Lett 104:123–126
Foulkes EC (1978) Renal tubular transport of cadmium-metallothionein. Toxicol Appl Pharmacol 45:505–512
Fowler BA, Goering PL, Squibb KS (1987) Mechanism of cadmium-metallothionein induced nephrotoxicity: relationship to renal calcium metabolism. Experientia [Suppl] 52:661–668
Fowler BA, Gandley RE, Akkerman M, Lipsky MM, Smith M (1991) Proximal tubule cell injury. In: Klaassen CD, Suzuki KT (eds) Metallothionein in biology and medicine. CRC Press, Boca Raton, FL, pp 311–321

Friberg L (1948) Proteinuria and emphysema among workers exposed to cadmium and nickel dust in a storage battery plant. Proc Int Cong Ind Med 9:641–644
Friberg L (1950) Health hazards in the manufacture of alkaline accumulators with special reference to chronic cadmium poisoning. Acta Med Scand 138:1–124
Friberg L, Kjellström T, Nordberg GF (1986) Cadmium. In: Friberg L, Nordberg G, Vouk VB (eds) Handbook on the toxicology of metals, vol 2, 2nd edn. Elsevier, New York, pp 130–184
Gabor S, Anca Z, Bordas E (1978) Cadmium-induced lipid peroxidation in kidney and testes: effect of zinc and copper. Rev Roum Biochim 15:113–117
Giavini E, Prati M, Vismara C (1980) Effects of cadmium, lead, and copper on rat preimplantation embryos. Bull Environ Contam Toxicol 25:702–705
Girod C, Chauvineau A (1964) Nouvelles observations concernant l'influence duchlorure de cadmium sur le testicule du Singe Macacus irus. F Cuv C R Soc Biol (Paris) 158:2113–2115
Glaser U, Hochrainer D, Otto FJ, Oldiges H (1990) Carcinogenicity and toxicity of four cadmium compounds inhaled by rats. Chem Environ Toxicol 27:153–162
Goering PL, Klaassen CD (1983) Altered subcellular distribution of cadmium following cadmium pretreatment: possible mechanism of tolerance to cadmium-induced lethality. Toxicol Appl Pharmacol 70:195–203
Goering PL, Klaassen CD (1984) Zinc-induced tolerance to cadmium hepatotoxicity. Toxicol Appl Pharmacol 74:299–307
Goering PL, Squibb KS, Fowler BA (1985) Calcuria and proteinuria during cadmium-metallothionein-induced proximal tubule cell injury. In: Hemphill DD (ed) Trace substances in environmental health, vol XIX. University of Missouri Press, Columbia, pp 22–35
Goering PL, Mistry P, Fowler BA (1987) Mechanisms of metal-induced cell injury. In: Haley TJ, Berndt WO (eds) Handbook of toxicology. Hemisphere, Washington DC, pp 384–425
Goyer RA (1991) Toxic effect of metals. In: Amdur MO, Doull J, Klaassen CD (eds) Casarett and Doull's toxicology, the basic science of poisons, 4th edn. McGraw-Hill Pergamon, New York, pp 623–680
Goyer RA, Cherian MG (1992) Role of metallothionein in human placenta and rats exposed to cadmium. In: Nordberg GF, Herber RFM, Alessio L (eds) Cadmium in the human environment. IARC, Lyon, pp 239–247
Gregus Z, Watkins JB, Thompson TN, Klaassen CD (1982) Resistance of some phase II biotransformation pathways to hepatotoxins. J Pharmacol Exp Ther 222:471–479
Gunn SA, Gould TC, Anderson WAD (1963a) The selective injurious response of testicular and epididymal blood vessels to cadmium and its prevention by zinc. Am J Pathol 42:685–702
Gunn SA, Gould TC, Anderson WAD (1963b) Cadmium-induced interstitial cell tumors in rats and mice and their prevention by zinc. J Natl Cancer Inst 31:745–753
Gunn SA, Gould TC, Anderson WAD (1964) Effect of zinc on cancerogenesis by cadmium. Proc Soc Exp Biol Med 115:653–657
Habermann E, Crowell K, Janicki P (1983) Lead and other metals can substitute for Ca^{2+} in calmodulin. Arch Toxicol 54:61–70
Haddow A, Roe FJC, Dukes CE, Mitchely BCV (1964) Cadmium neoplasia: sarcomata at the site of injection of cadmium in rats and mice. Br J Cancer 18:667–673
Hadley WM, Miya TS, Bousquet WF (1974) Cadmium inhibition of hepatic drug metabolism in the rat. Toxicol Appl Pharmacol 28:284–291
Harrison PTC, Heath JC (1986) Apparent synergy in lung carcinogenesis: interactions between N-nitrosoheptamethyleneimine, particulate cadmium and crocidolite asbestos fibers in rats. Carcinogenesis 7:1903–1908
Heath JC, Daniel MR, Dingle JT, Webb M (1962) Cadmium as a carcinogen. Nature 193:592–593

Heinrich U, Peters L, Ernst H, Rittinghausen S, Dasenbrock C, König H (1989) Investigation of the carcinogenic effects of various compounds after inhalation in hamsters and mice. Exp Pathol 37:253–258
Hildebrand CE, Griffith JK, Tobey RA, Walters RA, Enger MD (1982) Molecular mechanisms of cadmium detoxification in cadmium-resistant cultured cells: role of metallothionein and other inducible factors. In: Foulkes EC (ed) Biological roles of metallothionein. Elsevier/North-Holland, New York, pp 279–303
IARC (1976) Cadmium, nickel, some epoxides, miscellaneous industrial chemicals and general considerations on volatile anesthetics. Lyon, France, pp 39–74 (International Agency for Research on Cancer monographs, vol 11)
IARC (1993) Beryllium, cadmium, mercury, and exposures in the glass manufacturing industry. Intl Agency Res Cancer, vol 58. Lyon, France, pp 119–238
Ishizu S, Minami M, Suzuki A, Yamada M, Sato M, Yamamura K (1973) An experimental study on teratogenic effect of cadmium. Ind Health 11:127–139
Ito T, Sawauchi K (1966) Inhibitory effects on cadmium-induced testicular damage by pretreatment with smaller cadmium doses. Okajimas Fol Anat Jpn 42: 107–117
Izatt R, Christensen J, Rytting J (1971) Sites and thermodynamic quantities with proton and metal ion interactions with ribonucleic acid, deoxyribonucleic acid, and their constituent bases, nucleosides and nucleotides. Chem Rev 71:439–481
Jin P, Ringertz N (1990) Cadmium induces transcription of protooncogenes *c-jun* and *c-myc* in rat L6 myoblasts. J Biol Chem 265:14061–14064
Kägi JHR, Hapke H-J (1984) Biochemical interactions of mercury, cadmium, and lead. In: Nriagu JO (ed) Changing metal cycles and human health. Springer, Berlin Heidelberg New York, pp 237–250
Kang Y-J, Clapper JA, Enger MD (1989) Enhanced cadmium cytotoxicity in A549 cells with reduced glutathione levels is due to neither enhanced cadmium accumulation nor reduced metallothionein synthesis. Cell Biol Toxicol 5:249–260
Kasparzak KS (1990). Metal interactions in nickel, cadmium and lead carcinogenesis. In: Foulkes EC (ed) Biological effects of heavy metals, vol II: metal carcinogenesis. CRC Press, Boca Raton, FL, pp 173–189
Kazantzis G (1963) Induction of sarcoma in the rat by cadmium sulphide pigment. Nature 198:1213–1214
Kazantzis G (1986) Cadmium: sources, exposure and possible carcinogenicity. In: O'Neill IK, Schuller P, Fishbein L (eds) Environmental carcinogens selected methods of analysis, vol 8. IARC, Lyon, France, pp 93–100
Kazantzis G (1987) Cadmium. In: Fishbein L, Furst A, Mehlman MA (eds) Advances in modern toxicology, vol XI. Princeton Scientific, Princeton, NJ, pp 127–143
Kazantzis G, Lam TH, Sullivan K (1988) Mortality of cadmium exposed workers. Scand J Work Environ Health 14:220–223
Kido T, Honda R, Tsuritani I, Yamaya H, Ishizaki M, Yamada Y, Nogawa K (1988) Progress of renal dysfunction in inhabitants environmentally exposed to cadmium. Arch Environ Health 43:213–217
Kido T, Nogawa K, Honda R, Tsuritani I, Ishizaki M, Yamada Y, Nakagawa H (1990) The aassociation between renal dysfunction and osteopenia in environmental cadmium-exposed subjects. Environ Res 51:71–82
Kipling MD, Waterhouse JAH (1967) Cadmium and prostatic carcinoma. Lancet 1:730–731
Kjellström T (1986) Itai-itai disease. In: Friberg L, Elinder CG, Kjellström T, Nordberg GF (eds) Cadmium and health: a toxicological and epidemiological appraisal, vol II, effects and response. CRC Press, Boca Raton, FL, pp 257–290
Klaassen CD (1990) Heavy metals and heavy-metal antagonists. In: Gilman AG, Rall TW, Nies AS, Taylor P (eds) Goodman and Gilman's the pharmacological basis of therapeutics. McGraw-Hill, New York, pp 1592–1614

Koller LD (1980) Review/commentary: immunotoxicology of heavy metals. Intl J Immunopharmacol 2:269–279
Koller LD, Exon JH, Roan JG (1975) Antibody suppression by cadmium. Arch Environ Health 30:598–601
Kolonel LN (1976) Association of cadmium with renal cancer. Cancer 37:1782–1787
Kotsonis FN, Klaassen CD (1977) Toxicity and distribution of cadmium administered to rats at sublethal doses. Toxicol Appl Pharmacol 41:667–680
Kotsonis FN, Klaassen CD (1978) The relationship of metallothionein to the toxicity of cadmium after prolonged oral administration to rats. Toxicol Appl Pharmacol 46:39–54
Kurokawa Y, Matsushima M, Imazawa N, Takamura N, Takahashi M, Mayashi Y (1985) Promoting effect of metal compounds on rat renal tumorigenesis. J Am Coll Toxicol 4:321–327
Lancet (1986) Carcinogenicity of cadmium. II:931
Lauwerys R (1978) CEC criteria (dose/effect relationship) for cadmium. Pergamon, Oxford
Lauwerys R, Bernard A, Buchet JP, Roels H, Bruaux P, Claeys F, Ducoffre G, DePlaen P, Staessen J, Amery A et al. (1991) Does environmental cadmium represent a health risk? Conclusions from the Cadmibel study. Acta Clin Belg 46:219–225
Leber AP, Miya TS (1976) A mechanism for cadmium and zinc-induced tolerance to cadmium toxicity: involvement of metallothionein. Toxicol Appl Pharmacol 37:403–414
Lemen RA, Lee JS, Wagoner JK, Blejer HP (1976) Cancer mortality among cadmium production workers. Ann NY Acad Sci 271:273–279
Lewis GP, Jusko WJ, Coughlin LL, Hartz S (1972) Cadmium accumulation in man: influence of smoking, occupation, alcoholic habit and disease. J Chronic Dis 25:717–726
Loeb L, Sirover M, Wymouth L, Dube D, Seal G, Agarwal S, Katz E (1977) Infidelity of DNA synthesis as related to mutagenesis and carcinogenesis. J Toxicol Environ Health 2:1297–1304
Lohiya NK, Arya M, Shivapuri VS (1976) The effects of cadmium chloride on the testis and epididymis of the Indian hanuman langur, Presbytis entellus entellus Dufresne. Acta Eur Fertil 7:339–348
Maines MD, Kappas A (1977) Metals as regulators of heme metabolism. Science 198:1215–1221
Maitani T, Waalkes MP, Klaassen CD (1984) Distribution of cadmium after oral administration of cadmium-thionein to mice. Toxicol Appl Pharmacol 74: 237–243
Maitani T, Cuppage FE, Klaassen CD (1988) Nephrotoxicity of intravenously injected cadmium-metallothionein: critical concentration and tolerance. Fund Appl Toxicol 10:98–108
Makowski GS, Lin S-M, Brennan SM, Smilowitz HM, Hopfer SM, Sunderman FW Jr (1991) Detection of two Zn-finger proteins of Xenopus laevis, TFIIIA, and p43, by probing Western blots of ovary cytosol with $^{65}Zn^{2+}$, $^{63}Ni^{2+}$, or $^{109}Cd^{2+}$. Biol Trace Elem Res 29:93–109
Mason HJ, Davison AG, Wright AL, Guthrie CJ, Fayers PM, Venables KM, Smith NJ, Chettle DR, Franklin DM, Scott MC, Holden H, Gompertz D, Newman-Taylor AJ (1988) Relations between liver cadmium, cumulative exposure, and renal function in cadmium alloy workers. Br J Ind Med 45:793–802
McLellan JA, Flanagan PR, Chamberlain MJ, Valberg LS (1978) Measurement of dietary cadmium absorption in humans. J Toxicol Environ Health 4:131–138
Means JR, Carlson GP, Schnell RC (1979) Studies on the mechanism of cadmium-induced inhibition of the hepatic microsomal monooxygenase of the male rat. Toxicol Appl Pharmacol 48:293–304

Meek ES (1959) Cellular changes induced by cadmium in mouse testis and liver. Br J Exp Pathol 40:503–506
Mills JS, Johnson JD (1985) Metal ions as allosteric regulators of calmodulin. J Biol Chem 260:15100–15105
Miyake M, Murata I, Osabe M, Ono T (1979) Effect of metal cations on misincorporation by E. coli DNA polymerases. Biochem Biophys Res Commun 77: 854–860
Müller S, Gillert KE, Krause C, Jautzke G, Gross U, Diamantstein T (1979) Effects of cadmium on the immune system of mice. Experientia 35:909–910
Nakashima K, Wakisaka T, Fujiki Y (1988) Dose-response relationship of cadmium embryotoxicity in cultured mouse embryos. Reprod Toxicol 1:293–298
Niewenhuis KJ, Prozialeck WC (1987) Calmodulin inhibitors protect against cadmium induced testicular damage in mice. Biol Reprod 37:127–133
Nogawa K, Tsuritani I, Kido T, Honda R, Yamada Y, Ishizaki M (1987) Mechanism for bone disease found in inhabitants environmentally exposed to cadmium: decreased 1,25-dihydroxy-vitamin D level. Int Arch Occup Environ Health 59:21–30
Nogawa K, Honda R, Kido T, Tsuritani I, Yamada Y, Ishizaki M, Ymaya H (1989) A dose-response analysis of cadmium in the general environment with special reference to total cadmium intake limit. Environ Res 48:7–16
Nomiyama K (1982) Carcinogenicity of cadmium. Jpn J Ind Health 24:13–23
Nordberg GF (1989) Modulation of metal toxicity by metallothionein. Biol Trace Elem Res 21:131–135
Oberdörster G (1986) Airborne cadmium and carcinogenesis of the respiratory tract. Scand J Work Environ Health 12:523–537
Oberdörster G, Cox C (1990) Carcinogenicity of cadmium in animals: what is the significance for man? Chem Environ Toxicol 27:181–195
Ochi T, Ishiguro T, Ohsawa M (1983) Participation of active oxygen species in the induction of DNA single-strand scissions by cadmium chloride in cultured Chinese hamster cells. Mutat Res 122:169–175
Ochi T, Otsuka F, Takahashi K, Ohsawa M (1988) Glutathione and metallothioneins as cellular defense against cadmium toxicity in cultured Chinese hamster cells. Chem Biol Interact 65:1–14
Oldiges H, Hochrainer D, Glaser U (1989) Preliminary results from a long-term inhalation study with four cadmium compounds. Toxicol Environ Chem 23: 35–41
Paksy K, Varga B, Horváth E, Tátrai E, Ungváry G (1989) Acute effects of cadmium on preovulatory serum FSH, LH, and prolactin levels and on ovulation and ovarian hormone secretion in estrous rats. Reprod Toxicol 3:241–247
Parizek J (1957) The destructive effect of cadmium ion on testicular tissue and its prevention by zinc. J Endocrinol 15:56–63
Parizek J, Oskadolova I, Bemes I, Pitha J (1968) The effect of a subcutaneous injection of cadmium salts on the ovaries of adult rat in persistent oestrus. J Reprod Fertil 17:559–562
Parzyck DC, Shaw SM, Kessler WV, Vetter RJ, VanSickle RJ, Mayes RA (1978) Fetal effects of cadmium in pregnant rats on normal and zinc-deficient diets. Bull Environ Contam Toxicol 19:206–214
Poirier LA, Kasprzak KS, Hoover KL, Wenk ML (1983) Effects of calcium and magnesium acetates on the carcinogenicity of cadmium chloride in Wistar rats. Cancer Res 43:4575–4581
Rehm S, Waalkes MP (1988) Cadmium-induced ovarian toxicity in hamsters, mice, and rats. Fund Appl Toxicol 10:635–647
Richard G, Federolf G, Habermann E (1985) The interaction of aluminum and other metal ions with calcium-calmodulin-dependent phosphodiesterase. Arch Toxicol 57:257–259

Roberts SA, Miya TS, Schnell RC (1976) Tolerance development to cadmium-induced alteration of drug action. Res Commun Chem Pathol Pharmacol 14: 197–200

Roe FJC, Dukes CE, Cameron KM, Pugh RCD, Mitchley BCV (1964) Cadmium neoplasia: testicular atrophy and Leydig cell hyperplasia and neoplasia in rats and mice following the subcutaneous injection of cadmium salts. Br J Cancer 18:674–681

Rossi A, Manzo L, Orrenius S, Vahter M, Nicotera P (1991) Modifications of cell signaling in the cytotoxicity of metals. Pharmacol Toxicol 68:424–429

Sacco-Gibson N, Chaudhry S, Brock A, Sickles AB, Patel B, Hegstad R, Johnston S, Peterson D, Bhattacharyya M (1992) Cadmium effects on bone metabolism: accelerated resorption in ovarectomized, aged beagle dogs. Toxicol Appl Pharmacol 113:274–283

Samarawickrama GP (1979) Biological effects of cadmium in mammals. In: Webb M (ed) The chemistry, biochemistry and biology of cadmium. Elsevier/North-Holland, New York, pp 341–421

Shaikh Z (1991) Chronic cadmium exposure and metallothionein. In: Klaassen CD, Suzuki KT (eds) Metallothionein in biology and medicine. CRC Press, Boca Raton, FL, pp 383–391

Shiraishi N, Rehm S, Waalkes MP (1994) Effect of chlorpromazine pretreatment on cadmium toxicity in the male Wistar (WF/NCr) rat. J Toxicol Environ Health 42:123–138

Shukla GS, Hussain T, Chandra SV (1987) Possible role of regional superoxide dismutase activity and lipid peroxide levels in cadmium neurotoxicity: in vivo and in vitro studies in growing rats. Life Sci 41:2215–2221

Sikorski R, Radomanski T, Paszkowski T, Skoda J (1988) Smoking during pregnancy and the perinatal body burden. J Perinat Med 16:225–231

Singhal RK, Anderson ME, Meister A (1987) Glutathione, a first line of defense against cadmium toxicity. FASEB J 1:220–223

Smith JB, Dwyer SD, Smith L (1989) Cadmium evokes inositol polyphosphate formation and calcium mobilization. Evidence for a cell surface receptor that cadmium stimulates and zinc antagonizes. J Biol Chem 264:7115–7118

Sporn A, Dinu I, Stoenescu L, Cirstea A (1969) Beitrage zur Ermittlung der Wechselwirkungen zwischen Cadmium und Zink. Nahrung 13:461–469

Squibb KS, Pritchard JB, Fowler BA (1984) Cadmium metallothionein nephropathy: ultrastructural/biochemical alterations and intracellular cadmium binding. J Pharmacol Exp Ther 229:311–321

Stacey NH, Klaassen CD (1980) Cadmium uptake by isolated rat hepatocytes. Toxicol Appl Pharmacol 55:448–455

Stacey NH, Cantilena LR Jr, Klaassen CD (1980) Cadmium toxicity and lipid peroxidation in isolated rat hepatocytes. Toxicol Appl Pharmacol 53:470–480

Stayner L, Smith R, Thun M, Schorr T, Lemen R (1992) A dose-response analysis and quantitative assessment of lung cancer risk and occupational cadmium exposure. Ann Epidemiol 2:177–194

Stoeppler M (1991) Cadmium. In: Merian E (ed) Metals and their compounds in the environment. VCH, New York, pp 803–851

Stoll R, White J, Miya T, Bousquet W (1976) Effects of cadmium on nucleic acid and protein synthesis in rat liver. Toxicol Appl Pharmacol 37:61–74

Stowe HD, Wilson M, Goyer RA (1972) Clinical and morphological effects of oral cadmium toxicity in rabbits. Arch Pathol 94:389–405

Sunderman FW Jr (1978) Carcinogenic effects of metals. Fed Proc 37:40–46

Sunderman FW Jr, Barber AM (1988) Finger loops, oncogenes, and metals. Ann Clin Lab Sci 18:267–288

Suzuki CAM, Cherian MG (1989) Renal glutathione depletion and nephrotoxicity of cadmium-metallothionein in rats. Toxicol Appl Pharmacol 98:544–552

Suzuki Y, Chao S-H, Zysk JR, Cheung WY (1985) Stimulation of calmodulin by cadmium ion. Arch Toxicol 57:205–211
Takenaka S, Oldiges H, König H, Hochrainer D, Oberdörster G (1983) Carcinogenicity of cadmium chloride aerosols in Wistar rats. J Natl Cancer Inst 70: 367–373
Tang N, Enger D (1992) Cadmium's action on NRK-49F cells to produce responses induced also by TGFβ production or activation. Toxicology 71:161–171
Thun MJ, Schnorr TM, Smith AB, Halperin WE, Lemen RA (1985) Mortality among a cohort of U.S. cadmium production workers – an update. J Natl Cancer Inst 74:325–333
Thun MJ, Osorio AM, Schober S, Hannon WH, Lewis B, Halperin W (1989) Nephropathy in cadmium workers: assessment of risk from airborne occupational exposure to cadmium. Br J Ind Med 46:689–697
Trump BF, Berezesky IK (1987) Mechanisms of cell injury in the kidney: the role of calcium. In: Fowler BA (ed) Mechanisms of cell injury: implications for human health. Wiley, New York, pp 135–151
Tsuchiya K (1978) Cadmium studies in Japan – a review. Kodansha, Tokyo – Elsevier/North-Holland Biomedical, New York
Van der Velde WJ, Copius Peereboom-Stegeman JHJ, Treffers PE, James J (1983) Structural changes in the placenta of smoking mothers: a quantitative study. Placenta 4:231–240
Vallee BL, Ulmer DD (1972) Biochemical effects of mercury, cadmium and lead. Annu Rev Biochem 41:91–128
Vallee BL, Glades A (1974) The metallobiochemistry of zinc enzymes. Adv Enzymol 56:283–430
Varga B, Paksy K (1991) Toxic effects of cadmium on LHRH-induced LH release and ovulation in rats. Reprod Toxicol 5:199–203
Varga B, Paksy K, Náray M (1991) Distribution of cadmium in ovaries, adrenals and pituitary gland after chronic administration in rats. Acta Physiol Hungarica 78:221–226
Waalkes MP (1986) Effect of dietary zinc deficiency on the accumulation of cadmium and metallothionein in selected tissues of the rat. J Toxicol Environ Health 18:301–313
Waalkes MP, Goering PL (1990) Metallothionein and other cadmium-binding proteins: recent developments. Chem Res Toxicol 3:281–288
Waalkes MP, Oberdörster G (1990) Cadmium carcinogenesis. In: Foulkes EC (ed) Biological effects of heavy metals, vol II: metal carcinogenesis. CRC Press, Boca Raton, FL, pp 129–158
Waalkes MP, Poirier LA (1984) In vitro cadmium-DNA interactions: cooperativity of cadmium binding and competitive antagonism by calcium, magnesium, and zinc. Toxicol Appl Pharmacol 75:539–546
Waalkes MP, Poirier LA (1985) Interactions of cadmium with interstitial tissue of the rat testes: uptake of cadmium by isolated interstitial cells. Biochem Pharmacol 34:2513–2518
Waalkes MP, Rehm S (1992) Carcinogenicity of oral cadmium in the male Wistar (WF/NCr) rat: effect of chronic dietary zinc deficiency. Fund Appl Toxicol 19:512–520
Waalkes MP, Rehm S (1994) Chronic toxic and carcinogenic effects of cadmium in male DBA/2NCr and NFS/NCr mice: strain-dependent association with tumors of the hematopoietic system, injection site, liver and lung. Fund Appl Toxicol 23:21–31
Waalkes MP, Poisner AM, Wood GW, Klaassen CD (1984) Metallothionein-like proteins in human placenta and fetal membranes. Toxicol Appl Pharmacol 74:179–184
Waalkes MP, Rehm S, Riggs CW, Bare RM, Devor DE, Poirier LA, Wenk ML, Henneman JR, Balaschak MS (1988a) Cadmium carcinogenesis in the male

Wistar [Crl:(WI)BR] rats: dose-response analysis of tumor induction in the prostate and testes and at the injection site. Cancer Res 48:4656–4663
Waalkes MP, Ward JM, Konishi N (1988b) Toxicity and carcinogenicity of cadmium following repeated injections in rats: rapid induction of highly malignant tumors at the injection site. Proc Am Assoc Cancer Res 29:132
Waalkes MP, Rehm S, Riggs CW, Bare RM, Devor DE, Poirier LA, Wenk ML, Henneman JR (1989) Cadmium carcinogenesis in the male Wistar [Crl:(WI)BR] rats: dose-response analysis of effects of zinc on tumor induction in the prostate, in the testes, and at the injection site. Cancer Res 49:4282–4288
Waalkes MP, Rehm S, Sass B, Konishi N, Ward JM (1991a) Chronic carcinogenic and toxic effects of a single subcutaneous dose of cadmium in the male Fischer rat. Environ Res 55:40–50
Waalkes MP, Kovatch R, Rehm S (1991b) Effect of chronic dietary zinc deficiency on cadmium toxicity and carcinogenesis in the male Wistar [Hsd:(WI)BR] rat. Toxicol Appl Pharmacol 108:448–456
Waalkes MP, Diwan BA, Bare RM, Ward JM, Weghorst C, Rice JM (1991c) Anticarcinogenic effects of cadmium in B6C3F1 mouse liver and lung. Toxicol Appl Pharmacol 110:327–335
Waalkes MP, Wahba ZZ, Rodriguez RE (1991d) Cadmium. In: Sullivan JB Jr, Krieger GR (eds) Hazardous materials toxicology. Williams and Wilkens, Baltimore, MD, pp 845–852
Waalkes MP, Rehm S, Perantoni A, Coogan TP (1992a) Cadmium exposure in rats and tumors of the prostate. In: Nordberg GF, Alessio L, Herber RFM (eds) Cadmium in the human Environment: toxicity and carcinogenicity. IARC Scientific Publications, Lyon, France, pp 390–400
Waalkes MP, Rehm S, Sass B, Ward JM (1992b) Induction of tumors of the hematopoietic system in rats. In: Nordberg GF, Alessio L, Herber RFM (eds) Cadmium in the human environment: toxicity and carcinogenicity. IARC Scientific Publications, Lyon, France, pp 401–404
Waalkes MP, Coogan TP, Barter RA (1993a) Toxicological principles of metal carcinogenesis with special emphasis on cadmium. Crit Rev Toxicol 22:175–201
Waalkes MP, Diwan BA, Weghorst CM, Ward JM, Rice JM, Cherian MG, Goyer RA (1993b) Further evidence of the tumor suppressive effects of cadmium in the B6C3F1 mouse liver and lung: late stage vulnerability of tumors to cadmium and the role of metallothionein. J Pharmacol Exper Therap 266:1656–1663
Waalkes MP, Rehm S, Sass B, Kovatch R, Ward JM (1994) Chronic carcinogenic and toxic effects of a single subcutaneous dose of cadmium in male NFS and C57 mice and male Syrian hamsters. Toxic Sub J 13:15–28
Wacker W, Vallee B (1959) Nucleic acids and metals. I. Chromium, manganese, nickel, iron and other metals in ribonucleic acid from diverse biological sources. J Biol Chem 234:3257–3262
Wade GG, Mandel R, Ryser HJ-P (1987) Marked synergism of dimethylnitrosamine carcinogenesis in rats exposed to cadmium. Cancer Res 47:6606–6613
Wahba Z, Waalkes MP (1990) Effect of in vivo low-dose cadmium pretreatment on the in vitro interactions of cadmium with isolated interstitial cells of the rat testes. Fund Appl Toxicol 15:641–650
Webb M (1979) The chemistry, biochemistry, and biology of cadmium. Elsevier/North-Holland Biomedical, NY
Wisniewska-Knypl JM, Jablonska J, Myslak Z (1971) Binding of cadmium on metallothionein in man: an analysis of a fatal poisoning by cadmium chloride. Arch Toxicol 28:46–55
Wong KL, Klaassen CD (1980) Age difference in the susceptibility to cadmium induced testicular damage in rats. Toxicol Appl Pharmacol 55:456–466
Yoshikawa H (1973) Preventive effects of pretreatment with cadmium on acute cadmium poisoning in rats. Ind Health 11:113–119

Zavon MR, Meadow CD (1970) Vascular sequelae to cadmium fume exposure. Am Ind Hyg Assoc J 31:180–182
Zeng J, Heuchel R, Schaffner W, Kägi JHR (1991) Thionein (apometallothionein) can modulate DNA binding and transcription activation by zinc finger containing factor Sp1. FEBS Lett 279:310–312

CHAPTER 10

Chromium Toxicokinetics

E.J. O'FLAHERTY

A. Introduction

Chromium (Cr) toxicokinetics is the toxicokinetics of two different oxidation states, Cr(III) and Cr(VI), linked by reduction processes that are ubiquitous in body fluids and tissues. The kinetic behaviors of these two major oxidation states of chromium are very different. Reduction of Cr(VI) to Cr(III) in the body, the lung, and the gastrointestinal tract is sufficiently rapid that bulk chromium kinetics may be considered to be the kinetics of Cr(III). However, certain detectable differences in chromium disposition depend upon whether exposure is to a Cr(III) or a Cr(VI) salt. In addition, the reduction process itself is of interest relative to the carcinogenicity of Cr(VI) in the lung. Therefore, a comprehensive understanding of the toxicokinetics of chromium must include the disposition of both Cr(III) and Cr(VI).

The important features of Cr(III) and Cr(VI) toxicokinetics are discussed in this chapter. The picture they form is consistent with present understanding of chromium disposition and toxicity. However, great uncertainties remain. Data gaps and uncertainties are identified and discussed in Sect. D.

B. Chromium Actions and Kinetics

I. Local and Systemic Toxicity

Cr(VI) compounds are corrosive as a result of their acidity and oxidizing potential. Oral, pulmonary, and dermal exposure to Cr(VI) may involve local irritation and corrosive action. Dermal irritation with ulceration can lead to systemic uptake of chromium, but this is an extreme situation. Perforation of the nasal septum, formerly common in industries using chromium, is rarely seen today. Dermal contact can also cause delayed sensitization and allergic dermatitis; Cr(VI) appears to be more active than Cr(III) in this regard. These actions of Cr(VI) are local, not systemic, and do not involve any consideration of chromium kinetics.

Apart from their localized toxic actions, Cr(VI) compounds are nephrotoxic in humans (GOYER 1990) and animals (GUMBLETON and NICHOLLS 1988) following high-level acute exposure. Both tubular and glomerular damage occur (APPENROTH and BRÄUNLICH 1988). Evidence for human kidney damage from lower-level chronic exposure to Cr(VI) compounds is equivocal (VERSCHOOR et al. 1988). Chronic animal studies have afforded no evidence of Cr(VI) toxicity, either in the kidney or in other tissues (MACKENZIE et al. 1958).

It has been suggested that there may be a threshold for Cr(VI)-related renal tubular damage (FRANCHINI and MUTTI 1988). Indeed, the rapidity and ubiquity of Cr(VI) reduction processes have suggested to some investigators the existence of a threshold for Cr(VI) dose-response relationships in general (PETRILLI and DEFLORA 1988). In this view, Cr(VI) would be systemically toxic only when the capacity of the multiple mechanisms for its detoxification by reduction to Cr(III) had been exceeded.

Cr(III) has not been shown to be systemically toxic. It has been generally understood that the absence of Cr(III) toxicity is due partly to its inability to cross membranes and thus either to be absorbed or to reach peripheral tissues in significant amounts. However, Cr(III) is now known to be capable of crossing cell and other membranes when the ligand environment is favorable (BIANCHI and LEVIS 1988). Recently, the question has been raised whether Cr(III) may be toxic even though only very small amounts are able to penetrate into cells (DEBETTO and LUCIANI 1988). The possibility of Cr(III) toxicity remains entirely speculative at the present time.

II. Essentiality of Cr(III)

The question whether Cr(III) might be toxic is complicated by the fact that it is considered an essential metal. Biologically active Cr(III) facilitates the action of insulin. Chromium deficiency results in impaired glucose tolerance, which can be corrected by the administration of chromium (ANDERSON et al. 1983; ANDERSON 1986). Hypoglycemia and its associated symptoms can also be corrected by administration of chromium (ANDERSON et al. 1987). These effects of chromium supplementation are often associated with improvements in lipid levels, with a net decrease in total lipids and cholesterol and an increase in the ratio of HDL cholesterol to LDL cholesterol (RIALES and ALBRINK 1981; MOSSOP 1983; EVANS 1989; WANG et al. 1989).

Only very small amounts of chromium are required. The estimated safe and adequate adult daily dietary intake is 50–200 μg/day (National Research Council 1989). Below intakes of about 40 μg/day, fractional oral absorption of chromium increases as intake is decreased (ANDERSON and KOZLOVSKY 1985). Biologically active Cr(III) may be more efficiently absorbed from the gastrointestinal tract than other forms of soluble chromium (MERTZ and ROGINSKI 1971).

It is reasonable to speculate that the disposition kinetics of biologically active Cr(III) may also be different from the kinetics of other complexes of

Cr(III) that occur in blood plasma, but there are no data at the present time to support or refute this possibility. The kinetics of biologically active Cr(III) would not be expected to have a detectable impact on the kinetics of total chromium at the typical total chromium doses given in experimental animal toxicity studies. However, the essentiality of chromium is an important consideration in the development of a global understanding of chromium kinetics and toxicity.

III. Carcinogenicity of Cr(VI)

Industrial exposures to chromium are associated primarily with the manufacture of chromates, including chromate pigments; chromium electroplating, which generates aerosols of chromic acid containing chromic trioxide; production of chromium ferroalloys; and stainless steel welding. Chromate production, electroplating, and welding involve exposure to soluble Cr(VI) compounds. Chromium ferroalloy production involves exposure to insoluble compounds presumed to be largely Cr(III), although some exposure to Cr(VI) probably also occurs. Stainless steel welders are exposed to nickel in addition to chromium (RAITHEL et al. 1988), and chrome electroplaters may be exposed to nickel compounds as well (HATHAWAY 1989). Consideration of a large number of epidemiologic studies has led to the conclusion that exposure to Cr(VI) as chromates, dichromates, or chromic trioxide places the worker at risk for development of lung cancer (LANGÅRD 1990). The epidemiologic data for production of ferrochromium alloys, electroplating, and welding are weak or equivocal, and it is not clear whether associations exist in these trades between exposure to chromium and lung cancer, particularly since nickel in some chemical forms is an established lung carcinogen (STERN 1983; BECKER et al. 1985; LANGÅRD et al. 1990).

The mechanism of Cr(VI) carcinogenicity in the lung is believed to be its reduction to Cr(III) with accompanying generation of reactive intermediates (WETTERHAHN and HAMILTON 1989). Cr(III) is capable of reacting with isolated DNA, and of causing DNA-protein cross-linking in solution and in isolated nuclei (FORNACE et al. 1981). However, it is generally inactive in bacterial and mammalian cell test systems (DEFLORA et al. 1990). In contrast, soluble Cr(VI) salts are not ordinarily genotoxic in acellular test systems, but do test positive for mutagenicity in short-term tests using cellular systems (DEFLORA et al. 1990). Cr(III) binds to isolated native and synthetic DNA only at low levels compared with the amount of chromium bound when Cr(VI) is incubated with isolated DNA in the presence of a microsomal reducing system (TSAPAKOS and WETTERHAHN 1983). These observations have led to the understanding that electrophilic intermediates produced in the course of reduction of Cr(VI) to Cr(III) are the active genotoxic agents. In this view, reduction of Cr(VI) in the lung would be considered an activation rather than a detoxification process.

Cr(VI) is reduced intracellularly by a wide variety of enzyme systems: microsomal (ALEXANDER et al. 1986), mitochondrial (ROSSI and WETTERHAHN 1989), and cytosolic (BANKS and COOKE 1986). It also is reduced nonenzymatically. Reactive moieties generated in these reduction processes include Cr(V) (ROSSI and WETTERHAHN 1989), hydroxyl radicals, and singlet oxygen (KAWANISHI et al. 1986).

The relative efficiencies of Cr(VI) reduction in the lung and its clearance from the lung are important determinants of its local carcinogenic potency. Nonetheless, data that might fix the relative magnitudes of these pathways are scanty. In addition, it is likely that the balance among Cr(VI) elimination mechanisms in the lung is influenced by extrinsic factors such as particle size and solubility (see Sect. C.II.2). The behavior of chromium in the lung has not been quantified to an extent that would allow reliable simulation of a relationship between Cr(VI) inhalation exposure and carcinogenic action.

C. Key Features of Chromium Kinetics

The important features of chromium kinetics discussed in this section have been drawn from studies in the rat unless otherwise indicated. Table 1 contains a listing of the most important studies of the kinetics of Cr(III) and Cr(VI) in rats. The results of these studies form the core of the following discussion.

I. Solubility

Both Cr(III) and Cr(VI) form compounds whose water solubilities range from very high to very low. Many compounds of industrial importance, such as the sodium, potassium, and ammonium chromates and dichromates, chromium trioxide (CrO_3), and the hydrated Cr(III) nitrate, chloride,

Table 1. Key chromium kinetic studies in rats

Oxidation state	Administration	Reference
Cr(III)	Intravenous	MERTZ et al. (1965), HOPKINS (1965)
Cr(VI)	Intravenous	None
Cr(III)	Stomach tube[a]	MACKENZIE et al. (1959)
Cr(VI)	Stomach tube[a]	MACKENZIE et al. (1959)
Cr(III)	Drinking water; chronic	MACKENZIE et al. (1958)
Cr(VI)	Drinking water; chronic	MACKENZIE et al. (1958)
Cr(III)	Intratracheal	EDEL and SABBIONI (1985)
Cr(VI)	Intratracheal	WEBER (1983)
		BRAGT and VAN DURA (1983), EDEL and SABBIONI (1985)
Cr(III)	Inhalation	None
Cr(VI)	Inhalation	LANGÅRD et al. (1978)

[a] Plasma/red cell ratio only.

acetate, and sulfate salts, are soluble or very soluble in water. Some of the chromate pigments are poorly soluble. Poorly soluble or essentially insoluble salts include lead chromate, chromium oxide (Cr_2O_3), and chromic hydroxide [$Cr(OH)_3$]. This wide range of solubilities complicates efforts to model the kinetics of chromium and to rationalize its toxicity, especially its carcinogenicity.

Cr(VI) from soluble salts will be absorbed from the lung if it is not reduced to Cr(III). Insoluble salts may be transported to the gastrointestinal tract, where, if the Cr(VI) escapes reduction in the stomach, it can be absorbed from the intestine. However, Cr(VI) is generally largely reduced in the acid environment of the stomach, and subsequently handled like Cr(III). Cr(III) salts are poorly absorbed, in either the lung or the gastrointestinal tract, whether or not they are soluble (see Sect. C.II).

Solubility also influences carcinogenic action. Soluble Cr(VI) compounds may generate higher local concentrations of chromium in the lung than insoluble compounds. However, soluble chromium is also more rapidly cleared from the lung to the systemic circulation and to the gastrointestinal tract. The results of experimental animal studies and of epidemiologic studies of workers with industrial chromium exposure support the hypothesis that moderately soluble chromate salts (zinc, calcium, and strontium chromates) are the most potent pulmonary carcinogens, while the highly soluble salts (sodium, potassium, and ammonium chromates) are much less active (HATHAWAY 1989; GIBB and CHEN 1989). It is inferred from these observations that the less readily soluble salts generate lower but persistent local concentrations of Cr(VI) in the lung.

II. Membrane Permeability and Chromium Absorption

Reasonable solubility is a prerequisite for absorption of either Cr(III) or Cr(VI). Insoluble chromium compounds were used historically as markers of gastrointestinal transit time.

Cr(III) is very poorly transferred across membranes. It is absorbed only to a limited extent, usually less than 1%, from soluble Cr(III) salts in the gastrointestinal tract (MERTZ et al. 1965; MACKENZIE et al. 1959) and the lung (EDEL and SABBIONI 1985). Soluble salts of Cr(VI) are more readily absorbed from both the lung (BRAGT and VAN DURA 1983) and the gastrointestinal tract (MACKENZIE et al. 1959) than are soluble salts of Cr(III).

1. Gastrointestinal Absorption

As much as 20% of a dose of Cr(VI) injected directly into the small intestine was absorbed by rats (MACKENZIE et al. 1959). It must be remembered, however, that much of the Cr(VI) entering the stomach will actually be absorbed as Cr(III), because it will have been reduced to Cr(III) before entering the small intestine. In the MACKENZIE et al. (1959) study, 6% of an oral dose of a soluble Cr(VI) salt was found to be absorbed. By comparison of concentrations of radiolabeled chromium in the blood 4–10 days after intravenous or oral administration of a soluble Cr(III) salt, MERTZ et al. (1965) determined that 2%–3% of the oral dose had been absorbed, independent of dose in the range 0.15–10.0 μg/kg. MACKENZIE et al. (1959), however, estimated that Cr(III) was absorbed only one-tenth or less as well as Cr(VI).

The nature of the anion making up the Cr(III) salt, as well as the gastrointestinal tract contents and the nutritional status of the animal, are important determinants of efficiency of absorption from the gastrointestinal tract. For example, oxalates are better absorbed than phytates (CHEN et al. 1973). There is a suggestion that biologically active complexes of Cr(III) are even better absorbed, perhaps to the extent of 10%–25% (MERTZ and ROGINSKI 1971). Zinc status also influences chromium absorption (HAHN and EVANS 1975). Both Cr(III) and Cr(VI) salts are better absorbed in the fasting than in the fed state (MACKENZIE et al. 1959).

2. Pulmonary Absorption

In the lung, solubility is probably the single most important determinant of Cr(VI) absorption efficiency. Within a series of Cr(VI) salts of different solubilities, systemic availability tends to parallel solubility (BRAGT and VAN DURA 1983). The range of half-lives reported for clearance of chromium from the lung (BRAGT and VAN DURA 1983; WEBER 1983) suggests that chromium may be present in the lung in different states or compartments from which it is cleared at different rates. Particle size distribution may also be expected to be a critical factor determining localization and clearance of chromium in the lung. Unfortunately, little is known about these factors as they relate either to individual chromium compounds or to the mixture of salts that may be present in the lung after workplace or ambient exposure to chromium-containing compounds.

Transfer to the gastrointestinal tract is an important clearance mechanism for chromium in the lung. LANGÅRD et al. (1978) observed rapid increases in fecal chromium during 6-h inhalation exposures of rats to dusts of zinc chromate, a moderately soluble Cr(VI) salt. BRAGT and VAN DURA (1983) found that the fraction of administered chromium appearing in the feces increased sharply as the solubility of the Cr(VI) salt decreased, with as much as 80% of an intratracheal dose of insoluble lead chromate appearing in feces by the 9th day postadministration compared with 20% of an intratracheal dose of soluble sodium chromate. The figure of 20% for sodium chromate was confirmed by WEBER (1983) and by EDEL and SABBIONI (1985), who showed also that 36% of an intratracheal dose of a soluble Cr(III) salt had been excreted in the feces by day 7, presumably because the Cr(III) was poorly cleared by systemic absorption.

The kinetics of clearance of chromium salts from the lung suggest that systemic absorption is a rapid process that may be complete within 1 h after a single intratracheal dose (BRAGT and VAN DURA 1983). Subsequent clearance of chromium from the lung can be described as a sum of exponential terms with half-lives of the order of days up to 1 month (BRAGT and VAN DURA 1983; WEBER 1983). Pulmonary clearance of chromium is not dose dependent within a reasonable dose range (BRAGT and VAN DURA 1983; WEBER 1983).

III. Reduction of Cr(VI) to Cr(III)

In biologic milieu, Cr(VI) is rapidly reduced to Cr(III) (see Sect. B.III). Although Cr(VI) reduction is catalyzed intracellularly by a wide range of microsomal, mitochondrial, and cytosolic enzyme systems, it does not require enzyme or even biochemical mediation; it occurs, for example, very rapidly in the acid environment of the stomach (DEFLORA et al. 1987). The rate of reduction in the stomach has been shown to be influenced both by acidity (DONALDSON and BARRERAS 1966; work done in humans) and by the pattern of food intake (DEFLORA et al. 1987).

Given the variety of mechanisms involved, it is reasonable to think that the net reduction rate of Cr(VI) to Cr(III) in the body may not only be tissue specific but may also be influenced by the nutritional and physiologic status of the animal. Unfortunately, very little experimental information is available that would allow this premise to be tested.

CAVALLERI et al. (1985) found that the half-life of Cr(VI) in rat whole blood and plasma in vivo was less than 1 min. Chromium was administered intravenously as potassium dichromate in this study, and the observed half-life could have been influenced by specific physicochemical features of the salt (see Sect. C.IV). It is not known whether this is a representative half-life for systemic Cr(VI) administered by other routes.

IV. General Chromium Disposition

Cr(III) can penetrate into cells of peripheral tissues (See Sect. B.I), presumably in the form of diffusible complexes (BIANCHI and LEVIS 1988). It can also enter and leave the erythrocyte (EDEL and SABBIONI 1985) and be secreted into the bile (CIKRT and BENCKO 1979). However, these transfer processes are sluggish compared to Cr(III) elimination in the urine.

When chromium is administered by a "natural" route, it presumably reaches the plasma largely as a mix of stable Cr(III) organic complexes with amino acids, other low-molecular-weight organic acids, and proteins (GAD 1989). The composition of the mix of Cr(III) complexes in the plasma should vary to some extent with the nutritional and physiologic status of the subject, but should be essentially independent of the chemical form of the administered chromium. When Cr(III) is administered intravenously, however, its disposition is strongly dependent both on the chemical form of the chromium compound and on the medium in which it is administered. In a study of the disposition of chromium salts (VISEK et al. 1953), unbuffered Cr(III) chloride was found to be taken up principally by the liver, spleen, and bone marrow, suggesting that it acted as a colloid under these conditions. The results of electrophoretic studies supported this interpretation. Sodium chromite (Na_3CrO_2) was even more strongly trapped by tissues of the reticuloendothelial system, and was only minimally excreted during the first 4 days after administration. Buffering the Cr(III) chloride solution with either acetate or citrate resulted in a very different pattern of disposition, with less chromium in tissues and a much larger fraction excreted during the first 4 days after administration. Tissue distribution was similar to that seen after oral administration of Cr(III) chloride. Therefore, the tissue distribution of chromium in studies in which unbuffered Cr(III) has been administered intravenously may not be the same as it would have been following a more physiologic route of exposure. It is also possible that Cr(III) kinetics following intraperitoneal administration may be influenced either by the identity of the chemical compound or by toxicity at the injection site (VISEK et al. 1953).

Chromate salts appear to behave as ionic species in the plasma (VISEK et al. 1953; GAD 1989), while Cr(III) travels largely bound to plasma proteins, mostly globulins. Unless Cr(III) is present in excess, the majority of it travels with the iron-binding protein transferrin (HOPKINS and SCHWARZ 1964). With larger amounts of Cr(III), binding to other proteins, including albumin, occurs. A fraction of the Cr(III) in plasma is complexed with low-molecular-weight ligands (MERTZ 1969), and it is this small fraction that may be able to traverse membranes and thus diffuse out of the blood.

V. Chromium in the Red Cell

Cr(VI) readily enters the red cell, where it is reduced to Cr(III). It had at one time been thought that Cr(III) was trapped in the red cell for the remaining life of the cell. However, the half-life of chromium in the red cell is substantially lower than the cell turnover time (GRAY and STERLING 1950; BISHOP and SURGENOR 1964), and chromium has been shown to enter and leave the red cell in studies in which Cr(III) salts were administered intratracheally (EDEL and SABBIONI 1985). The picture that emerges from these observations is that Cr(VI) rapidly enters the red cell where it is reduced to Cr(III), in which form it is slowly lost from the cell. Cr(III) also diffuses into the red cell, but much more slowly.

In Table 2, a comparison is made of the red cell : plasma chromium ratio in rats given soluble Cr(VI) by three different routes of administration. The fraction of absorbed chromium entering the systemic circulation as Cr(VI) declines in the order: intravenous > intestinal > oral administration. If

Table 2. Red cell: plasma chromium after administration of Cr(VI) salts

Administration	Time post-administration	Red cell: plasma concentration ratio	Reference
Intratracheal instillation	48 h	2.7	WEBER (1983)
Intratracheal instillation	24 h	1.6	EDEL and SABBIONI (1985)
Intratracheal instillation	24 h	1.7	WEBER (1983)
Intratracheal instillation	6 h	0.87	WEBER (1983)
Intravenous injection	4 h	1.9	GRAY and STERLING (1950)
Intestinal injection	4 h	0.31	MACKENZIE et al. (1959)
Stomach tube	4 h	0.11	MACKENZIE et al. (1959)

Cr(VI) enters the red cell rapidly while Cr(III) does not, the red cell: plasma chromium ratio at any single time point after administration should decline in the same order. As Table 2 shows, this is in fact the case. In addition, Table 2 shows that the red cell: plasma chromium ratio increases with time after administration, and suggests that the fraction of an intratracheal dose of Cr(VI) entering the systemic circulation as Cr(VI) may fall between the values for fractional absorption from intravenous and intestinal doses.

After intratracheal administration of a radiolabeled soluble Cr(VI) salt to rats, WEBER (1983) found that the amount of radiolabel in the serum was greater than that in the red cells at 6 h after administration, but that by 2 days after administration the serum contained only 40% as much chromium as the red cells. By 40 days after administration, the percentage had declined to 1.0. Therefore, red cell chromium is lost more slowly than plasma chromium, and plasma chromium is the pool from which chromium is transferred into other tissues and from which it is excreted.

VI. Chromium in Bone

A significant fraction of the body burden of chromium is found in bone (WITMER and HARRIS 1991). WEBER (1983) found that radiolabel from an intratracheally administered chromate salt was located principally in the epiphyseal region of the long bones. The distribution of radiolabel was reported to be similar to that seen after administration of bone-seeking elements such as Ca^{45} or Sr^{89}. KRAINTZ and TALMAGE (1952) had also noted localization of radiolabel in the epiphyses of the long bones after intravenous administration of $Cr^{51}Cl_3$ to rabbits.

The fraction of a single dose of chromium that is incorporated into the bone is dependent on age in such a way as to suggest a link with bone formation rate. HOPKINS (1965) observed that bone of young, growing rats tended to concentrate chromium with time (7% of the dose 4h after an intravenous injection of $Cr^{51}Cl_3$, increasing to 12% of the dose at 24h postinjection) while bone of mature rats acquired a smaller fraction of the label (4% of the dose at 4h) that did not increase with time. These observations strongly suggest that a portion of the chromium incorporated into bone is associated with formation of new bone, and that around 4% of the intravenous dose was taken up by rapid surface exchange of plasma and bone chromium, while the additional incorporation of chromium into bone of the growing rats was attributable to bone growth.

VII. Chromium in Other Tissues

There is little reliable information on amounts of chromium in tissues following single oral or inhalation exposures, and some of these data, especially those from earlier studies, are contradictory. Retention of chromium by liver, kidney, and spleen is prolonged (MACKENZIE et al. 1959; BRAGT and VAN DURA 1983; EDEL and SABBIONI 1985). The relatively slow loss from these and other tissues (testis, epididymis) (HOPKINS 1965) requires the assumption that transfer of chromium into and out of tissues is diffusion-limited rather than flow-limited.

Previous dietary history, including feeding of a chromium-deficient diet, does not appear to affect chromium distribution or kinetics; in addition, no dose or sex dependence of chromium distribution in rats at single intravenous doses between 0.01 μg/100 g and 15 μg/100 g was observed (HOPKINS 1965; MERTZ et al. 1965). However, one study of chronic oral exposure in rats showed that after 1 year of exposure to a range of drinking water concentrations up to 25 ppm, liver concentrations of chromium were not linearly related to dose but were disproportionately high at drinking water concentrations above about 8–10 ppm (MACKENZIE et al. 1958). While concentrations of chromium in kidney and bone after this chronic exposure period were not clearly disproportional to dose, it is apparent that alterations in chromium kinetics may occur during chronic higher-level exposure.

VIII. Excretion

The percentage of a chromium dose excreted in the urine is independent of the oxidation state of the administered chromium. About 21%–22% is excreted in the 24h following intravenous injection of a soluble salt of either Cr(III) or Cr(VI) in rats (CIKRT and BENCKO 1979).

Not all chromium is excreted in the urine, however. CIKRT and BENCKO (1979) found 0.5% of an intravenous dose of $Cr^{51}Cl_3$ and 3.5% of an intravenous dose of $Na_2Cr^{51}O_4$ in the bile of rats during the first 24h after administration. The chief difference in the behaviors of the two oxidation states of chromium was a more rapid excretion of Cr(VI) during the first few hours, which is consistent with the different membrane transfer capabilities of Cr(III) and Cr(VI). CAVALLERI et al. (1985) administered $K_2Cr^{51}O_4$ to adult rats by intravenous injection in order to determine the amounts of total chromium and of Cr(VI) appearing in bile. They found that about 2% of the dose was excreted in the bile within 2 h; less than 1% of this amount was excreted as Cr(VI). These percentages were independent of dose from 0.1–1.0 mg chromium. Thus, while most of an intravenous Cr(VI) dose is reduced to Cr(III) before its excretion, a small fraction appears in the bile as Cr(VI). Since no special precautions were reported by these authors for avoiding reduction of Cr(VI) to Cr(III) in the isolated bile sample [Cr(VI) was found to be only slowly reduced in vitro in rat bile over a 1-h period], it is possible that the fraction of total chromium excreted as Cr(VI) in the bile might have been somewhat greater than 1%.

Interestingly, CIKRT and BENCKO (1979) found that the gastrointestinal tract contents and feces together accounted for 4.2% of an intravenous Cr(III) dose and 7.3% of an intravenous Cr(VI) dose in rats after 24 h, suggesting a fairly substantial gastrointestinal excretion.

D. Uncertainties and Research Needs

A qualitative outline of chromium kinetic behavior emerges from these studies. Chromium(VI) is absorbed, distributed, and excreted substantially more readily than Cr(III). At the same time, reduction of Cr(VI) to Cr(III) occurs so rapidly in the lung, the gastrointestinal tract, and the body that to a large extent the kinetics of Cr(VI) have been thought of as the kinetics of Cr(III). This is not precisely correct. The rapidity with which Cr(VI) is reduced, compared to the rapidity of its absorption and excretion processes, controls a sensitive balance that determines overall absorption, distribution, and excretion of chromium as well as the amounts absorbed, distributed, and excreted as Cr(VI). In addition, the nature and rate of the reduction process itself appear to be linked with Cr(VI) pulmonary carcinogenicity.

Several important data gaps relating to chromium kinetic behavior can be identified.

1. A fully adequate quantitative understanding of chromium kinetics as related to pulmonary toxicity requires additional information about the relative importance of the competing process integral to chromium kinetic behavior in the lung. Studies on local and total chromium kinetics in the lung should include investigation of the oxidation states and compartmentalization of chromium as well as of the relative rates of Cr(VI) reduction, chromium transfers within the lung, and elimination of chromium from the lung by systemic absorption and transfer to the gastrointestinal tract.

2. Only limited information regarding the absorbability of chromium from environmental sources is available. Since even soluble salts of chromium are not particularly well absorbed, either in the lung or in the gastrointestinal tract, it may be that availability of chromium to absorption pro-

cesses will prove to be the most important single characteristic of a chromium source determining its potential absorption and toxicity. The availability/absorbability of chromium from environmental sources is unknown for all but a few chemically defined salts.

3. Little is known about the importance of bone as a reservoir and continuing source of internal exposure to chromium. The mechanism(s) by which chromium is incorporated into bone, and the dependence of bone chromium uptake on age and physiologic status, are important features of any complete model of chromium kinetics.

4. Nearly all of our understanding of chromium kinetics is based on single-dose animal studies. However, humans are generally chronically exposed to chromium, perhaps at varying rates over time. The impact of chronic higher-level as well as lower-level exposure on chromium kinetics should be investigated.

References

Alexander J, Mikalsen A, Ryberg D (1986) Microsomal reduction of Cr VI. Acta Pharmacol Toxicol (Copenh) 59:267–269

Anderson RA (1986) Chromium metabolism and its role in disease processes in man. Clin Physiol Biochem 4:31–41

Anderson RA, Kozlovsky AS (1985) Chromium intake, absorption and excretion of subjects consuming self-selected diets. Am J Clin Nutr 41:1177–1183

Anderson RA, Polansky MM, Bryden NA, Roginski EE, Mertz W, Glinsmann W (1983) Chromium supplementation of human subjects: effects on glucose, insulin, and lipid variables. Metabolism 32:894–899

Anderson RA, Polansky MM, Bryden NA, Canary J (1987) Supplemental chromium effects on glucose, insulin, glucagon, and urinary chromium losses in subjects consuming controlled low-chromium diets. Am J Clin Nutr 54:909–916

Appenroth D, Bräunlich H (1988) Age dependent differences in sodium dichromate nephrotoxicity in rats. Exp Pathol 33:179–185

Banks RB, Cooke RT Jr (1986) Chromate reduction by rabbit liver aldehyde oxidase. Biochem Biophys Res Commun 137:8–14

Becker N, Claude J, Frenzel-Beyme R (1985) Cancer risk of arc welders exposed to fumes containing chromium and nickel. Scand J Work Environ Health 11:75–82

Bianchi V, Levis AG (1988) Review of genetic effects and mechanisms of action of chromium compounds. Sci Total Environ 71:351–355

Bishop C, Surgenor M (eds) (1964) The red blood cell: a comprehensive treatise. Academic, New York

Bragt PC, van Dura EA (1983) Toxicokinetics of hexavalent chromium in the rat after intratracheal administration of chromates of different solubilities. Ann Occup Hyg 27:315–322

Cavalleri A, Minoia C, Richelmi P, Baldi C, Micoli G (1985) Determination of total and hexavalent chromium in bile after intravenous administration of potassium dichromate in rats. Environ Res 37:490–496

Chen NSC, Tsai A, Dyer IA (1973) Effect of chelating agents on chromium absorption in rats. J Nutr 103:1182–1186

Cikrt M, Bencko V (1979) Biliary excretion and distribution of $^{51}Cr(III)$ and $^{51}Cr(VI)$ in rats. J Hyg Epidemiol Microbiol Immunol 23:241–246

DeBetto P, Luciani S (1988) Toxic effect of chromium on cellular metabolism. Sci Total Environ 71:365–377

DeFlora S, Badolati GS, Serra D, Picciotto A, Magnolia MR, Savarino V (1987) Circadian reduction of chromium in the gastric environment. Mutat Res 192: 169–174
DeFlora S, Bagnasco M, Serra D, Zanacchi P (1990) Genotoxicity of chromium compounds. A review. Mutat Res 238:99–172
Donaldson RM Jr, Barreras RF (1966) Intestinal absorption of trace quantities of chromium. J Lab Clin Med 68:484–493
Edel J, Sabbioni E (1985) Pathways of Cr(III) and Cr(VI) in the rat after intratracheal administration. Hum Toxicol 4:409–416
Evans GW (1989) The effect of chromium picolinate on insulin controlled parameters in humans. Int J Biosoc Med Res 11:163–180
Fornace AJ, Seres DS, Lechner JF, Harris CC (1981) DNA-protein cross-linking by chromium salts. Chem Biol Interact 36:345–354
Franchini I, Mutti A (1988) Selected toxicological aspects of chromium (VI) compounds. Sci Total Environ 71:379–387
Gad SC (1989) Acute and chronic systemic chromium toxicity. Sci Total Environ 86:149–157
Gibb H, Chen C (1989) Evaluation of issues relating to the carcinogen risk assessment of chromium. Sci Total Environ 86:181–186
Goyer RA (1990) Environmentally related diseases of the urinary tract. Environ Med 74:377–389
Gray SJ, Sterling K (1950) The tagging of red cells and plasma proteins with radioactive chromium. J Clin Invest 29:1604–1613
Gumbleton M, Nicholls PJ (1988) Dose-response and time-response biochemical and histological study of potassium dichromate-induced nephrotoxicity in the rat. Food Chem Toxicol 26:37–44
Hahn CJ, Evans GW (1975) Absorption of trace metals in the zinc-deficient rat. Am J Physiol 228:1020–1023
Hathaway JA (1989) Role of epidemiologic studies in evaluating the carcinogenicity of chromium compounds. Sci Total Environ 86:169–179
Hopkins LL Jr (1965) Distribution in the rat of physiological amounts of injected Cr^{51}(III) with time. Am J Physiol 209:731–735
Hopkins LL Jr, Schwarz K (1964) Chromium (III) binding to serum proteins, specifically siderophilin. Biochim Biophys Acta 90:484–491
Ivankovic S, Preussmann R (1975) Absence of toxic and carcinogenic effects after administration of high doses of chromic oxide pigment in sub-acute and long-term feeding experiments in rats. Food Cosmet Toxicol 13:347–351
Kawanishi S, Inoue S, Sano S (1986) Mechanism of DNA cleavage induced by sodium chromate (VI) in the presence of hydrogen peroxide. J Biol Chem 261:5952–5958
Kraintz L, Talmage RV (1952) Distribution of radioactivity following intravenous administration of trivalent chromium 51 in the rat and rabbit. Proc Soc Exp Biol Med 81:490–492
Langård S (1990) One hundred years of chromium and cancer: a review of epidemiological evidence and selected case reports. Am J Ind Med 17:189–215
Langård S, Gundersen N, Tsalev DL, Gylseth B (1978) Whole blood chromium level and chromium excretion in the rat after zinc chromate inhalation. Acta Pharmacol Toxicol 42:142–149
MacKenzie RD, Byerrum RU, Decker CF, Hoppert CA, Langham RF (1958) Chronic toxicity studies. II. Hexavalent and trivalent chromium administered in drinking water to rats. AMA Arch Ind Health 18:232–234
MacKenzie RD, Anwar RA, Byerrum RU, Hoppert CA (1959) Absorption and distribution of Cr^{51} in the albino rat. Arch Biochem Biophys 79:200–205
Mertz W (1969) Chromium occurrence and function in biological systems. Physiol Rev 49:163–239

Mertz W, Roginski EE (1971) Chromium metabolism: the glucose tolerance factor. In : Mertz W, Cornatzer WE (eds) Newer trace elements in nutrition. Dekker, New York, pp 123–153

Mertz W, Roginski EE, Reba RC (1964) Biological activity and fate of trace quantities of intravenous chromium (III) in the rat. Am J Physiol 209:489–494

Mossop RT (1983) Effects of chromium (III) on fasting glucose, cholesterol and cholesterol HDL levels in diabetics. Cent Afr J Med 29:80–82

National Research Council (1989) RDA (Recommended Dietary Allowance), 10th edn. National Academy Press, Washington

Norseth T, Alexander J, Aaseth J, Langård S (1982) Biliary excretion of chromium in the rat: a role of glutathione. Acta Pharmacol Toxicol (copenh) 51:450–455

Petrilli FL, De Flora S (1988) Metabolic reduction of chromium as a threshold mechanism limiting its in vivo activity. Sci Total Environ 71:357–364

Raithel HJ, Schaller KH, Reith A, Svenes KB, Valentin H (1988) Investigations on the quantitative determination of nickel and chromium in human lung tissue. Int Arch Occup Environ Health 60:55–66

Riales R, Albrink MJ (1981) Effect of chromium chloride supplementation on glucose tolerance and serum lipids including high density lipoprotein of adult men. Am J Clin Nutr 34:2670–2678

Rossi SC, Wetterhahn KE (1989) Chromium (V) is produced upon reduction of chromate by mitochondrial electron transport chain complexes. Carcinogenesis 10:913–921

Stern RM (1983) Assessment of risk of lung cancer for welders. Arch Environ Health 38:148–155

Tsapakos MJ, Wetterhahn KE (1983) The interaction of chromium with nucleic acids. Chem Biol Interact 46:265–277

Verschoor MA, Bragt PC, Herber RFM, Zielhuis RL, Zwennis WCM (1988) Renal function of chrome-plating workers and welders. Int Arch Occup Environ Health 60:67–70

Visek WJ, Whitney IB, Kuhn USG III, Comar CL (1953) Metabolism of Cr^{51} by animals as influenced by chemical state. Proc Soc Exp Biol Med 84:610–615

Wang MM, Fox EA, Stoecker BJ, Menendez CE, Chan SB (1989) Serum cholesterol of adults supplemented with brewer's yeast or chromium chloride. Nutr Res 9:989–998

Weber H (1983) Long-term study of the distribution of soluble chromate-51 in the rat after a single intratracheal administration. J Toxicol Environ Health 11:749–764

Wetterhahn KE, Hamilton JW (1989) Molecular basis of hexavalent chromium carcinogenicity: effect on gene expression. Sci Total Environ 86:113–129

Witmer CM, Harris R (1991) Chromium content of bone after oral and intraperitoneal (ip) administration of chromium (VI) to rats (Abstr). Toxicologist 11:41

World Health Organization (1988) Environmental health criteria 61: chromium. IPCS International Programme on Chemical Safety. World Health Organization, Geneva

CHAPTER 11

Metals and Stress Proteins

P.L. GOERING and B.R. FISHER

A. Introduction

Prokaryotic and eukaryotic cells respond to physical and chemical stressors or stress by increasing the transcription of specific genes that encode for a small class of proteins termed heat shock proteins (hsps). This response is believed to represent a transient reprogramming of gene expression and biological activity which serves to protect sensitive cellular components from irreversible damage and assists in the rapid recovery after the stress is removed or ceases. The changes in gene expression associated with this response following exposure to a stimulus are rapid, and result in both increased de novo synthesis and accumulation of stress proteins. Originally termed the heat shock response because of the induction of these proteins following hyperthermia (RITOSSA 1962), the signaling mechanism involved in its initiation is sensitive to a variety of physical and chemical insults, including metals (NOVER 1991). Because the response can be initiated by a variety of stressors it is generically referred to as the "stress response." In this review, for convention and to avoid confusion, we have adopted the use of the term "stress proteins" when referring to classic hsps and other stress proteins. Therefore, the hsps will be considered as a subset of the stress proteins, and reference to hsps will be used only when deemed necessary, e.g., when referring to specific references and studies describing specific hsps.

The inducing stimuli of stress proteins and the stress response are numerous and varied, and can be divided into three broad categories: (a) environmental stresses; (b) pathophysiological conditions, including viruses, microbial infection, tissue trauma, and disease states; and (c) particular physiological processes such as cell cycling, embryonic development, cell growth and differentiation, hormonal stimulation, and oncogene activation. Of particular interest to the field of toxicology are the numerous and diverse chemical and physical inducers of stress proteins. These inducers include many metals, hypoxia, ischemia/reperfusion, UV irradiation, ethanol, hydrogen peroxide, dioxins, pentobarbital, diphenylhydantoin, and several other carcinogens, mutagens, teratogens, sulfhydryl reagents, and pro-oxidants (SCHLESINGER 1990; NOVER 1991).

The purpose of this chapter is to review the impact of an important class of chemical inducers of stress proteins, the metals. The synthesis of stress proteins has been shown to be induced by a host of different metals, which includes arsenite, cadmium, mercury, copper, zinc, lead, iron, gold, and gallium. While the understanding of the relationships between metals and their capacity to induce the stress response is incomplete, these interactions are important to consider because they may reveal information regarding mechanisms of toxicity, cellular defense mechanisms against metal toxicity, and biochemical responses which can be exploited as biomarkers of exposure/toxicity for metals. These issues will be addressed in the following manner. First, the current knowledge of the physiologic roles and functions of several major stress proteins (classically referred to as the hsps) will be discussed to acquaint the metals toxicologist with this broad area of research. Although many proteins have been identified, this review will focus on the major ones and will not be an exhaustive review of the literature. Since the induction of specific proteins is dependent on the stressor, tissue, cell type, and species, the discussion will focus on the major stress protein families, including hsp70, hsp90, hsp110, hsp28, and hsp32. Second, the effects of specific metals that can alter the expression of various stress proteins will be discussed. Following this review, evidence will be presented which identifies an enzyme well known to be affected by metals, heme oxygenase, as a stress protein. Fourth, we will make a case that metallothionein, which is highly inducible and has a high affinity for several metals, is a stress protein. Finally, the potential for using stress proteins as biomarkers of exposure, toxicity, and environmental stress will be addressed.

B. Stress Proteins and Their Functions

Studies on stress protein genes have revealed that these genes contain highly conserved sequence homologies among widely divergent organisms and that several of the stress proteins are members of "multigene families." In general, eukaryotes possess at least two copies of most stress protein genes: a stress-inducible gene, and a constitutively expressed gene (or cognate gene). The constitutive proteins are present at lower levels in cells in the absence of stress and are involved in maintaining cellular homeostasis (for review, see Hendrick and Hartl 1993). In this function, these proteins readily bind with other proteins or polypeptides. Such protein-protein interactions are one mechanism whereby stress proteins aid in the transport of other proteins across intracellular membranes (Pelham 1986), assist in polypeptide assembly and prevent protein misfolding (Rothman 1989), and bind to a variety of intracellular receptors in the absence of ligand (Howard et al. 1990; Wilhelmsson et al. 1990).

The major stress proteins have been classified in protein families based on their sequence homology and their relative molecular mass: the small hsp family ranging from 15 kilodaltons (kDa) to 40 kDa (which now includes the heme catabolic enzyme, heme oxygenase), the hsp70 family comprising proteins from 66 kDa to 78 kDa, the hsp90 family with proteins ranging from 83 kDa to 90 kDa, and the large molecular weight stress proteins of 100 kDa to 110 kDa (MORIMOTO et al. 1990). A closely related stress protein family, the glucose-regulated proteins (grps), have masses of 78–94 kDa and share considerable sequence homology with several hsps.

Most hsp research in mammalian systems has concentrated on the characterization and function of the hsp70 proteins. There are two major forms of hsp70, a constitutively expressed 73-kDa protein and an inducible 72-kDa protein. In many cells, a rapid increase in 72-kDa protein expression occurs under physiologic stress conditions. In eukaryotic cells, the constitutive heat shock 73-kDa proteins, or hsc70, are associated with various organelles in the cytoplasm. These proteins are believed to be critical for maintenance of cellular homeostasis, and their proposed functions all involve the modulation of protein-protein interactions (SCHLESINGER 1990). The hsc70 proteins function in the transport of proteins from the cytoplasm into the endoplasmic reticulum microsomes (CHIRICO et al. 1988) and mitochondria (CHENG et al. 1989). The mechanism of translocation into organelles is believed to involve protein-protein interactions, whereby hsc70 proteins unfold and refold other proteins into conformations which are more readily translocated through membranes. These proteins have been termed "molecular chaperones" (HEMMINGSEN et al. 1988). BECKMANN et al. (1990) and SHEFFIELD et al. (1990) demonstrated that hsc70 proteins also interact with newly synthesized proteins. These proteins bind to nascent polypeptide chains in order to assist in proper protein assembly and prevent precursor aggregation.

During physiologic stress, newly synthesized hsp72 translocates from cytoplasm into the nucleus. Although less is known about the specific function of hsp72, it is believed to act in a fashion similar to the hsc70 proteins. HIGHTOWER (1991) has proposed that hsp72 binds to thermal-sensitive proteins to prevent irreversible denaturation and aggregation and facilitates the refolding and repair of proteins as cells recover from stress. It has been postulated that these protein-protein interactions may be one of the underlying mechanisms by which cells acquire thermotolerance, i.e., tolerance to an otherwise lethal heat stress after exposure to a sublethal heat shock sufficient to induce stress protein synthesis (WELCH 1987; JOHNSTON and KUCEY 1988; RIABOWOL et al. 1988).

The eukaryotic hsp90 proteins are constitutively expressed in cells under normal growth conditions, and the synthesis of this protein increases under physiologic stress conditions. The hsp90 proteins are localized in the cell cytoplasm and, like the hsp70 proteins, complex with a variety of cellular proteins, including protein kinases and cytoskeletal proteins (SCHLESINGER 1990). In addition, the hsp90 proteins have been shown to bind to steroid receptors (CARSON-JURICA et al. 1989; HOWARD et al. 1990) and the dioxin receptor (Wilhelmsson et al. 1990), stabilizing these receptors in inactive conformations in the absence of ligand. During thermal stress, hsp90 translocates to the nucleus, although the specific function of this protein has not been elucidated.

Similar to the proteins constituting the hsp90 family, proteins comprising the hsp110 family are constitutively expressed and upregulated during physiologic stress. This protein is localized within the nucleus and appears to be associated with the nucleolus. It is believed that hsp110 binds to and protects the nucleolar transcription machinery during physiologic stress and may be involved in the recovery of normal rRNA transcription following stress (Subjeck et al. 1983).

In mammalian cells, hsp28 is present in a relatively low steady-state concentration and increases many-fold in cells after stress. In addition to the increased synthesis of this protein following stress, Welch (1985) has shown that phosphorylation of hsp28 increased significantly in cells exposed to mitogens and tumor promoters, and postulated that the action of hsp28 is mediated by this change in phosphorylation state. Moretti-Rojas et al. (1988) have suggested that hsp28 is involved in estrogen and progesterone regulation, and recent data suggest that it may play a role in the acquisition of thermotolerance (Rollet et al. 1992). During thermal stress, much of the hsp28 relocalizes to the nucleus and only traces can be found dispersed throughout the cytoplasm. As the cells recover from heat shock, hsp28 leaves the nucleus and is localized in the perinuclear region (Arrigo et al. 1988)

Another major group of proteins which are expressed during physiologic stress is the glucose-regulated proteins (grps). The grps derived their name from studies originally demonstrating that inducers of grp synthesis included conditions or agents which interfered with glucose utilization, e.g., glucose deprivation, insulin, tunicamycin, and glucosamine (Nover 1991). Synthesis of grps is also induced by agents which perturb intracellular calcium homeostasis, e.g., calcium ionophore. The grps exhibit significant sequence homology with hsps. Although grps are functionally related to the hsps, the two groups of proteins differ in their intracellular compartmentalization. Both grps and hsps are induced independently by distinct agents but also share a common set of inducers, including several metals (Nover 1991; Goering et al. 1993a). In some cell types, a coordinate induction of synthesis of hsps and grps is observed (Welch 1992), while under other experimental conditions, regulation of the two protein groups appears to be inversely related (Nover 1991).

The preceding brief overview has illustrated the structural and functional diversity of the classical hsps and grps. The complexity of these stress proteins and the stress response can be further appreciated by the following review on the unique and diverse effects of several metals on the expression of stress proteins.

C. Metals and Their Effects on Expression of Stress Proteins

I. General

The enhanced expression of metal-induced stress proteins is controlled primarily at the transcriptional level similar to the induction of hsps by heat (Wu et al. 1986). Regulation of hsp genes in eukaryotic systems is mediated by a *cis*-acting heat shock control element (HSE) that is found in multiple copies upstream of the transcriptional start site (Pelham 1982). Transcriptional activation of the hsp genes is mediated by a *trans*-acting protein, known as the heat shock factor (HSF), which binds specifically to the HSE (Wu 1984a,b).

The enhanced synthesis of specific stress proteins by metals can vary considerably and is influenced by such factors as the specific metal tested, the chemical speciation or composition of the metal, the treatment protocol (dose and time), the tissue or cell type, the developmental stage of an organism, and the state of differentiation of a tissue (Shuman and Przybyla 1988; Lai et al. 1993). When compared to the proteins induced by heat, three basic patterns of metal-induced stress protein synthesis are observed: (a) analogous induction, (b) subset induction, and (c) metal-specific induction. A comparison of heat-induced proteins with metal-induced proteins which demonstrates these three patterns is presented in Table 1. In studies demonstrating analogous induction, the same proteins are induced by both heat and metals. In addition to the examples presented in Table 1, analogous induction has also been reported in rat lens (De Jong et al. 1986) and chicken macrophages (Miller and Qureshi 1992). Bauman et al. (1993) reported analogous induction in rat hepatocytes between heat and a variety of metals, although the total amount of protein induced and the kinetics of induction varied with specific metals. In this experiment, heat treatment resulted in the greatest increase in hsp70 and hsp90; cadmium produced a fivefold elevation in hsp70 and a twofold increase in hsp90; arsenite, a sixfold rise in hsp70 and a twofold elevation in hsp90; and nickel, a twofold increase in hsp70 and a 30% increase in hsp90.

In experiments demonstrating subset induction, the proteins induced by metals represent a subset of the proteins induced by heat (Table 1). Findings from these studies indicate that some of the proteins induced by heat are insensitive to metal induction. Additional accounts of subset induction have been reported in murine lymphocytes (Rodenhiser et al. 1986) and human melanoma cells (Delpino et al. 1992). In contrast, in metal-specific induction, specific proteins have been identified that are only sensitive to metals (Table 1). Other examples of metal-specific induction have been reported in human periodontal ligament cells (Sauk et al. 1988), rat aortic smooth muscle (Kohane et al. 1990), and Chinese hamster ovary cells (Lee et al.

Table 1. Patterns of heat-induced vs. metal-induced protein synthesis

Experimental system	Stressor	Induced proteins (kDa)
I. Analogous induction		
Human chorionic villus (HONDA et al. 1992)	Heat	70,73,85,105
	As	70,73,85,105,
	Cd	70,73,85,105
Guinea pig tracheal epithelial cells and alveolar macrophages (COHEN et al. 1991)	Heat	28,32,72,73,90,110
	As	28,32,72,73,90,110
II. Subset induction		
Rat hepatocytes (CAJONE and BERNELLI-ZAZZERA 1988)	Heat	24,27,31,47,80,95,100
	Fe	31,80,95
Chicken myoblasts (ATKINSON et al. 1983)	Heat	25,64,82,85
	As	25,64
	Cu	25,64
	Zn	25,64
Xenopus laevis epithelial cells (HEIKKILA et al. 1987)	Heat	30,31,54,70,73,87
	As	70,73
III. Metal-specific induction		
Human melanoma cells (CALTABIANO et al. 1986a)	Heat	72,90,100
	Cd	**32**,72,90,100
	Cu	**32**,72,90,100
	Zn	**32**,72,90,100
Tetrahymena pyriformis (AMARAL et al. 1988)	Heat	25–29,35,70–75,92
	As	25–29,35,**36**,**42**,**46** 70–75,**83**,92

Note, The table does not present an exhaustive comparison of the three patterns. In III, bold numbers represent molecular masses of metal-specific proteins which were not induced by heat. Abbreviation: kDa = kilodaltons.

1991). The findings summarized in these last two induction categories imply that the regulatory mechanism(s) involved in some heat-induced proteins is independent of the mechanisms responsible for some metal-induced stress proteins.

The proximal mechanism for induction of stress protein synthesis leading to the activation of HSF and gene activation is not completely understood, but evidence for several possibilities exists. Activation of HSF by prooxidants does not result in the accumulation of specific stress proteins (BRUCE et al. 1993). These results suggest that induction of stress proteins by specific metals, whose toxicity is mediated via oxidative damage to membranes or DNA, may be fundamentally different from that of the heat-induced activation of the stress response (KEYSE and TYRRELL 1987; BRUCE et al. 1993). Thus, metals such as cadmium, mercury, nickel, arsenite, copper, lead, and iron, which induce oxygen free radicals or promote formation of lipid peroxides (STACEY and KLAASSEN 1981; HALLIWELL and GUTTERIDGE 1984; CHRISTIE and COSTA 1984; KASPRZAK 1991; DONATI et al. 1991), may induce stress proteins via this mechanism. A second mechanism of metal-induced stress protein synthesis may involve depletion of intracellular thiol pools, since several sulfhydryl reactive agents increase synthesis of these proteins. A third mechanism of induction by metals may involve metal-induced proteotoxicity, since aberrant or denatured proteins are a stimulus for stress protein synthesis (ANANTHAN et al. 1986; HIGHTOWER 1991). Specific induction of grps by metals such as cadmium (GOERING et al. 1993a) and lead (SHELTON et al. 1986) may involve the property of these metals to disrupt cellular calcium homeostasis, since grps are induced by other agents which alter calcium homeostasis (NOVER 1991).

A comparison of stress proteins induced by various metals is presented in Table 2. This table does not represent an exhaustive list of all studies, but is intended to illustrate the degree of variability reported in the synthesis of specific proteins induced by different metals in various experimental models. Because of this variability, it is difficult to make generalizations about specific protein induction by metals. Although all these metals share the ability to induce synthesis of stress proteins in one or more experimental models, protein induction by a specific metal may have unique characteristics. The following review of the stress protein induction effects by specific metals will focus on those metals which illustrate some of the these unique characteristics. Other important or unique properties of specific metals will be discussed in Sect. G on biomarkers.

II. Arsenic

The majority of literature on metal-induced stress proteins results from studies employing the arsenic oxides, sodium arsenate and sodium arsenite (Table 2). Although these compounds are very effective inducers of stress proteins, arsenite has been shown to be the more potent inducer (BOURNIAS-VARDIABASIS et al. 1990; BAUMAN et al. 1993). Sodium arsenite has been shown to elicit thermotolerance, self-tolerance, and cross-tolerance in response to a variety of physiologic stresses (see Sect. D). The thermotolerance induced by arsenite treatment has been correlated with the increase in newly synthesized stress proteins (LEE and DEWEY 1987). Arsenite has been shown to induce the same proteins as heat in several experimental systems (DE JONG et al. 1986; BOURNIAS-VARDIABASIS et al. 1990; COHEN et al. 1991; HONDA et al. 1992). In contrast, there are reported differences in the stress protein responses produced by these two stressors.

Sodium arsenite and hyperthermia both induce a similar stress response in rat embryos treated in vitro as measured by stress protein synthesis and accumulation of hsp70 mRNA and hsp72 expression, although the abnormal morphologies produced by these embryotoxic agents appear to be different (MIRKES and CORNEL 1992). Studies conducted by AMARAL et al. (1988) have shown that arsenite does not inhibit synthesis of constitutive, nonstress proteins as dramatically as does hyperthermic shock. Several studies have

Table 2. Comparison of stress proteins induced by various metals

Metal	Induced proteins (kDa)	Experimental system	Reference
As	70,73,85,105	Human chorionic villus	Honda et al. 1992
	30,36,41–47,68–72,90	Human periodontal ligament cells	Sauk et al. 1988
	30,70,100	Human foreskin fibroblasts	Levinson et al. 1980
	28,116	Human keratinocytes and fibroblasts	Edwards et al. 1991
	70,90	HeLa cells	Duncan and Hershey 1987
	32,72,90,100	Human melanoma cells	Caltabiano et al. 1986a
	28,70,90	Human melanoma cells	Delpino et al. 1992
	16,71,85	Rat lens	De Jong et al. 1986
	30,70,90,110	Rat aortic smooth muscle	Kohane et al. 1990
	28,30,38,51,65,71,85	Rat kidney PCT epithelial cells	Aoki et al. 1990
	70,90	Rat primary hepatocyte culture	Bauman et al. 1993
	32,70,72,90,110	Rat brain tumor cells	Lai et al. 1993
	73,105	Mouse embryo	Honda et al. 1992
	65,70,90,110	Murine lymphocytes	Rodenhiser et al. 1986
	34,72,90,110	Murine melanoma cells	Caltabiano et al. 1986a
	32,70,90	Murine 3T3 cells	Hiwasa and Sakiyama 1986
	28,32,72,73,90,110	Guinea pig tracheal epithelial cells and alveolar macrophages	Cohen et al. 1991
	28,70,87,110	CHO cells	Lee et al. 1991
	70,90	CHO cells	Li 1983
	25,64	Chicken myoblasts	Atkinson et al. 1983
	70,73	*Xenopus laevis* epithelial cells	Heikkila et al. 1987
	25–29,35,36,42,46,70–75,83,92	*Tetrahymena pyriformis*	Amaral et al. 1988
	–	Rotifer *Brachionus plicatilis*	Cochrane et al. 1991
Cd	70,73,85,105	Human chorionic villus	Honda et al. 1992
	72,90	Human keratinocytes and fibroblasts	Edwards et al. 1991
	70,78	HeLa cells	Watowich and Morimoto 1988
	32,72,90,100	Human melanoma cells	Caltabiano et al. 1986a
	28,70,90	Human melanoma cells	Delpino et al. 1992
	72,94,110	Rat liver	Goering et al. 1993a

	70,90	Rat primary hepatocyte culture	BAUMAN et al. 1993
	32,70,72,78,90,110	Rat brain tumor cells	LAI et al. 1993
	73,105	Mouse embryo	HONDA et al. 1992
	30,68,71	Murine cardiac myocytes	LÖW-FRIEDRICH and SCHOEPPE 1991
	34,72,90,100	Murine melanoma cells	CALTABIANO et al. 1986a
	32,70,90	Murine 3T3 cells	HIWASA and SAKIYAMA 1986
	22,23,26,27,70	*Drosophila* cells	BOURNIAS-VARDIABASIS et al. 1990
	135	*Candida albicans*	ZEUTHEN and HOWARD 1989
Zn	70,90	HeLa cells	DUNCAN and HERSHEY 1987
	32,72,90,10	Human melanoma cells	CALTABIANO et al. 1986a
	–	Rat lens	DE JONG et al. 1986
	70,90	Rat primary hepatocyte culture	BAUMAN et al. 1993
	32,70,72,78,90,110	Rat brain tumor cells	LAI et al. 1986a
	34,72,90,100	Murine melanoma cells	CALTABIANO et al. 1986a
	32,70,90	Murine 3T3 cells	HIWASA and SAKIYAMA 1986
	25,64	Chicken myoblasts	ATKINSON et al. 1983
	22,23	*Drosophila* cells	BOURNIAS-VARDIABASIS et al. 1990
Cu	30,70,100	Human foreskin fibroblasts	LEVINSON et al. 1980
	32,72,90,100	Human melanoma cells	CALTABIANO et al. 1986a
	32,70,72,90,110	Rat brain tumor cells	LAI et al. 1993
	34,72,90,100	Murine melanoma cells	CALTABIANO et al. 1986a
	32	Murine 3T3 cells	HIWASA and SAKIYAMA 1986
	25,64	Chicken myoblasts	ATKINSON et al. 1983
	58	Rotifer *Brachionus plicatilis*	COCHRANE et al. 1991
	25,35,70,100	Chicken embryonic fibroblasts	LEVINSON et al. 1980
	22,23,26,27,70	*Drosophila* cells	BOURNIAS-VARDIABASIS et al. 1990
Hg	43,72,90,110	Rat kidney	GOERING et al. 1992
	70,90	Rat primary hepatocyte culture	BAUMAN et al. 1993
	22,23,26,27,70	*Drosophila* cells	BOURNIAS-VARDIABASIS et al. 1990
	–	Rotifer *Brachionus plicatilis*	COCHRANE et al. 1991
	81,84,88	*Candida albicans*	ZEUTHEN and HOWARD 1989

Table 2. *Continued*

Metal	Induced proteins (kDa)	Experimental system	Reference
Pb	–	Rat primary hepatocyte culture	BAUMAN et al 1993
	32	Rat fibroblast, kidney epithelial cells	SHELTON et al. 1986
	22,32,70,90	Chicken macrophages	MILLER and QURESHI 1992
	–	*Drosophila* cells	BOURNIAS-VARDIABASIS et al. 1990
	70	Crustacean *Oniscus asellus*	KÖHLER et al. 1992
Ga	28,30,38,51,65,71,85	Rat kidney PCT epithelial cells	AOKI et al. 1990
	12,26,39,50,89	*Pseudomonas fluorescens*	AL-AOUKATY et al. 1992
Ni	70,90	Rat primary hepatocyte culture	BAUMAN et al. 1993
	22,23	*Drosophila* cells	BOURNIAS-VARDIABASIS et al. 1990
Fe	31,80,95	Rat hepatocytes	CAJONE and BERNELLI-ZAZZERA 1988
	–	Rat primary hepatocyte culture	BAUMAN et al. 1993
Au	32,72,90,100	Human monocytes	CALTABIANO et al. 1986b
	90	Human monocytes	SCHMIDT and ABDULLA 1988
	34,72,90	Murine melanoma cells	CALTABIANO et al. 1986b

Note, The apparent molecular weights indicated in the table were determined on the basis of relative migration on electrophoretic gels. The symbol "–" indicates that no proteins were induced in response to the specific metal, although proteins were induced in this experimental model in response to other metals listed in the table. Abbreviation: kDa = kilodaltons.

reported differences in the kinetics of synthesis and cellular redistribution of specific stress proteins following treatment with either heat or arsenite (DARASCH et al. 1988; LEE et al. 1991). Although transcripts of these proteins may accumulate at the same rate after arsenite or heat treatments, decay of stress protein message and detection of stress proteins after arsenite exposure appear to be protracted (SHUMAN and PRZYBYLA 1988). This and similar findings have led to the suggestion that stress protein mRNA induced by arsenite may be more stable than the corresponding heat-shock-induced mRNA transcript (AMARAL et al. 1988). This difference in kinetics of the response by these two agents may be due, at least in part, to the differences in the nature of the insult. Heat exposure is an on/off type of exposure, whereas the time that tissues would be exposed to a metal would depend upon the plasma and whole-body half-lives of the metal.

III. Cadmium

With the exception of the arsenic compounds, cadmium has been one of the most studied inducers of stress proteins (Table 2). Analysis of cadmium-induced stress protein synthesis has occurred in studies using various mammalian and nonmammalian systems in vitro. Cadmium induced stress protein synthesis in primary hepatocyte cultures and *Drosophila* cells prior to cytotoxicity (COURGEON et al. 1984; BAUMAN et al. 1993; GOERING et al. 1993c). Examples of stressor-specific and cell-type-specific induction of stress proteins in vitro have been observed in studies using cadmium. MISRA et al. (1989) demonstrated enhanced synthesis of hsp70 in rainbow trout hepatoma cells and chinook salmon embryonic cells exposed to heat shock and zinc, but the embryonic cells were additionally responsive to cadmium. Cadmium induced synthesis of metallothionein (MT) in the hepatoma cells but heat shock was negative, and the embryonic cells did not synthesize MT after cadmium exposure. Exposure of human or murine melanoma cells to cadmium, as well as copper, zinc, and other thiol-reactive agents, elicited the synthesis of 100-, 90-, 72-, and 32- (human) or 34- (murine) kDa proteins (CALTABIANO et al. 1986a). Heat shock and calcium ionophore induced synthesis of the major stress proteins, but failed to enhance synthesis of the 32- and 34-kDa proteins. Thus, the 32- and 34-kDa proteins represent a subset of stress-inducible gene products which are sensitive to thiol-reactive chemicals but not heat shock. Treatment of neuroblastoma cells with cadmium resulted in an increased expression of mRNAs for hsp70, hsp90, hsp32, MT, N-*myc*, and the multidrug-resistance (*MDR1*) gene (MURAKAMI et al. 1991). In a laboratory evaluating the role of stress proteins in cutaneous inflammatory diseases, cadmium induced synthesis of hsp72 and hsp90 in human keratinocytes, skin fibroblasts, and human epithelial tumor cells 3 h after exposure (EDWARDS et al. 1991).

Cadmium has been shown to elicit stress protein synthesis in tissues in vivo. Within 2–4 h after acute dosing of rats with cadmium, an enhanced de novo synthesis of 70-, 90-, and 110-kDa proteins was observed in liver but not kidney. Synthesis of these proteins was enhanced prior to detection of cell injury. Two of these proteins were identified as hsp72 and grp94 immunochemically using monoclonal antibodies, and were accumulating up to 24 h after exposure to cadmium (Goering et al. 1993a).

IV. Mercury

Mercury elicits the synthesis of stress proteins in vitro and in vivo. A comparison of mercury-induced stress proteins shows they vary from heat-induced proteins and these differences may partially be determined by the experimental system. Bournias-Vardiabasis et al. (1990) reported that the spectrum of proteins induced in *Drosophila* embryonic cells were identical after heat or mercury treatment. In contrast, yeast cells (*Candida albicans*) exposed to mercury induced the synthesis of three stress proteins, only one of which corresponded in molecular mass to a major hsp.

Goering et al. (1992) demonstrated that mercury in vivo produced dose- and time-dependent alterations in expression of renal gene products as evidenced by enhanced synthesis of stress proteins and inhibition of the synthesis of constitutive proteins. Enhanced de novo synthesis of 43-, 70-, 90-, and 110-kDa proteins was detected in kidney but not liver from mercury-injected rats. The synthesis of the three proteins was discoordinate, suggesting that the synthesis of these proteins may be independently regulated. The changes in renal protein synthesis were rapid (2–4 h), and occurred prior to overt renal injury as assessed by functional and histopathological evidence.

V. Copper

In addition to inducing hsps in a variety of experimental systems, copper has been shown to induce MT and grps (Bunch et al. 1988; Hatayama et al. 1991; Lai et al. 1993). Bunch et al. (1988) demonstrated a 30- to 100-fold induction of MT mRNA levels in copper-treated *Drosophila* cells. The level of induction was found to be dose dependent and MT mRNA levels remained elevated for at least 4 days. Similar results were reported for cadmium, although copper was found to be less toxic than cadmium. Hatayama et al. (1991) found that MT was induced at lower copper concentrations than hsp70 in HeLa cells. Incubation of HeLa cells with 100 μM cupric sulfate induced the synthesis of MT with no observed changes in hsp70 synthesis or cell growth. In contrast, 200 μM cupric sulfate was required to induce hsp70 and was associated with a concomitant inhibition of cell growth. This concentration-dependent difference in induction of MT and stress protein synthesis may be explained as follows: MT induction is more sensitive to lower concentrations of copper or zinc (vide infra); however, higher metal concentrations are proteotoxic and the resultant denatured proteins initiate the cascade of cellular and genetic events which results in the synthesis of stress proteins.

SANCHEZ et al. (1992) have reported that the induction of hsp104 in *Saccharomyces cerevisiae* by heat is of critical importance in establishing tolerance to ethanol and of moderate importance in establishing tolerance to sodium arsenite, but was not sufficient to confer tolerance to copper. These investigators conclude that the lethal lesions produced by copper in yeast are fundamentally different from those produced by heat.

VI. Zinc

BOURNIAS-VARDIABASIS et al. (1990) exposed *Drosophila* embryonic cells to a number of metal ions that had previously been reported to act as teratogens in mammalian systems. Although the induced protein patterns varied between different metals, zinc induced a pattern similar to that produced by "classical" teratogens. ROCCHERI et al. (1988) reported that treatment of sea urchin embryos with zinc induced the same stress proteins as those observed in heat-treated embryos. These investigators found that zinc treatment induced a protracted synthesis of the stress proteins relative to heat. This finding might be explained by the prolonged exposure of the sea urchin embryos to zinc in the culture medium. Unlike heat shock, zinc treatment did not inhibit overall protein synthesis or induce thermotolerance in these zinc-treated sea urchin embryos.

Similar to the data reported for copper, HATAYAMA et al. (1992) found that MT was induced at lower zinc concentrations than hsp70. Incubation of HeLa cells incubated with $100\,\mu M$ zinc sulfate induced the synthesis of MT with no observed changes in hsp70 synthesis or cell growth. In contrast, $200\,\mu M$ zinc sulfate induced hsp70 but was also associated with an inhibition of cell growth. These investigators concluded that since the zinc mediated induction of hsp70 synthesis and inhibition of cell growth occurred concomitantly, hsp70 would appear not to be involved with the detoxification of zinc, but may participate in the repair of zinc-induced cellular damage.

VII. Lead

Data generated from a study of lead-induced synthesis of stress proteins illustrated that different metals have both common and individual effects on expression of stress proteins. Lead-induced stress protein synthesis was unusual in that it did not share the properties of metals, such as cadmium and arsenite, to mimic the effects of heat (SHELTON et al. 1986). Rather, lead induced two minor classes of proteins. In this study, exposure of primary rat kidney epithelial cells and rat fibroblasts to lead glutamate induced synthesis of two grps and a protein of 32 kDa, which was shown to be distinct from the lead nuclear inclusion body protein. The treatment did not affect synthesis of hsp70. In contrast, cadmium exposure increased expression of hsp70 and the 32-kDa protein, but did not affect grp synthesis. The three proteins induced by lead were distinct from the major group of stress proteins because of the relative lack of synthesis observed following heat shock. The induction of grps by metals such as cadmium and lead might result from interfering with glucose utilization, disrupting calcium homeostasis, or binding sulfhydryl group targets. It should be noted that in another study exposure of avian macrophages to lead acetate in vitro resulted in the increased expression of a hsp70, a hsp90, and a hsp32 and hsp23; these same proteins were induced after exposure to heat (MILLER and QURESHI 1992).

VIII. Iron

Iron has been shown to increase levels of stress proteins, including ubiquitin, induce the synthesis of heme oxygenase, and increase superoxide dismutase activity (KANTENGWA and POLLA 1993; UNEY et al. 1993). Oxidative stress produced by ADP-iron induced stress proteins (31, 80, and 95 kDa) in isolated rat hepatocytes and cultivated rat hepatoma cells (CAJONE and BERNELLI-ZAZZERA 1988). These iron-induced proteins are a subset of the proteins induced in these cells by hyperthermia. Treatment of mouse neuroblastoma cells with iron increased mRNA levels of hsp70 and ubiquitin. This response was blocked by the chelating agents α-tocopherol or desferroxamine, which demonstrates that the induction of these transcripts was mediated by the presence of iron (UNEY et al. 1993). These results suggest that free radical generation by iron is potentially responsible for the initiation of iron-induced stress protein synthesis and iron-mediated cell injury. Oxygen free radicals generated in the presence of hemoglobin-released iron are also believed to initiate stress protein induction by human monocytes-macrophages during erythrophagocytosis (CLERGET and POLLA 1990; DONATI et al. 1991). In contrast, iron limitation resulted in increased expression of several proteins in bacteria, one of which demonstrated 65% homology in amino acid sequence with members of the hsp60 protein family (PANNEKOEK et al. 1992; VALONE et al. 1993). Iron was also an ineffective inducer of hsp70 and hsp90 in primary cultures of rat hepatocytes (BAUMAN et al. 1993).

IX. Gold

Auranofin, a gold salt currently used for therapy of chronic inflammation associated with rheumatoid arthritis, has been reported to induce several stress proteins and inhibit interleukin 1β synthesis in human peripheral blood monocytes (CALTABIANO et al. 1986b; SCHMIDT and ABDULLA 1988).

Of the several stress proteins induced, CALTABIANO et al. (1986b) observed that the enhanced synthesis of two polypeptides, p32 and p34, was most prominent. In addition, oral administration of auranofin to rats enhanced the synthesis of the 32-kDa stress protein in peritoneal exudate cells. To establish a structure-activity relationship, these investigators tested a series of gold analogs and other anti-arthritic agents for their abilities to stimulate this low molecular weight stress protein (CALTABIANO et al. 1988). These studies demonstrated that the enhanced expression of the 32-kDa polypeptide was dependent on the presence of gold atoms. Gold complexes bearing several phosphine or thiosugar groups were the most effective inducers of this stress protein, and most likely reflected the ability of these analogs to more readily cross cell membranes.

D. Tolerance Induction and Stress Proteins

The stress, or heat shock, response is believed to serve an adaptive or survival function which involves a rapid but transient reprogramming of cellular metabolic activity to protect critical cellular macromolecules against stressors and to facilitate rapid resumption of physiologic functions during a recovery or repair phase. Although the specific stress proteins induced vary with the organism studied and the type and dose of the stressor (Table 2), the protective effect of stress proteins is believed to occur via protecting the genome through interactions with the nuclear matrix or via protein-protein interactions, i.e., stress proteins bind to proteins damaged or denatured by stressors and aid in their solubilization, refolding, and stabilization (WELCH 1990).

GERNER and SCHNEIDER (1975) were the first to demonstrate that heat shock induces a transient state of heat resistance in mammalian cells. Subsequent research has demonstrated that cells or embryos given a sublethal heat shock sufficient to induce hsp synthesis exhibit tolerance to an otherwise lethal heat shock (LI and LASZLO 1985; MIRKES 1987; JOHNSTON and KUCEY 1988; RIABOWOL et al. 1988). This phenomenon has been termed "acquired thermotolerance," and is defined for the purposes of this review as self-tolerance. Data from numerous studies demonstrate that the acquisition of thermotolerance is attributable to the induction and expression of stress proteins (LANDRY et al. 1982; WELCH and MIZZEN 1988). Many investigations, including those of LANDRY et al. (1982), LI and WERB (1982), and LAVOIE et al. (1993), have demonstrated that the kinetics of thermotolerance induction and decay are correlated with stress protein synthesis and degradation, respectively.

A considerable number of stress protein inducers, such as cadmium, ethanol, and sodium arsenite, are as effective as heat in conditioning the cell to become thermotolerant. This type of tolerance, where the inducing and stress stimulus are different, is termed cross-tolerance. Induction of cross-tolerance in cells is for the most part unilateral, i.e., some metals readily confer thermotolerance, but heat shock generally does not confer tolerance to the adverse effects of chemical stressors, although there are exceptions (NOVER 1991). All chemical stressors reported to evoke thermotolerance are known inducers of stress proteins. Analysis of studies in vertebrate cells reveals that the amount of hsp70, but not of hsp27, hsp90, and hsp110, correlates with the degree of thermotolerance; other studies implicate a similar role for hsp27 (NOVER 1991).

The studies described below demonstrate a strong correlation between the induction of stress proteins and the development of self-tolerance and cross-tolerance. Numerous studies have demonstrated that sodium arsenite can induce thermotolerance in various experimental systems (WIEGANT et al. 1987; LEE and DEWEY 1988; LEE and HAHN 1988; BURGMAN et al. 1993). LEE and HAHN (1988) have shown that cells treated with heat or sodium arsenite induce the synthesis of stress proteins which confers self-tolerance, as well as cross-tolerance, in response to these different types of stressors. In addition, pretreatment of rat neuroblastoma and hepatoma cells with arsenite or heat ameliorated heat-induced injury to the cytoskeleton (WIEGANT et al. 1987). The kinetics of cytoskeletal thermotolerance in these cells paralleled the kinetics of accumulation and decay of the stress proteins induced by sodium arsenite. Conversely, studies conducted using the protein synthesis inhibitor cycloheximide have demonstrated that, while heat-induced thermotolerance is partly diminished by inhibition of stress protein synthesis, sodium arsenite-induced thermotolerance is abolished, indicating that it is dependent on the synthesis of new proteins (KAMPINGA et al. 1992; BURGMAN et al. 1993). Collectively, these results suggest that the reciprocal cross-tolerance induced between sodium arsenite and heat may be mediated by slightly different mechanisms.

In contrast to the cross-tolerance studies described above employing arsenic, similar experiments reported that cadmium induces self-tolerance (CERVERA 1985), but pretreatment with zinc or cadmium generally does not confer thermotolerance in cultured cells (CERVERA 1985) and sea urchin embryos (ROCCHERI et al. 1988). Heat pretreatment does not confer cross-tolerance to cadmium in cultured cells (CERVERA 1985; CIAVARRA and SIMEONE 1990a,b), although a few exceptions have been reported. KAPOOR and SVEENIVASAN (1988) demonstrated that cadmium chloride treatment led to the induction of high levels of peroxidase activity and induced thermotolerance in the mold *Neurospora crassa*, although no correlation between levels of superoxide dismutase and the development of thermotolerance could be established.

The genetic differences in heat-induced cross-tolerance to cadmium were investigated in two inbred strains of mice (KAPRON-BRAS and HALES 1992). Although pre-exposure to heat provided no cross-tolerance to cadmium in BALB/c mouse embryos, heat treatment of SWV mouse embryos essentially eliminated the toxic effects of cadmium on embryonic growth and development. Mouse embryos from both strains displayed a comparable rapid increase in the synthesis of hsp68 following heat shock. Thus, the investigators concluded that cross-tolerance in the embryos was not dependent on induction of hsp68, and that other hsps or physiologic mechanisms must be involved. SANCHEZ et al. (1992) conducted mutation experiments in *Saccharomyces cerevisiae* and demonstrated that hsp104 is of critical importance in the development of thermotolerance and of moderate importance in tolerance to sodium arsenite. These investigators found that hsp104 was ineffective in reducing the toxic effects of copper or cadmium.

Because of the diversity of the stress response in various organisms in response to different stressors, it is difficult to make generalities regarding the involvement of any specific stress protein in the establishment of tolerance to metals and other xenobiotics. Additional research is needed to determine which of these stress proteins, if any, are essential for the induction of self- and cross-tolerance to metals and other stressors, and to elucidate other physiologic mechanisms which may be involved in this phenomenon.

E. Heme Oxygenase Is a Stress Protein

Recently, a stress protein with a relative molecular mass of approximately 32 kDa has been identified as heme oxygenase (HO). HO, localized in the endoplasmic reticulum or microsomal cellular fraction, is the enzyme which initiates the catabolism of heme. Thermal induction of HO mRNA and protein, as determined by cDNA probes and specific antibodies, respectively, and the presence of a heat shock gene regulatory element (HSE) on the HO gene promoter, is evidence that investigators have used to classify HO as a hsp or stress protein (SHIBAHARA et al. 1987; TAKETANI et al. 1988; EWING and MAINES 1991). In addition to being expressed in response to a number of hsp-inducing chemicals, HO is induced by a number of tumor-promoting agents (HIWASA et al. 1982; HIWASA and SAKIYAMA 1986). Rodent HO has been localized in brain, kidney, liver, heart, and skin cells using these techniques (EWING et al. 1992; TACCHINI et al. 1993; MAINES et al. 1993; KATAYOSE et al. 1993). Human HO is not readily heat-inducible and therefore may not be classified as a hsp (YOSHIDA et al. 1988; KEYSE and TYRRELL 1989), but may still be classified as a stress protein due to its responsiveness to other stressors, such as UVA radiation, hydrogen peroxide, sodium arsenite, and cadmium (KEYSE and TYRRELL 1989; SATO et al. 1990). HO cleaves the heme tetrapyrrole ring structure to form the intermediate biliverdin, which is subsequently converted to bilirubin via the action of biliverdin reductase. Two HO isozymes, HO-1 and HO-2, are encoded by separate genes, and vary in their tissue distribution and regulation. HO-1 is the highly inducible form of the enzyme, with inducers as diverse as the substrate heme, heat shock, and other chemicals, including metals and oxygen free radicals (Maines 1988; Keyse and Tyrell 1989; Ewing et al. 1992).

Abnormal elevation of HO activity in response to metals and other xenobiotics has both negative and positive consequences for cells. While the physiological role of HO involves the turnover of heme compounds, a protracted increase in the de novo synthesis of HO has been generally associated with perturbation of cellular processes which may lead to cell injury or death. Reduction of heme via HO can deplete cytochrome P-450 levels, inhibit mixed function oxidase activities, and deplete mitochondrial respiratory cytochromes (Maines and Kappas 1977). Cellular respiration and biotransformation of endogenous compounds and xenobiotics may become compromised.

Recently, it has been shown that biliverdin and bilirubin, the products of heme degradation, possess antioxidant activity and may play a role in protecting the cell against damage induced by free radicals (Stocker et al. 1987a,b). Many metals, such as cadmium, arsenite, iron, and copper, which are known to induce HO and other stress proteins, have been shown to promote production of free radicals. Thus, HO may represent a physiologic defense mechanism against stressors which produce cell injury by promoting oxidative stress, e.g., UVA, hydrogen peroxide, and metals. Since many prooxidants elevate stress protein levels, this type of cell injury may represent a common mechanism by which these types of compounds increase stress protein synthesis. The HO-bile pigment system may be particularly important in tissues which are deficient in other antioxidant systems, e.g., brain is deficient in glutathione, but possesses the capacity to markedly induce HO-1 (Ewing and Maines 1991; Ewing et al. 1992). Some metals, such as cadmium and nickel, are capable of reducing GSH concentrations, and GSH depletion is a known stimulus for HO-1 induction (Saunders et al. 1991; Ewing and Maines 1993). Thus, GSH and HO may constitute a closely integrated cellular defense mechanism against toxic insults, including metals.

Early in vivo studies examining the effect of metals on HO demonstrated that metal induction of HO is tissue specific with respect to the inducing metal. Most of these studies examined HO induction by assaying enzymatic activity. Cadmium, mercury, lead, chromium, manganese, iron, cobalt, nickel, copper, tin, indium, thallium, platinum, silver, and selenium have been shown to induce HO in liver (Maines and Kappas 1977; Woods et al. 1979, 1984; Eaton et al. 1980); however, cobalt and cadmium are the most effective inducers. In contrast, nickel, tin, and platinum represent the best inducers of the enzyme in kidney; in heart, mercury is the most effective. While the usual response of HO to metals is induction of synthesis and activity, cadmium, a potent inducer in liver, causes a marked depression of enzyme activity in testes (MAINES et al. 1982).

More recent studies have demonstrated that metals and heat can induce a protein whose M_r ranges from 30 to 35 kDa. This protein has been identified as HO in rodent, human, and avian cells via immunoblot techniques using specific HO antibodies; its mRNA has been identified using cDNA probes. In a variety of cell systems, sodium arsenite is the most common and effective inducer of HO. Sodium arsenite and cadmium are strong inducers of HO mRNA in human skin fibroblasts, with maximal responses observed within 2–4 h after exposure (KEYSE and TYRRELL 1989). Arsenite induced hsp70 and HO mRNAs in three human hepatoma cell lines, while cadmium increased the hsp70 and HO mRNAs in only two of the cell lines (MITANI et al. 1990). Heat induced hsp70 mRNA in all three cell lines but only induced HO mRNA in one cell line. Treatment of rat hepatoma cells, HeLa cells, and mouse peritoneal macrophages with sodium arsenite and cadmium resulted in marked induction of HO within 7 h after exposure (TAKETANI et al. 1988, 1989, 1990). In these studies, cobalt and heat induced HO in the rat hepatoma cells only, but not in the other cell types. Cadmium induced HO-1, but not the HO-2 isozyme, in CHO cells (SAUNDERS et al. 1991). These results illustrate the cell-type specificity in the stress protein response in general and caution the reader in forming conclusions and extrapolating results to higher levels of biological complexity.

Based on the evidence presented above, HO can be classified as a stress protein. To date, it is the only stress protein covered in this review which possesses enzymatic activity. In a manner similar to other well-known stress proteins, HO induction is dependent on the type of stressor, tissue, and species. Further research efforts should focus on the role of HO as a player in cellular defense mechanisms against injury by stressors.

F. Is Metallothionein a Stress Protein?

Metallothionein (MT) is a low-molecular-weight metal-binding protein well known to toxicologists who study metals. The protein is believed to play an important role in the homeostasis of the essential micronutrients zinc and copper, but also influences the metabolism and toxicity of other metals, such as cadmium, mercury, and copper.

Is MT a stress protein? In this section, the following common properties of MTs and stress proteins will be compared: evolutionary conservation, common inducers, protective roles and cross-tolerance, gene regulation, increased expression of these proteins in neoplasms and other disease states, and as adjuncts in chemotherapy. Since the two classes of proteins represent many proteins, we will speak of them in a generic sense for the purposes of this discussion. The discussion will include many generalizations and is intended to stimulate debate, rather than represent an exhaustive comparison.

I. Evolutionary Conservation

Metallothioneins and stress proteins have been highly conserved through evolution. The stress proteins have been identified in organisms ranging from "bacteria to humans," and MTs, or MT-like proteins, share a similar phylogenetic distribution. Many gene families exist for MTs and stress proteins. Of all mammalian MTs compared thus far, 35 of 61 residues are invariant (KÄGI and KOJIMA 1987). The hsp70 gene in humans is approximately three-fourths homologous with the hsp70 gene in *Drosophila*; distant prokaryotic and eukaryotic species have at least 50% hsp gene homology (MORIMOTO et al. 1992). In most species examined, MTs display genetic polymorphism. In mammals and crustaceans, MTs generally consist of several isoforms, most notably MT-1 and MT-2, which are coded by nonallelic genes. The stress proteins comprise both constitutive and inducible forms.

II. Common Inducers

Many cells possess the remarkable capacity to induce the transcription of MT and stress protein mRNAs and synthesis of their respective proteins to levels as much as 10–30 times higher than control levels. The stimuli for this marked induction represent a diverse group of chemical and physical agents. The most effective inducers of both MTs and stress proteins are metals or metalloids, and include cadmium, sodium arsenite, mercury, zinc, and copper. Other common inducers of stress proteins and MTs include heat, several alkylating agents and carcinogens, UV irradiation, interleukin-1, ethanol, iodoacetamide, phthalate esters, dexamethasone, and estradiol. Several hepatotoxicants and hepatocarcinogens are capable of inducing MT synthesis in liver (WAALKES and WARD 1989), and these compounds also induce stress protein synthesis (CARR et al. 1986). Synthesis of MTs and stress proteins is also induced in regenerating liver, e.g., following partial hepatectomy, or in proliferating tissue. Other chemical and physical inducers of MT include carbon tetrachloride, endotoxin, hypothermia, infection, hyperbaric oxygen, and X-irradiation. Other inducers of stress proteins include ischemia/reperfusion, dioxins, diphenylhydantoin, and cigarette smoke (WAALKES and GOERING 1990; NOVER 1991).

III. Protective Roles and Cross-tolerance

A large body of evidence supports the role of both MTs and stress proteins in protecting cells from injury. While the constitutive forms have important functions in maintaining cellular homeostasis, prior induction of synthesis of these proteins by a relatively nontoxic stimulus appears to represent a critical step in an organism's ability to develop tolerance to subsequent exposure to adverse stimuli. Cross-tolerance, whereby the cell can be made tolerant to a toxic insult, such as heat, after induction of protein synthesis by a different stimulus, such as cadmium, has been observed for both MTs and stress proteins (see Sect. D). The protective function of both MTs and stress proteins can exceed a threshold whereby this capacity is overridden and cell injury will ensue.

Prior induction of stress proteins by heat or metals has been shown to protect cells or organisms against toxic injury by metals, such as arsenite or cadmium, as was discussed in more detail in Sect. D. The mechanism for this tolerance is believed to occur via protein-protein interactions, which involves the salvaging of damaged proteins or targeting damaged proteins for proteolysis.

Treatment with low, nontoxic doses of cadmium sufficient to induce MT synthesis will reduce or prevent the toxicity of a subsequent toxic cadmium dose. Cross-tolerance also occurs; induction of MT by zinc or other xenobiotics will confer tolerance to cadmium or other insults, such as ionizing radiation. MTs contain numerous thiol ligands (33% of all amino acid residues are cysteines) which serve to detoxify specific chemical and physical insults by at least two mechanisms. In the case of metals, such as cadmium, the mechanism of detoxication involves the high-affinity sequestration of metal ions into a relatively inert intracellular complex, thus reducing the interaction of toxic cations with target molecules (for review, see WAALKES and GOERING 1990). Tolerance develops to cadmium-induced lethality, genotoxicity, inhibition of hepatic drug metabolism, testicular necrosis, hepatotoxicity, Cd-MT nephrotoxicity, carcinogenesis, fetotoxicity, and teratogenicity (ITO and SAWAUCHI 1966; YOSHIKAWA 1973; LEBER and MIYA 1976; ROBERTS et al. 1976; FERM and LAYTON 1979; GOERING and KLAASSEN 1983, 1984; SQUIBB et al. 1984; COOGAN et al. 1992, 1994). A second cellular defense mechanism involving MT may be its role as an antioxidant (THORNALLEY and VASAK 1985; TAMAI et al. 1993). These protective effects may be related to the disproportionately high level of cysteine residues which provide a significant pool of reduced sulfhydryl groups. The toxic effects of ionizing radiation and free-radical generating compounds may be reduced via this mechanism (MATSUBARA et al. 1987; IMURA et al. 1991).

Strong experimental evidence exists which demonstrates that both MTs and stress proteins play a role in tolerance to toxic metal insults, albeit by different mechanisms. Perhaps both MTs and stress proteins are operating in concert or sequentially, along with other cellular defense mechanisms, e.g., GSH, in a multitiered system to protect cells from toxic injury.

IV. Gene Regulation

The transcriptional regulation of MT and stress protein genes share similar properties. Recently, SILAR et al. (1991) have shown that transcription of the copper-inducible *S. cerevisiae* yeast MT gene, CUP1, is controlled by the heat shock *trans*-acting factor (HSF) which activates hsp genes. MT inducibility is regulated at the transcriptional level by various *cis*-acting DNA sequences located in the promoter region of MT genes. Such sequences are responsible for induction of MT synthesis by metals (metal regulatory, or responsive, elements; MRE) and glucocorticoids (glucocorticoid regulatory elements; GRE). The human hsp70 promoter contains a sequence that is homologous to the human MT promoter MRE (WU et al. 1986). Soluble *trans*-acting DNA-binding proteins, which bind to MT-inducing metals, facilitate the interaction of the metal with the MREs to promote MT gene transcription (HAMER 1986; PALMITER 1987). The promoter region of the human hsp70 gene contains multiple heat shock elements (HSE) which confer inducibility by a variety of stressors. In eukaryotes, stress gene transcription is mediated by the activation of soluble HSFs. Studies have revealed multiple HSFs, and the HSF responsible for activation of gene transcription in response to physical and chemical stress, including metals, has been designated HSF1(MORIMOTO et al. 1992). Under stress conditions, HSF1 displays such activities as oligomerization, translocation to the nucleus, and DNA-binding in order to activate stress genes.

V. Increased Expression in Neoplasms and Other Diseases

Elevated expression of MT and stress proteins has been observed in neoplastic tissue. MT has been detected in various types of human tumors, including thyroid tumors, testicular embryonal carcinomas, adenomas of the breast and colon, and skin papillomas (NARTLEY et al. 1987; KONTOSOGLOU et al. 1989; HASHIBA et al. 1989). MT is also elevated in liver during regenerative proliferation induced by the hepatotoxicants and tumor promoters acetaminophen and bis(2-ethylhexyl) phthalate (DEHP) (WAALKES and WARD 1989). The tumor-promoting phorbol esters, such as TPA, enhance expression of MT and hsp32 (heme oxygenase) (HIWASA and SAKIYAMA 1986; HASHIBA et al. 1989). MORTON et al. (1988) reported that MT is elevated in mouse tumors in a manner related to their degree of malignancy.

Increased levels of members of the hsp70 and hsp90 families are found in mammalian cells transformed by DNA tumor viruses. Mouse ascites tumor cells and human mammary tumor cells express hsp25–27. Formation of complexes of hsp70 and hsp90 with oncogene products, including p53 and c-*myc*, has been observed. It is speculated that stress proteins may play a role in malignant transformation as a result of these complexes increasing the half-life and concentrations of particular immortalizing nuclear oncogene products, such as p53 (MORIMOTO et al. 1990; NOVER 1991). Both MT and stress protein expression are cell cycle dependent and can be used as a marker of cell proliferation. The exact role of MTs and stress proteins in tumor cell pathobiology is not known but is important to define.

Abnormal regulation and synthesis of MTs and stress proteins may play roles in the etiology of other diseases. MT as been implicated in metal storage diseases, such as Menkes' and Wilson's diseases, familial hyperzincemia, and diabetes. Several of the stress proteins have been implicated to play roles in the development of certain autoimmune and infectious diseases, immune dysfunction, osteoarthritis, hypertension and atherosclerosis, and cutaneous inflammatory diseases (KAUFMANN 1990; VAN EDEN 1991; WINFIELD and JARJOUR 1991; HAMET 1992). The hsp70 gene is located between the tumor necrosis factor α and β genes, thus suggesting a relationship between stress proteins and neoplasia and infection, since these cytokines evoke a range of actions in inflammation, immunity, and cell injury (MORIMOTO et al. 1990).

VI. Adjuncts to Chemotherapy

The ubiquitous distribution of both MTs and stress proteins may be exploited in cancer therapy regimens. MT may serve as an adjunct in cancer chemotherapy (IMURA et al. 1991; CHERIAN et al. 1994). The selective induction of MT by bismuth in noncancer cells protects these cells from harmful side effects of a number of chemotherapeutics, including alkylating agents, antimetabolites, and mitotic inhibitors, but does not interfere with the efficacy of these compounds. Conversely, the presence of MT in tumors may act as a multidrug resistance factor in cancer chemotherapy (CHERIAN et al. 1994). While the role of stress proteins is not clearly defined, hyperthermia potentiates the antitumor effects of chemotherapeutics and radiotherapy (NOVER 1991).

While MTs are structurally and mechanistically distinct from the classical stress proteins, MTs share many similar functions with this broad class of proteins. MT appears to be a stress protein, based on similarities with stress proteins, with regard to ubiquitous distribution, high evolutionary conservation, protective functions, and gene regulation properties.

G. Stress Proteins as Biomarkers of Metal Exposure and Toxicity

The development of more sensitive and predictive test methods to characterize the hazards associated with xenobiotic exposure is an area of intense research in toxicology. One approach is to develop methodologies which would define biomarkers of exposure/toxicity. A biomarker can be defined broadly as a change which occurs in a biological system that is quantitatively or qualitatively predictive of disease or potential disease development resulting from toxicant exposure (Committee on Biological Markers 1987). Based on findings in recent studies, an increasing number of investigators is postulating that stress proteins, or altered patterns of protein synthesis, may serve as markers of exposure, toxicity, or general organismal stress for metals and other xenobiotics (DEATON et al. 1990; GLAVEN et al. 1991; COCHRANE et al. 1991; STEGEMAN et al. 1992; GOERING et al. 1993a,b).

I. Rationale and Criteria

Chemical-specific or target-tissue-specific alterations in protein synthesis may represent "biochemical fingerprints" or "biochemical signatures" of exposure and toxicity, respectively (FOWLER and SILBERGELD 1989; STEGEMAN et al. 1992). Since in the "real world" humans are more likely exposed to mixtures of xenobiotics rather than to a single agent, analysis of tissue-specific differences in stress protein synthesis, or agent-specific responses within a tissue, would enable toxicologists to identify which contaminant elicited the response. Several properties of the stress protein response make it a viable system to exploit as a biomarker. First, induction of synthesis is rapid; changes in expression of mRNA and their protein products occur within a few hours after exposure (BLAKE et al. 1990). Second, these changes in gene expression have been demonstrated at chemical concentrations which are below toxicity thresholds or occur prior to the onset of overt clinical toxicity (AOKI et al. 1990; SANDERS et al. 1991; GOERING et al. 1992, 1993a). Finally, interspecies comparisons are more meaningful when a response such as stress protein induction is highly conserved across a broad range of species, i.e., from bacteria to humans (SCHLESINGER 1990; NOVER 1991). These features make the stress protein response attractive for development as a biomarker of exposure and toxicity, and may help toxicologists more completely understand the early cellular responses which play a role in mechanisms of cell injury. Since stress proteins are localized in specific intracellular compartments, changes in synthesis may reflect ensuing damage to specific organelles.

There are important criteria for biomarkers: (a) the response should occur prior to the onset of overt disease symptomology, (b) for biomarkers of exposure, the response should demonstrate a level of specificity so as to allow the detection of single toxicants or classes of toxicants, and (c) for markers of toxicity or cellular stress the response should demonstrate a mechanistic correlation to intracellular changes which ultimately lead to overt toxicities (FOWLER et al. 1984; GLAVEN et al. 1991).

II. Exposure and Toxicity

Several reports indicate that the synthesis and accumulation of stress proteins meet the criteria of sensitivity and specificity described above for biomarkers of exposure and toxicity for metals. A concomitant decrease in synthesis of constitutive, non-stress proteins may also contribute to the specificity of the response. For example, AOKI et al. (1990) found that noncytotoxic concentrations of the semiconductor metal gallium stimulated the synthesis of several proteins and inhibited the synthesis of others in cultured kidney tubules. The pattern of synthesis and inhibition was distinct from that which occurred in the induction response to arsenite. In another study, noncytotoxic concentrations of arsenite induced the synthesis of hsp70 and hsp90, while phenyldichloroarsine, a skin vesicant, stimulated synthesis of hsp70, but not hsp90, in cultured human keratinocytes (DEATON et al. 1990). Cadmium, zinc, arsenite, mercury, and nickel induced hsp70, hsp90 and MT synthesis in primary cultures of hepatocytes (BAUMAN et al. 1993; GOERING et al. 1993c) prior to cell damage as assessed by increased release of intracellular potassium or decreased cell survival. The response exhibited some specificity towards the inducing metals; iron, lead, and vanadium were ineffective inducers of these three proteins (BAUMAN et al. 1993). Exposure of the rotifer *B. plicatilis* to copper and tributyl tin resulted in induction of a 58-kDa protein at concentrations equal to 0.1%–10% of the metal concentrations which were lethal to 50% of the organisms (LC_{50}). These studies indicate that changes in stress protein abundance may prove useful as a sensitive biomarker of metal exposure (COCHRANE et al. 1991). Specificity was observed with regard to the inducing metal, e.g., copper and tributyl tin induced p58, while aluminum, mercury, arsenic, and zinc had no effect.

In studies using cadmium as a well-known hepatotoxicant and mercuric chloride as a well-known nephrotoxicant, GOERING et al. (1992, 1993a) characterized changes in stress protein synthesis in target and nontarget tissues in adult rats (Fig. 1). Dose- and time-dependent alterations in gene expression in kidney and liver were demonstrated after exposure to mercury and cadmium, respectively, as evidenced by enhanced de novo synthesis of the 70-, 90-, and 110-kDa stress proteins 2–4 h after exposure. A concomitant inhibition of the synthesis of constitutive proteins of 68 and 38 kDa was observed. The changes in protein synthesis were target-organ specific, i.e., the nephrotoxicant mercury produced changes only in kidney, but not in liver, and the hepatotoxicant cadmium produced changes in liver, but not in kidney. The effects on protein synthesis occurred prior to the detection of liver and renal injury using histopathologic, functional, and clinical indices. An important step in evaluating the stress protein response as a marker to assess exposures and predict potential adverse effects is to validate in vivo results with data generated in vitro. The dose- and time-dependent changes in protein synthesis observed in vivo by GOERING et al. (1993a) after exposure to cadmium were confirmed in primary cultures of hepatocytes, and also occurred prior to detection of cytotoxicity (GOERING et al. 1993c). It should be noted that the utility of stress proteins as markers of exposure and toxicity is not limited to metals. Other studies have demonstrated that stress proteins may be useful markers of cellular injury or toxicity for other classes

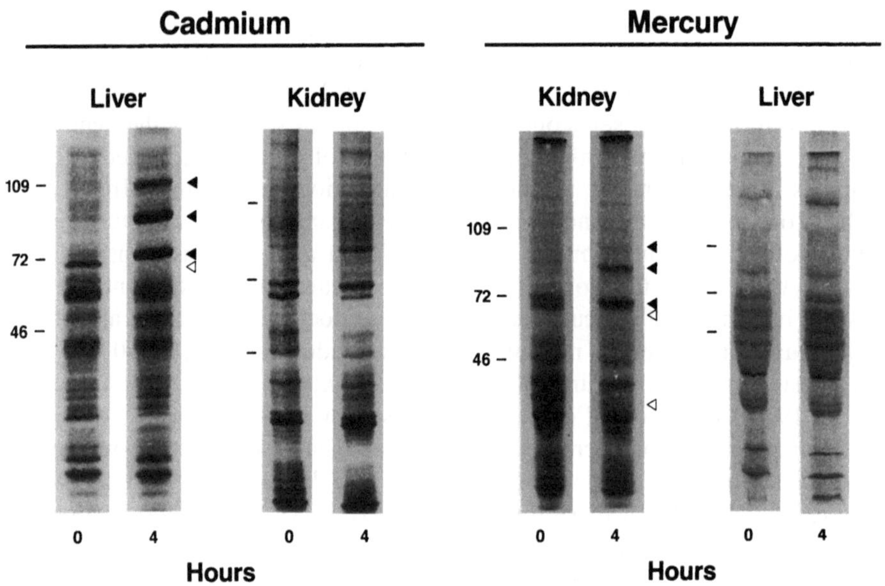

Fig. 1. Representative SDS-PAGE profiles of ^{35}S-methionine-labeled rat kidney and liver proteins from controls (0 h) and 4 h after exposure to $CdCl_2$ (2 mg Cd/kg, i.v.) and $HgCl_2$ (1 mg Hg/kg, i.v.). Samples were applied to gels at 60 000 cpm per lane. Molecular weights of protein standards are indicated (×1000). All gels were 12.5% acrylamide except the mercury-liver gel, which was an 8%–25% gradient gel. De novo synthesis of 70-, 90-, and 100-kDa proteins (*solid arrows*) and inhibition of synthesis of 68- and 38-kDa proteins (*hollow arrows*) were observed. Changes in protein synthesis were observed in liver, but not kidney, after cadmium treatment, and in kidney, but not liver, after mercury treatment

of xenobiotics and various disease states such as arthritis, autoimmune disease, and neurologic injury.

III. Toxicity Screening Assays

Several investigators have suggested that the stress protein response may be applied to in vitro and in vivo screening assays for various toxicity endpoints resulting from metal exposure. The use of recombinantly constructed "stress reporter" cells, which link the hsp DNA promoter to a reporter gene such as β-galactosidase, could be developed into a short-term in vitro screen in order to evaluate the hazardous potential of particular metal species based on the extent of the reporter gene response (WELCH 1992). Transgenic "stress reporter" organisms may also represent tools to evaluate environmental contamination from metals and other pollutants. Transgenic worms (*C. elegans*), developed with an enzyme reporter gene under hsp promoter control, could be transplanted to ecosystems to determine the extent of pollution based on the activity of the reporter gene (WELCH 1993).

Differential induction of many stress proteins, with an M_r ranging from 10 to 100 kDa, was observed after exposure of *E. coli* to cadmium, nickel, and hexavalent chromate (ÖDBERG-FERRAGUT et al. 1991). Six proteins were induced by all three metals, another group of proteins was selectively induced by cadmium, another group was selectively induced by chromate, and some proteins were induced by two of the metals but not the third. These effects were a more sensitive marker of toxicity than growth inhibition. In another study in which *E. coli* was exposed to nine diverse "model pollutants" including cadmium, 13–39 stress proteins were induced at concentrations that had no effect on growth (BLOM et al. 1992). At least 50% of these proteins were unique for a given chemical and can be potentially used to assess the degree of environmental stress and identify the nature of pollutants.

The possibility of developing the cellular reponse to stress with regard to gene expression into an in vitro test method for general metal toxicity screening has been explored by FISCHBACH et al. (1993). The basic tool of the assay is a transformed NIH/3T3 mouse cell line transfected with a plasmid construct carrying the human hsp70 promoter sequence linked to a human growth hormone "reporter" gene. Thus, by assaying cell cultures for growth hormone, one can quantitate hsp70 gene promoter activity. Of 31 metals or metalloids tested, approximately one-half were effective inducers of the hsp70 promoter, and could be ranked in groups as strong, intermediate, or weak inducers. Evidence for induction of the hsp70 promoter occurred at concentrations several orders of magnitude below those which resulted in cell lysis. These results demonstrated a high degree of concordance with relative toxicity results generated with other in vitro and in vivo metal toxicity assays. This hsp70 test identified correctly approximately 80% of the positive and negative compounds for several endpoints, such as LD_{50} in mice, carcinogenesis/mutagenesis, in vitro cell transformation, and sister chromatid exchange. The results obtained with this reporter gene assay confirm a growing body of evidence that the stress response in general has great potential to be exploited as an in vitro marker of metal toxicity.

Several assays have been reported which may prove valuable in identifying the potential teratogenic properties of chemicals. Results from studies using *Drosophila* embryonic cells exposed to teratogenic metals (arsenite, cadmium, mercury) demonstrated a strong correlation between induction of specific stress proteins, inhibition of cell differentiation, and the known animal/human teratogenic effects associated with these metals (BOURNIAS-VARDIABASIS et al. 1990). Based on results from a number of studies, these authors concluded that any agent which interferes with normal cellular morphogenesis could be identified in this assay. In another study, human chorionic villus tissue, obtained from women undergoing therapeutic termination of a normal pregnancy at 10–17 weeks of gestation, demonstrated an increased expression of hsp70, hsp85, and hsp105 when exposed in vitro to the teratogens heat shock, arsenite, and cadmium (HONDA et al. 1992).

These proteins were also induced in mouse embryo neuroepithelial tissue on day 9 of gestation after injection of pregnant dams with arsenite and cadmium. Induction of stress proteins by metals in human chorionic villus cells may prove useful for evaluating the potential environmental risk for the embryo to these agents. While not tested with metals yet, an experimental model using rat embryos exposed to teratogens from gestation days 9 to 11 in culture or through treating pregnant dams may be developed into a short-term screen for potential teratogens. Analysis of stress protein patterns which have been correlated with specific terata would reveal information regarding the potential teratogenicity of a given test compound (HANSEN et al. 1988; MIRKES and DOGGETT 1992; GOERING et al. 1993b).

IV. Environmental Monitoring

Another fertile area for the application of altered protein synthesis patterns as a biomarker is to assess degrees of environmental stress or contamination by metals. Such monitoring may enable toxicologists to detect relative levels of pollutant metals or to assess the general status of a particular ecosystem by measuring protein changes in native or transplanted (sentinel) species (see STEGEMAN et al. 1992). For this application, changes in stress protein patterns must be protracted and sensitive enough to be detected at relevant environmental exposures. Current evidence suggests that, although de novo synthesis is transient, various stress proteins accumulate in tissues for relatively longer periods (SANDERS 1990; GOERING et al. 1992, 1993a). For example, accumulation of hsp60 was detected in mantle tissue of the blue mussel *Mytilis edulis* after 7 days of exposure to various concentrations of copper (SANDERS et al. 1991). Accumulation of hsp60 was found to be a sensitive marker with elevated levels detected at copper concentrations one order of magnitude lower than that which caused a decrease in scope-for-growth, an index derived from several parameters which reflects the general physiologic status of the organism.

Isoforms of various stress proteins have been shown to be expressed under environmental stress in different tissues in many field species such as fish, mollusks, minnows, sea urchins, insects, and nematodes (see review by SANDERS 1993). Through carefully controlled laboratory studies, "signature" or "fingerprint" changes could be identified for sentinel species such that a particular metal contaminant could be identified in the field based on a similar response. For example, laboratory studies with the crustacean isopod *Oniscus asellus* demonstrated hsp70 expression in response to heat treatment and lead (100 mg/kg). In field studies, *O. asellus* removed from a polluted site in the vicinity of a lead/zinc smelter similarly demonstrated elevated hsp70 expression (KÖHLER et al. 1992).

V. Human Applications

The intracellular localization of stress proteins is problematic for the evaluation of the response in humans. Because the cells for the assay of stress proteins are not readily available through noninvasive procedures, the application of this response to human monitoring is limited. Recently, however, enhanced synthesis of stress proteins was demonstrated in primary cultures of human lymphocytes exposed to several metals (YAMADA and KOIZUMI 1993). The specificity of the response was dependent on the metal to which the cultures were exposed. For example, cadmium and zinc induced both hsp70 and MT, while cobalt and triphenyltin induced only hsp70. Conversely, copper, mercury, nickel, and silver all induced synthesis of MT, but not of hsp70. Enhanced synthesis of stress proteins has also been demonstrated in vivo in lymphocytes and spleen cells excised from mice exposed to hyperthermia (RODENHISER et al. 1985).

References

Al-Aoukaty A, Appanna VD, Falter H (1992) Gallium toxicity and adaptation on *Pseudomonas fluorescens*. FEMS Microbiol Lett 71:265–272

Amaral MD, Galego L, Rodrigues-Pousada C (1988) Stress response of *Tetrahymena pyriformis* to arsenite and heat shock: differences and similarities. Eur J Biochem 171:463–470

Ananthan J, Goldberg AL, Voellmy R (1986) Abnormal proteins serve as eukaryotic stress signals and trigger the activation of heat shock genes. Science 232: 522–524

Aoki Y, Lipsky MM, Fowler BA (1990) Alterations in protein synthesis in primary cultures of rat kidney proximal tubule epithelial cells by exposure to gallium, indium, and arsenite. Toxicol Appl Pharmacol 106:462–468

Arrigo AP, Suhan JP, Welch WJ (1988) Dynamic changes in the structure and intracellular locale of the mammalian low-molecular-weight heat shock protein. Mol Cell Biol 8:5059–5071

Atkinson BG, Cunningham T, Dean RL, Somerville M (1983) Comparison of the effects of heat shock and metal-ion stress on gene expression in cells undergoing myogenesis. Can J Biochem Cell Biol 61:404–413

Bauman JW, Liu J, Klaassen CD (1993) Production of metallothionein and heat-shock proteins in response to metals. Fundam Appl Toxicol 21:15–22

Beckmann RP, Mizzen LA, Welch WJ (1990) Interaction of hsp70 with newly synthesized proteins: implications for protein folding and assembly. Science 248:850–854

Blake MJ, Gershom D, Fargnoli J, Holbrook NJ (1990) Discordant expression of heat shock protein mRNAs in tissues of heat-stressed rats. J Biol Chem 265: 15275–15279

Blom A, Harder W, Matin A (1992) Unique and overlapping pollutant stress proteins of *Escherichia coli*. Appl Environ Microbiol 58:331–334

Bournias-Vardiabasis N, Buzin C, Flores J (1990) Differential expression of heat shock proteins in *Drosophila* embryonic cells following metal ion exposure. Exp Cell Res 189:177–182

Bruce JL, Price BD, Coleman CN, Calderwood SK (1993) Oxidative injury rapidly activates the heat shock transcription factor but fails to increase levels of heat shock proteins. Cancer Res 53:12–15

Bunch TA, Grinblat Y, Goldstein LS (1988) Characterization and use of the *Drosophila* metallothionein promoter in cultured *Drosophila melanogaster* cells. Nucleic Acids Res 16:1043–1061

Burgman PW, Kampinga HH, Konings AW (1993) Possible role of localized protein denaturation in the mechanism of induction of thermotolerance by heat, sodium-arsenite and ethanol. Int J Hyperthermia 9:151–162

Cajone F, Bernelli-Zazzera A (1988) Oxidative stress induces a subset of heat shock proteins in rat hepatocytes and MH1C1 cells. Chem Biol Interact 65:235–246

Caltabiano MM, Koestler TP, Poste G, Greig RG (1986a) Induction of 32- and 34-kDa stress proteins by sodium arsenite, heavy metals, and thiol-reactive agents. J Biol Chem 261:13381–13386

Caltabiano MM, Koestler TP, Poste G, Greig RG (1986b) Induction of mammalian stress proteins by a triethylphosphine gold compound used in the therapy of rheumatoid arthritis. Biochem Biophys Res Commum 138:1074–1080

Caltabiano MM, Poste G, Greig RG (1988) Induction of the 32-kD human stress protein by auranofin and related triethylphosphine gold analogs. Biochem Pharmacol 37:4089–4093

Carr BI, Huang TH, Buzin CH, Itakura K (1986) Induction of heat shock gene expression without heat by hepatocarcinogens and during hepatic regeneration in rat liver. Cancer Res 46:5106–5111

Carson-Jurica MA, Lee AT, Dobson AW, Conneely OM, Schrader WT, O'Malley BW (1989) Interaction of the chicken progesterone receptor with heat shock protein (hsp) 90. J Steroid Biochem 34:1–9

Cervera J (1985) Induction of self-tolerance and enhanced stress protein synthesis in L-132 cells by cadmium chloride and by hyperthermia. Cell Biol Int Rep 9:131-141

Cheng M, Hartl F, Martin J, Pollock R, Kalousek F, Neupert W, Hallberg E, Hallberg R, Horwich A (1989) Mitochondrial heat-shock protein hsp60 is essential for assembly of proteins imported into yeast mitochondria. Nature 337: 620–625

Cherian MG, Howell SB, Imura N, Klaassen CD, Koropatnick J, Lazo JS, Waalkes MP (1994) Role of metallothionein in carcinogenesis. Toxicol Appl Pharmacol 126:1–5

Chirico W, Waters MG, Blobel G (1988) 70K heat shock related proteins stimulate protein translocation into microsomes. Nature 332:805–810

Christie NT, Costa M (1984) In vitro assessment of the toxicity of metal compounds. Biol Trace Elem Res 6:139–158

Ciavarra RP, Simeone A (1990a) T lymphocyte stress response: I. Induction of heat shock protein synthesis at febrile temperatures is correlated with enhanced resistance to hyperthermic stress but not to heavy metal toxicity or dexamethasone-induced immunosuppression. Cell Immunol 129:363–376

Ciavarra RP, Simeone A (1990b) T lymphocyte stress reponse. II. Protection of translation and DNA replication against some forms of stress by prior hyperthermic stress. Cell Immunol 131:11–26

Clerget M, Polla BS (1990) Erythrophagocytosis induces heat shock protein synthesis by human monocytes-macrophages. Proc Natl Acad Sci USA 87:1081–1085

Cochrane BJ, Irby RB, Snell TW (1991) Effects of copper and tributyltin on stress protein abundance in the rotifer *Brachionus plicatilis*. Comp Biochem Physiol 98C:385–390

Cohen DS, Palmer E, Welch WJ, Sheppard D (1991) The response of guinea pig airway epithelial cells and alveolar macrophages to environmental stress. Am J Respir Cell Mol Biol 5:133–143

Committee on Biological Markers (1987) Biological markers in environmental health research. Environ Health Perspect 74:3–9

Coogan TP, Bare RM, Waalkes MP (1992) Cadmium-induced DNA damage: effects of zinc pretreatment. Toxicol Appl Pharmacol 113:227–233

Coogan TP, Bare RM, Bjornson EJ, Waalkes MP (1994) Enhanced metallothionein gene expression protects against cadmium genotoxicity in cultured rat liver cells. J Toxicol Environ Health (in press)
Courgeon A-M, Maisonhaute C, Best-Belpomme M (1984) Heat shock proteins are induced by cadmium in *Drosophila* cells. Exp Cell Res 153:515–521
Darasch S, Mosser DD, Bols NC, Heikkila JJ (1988) Heat shock gene expression on *Xenopus laevis* A6 cells in response to heat shock and sodium arsenite treatments. Biochem Cell Biol 66:862–870
Deaton MA, Bowman PD, Jones GP, Powanda MC (1990) Stress protein synthesis in human keratinocytes treated with sodium arsenite, phenyldichloroarsine, and nitrogen mustard. Fundam Appl Toxicol 14:471–476
De Jong WW, Hoekman WA, Mulders JW, Bloemendal H (1986) Heat shock response in the rat lens. J Cell Biol 102:104–111
Delpino A, Spinsanti P, Mattei E, Mileo AM, Vismara D, Ferrini U (1992) Identification of a 66 kD heat shock protein (HSP) induced in M-14 human melanoma cells by severe hyperthermic treatment. Melanoma Res 2:369–375
Donati YR, Kantengwa S, Polla BS (1991) Phagocytosis and heat shock response in human monocytes-macrophages. Pathobiology 59:156–161
Duncan RF, Hershey JWB (1987) Translational repression by chemical inducers of the stress response occurs by different pathways. Arch Biochem Biophys 256: 651–661
Eaton DL, Stacey NH, Wong K-L, Klaassen CD (1980) Dose-response effects of various metal ions on rat liver metallothionein, glutathione, heme oxygenase, and cytochrome P-450. Toxicol Appl Pharmacol 55:393–402
Edwards MJ, Marks R, Dykes PJ, Merrett VR, Morgan HE, O'Donovan MR (1991) Heat shock proteins in cultured human keratinocytes and fibroblasts. J Invest Dermatol 96:392–396
Ewing JF, Maines MD (1991) Rapid induction of heme oxygenase 1 mRNA and protein by hyperthermia in rat brain: heme oxygenase 2 is not a heat shock protein. Proc Natl Acad Sci USA 88:5364–5368
Ewing JF, Maines MD (1993) Glutathione depletion induces heme oxygenase-1 (HSP32) mRNA and protein in rat brain. J Neurochem 60:1512–1519
Ewing JF, Haber SN, Maines MD (1992) Normal and heat-induced patterns of expression of heme oxygenase-1 (HSP32) in rat brain: hyperthermia causes rapid induction of mRNA and protein. J Neurochem 58:1140–1149
Ferm VH, Layton WM Jr (1979) Reduction in cadmium teratogenesis by prior cadmium exposure. Environ Res 18:347–350
Fischbach M, Sabbioni E, Bromley P (1993) Induction of the human growth hormone gene placed under human hsp70 promoter control in mouse cells: a quantitative indicator of metal toxicity. Cell Biol Toxicol 9:177–188
Fowler BA, Silbergeld EK (1989) Occupational diseases – new workforces, new workplaces. Ann NY Acad Sci 572:46–54
Fowler BA, Abel J, Elinder CG, Hapke HJ, Kagi JHR, Kleiminger J, Kojima Y, Schoot-Uiterkamp AJM, Silbergeld EK, Silver S, Summer KH, Williams RJP (1984) Structure, mechanism, and toxicity. In: Nriagu JU (ed) Changing metal cycles and human health. Springer, Berlin Heidelberg New York, pp 391–404
Gerner EW, Schneider MJ (1975) Induced thermal resistance in HeLa cells. Nature 256:500–502
Glaven JA, Gandley RE, Fowler BA (1991) Biological indicators of cadmium exposure, Chap 67. Methods Enzymol 205:592–599
Goering PL, Klaassen CD (1983) Altered subcellular distribution of cadmium following cadmium pretreatment: possible mechanism of tolerance to cadmium-induced lethality. Toxicol Appl Pharmacol 70:195–203
Goering PL, Klaassen CD (1984) Zinc-induced tolerance to cadmium hepatotoxicity. Toxicol Appl Pharmacol 74:299–307

Goering PL, Fisher BR, Chaudhary PP, Dick CA (1992) Relationship between stress protein induction in rat kidney by mercuric chloride and nephrotoxicity. Toxicol Appl Pharmacol 113:184–191

Goering PL, Fisher BR, Kish CL (1993a) Stress protein synthesis induced in rat liver by cadmium precedes hepatotoxicity. Toxicol Appl Pharmacol 122:139–148

Goering PL, Fisher BR, Kimmel CA, Kimmel GL (1993b) Stress proteins as biomarkers of toxicity. In: Travis CC (ed) Use of biomarkers in assessing health and environmental impacts of chemical pollutants. Plenum, New York, pp 95–99

Goering PL, Kish CL, Dick SE (1993c) Stress protein synthesis induced by cadmium in rat hepatocyte primary cultures. Toxicologist 13:160 (abstract)

Halliwell B, Gutteridge JMC (1984) Oxygen toxicity, oxygen radicals, transition metals and disease. Biochem J 219:1–14

Hamer DH (1986) Metallothionein. Annu Rev Biochem 55:913–951

Hamet P (1992) Abnormal hsp70 gene expression: its potential key role in metabolic defects in hypertension. Clin Exp Pharmacol Physiol [Suppl] 20:53–59

Hansen DK, Anson JF, Hinson WG, Pipkin JL Jr (1988) Phenytoin-induced stress protein synthesis in mouse embryonic tissue. Proc Soc Exp Biol Med 189: 136–140

Hashiba H, Hosoi J, Karasawa M, Yamada S, Nose K, Kuroki T (1989) Induction of metallothionein mRNA by tumor promoters in mouse skin and its constitutive expression in papillomas. Mol Carcinog 2:95–100

Hatayama T, Tsukimi Y, Wakatsuki T, Kitamura T, Imahara H (1991) Different induction of 70000-Da heat shock protein and metallothionein in HeLa cells by copper. J Biochem 110:726–731

Hatayama T, Tsukimi Y, Wakatsuki T, Kitamura T, Imahara H (1992) Characteristic induction of 70000-Da heat shock protein and metallothionein by zinc in HeLa cells. Mol Cell Biochem 112:143–153

Heikkila JJ, Darasch SP, Mosser DD, Bols NC (1987) Heat and sodium arsenite act synergistically on the induction of heat shock gene expression in *Xenopus laevis* A6 cells. Biochem Cell Biol 65:310–316

Hemmingsen SM, Woolford C, van de Vies SM, Tilly K, Dennis DT, Georgopoulos CP, Hendrix RW, Ellis RJ (1988) Homologous plant and bacterial proteins chaperone oligomeric protein assembly. Nature 33:330–334

Hendrick JP, Hartl F-U (1993) Molecular chaperone functions of heat-shock proteins. Annu Rev Biochem 62:349–384

Hightower LE (1991) Heat shock, stress proteins, chaperones, and proteotoxicity. Cell 66:191–197

Hiwasa T, Sakiyama S (1986) Increase in the synthesis of a 32000 Mr protein in BALB/C 3T3 cells after treatment with tumor promoters, chemical carcinogens, metal salts and heat shock. Cancer Res 46:2474–2481

Hiwasa T, Fujimura S, Sakiyama S (1982) Tumor promoters increase the synthesis of a 32000-dalton protein in BALB/c 3T3 cells. Proc Natl Acad Sci USA 79: 1800–1804

Honda K, Hatayama T, Takahashi K, Yukioka M (1992) Heat shock proteins in human and mouse embryonic cells after exposure to heat shock or teratogenic agents. Teratogenesis Carcinog Mutagen 11:235–244

Howard KJ, Holley SJ, Yamamoto KR, Distelhorst CW (1990) Mapping the HSP90 binding region of the glucocorticoid receptor. J Biol Chem 265:11928–11935

Imura N, Satoh M, Naganuma A (1991) Possible application of metallothionein in cancer therapy. In: Klaassen CD, Suzuki KT (eds) Metallothionein in biology and medicine. CRC Press, Boca Raton, pp 375–382

Ito T, Sawauchi K (1966) Inhibitory effects on cadmium-induced testicular damage by pretreatment with smaller cadmium doses. Okajimas Folia Anat Jpn 42: 107–117

Johnston RN, Kucey BL (1988) Competitive inhibition of hsp70 gene expression causes thermosensitivity. Science 242:1551–1554
Kägi JHR, Kojima Y (1987) Chemistry and biochemistry of metallothionein. Experientia [Suppl] 52:25–61
Kampinga HH, Brunsting JF, Konings AW (1992) Acquisition of thermotolerance induced by heat and arsenite in HeLa S3 cells: multiple pathways to induce tolerance? J Cell Physiol 150:406–416
Kantengwa S, Polla BS (1993) Phagocytosis of *Staphylococcus aureus* induces a selective stress response in human monocytes-macrophages (M phi): modulation by M phi differentiation and by iron. Infect Immunol 61:1281–1287
Kapoor M, Sveenivasan GM (1988) The heat shock response of *Neurospora crassa*: stress-induced thermotolerance in relation to peroxidase and superoxide dismutase levels. Biochem Biophys Res Commun 156:1097–1102
Kapron-Bras CM, Hales BF (1992) Genetic differences in heat-induced tolerance to cadmium in cultured mouse embryos are not correlated with changes in a 68-kD heat shock protein. Teratology 46:191–200
Kasprzak KS (1991) The role of oxidative damage in metal carcinogenicity. Chem Res Toxicol 4:604–615
Katayose D, Isoyama S, Fujita H, Shibahara S (1993) Separate regulation of heme oxygenase and heat shock protein 70 mRNA expression in the rat heart by hemodynamic stress. Biochem Biophys Res Commun 191:587–594
Kaufmann SHE (1990) Heat shock proteins and the immune response. Immunol Today 11:129–136
Keyse SM, Tyrrell RM (1987) Both near ultraviolet radiation and the oxidizing agent hydrogen peroxide induce a 32-kDa stress protein in normal human skin fibroblasts. J Biol Chem 262:14821–14825
Keyse M, Tyrrell RM (1989) Heme oxygenase is the major 32-kDa stress protein induced in human skin fibroblasts by UVA radiation, hydrogen peroxide, and sodium arsenite. Proc Natl Acad Sci USA 86:99–103
Kohane DS, Sarzani R, Schwartz JH, Chobanian AV, Brecher P (1990) Stress-induced proteins in aortic smooth muscle cells and aorta of hypertensive rats. Am J Physiol 258:H1699–H1705
Köhler HR, Triebskorn R, Stöcker W, Kloetzel PM, Alberti G (1992) The 70 kD heat shock protein in soil invertebrates: a possible tool for monitoring environmental toxicants. Arch Environ Contam Toxicol 22:334–338
Kontosoglou TE, Banerjee D, Cherian MG (1989) Immunohistochemical localization of metallothionein in human testicular embryonal carcinoma cells. Virchows Arch [A] 415:545–549
Lai Y, Shen C, Cheng T, Hou M, Lee W (1993) Enhanced phosphorylation of a 65 kDa protein is associated with rapid induction of stress proteins in 9L rat brain tumor cells. J Cell Biochem 51:369–379
Landry J, Bernier D, Chretien P, Nicole LM, Tanguay RM, Marceau N (1982) Synthesis and degradation of heat shock proteins during development and decay of thermotolerance. Cancer Res 42:2457–2461
Lavoie JN, Gingras-Breton G, Tanguay RM, Landry J (1993) Induction of Chinese hamster HSP27 gene expression in mouse cells confers resistance to heat shock. HSP27 stabilization of the microfilament organization. J Biol Chem 268:3420–3429
Leber AP, Miya TS (1976) A mechanism for cadmium and zinc-induced tolerance to cadmium toxicity: involvement of metallothionein. Toxicol Appl Pharmacol 37:403–414
Lee KJ, Hahn GM (1988) Abnormal proteins as the trigger for the induction of stress responses: heat, diamide, and sodium arsenite. J Cell Physiol 136:411–420
Lee YJ, Dewey WC (1987) Effect of cycloheximide or puromycin on induction of thermotolerance by sodium arsenite in Chinese hamster ovary cells: involvement of heat shock proteins. J Cell Physiol 132:41–48

Lee YJ, Dewey WC (1988) Thermotolerance induced by heat, sodium arsenite, or puromycin: its inhibition and differences between 43 degrees C and 45 degrees C. J Cell Physiol 135:397–406

Lee YJ, Curetty L, Corry PM (1991) Differences in preferential synthesis and redistribution of HSP70 and HSP28 families by heat shock or sodium arsenite in Chinese hamster ovary cells. J Cell Physiol 149:77–87

Levinson W, Oppermann H, Jackson J (1980) Transition series metals and sulfhydryl reagents induce the synthesis of four proteins in eukaryotic cells. Biochim Biophys Acta 606:170–180

Li GC (1983) Induction of thermotolerance and enhanced heat shock protein synthesis in Chinese hamster fibroblasts by sodium arsenite and by ethanol. J Cell Physiol 115:116–122

Li GC, Laszlo A (1985) Animo acid analogs while inducing heat shock proteins sensitize CHO cells to thermal damage. J Cell Physiol 122:9197

Li GC, Werb Z (1982) Correlation between synthesis of heat shock proteins and development of thermotolerance in chinese hamster fibroblast. Proc Natl Acad Sci USA 79:3218–3222

Löw-Friedrich I, Schoeppe W (1991) Effects of calcium channel blockers on stress protein synthesis in cardiac myocytes. J Cardiovasc Pharmacol 17:800–806

Maines MD (1988) Heme oxygenase: function, Multiplicity, regulatory mechanisms, and clinical applications. FASEB J 2:2557–2568

Maines MD, Kappas A (1977) Metals as regulators of heme metabolism: physiological and toxicological implications. Science 198:1215–1221

Maines MD, Chung A-S, Kutty RK (1982) Inhibition of testicular heme oxygenase activity by cadmium: a novel cellular response. J Biochem 257:14116–14121

Maines MD, Mayer RD, Ewing JF, McCoubrey WK Jr (1993) Induction of kidney heme oxygenase-1 (HSP32) mRNA and protein by ischemia/reperfusion: possible role of heme as both promoter of tissue damage and regulator of HSP32. J Pharmacol Exp Ther 264:457–462

Matsubara J, Tajima Y, Karasawa M (1987) Promotion of radioresistance by metallothionein induction prior to irradiation. Environ Res 43:66–74

Miller L, Qureshi MA (1992) Heat-shock protein synthesis in chicken macrophages: influence of in vivo and in vitro heat shock, lead acetate, and lipopolysaccharide. Poult Sci 71:988–998

Mirkes PE (1987) Hyperthermia-induced heat shock response and thermotolerance in postimplantation rat embryos. Dev Biol 119:115–122

Mirkes PE, Cornel L (1992) A comparison of sodium arsenite- and hyperthermia-induced stress responses and abnormal development in cultured postimplantation rat embryos. Teratology 46:251–259

Mirkes PE, Doggett B (1992) Accumulation of heat shock protein 72 in post-implantation rat embryos after exposure to various periods of hyperthermia in vitro: evidence that heat shock protein 72 is a biomarker of heat-induced embryotoxicity. Teratology 46:301–309

Misra S, Zararullah M, Price-Haughey J, Gedamu L (1989) Analysis of stress-induced gene expression in fish cell lines exposed to heavy metals and heat shock. Biochim Biophys Acta 1007:325–333

Mitani K, Fujita H, Sassa S, Kappas A (1990) Activation of heme oxygenase and heat shock protein 70 genes by stress in human hepatoma cells. Biochem Biophys Res Commun 166:1429–1434

Moretti-Rojas I, Fugua SA, Montgomery RA, McGuire WL (1988) A cDNA for estradiol-regulated 24k protein: control of mRNA levels in MCF-7 cells. Breast Cancer Treat 11:155–163

Morimoto RI, Tissières A, Georgopoulos C (1990) The stress response, function of the proteins, and perspectives. In: Morimoto RI, Tissières A, Georgopoulos C (eds) Stress proteins in biology and medicine. Cold Spring Harbor Laboratory Press, Cold Spring Harbor, New York, pp 1–36

Morimoto RI, Sarge KD, Abravaya K (1992) Transcriptional regulation of heat shock genes. J Biol Chem 267:21987–21990
Morton KA, Alazaraki NP, Datz FL, Taylor AT (1988) Uptake of cadmium-109, a metallothionein-binding radiometal, by tumors in mice as a function of the transformed phenotype. Invest Radiol 23:200–204
Murakami T, Ohmori H, Katoh T, Abe T, Higashi K (1991) Cadmium causes increases of N-myc and multidrug-resistance gene mRNA in neuroblastoma cells. Sangyo Ika Daigaku Zasshi 13:271–278
Nartley N, Cherian MG, Banerjee D (1987) Immunohistochemical localization of metallothionein in human thyroid tumors. Am J Pathol 129:177–182
Nover L (1991) Heat shock response. CRC Press, Boca Raton
Ödberg-Ferragut C, Espigares M, Dive D (1991) Stress protein synthesis, a potential toxicity marker in *Escherichia coli*. Ecotoxicol Environ Safety 21:275–282
Palmiter R (1987) Molecular biology of metallothionein gene expression. Experientia [Suppl] 52:63–80
Pannekoek Y, van Putten JP, Dandert J (1992) Identification and molecular analysis of a 63-kilodalton stress protein from *Neisseria gonorrhoeae*. J Bacteriol 174: 6928–6937
Pelham HRB (1982) A regulatory upstream promoter element in the *Drosophila* hsp70 heat shock gene. Cell 30:517–528
Pelham HRB (1986) Speculations on the functions of the major heat shock and glucose-regulated proteins. Cell 46:959–961
Riabowol KT, Mizzen LA, Welch WJ (1988) Heat shock is lethal to fibroblasts microinjected with antibodies against hsp70. Science 242:433–436
Ritossa FM (1962) A new puffing pattern induced by heat shock and DNP in *Drosophila*. Experientia 18:571–573
Roberts SA, Miya TS, Schnell RC (1976) Tolerance development to cadmium-induced alteration of drug action. Res Commun Chem Pathol Pharmacol 14: 197–200
Roccheri MC, La Rosa M, Ferraro MG, Cantone M, Cascino D, Giudice G, Sconzo G (1988) Stress proteins by zinc ions in sea urchin embryos. Cell Differ 24: 209–213
Rodenhiser D, Jung JH, Atkinson BG (1985) Mammalian lymphocytes: stress-induced synthesis of heat-shock proteins in vitro and in vivo. Can J Biochem Cell Biol 63:711–722
Rodenhiser DI, Jung JH, Atkinson BG (1986) The synergistic effect of hyperthermia and ethanol on changing gene expression of mouse lymphocytes. Can J Genet Cytol 28:1115–1124
Rollet E, Lavoie JN, Landry L, Tanguay RM (1992) Expression of *Drosophila's* 27 kDa heat shock protein into rodent cells confers thermal resistance. Biochem Biophys Res Commun 185:116–120
Rothman JE (1989) Polypeptide chain binding proteins: catalysts of proteins folding and related processes in cells. Cell 59:591–601
Sanchez Y, Taulien J, Borkovich KA, Lindquist S (1992) Hsp104 is required for tolerance to many forms of stress. EMBO J 11:2357–2364
Sanders BM (1990) Stress proteins: potential as multitiered biomarkers. In: McCarthy J, Shugart L (eds) Biological markers of environmental contamination. Lewis, Boca Raton, pp 165–191
Sanders BM (1993) Stress proteins in aquatic organisms: an environmental perspective. Crit Rev Toxicol 23:49–75
Sanders BM, Martin LS, Nelson WG, Phelps DK, Welch W (1991) Relationships between accumulation of a 60 kDa stress protein and scope-for-growth in *Mytilus edulis* exposed to a range of copper concentrations. Marine Environ Res 31: 81–97
Sato M, Ishizawa S, Yoshida T, Shibahara S (1990) Interaction of upstream stimulatory factor with the human heme oxygenase gene promoter. Eur J Biochem 188:231–237

Sauk JJ, Norris K, Foster R, Moehring J, Somerman MJ (1988) Expression of heat stress proteins by human periodontal ligament cells. J Oral Pathol 17:496–499
Saunders EL, Maines MD, Meredith MJ, Freeman ML (1991) Enhancement of heme oxygenase-1 synthesis by glutathione depletion in Chinese hamster ovary cells. Arch Biochem Biophys 288:368–373
Schlesinger MJ (1990) Heat shock proteins. J Biol Chem 265:12111–12114
Schmidt JA, Abdulla E (1988) Down-regulation of IL-1 beta biosynthesis by inducers of the heat-shock response. J Immunol 141:2027–2034
Sheffield WP, Shore GC, Randall SK (1990) Mitochondrial precursor protein: effects of 70-kilodalton heat shock protein on polypeptide folding, aggregation, and import competence. J Biol Chem 265:11069–11076
Shelton KR, Todd JM, Egle PM (1986) The induction of stress-related proteins by lead. J Biol Chem 261:1935–1940
Shibahara S, Müller RM, Taguchi H (1987) Transcriptional control of rat heme oxygenase by heat shock. J Biol Chem 262:12889–12892
Shuman J, Przybyla A (1988) Expression of the 31-kDa stress protein in rat myoblasts and hepatocytes. DNA 7:475–482
Silar P, Butler G, Thiele DJ (1991) Heat shock transcription factor activates transcription of the yeast metallothionein gene. Mol Cell Biol 11:1232–1238
Squibb KS, Pritchard JB, Fowler BA (1984) Cadmium metallothionein nephropathy: ultrastructural/biochemical alterations and intracellular cadmium binding. J Pharmacol Exp Ther 229:311–321
Stacey NH, Klaassen CD (1981) Comparison of the effects of metals on cellular injury and lipid peroxidation in isolated rat hepatocytes. J Toxicol Environ Health 7:139–147
Stegeman JJ, Brouwer M, Di Giulio RT, Forlin L, Fowler BA, Sanders BM, Van Veld PA (1992) Molecular responses to environmental contamination: enzyme and protein systems as indicators of chemical exposure and effect. In: Huggett RJ, Kimerle RA, Mehrle PM, Bergman HL (eds) Biomarkers: biochemical, physiological, and histological markers of anthropogenic stress. Lewis, Boca Raton, pp 235–335
Stocker R, Glazer AN, Ames BN (1987a) Antioxidant activity of albumin-bound bilirubin. Proc Natl Acad Sci USA 84:5918–5922
Stocker R, Yamamoto Y, McDonagh AF, Glazer AN, Ames BN (1987b) Bilirubin is an antioxidant of possible physiological importance. Science 235:1043–1047
Subjeck JR, Shyy T, Shen J, Johnson RJ (1983) Association between mammalian 110000 dalton heat shock protein and nucleoli. J Cell Biol 97:1389–1398
Tacchini L, Schiaffonati L, Pappalardo C, Gatti S, Bernelli-Zazzera A (1993) Expression of HSP 70, immediate-early response and heme oxygenase genes in ischemic-reperfused rat liver. Lab Invest 68:465–471
Taketani S, Kohno H, Yoshinaga T, Tokunaga R (1988) Induction of heme oxygenase in rat hepatoma cells by exposure to heavy metals and hyperthermia. Biochem Int 17:665–672
Taketani S, Kohno H, Yoshinaga T, Tokunaga R (1989) The human 32-kDa stress protein induced by exposure to arsenite and cadmium ions is heme oxygenase. FEBS Lett 245:173–176
Taketani S, Sato H, Yoshinage T, Tokunaga R, Ishii T, Bannai S (1990) Induction in mouse peritoneal macrophages of 34 kDa stress protein and heme oxygenase by sulfhydryl-reactive agents. J Biochem 108:28–32
Tamai KT, Gralla EB, Ellerby LM, Valentine JS, Thiele DJ (1993) Yeast and mammalian metallothioneins functionally substitute for yeast copper-zinc superoxide dismutase. Proc Natl Acad Sci USA 90:8013–8017
Thornalley PJ, Vasak M (1985) Possible role for metallothionein in protection against radiation-induced oxidative stress. Kinetics and mechanism of its reaction with superoxide and hydroxyl radicals. Biochim Biophys Acta 827:36–44

Uney JB, Anderson BH Thomas SM (1993) Changes in heat shock protein 70 and ubiquitin mRNA levels in C1300 N2A mouse neuroblastoma cells following treatment with iron. J Neurochem 60:659–665

Valone SE, Chikami GK, Miller VL (1993) Stress induction of the virulence proteins (SpvA, -B, and -C) from native plasmid pSDL2 of *Salmonella dublin*. Infect Immun 61:705–713

Van Eden W (1991) Heat-shock proteins as immunogenic bacterial antigens with the potential to induce and regulate autoimmune arthritis. Immunol Rev 121:5–28

Waalkes MP, Goering PL (1990) Metallothionein and other cadmium-binding proteins: recent developments. Chem Res Toxicol 3:281–288

Waalkes MP, Ward JM (1989) Induction of hepatic metallothionein in male B6C3F1 mice exposed to hepatic tumor promoters: effects of phenobarbital, acetaminophen, sodium barbital, and di(2-ethylhexyl) phthalate. Toxicol Appl Pharmacol 100:217–226

Watowich SS, Morimoto RI (1988) Complex regulation of heat shock- and glucose-responsive genes in human cells. Mol Cell Biol 8:393–405

Welch WJ (1985) Phorbol ester, calcium ionophore, or serum added to quiescent rat embryo fibroblast cells all result in the elevated phosphorylation of two 28 000 dalton mammalian stress proteins. J Biol Chem 260:3058–3065

Welch WJ (1987) The mammalian heat shock (or stress) response: a cellular defense mechanism. Adv Exp Med Biol 225:287–304

Welch WJ (1990) The mammalian stress response: cell physiology and biochemistry of stress proteins. In: Morimoto RI, Tissieres A, Georgopoulos C (eds) Stress proteins in biology and medicine. Cold Spring Harbor Laboratory Press, Cold Spring Harbor, New York, pp 223–278

Welch WJ (1992) Mammalian stress response: cell physiology, structure/function of stress proteins, and implications for medicine and disease. Physiol Rev 72: 1063–1081

Welch WJ (1993) How cells respond to stress. Sci Am 268(5):56–64

Welch WJ, Mizzen LA (1988) Characterization of the thermotolerant cell. II. Effects on the intercellular distribution of heat-shock protein 70, intermediate filaments, and small nuclear ribonucleoprotein complexes. J Cell Biol 106:1117–1130

Wiegant FA, van Bergen en Henegouwen PM, van Dongen G, Linnemans WA (1987) Stress-induced thermotolerance of the cytoskeleton of mouse neuroblastoma N2A cells and rat Reuber H35 hepatoma cells. Cancer Res 47: 1674–1680

Wilhelmsson A, Cuthill S, Denis M, Wikstrom AC, Gustafsson JA, Poellinger L (1990) The specific DNA binding activity of the dioxin receptor is modulated by the 90 kD heat shock protein. EMBO J 9:69–76

Winfield JB, Jarjour WN (1991) Stress proteins, autoimmunity, and autoimmune disease. In: Kaufmann SH (ed) Heat shock proteins and immune response. Springer, Berlin Heidelberg New York, pp 161–189 (Current topics in microbiology and immunology, vol 167)

Woods JS, Carver GT, Fowler BA (1979) Altered regulation of hepatic heme metabolism by indium chloride. Toxicol Appl Pharmacol 49:455–461

Woods JS, Fowler BA, Eaton DL (1984) Studies on the mechanisms of thallium mediated inhibition of hepatic mixed-function oxidase activity: correlation with inhibition of NADPH-cytochrome c (P-450) reductase. Biochem Pharmacol 33:571–576

Wu BJ, Kingston RE, Morimoto RI (1986) Human HSP70 promoter contains at least two distinct regulatory domains. Proc Natl Acad Sci USA 83:629–633

Wu C (1984a) Two protein-binding sites in chromatin implicated in the activation of heat shock genes. Nature 309:229–234

Wu C (1984b) Activating protein factor binds in vitro to upstream control sequences in heat shock gene chromatin. Nature 311:81–84

Yamada H, Koizumi S (1993) Induction of a 70-kDa protein in human lymphocytes exposed to inorganic heavy metals and toxic organic compounds. Toxicology 79:131–138

Yoshida T, Biro P, Cohen T, Müller RM, Shibahara S (1988) Human heme oxygenase cDNA and induction of its mRNA by hemin. Eur J Biochem 171: 457–461

Yoshikawa H (1973) Preventive effects of pretreatment with cadmium on acute cadmium poisoning in rats. Ind Health 11:113–119

Zeuthen ML, Howard DH (1989) Thermotolerance and the heat-shock response in *Candida albicans*. J Gen Microbiol 135:2509–2518

CHAPTER 12

Metals and Anticancer Drug Resistance

J.S. LAZO

A. Introduction

Metals have historically held a distinguished position in the Pharmacopeia. Metals, such as mercury and arsenic, were the mainstays of early physicians (WOOD 1877). Today more than 80 of the known metals in the periodic table have been shown to have biological activity. Most, however, are toxic and do not have desirable therapeutic activity. Exceptions include aluminum, bismuth, gold, lithium, and platinum. Aluminum is used as an antacid, while bismuth finds employment for gastrointestinal disturbances. Gold salts are still useful for the treatment of rheumatism and lithium has become an important treatment for depression. Metals, such as zinc and iron, are also important as nutritional supplements. A number of drugs also require metals for their biological activity. For example, the anticancer agent bleomycin requires iron for its DNA cleaving and antitumor activity (LAZO and SEBTI 1989). Metals can also induce gene transcription and alter protein function. Many transcription factors and enzymes require metals.

The prevalence of metals as therapeutic, nutritional, and environmental substances makes a discussion of metals and drug pharmacodynamics warranted. As with drug-drug interactions, metal-drug interactions are particularly important for drugs with low therapeutic indices. One class of compounds with an especially poor therapeutic index is the antineoplastics. Curative treatment of many malignancies is thwarted because of drug resistance; significant increases in the drug dosage are not possible because of untoward effects. Thus, even modest changes in the toxicity or efficacy of anticancer drugs can markedly influence their therapeutic index. In this review I will focus primarily on the effects metals have on the biological activity and toxicity of antineoplastic chemotherapeutic agents.

B. Metal-Binding and Metal-Based Anticancer Agents

While most drugs do not directly interact with endogenous metals, two clinically used antineoplastic agents are distinguished by this attribute: bleomycin and doxorubicin. Metals also have a role in the treatment of cancer. The platinum-based antineoplastic agents, *cis*-diamminedichloroplatinum(II) (cisplatin) and *cis*-diamminedicarboxylatocyclobutaneplatinum

(II) (carboplatin), represent two highly effective drugs for the treatment of solid tumors.

I. Bleomycin

Bleomycin used throughout most of the world is actually a complex mixture of small glycopeptides isolated from *Streptomyces verticillus* as a copper complex. The copper is removed to avoid phlebitis associated with intravenous injection, and the resultant colorless bleomycins are metal free. It is generally believed that once injected bleomycin chelates copper in the blood (LAZO and SEBTI 1989). After cellular internalization by poorly defined mechanisms, copper is thought to be removed and replaced rapidly with iron(II). This iron(II)-bleomycin complex binds to molecular dioxygen and DNA, which results in sequence-specific single- and double-strand DNA damage that is lethal to cells (LAZO and SEBTI 1989). This redox-active bleomycin-iron-oxygen ternary complex is catalytic and capable of multiple cleavages once inside cells. Copper(I)- and copper(II)-bleomycin are also capable of cleaving DNA under experimental conditions and it is also possible that such a complex may contribute to the cytotoxicity of bleomycin (EHRENFELD et al. 1987).

The role of metal speciation in bleomycin uptake and cytotoxicity has been studied extensively. The relatively high amounts of copper in the blood, however,suggest that exogenous or endogenous metals in the blood do not markedly alter the uptake or activity of bleomycin. Nonetheless, a number of other metals can bind to bleomycin including indium and cobalt, and formulating bleomycin with these ex vivo can significantly reduce the cytotoxicity of bleomycin (LYMAN et al. 1986). Complexes of this type have been used, however, to examine tumor cell uptake of bleomycin compounds.

II. Doxorubicin

Doxorubicin and other anthracyclines, such as daunorubicin, are believed to exhibit their antineoplastic activity through inhibition of topoisomerase II activity. This appears to occur in the absence of significant metal binding by the anthracyclines. Doxorubicin can, however, bind with iron(III) and this results in the formation of a redox-active complex that may cause untoward effects (MYERS et al. 1986). Doxorubicin-iron(III) complexes have been shown to oxidatively damage membranes and inactivate protein kinase C, but the biological signficance of this is not well understood (HANNUN et al. 1989).

III. Cisplatin and Carboplatin

Cisplatin and a new analog carboplatin are coordinated complexes of platinum, which exhibit significant clinical antitumor activity as single agents and in combination with other antitumor compounds. It is widely accepted that the mechanism of action for this class of drugs is covalent bifunctional interactions with chromosomal DNA, especially intrastrand binding to the N^7 nitrogens of adjacent guanine residues (SHERMAN and LIPPARD 1987). The mechanism by which cisplatin or carboplatin enter cells is not firmly established, although many believe these compounds enter cells by passive diffusion. The cellular parameters that determine tumor responsiveness to cisplatin or carboplatin remain controversial. Drug entry and efflux, DNA repair, and alternate targets have been commonly suggested mechanisms but the relative importance of each of these mechanisms continues to be debated (ANDREWS and HOWELL 1990). It is interesting that cells made resistant to heavy metals, such as Cd, always seem to be cross-resistant to cisplatin and electrophilic anticancerr drugs (LAZO and BASU 1991). This has led some investigators to deduce that proteins induced by heavy metals, most notably metallothioneins, may participate in resistance to electrophilic anticancer drugs such as cisplatin and carboplatin (LAZO and BASU 1919). Because electrophilic anticancer drugs are frequently used anticancer drugs and because electrophilic mutagens are among the most common carcinogenic agents, the role of metals and metal-induced proteins could be of great biological importance.

C. Metal-Induced Anticancer Drug Resistance in Cell Culture

I. Cadmium

BAKKA et al. (1981) were among the first to suggest a role for metals in resistance to electrophilic anticancer drugs when they noted mammalian cells made chronically resistant to Cd were cross-resistant to the toxic effects of cisplatin. In subsequent studies they found the sensitivity of Cd-resistant cells was reduced by 1.5- to 3-fold not only to cisplatin but also to chlorambucil and prednimustine (ENDRESEN et al. 1983; ENDRESEN and RUGSTAD 1987). ANDREWS et al. (1987) and KELLEY et al. (1988) confirmed that Cd-resistant human tumor cells were cross-resistant to cisplatin. Moreover, ANDREWS et al. (1987) demonstrated that Cd-resistant cells did not have a decreased uptake of radiolabeled cisplatin. WEBBER et al. (1988) examined the sensitivity of Cd-resistant human DU-145 prostate cells to the anthracycline doxorubicin. A very modest (1.5-fold) resistance to doxorubicin was observed but the specificity of this resistance with respect to other antineoplastic agents was not determined.

Chronic exposure of mammalian cells to Cd generally increases the synthesis of the metal-binding protein metallothionein; often this is associated with an increase in metallothionein gene copy number. In all of the above-mentioned studies increases in metallothionein were noted.

II. Zinc

Zinc is a ubiquitous component of chromatin and proteins that interact with DNA. TOBEY et al. (1982) were among the first to investigate the effects of Zn on cellular sensitivity to electrophilic anticancer drugs. They found Zn pretreatment of Cd-resistant clones of Chinese hamster ovary cells induce resistance to alkylating agents such as melphalan. Interestingly parental Chinese hamster ovary cells are also protected by Zn pretreatment despite an inability to synthesize metallothionein after Zn treatment. They concluded that Zn-induced proteins other than metallothionein were responsible for the resistance. KOROPATNICK and PEARSON (1990) examined the effect of Zn treatment on cisplatin sensitivity of three mouse B16 melanoma cell strains. The three strains differed in their constitutive levels of metallothionein but after Zn pretreatment had similar metallothionein levels. A 4-h pretreatment with 37 μM Zn produced a twofold or greater resistance to cisplatin in all three cell strains. Interestingly, some but not all of the three cell strains displayed modest resistance to Cd and methotrexate after Zn pretreatment. The authors concluded resistance to cadmium was directly associated with metallothionein induction but suggested the Zn-mediated methotrexate and cisplatin resistance was produced by other effects. Zn has been shown to affect other cellular factors associated with anticancer drug action. For example, Zn inhibits both etoposide-induced DNA fragmentation and activation of poly(ADP-ribose) synthesis in human promylocytic leukemic cells (SHIMIZU et al. 1990) presumably by inhibiting an endonuclease.

D. Metallothionein and Anticancer Drug Resistance

A role for metallothionein in anticancer drug resistance is supported by the observation that every cell line made resistant to Cd, which overexpresses metallothionein, is cross-resistant to cisplatin. Some have hypothesized a scavenger role for metallothionein, which is a highly nucleophilic protein, with electrophilic antineoplastic drugs. Metals induce a number of proteins and affect many enzyme activities. This hypothesis has been addressed by several groups during the past 10 years using anticancer drug-resistant cell lines, in vitro assays, and metallothionein genes.

I. In Vitro Metallothionein-Drug Interactions

Several investigators (SHARMA and EDWARDS 1983; ZELAZOWSKI et al. 1984; KRAKER et al. 1985) reported that cisplatin reacts with metallothionein in vitro. Cisplatin binds stoichiometrically to the protein sulfhydryl groups in metallothionein and Zn, if present, is completely replaced with 10 ± 2 g atoms Pt/mol protein (PATTANAIK et al. 1992). The amine ligands in cisplatin are also lost, suggesting that Pt(II) has tetrathiolate coordination in metallothionein. The kinetics of the reaction are biphasic with the loss of Zn being principally associated with the second phase. Cisplatin reacts more rapidly with the thiols of Zn-metallothionein than with free thiols of glutathione. Interestingly trans-diamminedichloroplatinum(II), which is biologically much less active that cisplatin, also reacts with metallothionein. Thus, in vitro results suggest cisplatin can not only react with metallothionein but also can release Zn within cells. These in vitro results may reflect a relevant in vivo process: almost 10% of the radiolabel found in renal extracts of mice treated with ^{195m}Pt-labeled cisplatin was associated with metallothionein (ZELAZOWSKI et al. 1984). Using atomic absorption spectrophotometry and pharmacologically relevant concentrations of cisplatin, it has been reported that most platinum is in the cytosol of Ehrlich cells and that 30% of platinum is bound to metallothionein (KRAKER et al. 1985), supporting the notion that metallothionein is an important binding site within at least some cells. Others have questioned whether an interaction between metallothionein and cisplatin could be important based on the predicted kinetics of the interaction (ANDREWS and HOWELL 1990).

II. Metallothionein in Drug-Resistant Cells

Metals can alter the expression and activities of many proteins (see Chap. 10). Thus, several groups have examined if anticancer drug-resistant cells have elevated levels of metallothionein or if anticancer agents could alter metallothionein expression. Elevated levels of metallothionein protein have been reported in four of five human cells selected for cisplatin resistance (KELLEY et al. 1988) (Table 1). Murine L1210 cells that were 44-fold resistant to cisplatin contained 13-fold more metallothionein than the parental cells. Moreover, the revertant cells had reduced metallothionein. Cells with resistance to another anticancer drug, bleomycin, were not cross-resistant to cisplatin nor did they overexpress metallothionein. One cisplatin-resistant cell line (SCC25/CP) was tested and found to have increased levels of metallothionein mRNA. KASAHARA et al. (1991) examined two human small cell lung cancer cell lines that were 6- and 11-fold resistant to cisplatin and overexpressed metallothionein. These cells were also cross-resistant to cadmium chloride and several other electrophilic anticancer agents such as 4-hydroperoxycyclophosphamide, cis-diammine(glycolate) platinum, and a nitrosourea. Mouse fibrosarcoma cells with sevenfold resistance to cisplatin appeared to exhibit elevated levels of metallothionein, although the increase was not quantified (EICHHOLTZ-WIRTH et al. 1993). These cells were not crossresistant to radiation. The majority of evidence supports the hypothesis that cisplatin does not induce metallothionein directly, rather the increase reflects selection of protected cells. Increases in metallothionein mRNA and protein are not always seen in cisplatin-resistant cells (SCHILDER et al. 1990; ANDREWS and HOWELL 1990). It now seems likely that a number of different factors can control cellular sensitivity to electrophilic anticancer agents, such

Table 1. Relationship between metallothionein content and drug resistance[a]

Cell line	CP resistance ratio	MT content (fold increase ± SEM)
SCC-25/CP	7.1	4.37 ± 0.50
G3361/CP	6.7	2.00 ± 0.10
SW2/CP	4.5	5.10 ± 0.50
SL6/CP	2.5	3.37 ± 0.44
MCF-7/CP	2.5	0.95 ± 0.11
A-253/Cd	2.7	3.27 ± 0.37
A-253/C-10/BLM	0.88	0.72 ± 0.04
L1210/CP	44	13.3 ± 1.50
L1210/CP-R	5.8	2.18 ± 0.15
L1210/DACH	2.7	2.94 ± 0.18

[a] The cell lines used were SCC-25 (human head and neck carcinoma), G3361 (human melanoma), SW2 (human small cell carcinoma), SL6 (human large cell carcinoma), MCF-7 (human breast carcinoma), A-253 (human head and neck carcinoma), and L1210 (murine lymphocytic leukemia). The drugs to which cells were made resistant were cisplatin (CP), bleomycin (BLM), and 1,2-diaminocyclohexane platinum (DACH). L1210/CP-R is a partial revertant of L1210/CP. LCP resistance ratio is the relative concentration of drug required to inhibit the growth by 50% of the drug-resistant cells compared to the wild-type cells. (From KELLEY et al. 1988, Overexpression of metallothionein confers resistance to anticancer drugs. Science 241:1813–1815)

as cisplatin. These include not only metallothionein but also glutathione, DNA repair systems, and cellular drug content (ANDREWS and HOWELL 1990; GODWIN et al. 1992). Nonetheless, the ability of metals and other endogenous and exogenous substances to elevated metallothionein levels makes this protein family of significant interest.

III. Nonmetal Induction of Metallothionein

Metallothionein is also induced by a variety of nonmetals, and exposure to these agents often can cause tesistance to anticancer drugs such as cisplatin (BASU and LAZO 1991). Expression of v-*mos* can attenuate the glucocorticoid-induced metallothionein expression and cisplatin resistance. Transcriptional activation of c-Ha-ras oncogene expression in murine NIH 3T3 cells with dexamethasone produces an increase in metallothionein content and a decrease in cisplatin accumulation (ISONISHI et al. 1991). The ability of dexamethasone to affect the sensitivity of cells to electrophilic antineoplastic agents could be of some clinical interest because of its frequent usage as an antiemetic.

IV. Metallothionein Gene Transfer

The role of metallothionein in electrophilic anticancer drug resistance has been studied further using gene transfer methods. High levels of metallothionein expression have been obtained using a bovine papillomavirus vector containing human metallothionein IIA (KELLEY et al. 1988). The resulting transduced cells are resistant to cisplatin, chlorambucil, and melphalan but not to functionally unrelated agents like 5-fluorouracil or vincristine (Fig. 1). Using a similar approach, KAINA et al. (1990) observed significant resistance to mutagenic alkylating agents, such as *N*- methyl-*N*-nitrosourea, as well as some resistance to cisplatin. SCHILDER et al. (1990), however, reported no significant cisplatin resistance after transfer of the human metallothionein IIA gene, and others (KOROPATNICK and PEARSON

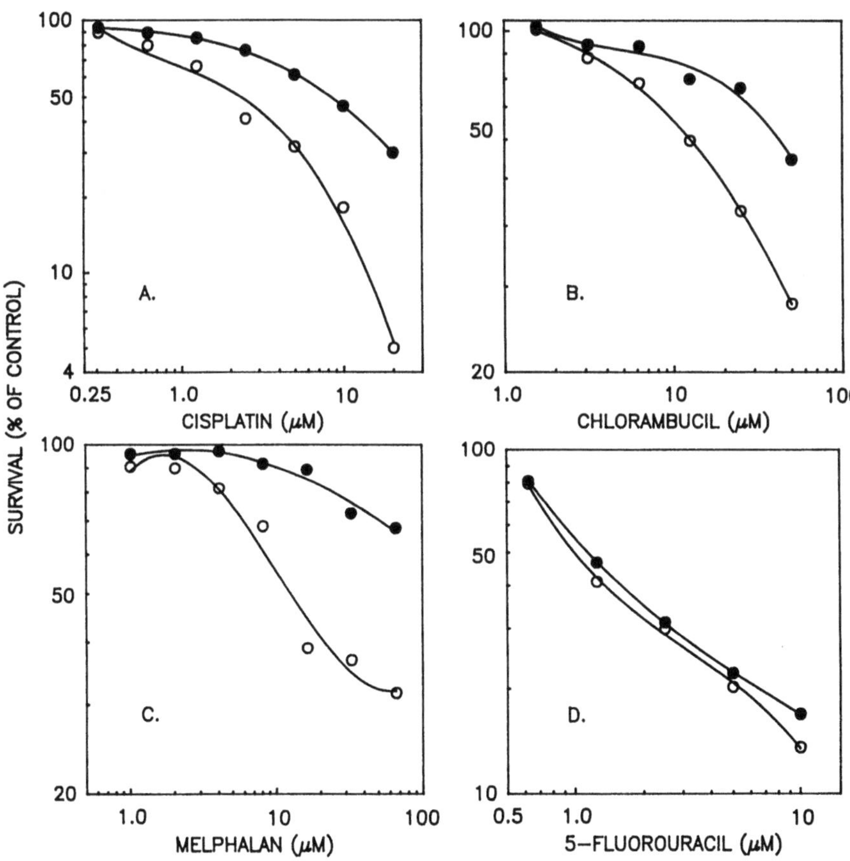

Fig. 1. Survival curves of C127 cells transduced with a bovine papilloma virus alone (○) or bovine papillomavirus containing an hMT IIA gene construct (●). Cells were treated for 72 h with the drugs and the number of cells determined by a colorimetric assay. (From KELLEY et al. 1988, Overexpression of metallothionein confers resistance to anticancer drugs. Science 241:1813–1815)

1990) have noted sensitization of transduced cells that is not easily reconciled with a scavenger role for metallothionein. It appears the drug-resistant phenotype depends on the cellular context in which metallothionein is expressed. Thus, the mechanism by which metallothionein modifies cellular sensitivity to anticancer drugs remains obscure.

V. Human Tumor Expression of Metallothionein and Drug Sensitivity

Several groups (NARTEY et al. 1987; BAHNSON et al. 1991) have shown human tumors contain metallothionein. Although it has been suggested that correlations may exist between metallothionein levels and tumor resistance to anticancer drugs (MATTERN and VOLM 1992), in other studies correlations are absent (MURPHY et al. 1991). Such attempts to relate the levels of metallothionein to in vivo drug sensitivity seem likely to fail, however, because of tumor heterogeneity and the multiple factors that seem to be involved in the drug-resistant phenotype. This problem has plagued attempts to address the clinical importance of other putative drug-resistant factors, such a *p*-glycoprotein or glutathione-*S*-transferase. Furthermore, previous metallothionein studies may be flawed because they failed to consider the metallothionein isotype or its subcellular distribution. It is now recognized there are at least seven different human metallothioneins and the functional significance of these is not known. Moreover, metallothionein can be located in both the cytoplasm and the nucleus of malignant cells (BANERJEE et al. 1982). The functional significance of this differential distribution all the factors that control it have not been clarified. We have found human metallothionein IIA most commonly overexpressed in the human tumor cell lines we have that are cisplatin resistant (YANG et al. 1994); we have also noted in our cisplatin-resistant cells a predominance of nuclear metallothionein compared to the drug-sensitive parental cells.

E. Metal-Mediated Changes in Drug Sensitivity In Vivo

Despite the lack of a clear understanding of the mechanism by which metallothionein engenders drug resistance, the isotypes involved, and the subcellular location of metallothionein, several investigators have begun to probe whether the successful cell culture observations can be applied to preclinical tumor models or patients.

I. Zinc

The protective effect of Zn in addition to several metals, such as Cu, Hg, Bi, Cd, Hg, Ni, and Co, against the toxic effects of cisplatin were first demonstrated by NAGANUMA et al. (1985). By pretreating mice with a single subcutaneous injection of 100 μmol/kg $ZnCl_2$, mice were protected against the lethal effects of cisplatin. Similar results were also seen with all of the above-mentioned metals but this protection was not firmly linked to the induction of metallothionein. Subsequently, SATOH et al. (1993) observed protection from the nephrotoxicity of cisplatin when mice were pretreated for 2 days with $ZnSO_4$ (200 μmol/kg per day) subcutaneously. Unfortunately, Zn pretreatment also reduced the antitumor activity. Coadministration of an inhibitor of γ-cystathionase selectively prevented tumor metallothionein synthesis and enhanced the therapeutic index of cisplatin in this model. Doz et al. (1992) capitalized on an observation that only extracerebral metallothionein synthesis is induced by a high-Zn diet and attempted to use Zn treatment to protect mice and rats against the hematotoxicity of carboplatin. A 2-week high-Zn diet protected these animals against monocyte and polymorphonuclear bone marrow progenitor depletion. The antitumor activity of carboplatin against an intracerebrally implanted 9L gliosarcoma was not altered by this treatment. Thus, it was suggested that increasing the dietary intake of Zn could increase the therapeutic index of carboplatin in the treatment of brain malignancies.

II. Bismuth

The observation of NAGANUMA et al. (1985) that metals could protect mice against the lethal effects of an anticancer drug stimulated additional studies with bismuth salts since they are relatively nontoxic antidiarrheal agents that induce metallothionein. Preclinical studies showing protection against bleomycin (IMURA et al. 1992), tumor nectrosis factor (SATOMI et al. 1988), and γ-irradiation (SATOH et al. 1989) toxicity with Bi pretreatment have been published. The extensive data, however, are with cisplatin or doxorubicin. Oral pretreatment with bismuth subnitrate daily for 5 days protects mice against the renal toxicity of cisplatin and the cardiotoxicity of doxorubicin (SATOH et al. 1988; NAGANUMA et al. 1988). This protection was attributed to increases in renal and cardiac metallothionein. Interestingly, bismuth subnitrate pretreatment does not increase tumor metallothionein nor did it alter the antitumor activity of cisplatin (NAGANUMA et al. 1987). Pretreatment of nude mice with bismuth nitrate also depresses the renal toxicity of cisplatin without compromising its activity against a human bladder tumor (KONDO et al. 1991). Renal metallothionein levels are increased without an increase in tumor metallothionein. The lack of tumor metallothionein induction with bismuth subnitrate or nitrate treatment may reflect the pharmacokinetics of Bi or a difference in metallothionein gene regulation in tumor tissue. BOOGAARD et al. (1991) demonstrated lower Pt levels in the kidneys of Bi-pretreated rats and suggested the protection afforded by Bi pretreatment was not due to increased binding of cisplatin to renal metallothionein. They proposed that cisplatin-induced nephrotoxicity is mediated by lipid peroxidation and that the antioxidant properties of metallothionein rather than its binding protected renal tissue. The results of a preliminary clinical trial have been reported (SOMMER et al. 1989) in which bismuth subsalicylate (1 g × 5) was administered during a 30-h period to women with ovarian cancer prior to cisplatin- and doxorubicin-containing therapy. A small protective effect was reported, although the patient sample size was very small (ten patients) and only one dose and schedule was tested. Thus, additional studies would be required to demonstrate any clinical benefit of Bi pretreatment.

F. Summary

Prior exposure to metals can have a pronounced effect on the responsiveness of cells or tissues to several anticancer drugs, most notably the electrophilic agents. While preclinical data suggest metal pretreatment may elevate the therapeutic index of these toxic drugs, there has been no convincing demonstration of the clinical efficacy of this approach. The mechanism by which metals protect against anticancer agents appears to reside in part in their ability to induce metallothionein, although other metal-induced effects may participate in drug resistance. The existence of different isotypes of metallothionein in cells adds to the complexity of the question. Metallothionein segregation to the nucleus within cells could also affect functionality. The availability of new reagents and methods should permit a critical analysis of these issues.

References

Andrews PA, Howell SB (1990) Cellular pharmacology of cisplatin: perspectives on mechanisms of acquired resistance. Cancer Cells 2:35–43

Andrews PA, Murphy MP, Howell SB (1987) Metallothionein-mediated cisplatin resistance in human ovarian carcinoma cells. Cancer Chemother Pharmacol 19:149–154

Bahnson RR, Banner BF, Ernstoff MS, Lazo JS, Cherian MG, Banerjee D, Chin JL (1991) Immunohistochemical localization of metallothionein in transitional cell carcinoma of the bladder. J Urol 146:1518–1520

Bakka A, Endresen L, Johnsen AB, Edminson PD, Rugstad HE (1981) Resistance against cis-dichlorodiammineplatinum in cultured cells with a high content of metallothionein. Toxicol Appl Pharmacol 61:215–226

Banerjee D, Onosaka S, Cherian MG (1982) Immunohistochemical localization of metallothionein in cell nucleus and cytoplasm of rat liver and kidney. Toxicology 24:95–105

Basu A, Lazo JS (1991) Suppression of dexamethasone-induced metallothionein expression and cis-diamminedichloroplatinum(II) resistance by v-mos. Cancer Res 51:893–896

Boogaard PJ, Slikkerveer A, Nagelkerke JF, Mulder GJ (1991) The role of metallothionein in the reduction of cisplatin-induced nephrotoxicity by Bi^{3+}-pretreatment in the rat in vivo and in vitro. Are antioxidant metallothionein more relevant than platinum binding? Biochem Pharmacol 41:369–375

Doz F, Berens ME, Deschepper CF, Dougherty DV, Bigornia V, Barker M, Rosenblum ML (1992) Experimental basis for increasing the therapeutic index of cis-diamminedicarboxylatocyclobutaneplatinum(II) in brain tumor therapy by a high-zinc diet. Cancer Chemother Pharmacol 29:219–226

Ehrenfeld GM, Shipley JB, Heimbrook DC, Sugiyama H, Long EC, van Boom JH, van der Marel GA, Oppenheimer NJ, Hecht SM (1987) Copper-dependent cleavage of DNA by bleomycin. Biochemistry 26:931–942

Eichholtz-Wirth H, Born R, Reidel G, Hietel B (1993) Transient cisplatin-resistant murine fibrosarcoma cell characterized by increased metallothionein content. J Cancer Res clin Oncol 119:227–233

Endresen L, Rugstad HE (1987) Protective function of metallothionein against certain anticancer agents. Experientia 52:595–602

Endresen L, Bakka A, Rugstad HE (1983) Increased resistance to chlorambucil in cultured cells with a high concentration of cytoplasmic metallothionein. Cancer Res 43:2918–2926

Godwin AK, Meister A, O'Dwyer PJ, Huang CS, Hamilton TC, Anderson ME (1992) High resistance to cisplatin in human ovarian cancer cell lines is associated with marked increase of glutathione synthesis. Proc Natl Acad Sci USA 89: 3070–3074

Hagenbeek A, Philip T, Bron D, Guglielmi C, Coiffier B, Gisselbrecht C, Kluin Nelemans JC, Somers T, Misset JL, Van der Lely J et al. (1991) The Parma international randomized study in relapsed non Hodgkin lymphoma: lst interim analysis of 128 patients (as 15 January 1991: 153 patients). Bone Marrow Transplant 7 [Suppl] 2:142

Hannun YA, Foglesong RJ, Bell RM (1989) The adriamycin-iron(III) complex is a potent inhibitor of protein kinase C. J Biol Chem 264:9960–9966

Imura N, Satoh M, Nagaruma A (1992) Possible application of metallothionein in cancer therapy. In: Klassen CD, Suzuki KT (eds) Metallothioneins in biology and medicine. CRC, Boca Raton, pp 375–382

Isonishi S, Hom DK, Thiebaut FB, Mann SC, Andrews PA, Basu A, Lazo JS, Eastman A, Howell SB (1991) Expression of the c-Ha-ras oncogene in mouse NIH3T3 cells induces resistance to cisplatin. Cancer Res 51:5903–5909

Kaina B, Lohrer H, Karin M, Herrlich P (1990) Overexpressed human metallothionein IIA gene protects Chinese hamster ovary cells from killing by alkylating agents. Proc Natl Acad Sci USA 87:2710–2714

Kasahara K, Fujiwara Y, Nishio K, Ohmori T, Sugimoto Y, Komiya K, Matsuda T, Saijo N (1991) Metallothionein content correlates with the sensitivity of human small cell lung cancer cell lines to cisplatin. Cancer Res 51:3237–3242

Kelley SL, Basu A, Teicher BA, Hacker MP, Hamer DH, Lazo JS (1988) Overexpression of metallothionein confers resistance to anticancer drugs. Science 241:1813–1815

Kelley J, Kovacs EJ, Nicholson K, Fabisiak JP (1991) Transforming growth factor-beta production by lung macrophages aand fibroblasts. Chest 99:85S–86S

Kondo Y, Satoh M, Imura N, Akimoto M (1991) Effect of bismuth nitrate given in combination with cis-diamminedichloroplatinum(II) on the antitumor activity and renal toxicity of the latter in nude mice inoculated with human bladder tumor. Cancer Chemother Pharmacol 29:19–23

Koropatnick J, Pearson J (1990) Zinc treatment, metallothionein expression, and resistance to cisplatin in mouse melanoma cells. Somat Cell Mol Genet 16: 529–537

Kraker A, Schmidt J, Krezoski S, Petering DH (1985) Binding of cis-dichlorodiammine platinum(II) to metallothionein in Ehrlich cells. Biochem Biophys Res Commun 130:786–792

Lazo JS, Basu A (1991) Metallothionein expression and transient resistance to electrophilic antineoplastic drugs. Semin Cancer Biol 2:267–271

Lazo JS, Sebti SM (1989) Malignant cell resistance to bleomycin to bleoymycin-group antibiotics. In: Kessel D (ed) Anticancer drug resistance. CRC, Boca Raton, pp 276–279

Lyman S, Ujjani B, Renner K, Antholine W, Petering DH, Whetstone JW, Knight JM (1986) Properties of the initial reaction of bleomycin and several of its metal complexes with Ehrlich cells. Cancer Res 46:4472–4478

Mattern J, Volm M (1992) Increased resistance to doxorubicin in human non-small cell lung carcinomas with metallothionein expression. Int J Oncol 1:687–689
Murphy D, McGown AT, Crowther D, Mander A, Fox BW (1991) Metallothionein levels in ovarian tumours before and after chemotherapy. Br J Cancer 63: 711–714
Myers C, Gianni L, Zweier J, Muindi J, Sinha BK, Eliot H (1986) Role of iron in adriamycin biochemistry. Fed Proc 45:2792–2797
Naganuma A, Satoh M, Koyama Y, Imura N (1985) Protective effect of metallothionein inducing metals on lethal toxicity of cis-diamminedichloroplatinum in mice. Toxicol Lett 24:203–207
Naganuma A, Satoh M, Imura N (1987) Prevention of lethal and renal toxicity of cis-diamminedichloroplatinum(II) by induction of metallothionein synthesis without compromising its antitumor activity in mice. Cancer Res 47:983–987
Naganuma A, Satoh M, Imura N (1988) Specific reduction of toxic side effects of adriamycin by induction of metallothionein in mice. Jpn J Cancer Res 79: 406–411
Nartey N, Cherian MG, Banerjee D (1987) Immunohistochemical localization of metallothionein in human thyroid tumors. Am J Pathol 129:177–182
Pattanaik A, Bachowski G, Laib J, Lemkuil D, Shaw CF III, Petering DH, Hitchcock A, Saryan L (1992) Properties of the reaction of cis-dichlorodiammineplatinum(II) with metallothionein. J Biol Chem 267:16121–16128
Satoh M, Naganuma A, Imura N (1988) Metallothionein induction prevents toxic side effects of cisplatin and adriamycin used in combination. Cancer Chemother Pharmacol 21:176–178
Satoh M, Kloth DM, Kadhim SA, Chin JL, Naganuma A, Imura N, Cherian MG (1993) Modulation of both cisplatin nephrotoxicity and drug resistance in murine bladder tumor by controlling metallothionein synthesis. Cancer Res 53:1829–1832
Satomi N, Sakurai A, Haranaka R, Haranaka K (1988) Preventive effects of several chemicals against lethality of recombinant human tumor necrosis factor. J Biol Response Mod 7:54–64
Schilder RJ, Hall L, Monks A, Handel LM, Fornace AJ Jr, Ozols RF, Fojo AT, Hamilton TC (1990) Metallothionein gene expression and resistance to cisplatin in human ovarian cancer. Int J Cancer 45:416–422
Sharma RP, Edwards IR (1983) Cisplatinum: subcellular distribution and binding to cytosolic ligands. Biochem Pharmacol 32:2665–2669
Sherman SE, Lippard SJ (1987) Structural aspects of platinum anticancer drug interactions with DNA. Chem Rev 87:1153–1181
Shimizu T, Kubota M, Tanizawa A, Sano H, Kasai Y, Hashimoto H, Akiyama Y, Mikawa H (1990) Inhibition of both etoposide-induced DNA fragmentation and activation of poly (ADP-ribose) synthesis by zinc ion. Biochem Biophys Res Commun 169:1172–1177
Sommer S, Thorling EB, Jakobsen A, Steiness E, Ostergaard K (1989) Can bismuth decrease the kidney toxic effect of cis-platinum? Eur J Cancer Clin Oncol 25:1903–1904
Tobey RA, Enger MD, Griffith JK, Hildebrand CE (1982) Zinc-induced resistance to alkylating agent toxicity. Cancer Res 42:2980–2984
Webber MM, Rehman SM, James GT (1988) Metallothionein induction and deinduction in human prostatic carcinoma cells: relationship with resistance and sensitivity to adriamycin. Cancer Res 48:4503–4508
Wood HC Jr (1877) A treatise on therapeutics comprising materia medica and toxicology, 2nd edn. JB Lipincott, Philadelphia, pp 351–384
Yang YY, Kuo SM, DeFilippo JM, Saijo N, Lazo JS (1994) Metallothionein isoform content and gene expression in cisplatin sensitive and resistant human small cell lung carcinomas. Proc Am Assoc Cancer Res 33:2766
Zelazowski AJ, Garvey JS, Hoeschele JD (1984) In vivo and in vitro binding of platinum to metallothionein. Arch Biochem Biophys 229:246–252

CHAPTER 13

Chemistry of Chelation: Chelating Agent Antagonists for Toxic Metals

M.M. JONES

A. Chelation: Its Basic Chemistry and Advantages as a Metal Complexation Process

The chemistry of chelation is involved in some very important aspects of the toxicology of metals. The first is that toxic metals exert many of their adverse biological effects by forming metal complexes with enzymes, DNA, or other molecules found within cells. Such metal chelate complexes have properties which differ in some important respects from the enzyme or DNA that they contain. This has the result that metal binding can inactivate enzymes. We also find that such chelate complexes are probably responsible for the carcinogenicity of certain chromium compounds which react with DNA. Another quite different aspect is the use of exogenous chelating agents as drugs in order to remove toxic metals from the sites to which they are bound in vivo. While it is this latter aspect which will be examined in detail in this chapter, it must be borne in mind that the successful use of a chelating agent as a drug for the treatment of metal intoxication is dependent upon the ability of that chelating agent to effectively compete with the in vivo binding site for the possession of the toxic metal ion (M). This basic detoxification process can be written as:

$$\text{M (at in vivo binding site)} + \text{chelating Agent} \rightarrow \\ \text{in vivo binding site} + \text{M–chelating agent complex}$$

Efforts to use chelating agents as antagonists for toxic metals began in the 1940s and since that time the literature in this area has grown considerably. Comprehensive reviews of this field are available (CATSCH 1968; CATSCH and HARMUTH-HOEHNE 1979; MAY and BULMAN 1983; BULMAN 1987; WILLIAMS and HALSTEAD 1982/1983; JONES 1991).

The word chelation was devised by the British chemist Gilbert Morgan about 1930 to describe the bonding situation which results when a given molecule (the ligand) is bound to a metal atom by two or more of its atoms. The molecules binding the metal are called chelating agents or chelators, and the ones of most interest as drugs are organic compounds. The resultant metal complexes often have unusual properties which had been noticed even earlier, such as the ability to be resolved into optically active (right- and left-hand) forms, and a greater stability than analogous complexes in which the

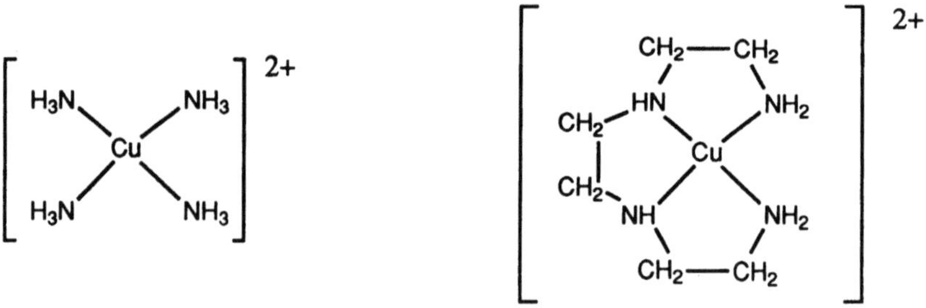

$[Cu(NH_3)_4]^{2+}$ $[Cu(TRIEN]^{2+}$

Fig. 1. The complex of Cu^{2+} with TRIEN, in which the nitrogen atoms are bonded together in a chain, is much more stable than the complex of Cu^{2+} with four separate NH_3 molecules, though each complex is held together by four Cu-N bonds. This is called the chelate effect

groups which bond to the metal are not bonded to each other (HANCOCK and MARTELL 1988, 1989). This difference in stability becomes more pronounced as we consider increasingly dilute solutions. This factor is of special importance as the main goal of the use of chelating agents in the treatment of metal intoxications is to complex such toxic metal ions under conditions where both the toxic metal ion and the chelating agent are present in dilute solution in the serum or various tissues. This can be seen in a comparison of the behavior of ammonia (NH_3) and triethylenetetramine (TRIEN), in bonding to copper(II) at increasing dilutions. The stability constants for these complexes are for the reactions:

$$Cu^{2+} + 4NH_3 = [Cu(NH_3)_4]^{2+}$$
$$[Cu(NH_3)_4^{2+}]/[Cu^{2+}][NH_3]^4 = K = 10^{13}$$
$$Cu^{2+} + \text{TRIEN} = [Cu(\text{TRIEN})]^{2-}$$
$$[Cu(\text{TRIEN})]/[Cu^{2+}][\text{TRIEN}] = K = 10^{20}$$

Chelating agents, such as TRIEN, are able to tie up metal ions much more effectively than simple ligands such as NH_3 (Fig. 1). Thus a $0.1\,M$ solution of $[Cu(NH_3)_4]^{2+}$ is dissociated to the extent of about 1%, while a $0.1\,M$ solution of $[Cu(TRIEN)]^{2+}$ is dissociated only to the extent of about 1×10^{-8}%!

B. Chemistry of Chelation in Biological Systems

One of the most important consequences of chelation is that the biochemical, chemical, and physical properties of the metal chelate complexes are quite different from those of the uncomplexed metal ion. In almost all environments a metal cation will form bonds to other species which can share electron pairs with it. The assembly of the metal ion and the atoms bonded to it constitute the "coordination-sphere" of the metal ion. To a first approximation each metal ion can be considered to have a characteristic number of bonded atoms in its coordination sphere. For the typical metal ion the actual atoms present in the coordination sphere are dependent upon the ligands present in the environment. These atoms in the coordination sphere commonly undergo a change when a typical toxic metal salt is ingested and absorbed by a person or other organism. Metal ions in biological systems reform a network of bonds to achieve a more stable characteristic coordination sphere. This is accomplished by forming bonds to various types of donor atoms which are present. Donor atoms found most commonly in living systems include oxygen atoms (in water, in carbonyl, carboxyl, and phenolic groups of amino acids, in phosphate groups, etc.), nitrogen atoms (in amino acids, porphyrins, nucleotides, etc.), and sulfur atoms (in cysteine and other compounds which contain thiol groups such as lipoic acid, as well as with thioether sulfur atoms such as those in methionine). In the typical process, the toxic metal atom incorporates atoms into its coordination sphere which are parts of an essential molecule such as an enzyme, a messenger molecule, DNA, or a similar molecule which has a definite function to perform in the cell. The reactivity of the resultant compound is altered so that it will not perform its usual tasks in the normal manner and the result is a decrease in the viability of the organism. These changes in the normal reactivity patterns of essential molecules are the basis for most of the biological effects of toxic metals.

Toxic metals usually do not react with just a single enzyme. Rather, they react with a wide variety of molecules which contain appropriate donor atoms. A given toxic metal is found to have an effect on any part of an organism where it reacts with, and changes the reactivity of, molecules which are critical for the normal functioning of the organism. Lead is selectively toxic towards the kidneys, the bone marrow, and the nervous system and can affect these at low lead concentrations, but has a much smaller effect on the action of the liver or the pancreas. In the bone marrow, Pb^{2+} reacts with and inactivates enzymes involved in the synthesis of the heme unit of hemoglobin (Moore et al. 1980), as shown in Fig. 2. In each of the organs subject to its action, lead apparently reacts with different types of molecules.

The clinical use of chelation is to transform the toxic metal complex with, say, an enzyme into a toxic metal complex with the administered chelating agent. Subsequently we want the enzyme to be reactivated and the toxic metal chelate complex to be excreted from the body in either the urine or the feces. When we use a chelating agent to treat metal intoxication, we transform the coordination sphere of the toxic metal ion and form a new complex with very different properties from those of the complex from which it was formed. One result of complexing a toxic metal ion with the therapeutically useful chelating agents is a considerable decrease in the

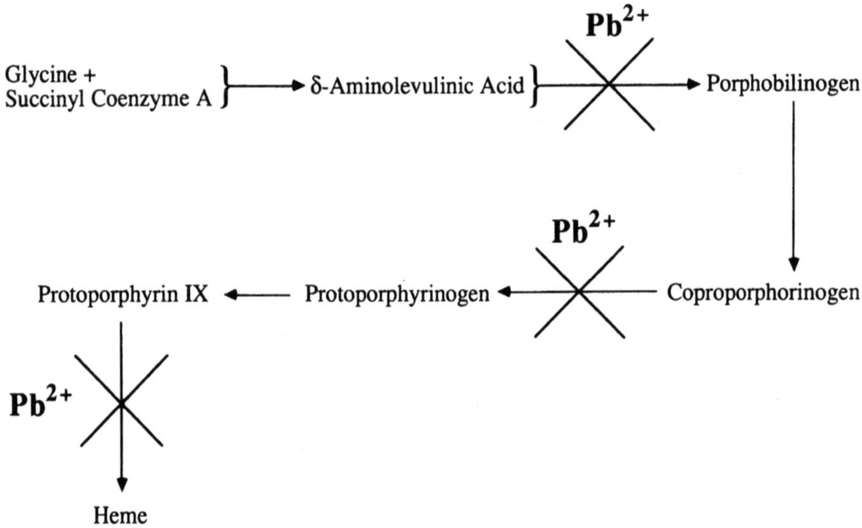

Fig. 2. Steps in the enzymatic synthesis of heme with which Pb^{2+} interferes. This interference can lead to the production of defective red blood cells

toxicity of the metal when it is present in the form of a complex in comparison to the uncomplexed ion. This is seen clearly in the data on the reduction of the toxicity of the cadmium ion when it is complexed with ethylenediaminetetraacetic acid (EDTA) and some of its analogs (Eybl and Sýkora 1966). Under conditions where the 10-day LD_{50} of cadmium chloride subcutaneously administered to mice was 4.9 mg Cd/kg for cadmium chloride, the simultaneous i.p. injection of EDTA (to give the Cd-EDTA complex) altered the LD_{50} to 18.4 mg Cd/kg, and for the Cd-diethylenetriamine-pentraacetic acid (DTPA) complex the LD_{50} was raised to 48.4 mg Cd/kg.

If we examine the processes which actually occur during the chelating agent-induced excretion of a toxic metal, we find a dependence upon whether the toxic metal occupies an extracellular or an intracellular site, and whether the chelating agent itself is restricted to the extracellular spaces or can gain access to the particular intracellular sites at which the toxic metal is present (Fig. 3). The use of a chelating agent restricted to the extracellular space can cause a large reduction of the toxic metal concentration in that space. This in turn will usually favor the diffusion of some of the toxic metal from intracellular sites to extracellular sites. The repeated administration of such a chelating agent will result in a gradual reduction of the total amount of toxic metal present. The removal of the toxic metal will occur more rapidly if we use an appropriate chelating agent which can penetrate the membranes surrounding the intracellular sites and gain direct access to the toxic metal ion in the cytosol and organelles inside the cell, as shown in Fig. 4. We find that chelating agents can be sorted out on the basis of the types of intra-

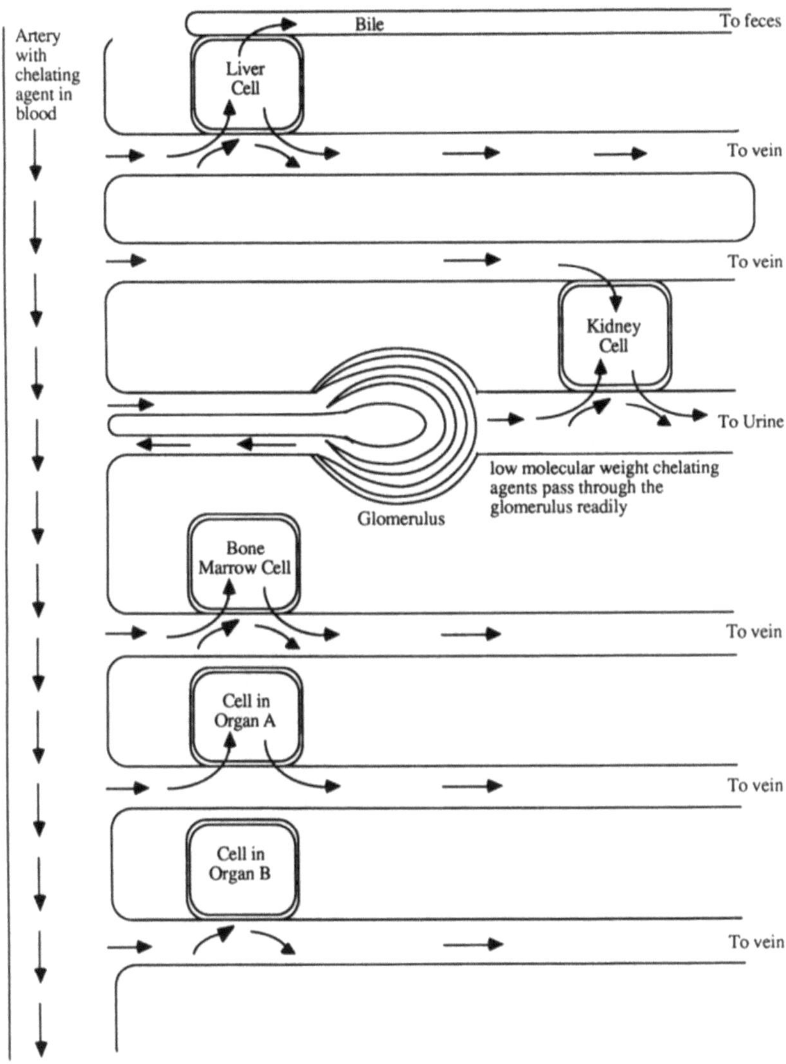

Fig. 3. Schematic view of the action of low molecular weight, water-soluble chelating agents in the mobilization of toxic metal ions

cellular sites to which they can readily gain access. This is summarized in Table 1.

Sites which actively transport anions are present in those two organs in which toxic metals frequently concentrate: the liver (MEIER 1988) and the kidneys (MØLLER and SHEIKH 1983; PRITCHARD and MILLER 1992). Using these criteria we can sort out those chelating agents used in the clinic (Fig. 5). BAL (British Anti-Lewisite or 2,3-dimercapto-1-propanol) is electrically

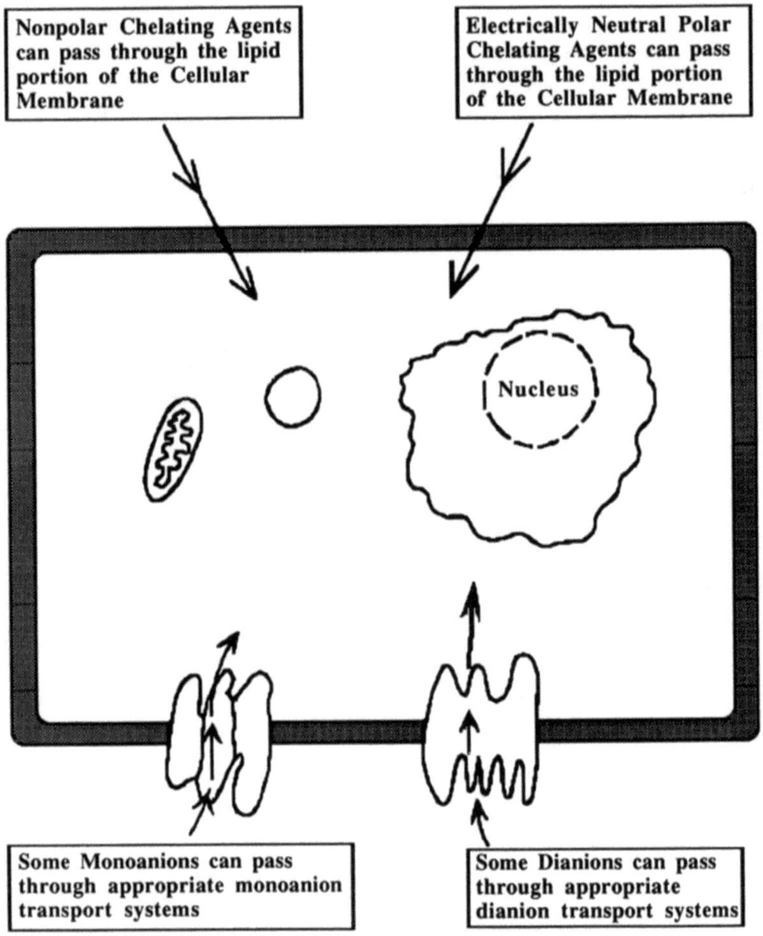

Fig. 4. Some pathways by which chelating agents may gain access to intracellular deposits of toxic metals

Table 1. Sites accessible to various types of chelating agents

Chelating agent type	Accessible sites
Neutral (uncharged)	
Polar	Many intra- and extracellular sites
Highly hydrophobic	Extracellular sites, most cells, brain, fatty tissue, etc.
Single negative charge	Extracellular spaces, cells with appropriate monoanionic transport systems in their membranes (e.g., kidneys and liver)
Double negative charge	Extracellular spaces, cells with appropriate transport systems
More than a −3 charge	Predominantly extracellular spaces
High molecular weight	Liver cells if compound is excreted in the bile

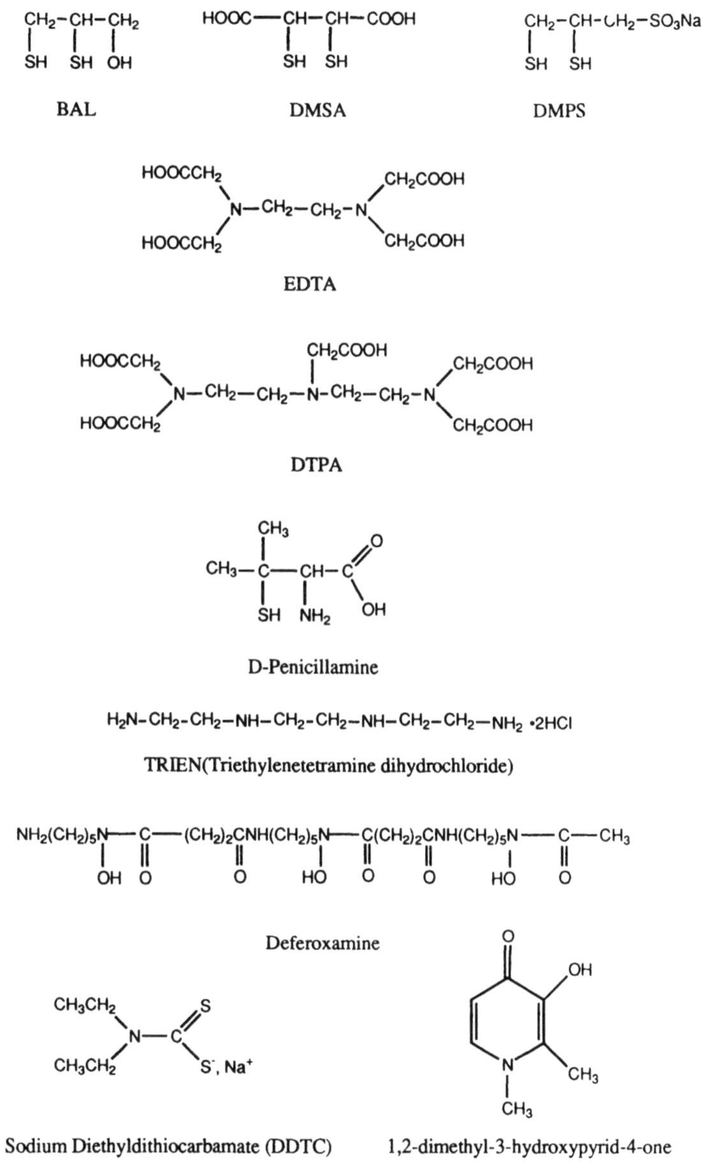

Fig. 5. Structures of chelating agents used in the clinical treatment of metal intoxications

uncharged and has a solubility in lipids which is sufficient to allow it to penetrate into most organs. DMPS (sodium 2,3-dimercaptopropane-1-sulfonate) has a single negative charge at physiological pH values and can be transported into intracellular sites in organs which have suitable transport systems, such as the kidneys (STEWART and DIAMOND 1987). DMSA (*meso-*

2,3-dimercaptosuccinic acid) has a −2 charge at physiological pH and indirect evidence suggests that it can apparently be transported into the kidney cells via the succinate transport system. EDTA and DTPA carry negative charges of −2 to −3 at physiological pH and are almost completely confined to the extracellular spaces. A small fraction of EDTA is, however, secreted by the kidneys (FOREMAN and TRUJILLO 1954) and a small fraction of DTPA is excreted in the bile (BHATACHARYYA and PETERSON 1979). D-Penicillamine is rapidly absorbed from the gastrointestinal tract, undergoes a complex metabolism, is rapidly excreted in the urine, and, in the rat, concentrates in organs which have a high collagen content (JOYCE 1989; PLANAS-BOHNE 1981a). TRIEN·2HCl (trientine or triethylenetetramine dihydrochloride) is absorbed up to about 20% in the gut of the rat. When given i.v., most passes rapidly into the urine, but a smaller fraction is presumably excreted in the bile and appears in the feces. Organ distribution of radioactive trientine in the rat indicates that it passes into the liver, spleen, kidneys, muscle, and gut wall (WALSHE 1982; GIBBS and WALSHE 1985). Deferoxamine is generally assumed to be largely confined to the extracellular space and the enhancement of urinary iron excretion which it induces is quite notable, but it increases the biliary excretion of iron (BERGERON et al. 1992), so it must enter liver cells also. Sodium diethyldithiocarbamate is a monoanion and can presumably use the monoanion transport systems in the kidneys and liver to attain intracellular sites (GALE et al. 1992).

C. Toxic Metal Excretion and Its Acceleration

I. Toxic Metal Half-Lives, Organ Distribution, and Normal Rates of Excretion

There are a variety of incompletely characterized processes which operate to limit the access of toxic metals to the body and to remove them from the body. The limited absorption of some toxic metals from the gastrointestinal tract is the first barrier. Further, of the fraction of metal ions which are absorbed, some will be rapidly excreted in the urine or feces. The ones which have been absorbed and incorporated into cells or other structures will be turned over and excreted in part during the normal metabolic processes. For certain metals in some structures, such as cadmium in the liver or kidney, mercury in the brain, or lead in adult bone, this is a rather slow process. For many other structures which turn over rapidly, such as the cells lining the intestines, the incorporated toxic metal ion may be excreted after a relatively brief stay in the body. For many toxic metals there are normal processes (involving endogenous chelating agents such as glutathione) operating for their excretion via both urinary and biliary excretion (FOULKES 1993).

The half-lives of toxic metals vary over an enormous range and the value for a given toxic metal depends upon the site at which it is deposited. Thus the half-life of mercury in the whole blood of men following exposure to mercury vapor was found to be 3.1 days for a fast phase and 18 days for a slow phase. In this same study the half-life for the mercury in the urine was found to be 40 days (BARREGÅRD et al. 1992). In a similar study of a more severe human exposure a half-life of 45 days was estimated (BLUHM et al. 1992). In this latter group of workers the administration of DMSA produced a threefold increase in the amount of mercury excreted in the urine, while the administration of *N*-acetyl-D,L-penicillamine produced a twofold increase. The half-life of mercury in the human brain may be much longer than this (CAVANAGH 1988). The half-life of elements which deposit in the bone, such as lead, can be of the order of many years (RABINOWITZ et al. 1976).

II. Acceleration of Rates of Excretion of Toxic Metal Ions Subsequent to Chelation

The most direct indication of the action of a chelating agent in accelerating the removal of a toxic metal from an organism is obtained via measurements of the toxic metal concentrations excreted in the urine or the feces. The main route of excretion of the toxic metal varies with both the toxic metal and the nature of the chelating agent used. The extent of the increase in toxic metal excretion is determined by both the degree of exposure to the toxic metal and the regimen used to administer the chelating agent. Selected data collected for lead illustrate these points.

1. Lead Intoxication

a) $Na_2CaEDTA$

This chelating agent is almost exclusively confined to the extracellular spaces and undergoes predominantly urinary excretion. The processes by which a complex such as $[CaEDTA]^{2-}$ exchanges its central calcium ion for a lead ion to give $[PbEDTA]^{2-}$ are probably similar to other processes of this type which have ben examined in vitro (MARGERUM et al. 1978). In these, one or more of the donor groups on the original complex breaks free and then forms a bond with the other metal ion. As the reaction proceeds, the chelating agent unwraps from one metal ion and wraps around the other. In this particular case, the process is assisted by the fact that the calcium complex is much less stable than the lead complex. Lead is removed from a binding site on serum albumin or an enzyme and transformed into a low molecular weight complex which can be excreted via glomerular filtration. A schematic representation of this process is shown in Fig. 6. EDTA itself undergoes very little metabolic change in the mammalian body (FOREMAN

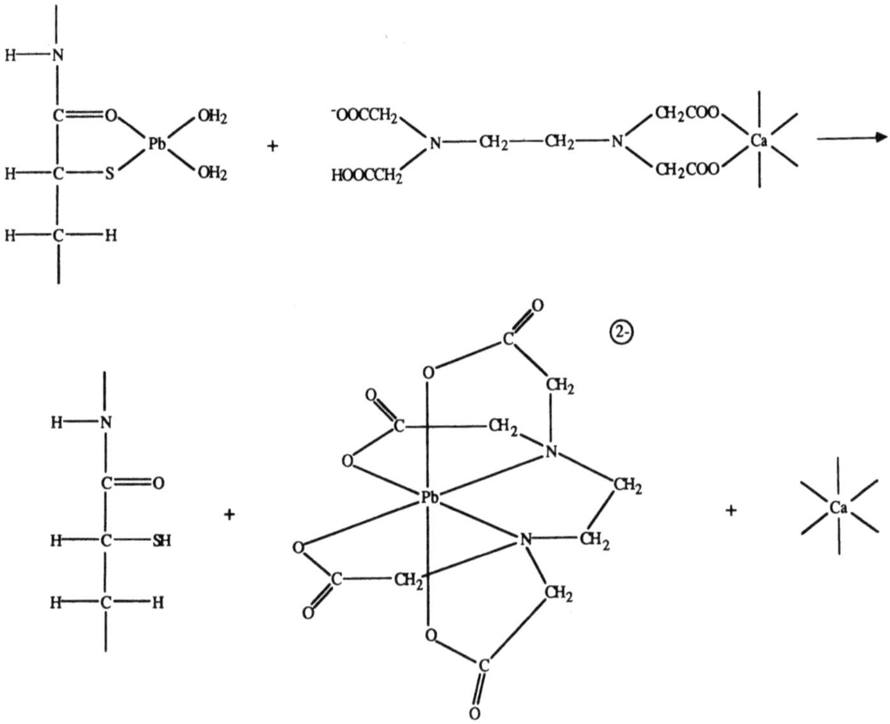

Fig. 6. Schematic representation of the removal of lead from a binding site on serum albumin by $Na_2CaEDTA$

and TRUJILLO 1954). The intravenous administration of solutions of $Na_2CaEDTA$ to individuals with lead poisoning usually results in a considerable enhancement of the urinary excretion of lead and this is used as a provocative test to gain information of the extent of lead exposure (FRIEDMAN and WEINBERGER 1990). Since $CaEDTA^{2-}$ is confined almost exclusively to the extracellular compartment, only a fraction of the lead is removed by such a treatment, even though a considerable molar excess of the chelating agent may be given. It is typical of chelating agent treatments that they are processes of relatively low efficiency. This is due in part to the fact that the distribution of the chelating agent and that of the toxic metal generally do not coincide. For example, in a study of 20 lead workers, it was found that the amount of lead excreted in the urine following the i.v. administration of 1 g $Na_2CaEDTA$ correlated well ($r = 0.86$) with the blood lead levels (TELL et al. 1992). However, while chelated lead correlated with bone lead levels in active workers, it did not if one included retired and no longer exposed workers in the group. No noticeable reduction in bone lead levels was observed, as would be expected if the lead removed by chelation came from the lead in the blood and in soft tissues.

D. Alteration of Metal Reactivity, Toxicity, and Distribution by Chelation

The properties of the toxic metal are significantly changed subsequent to its in vivo reaction with the exogenous chelating agent. Its reactivity with enzymes and other tissues is reduced, the reactions which give rise to its typical toxic symptoms are suppressed, and its distribution is drastically altered. The reaction of a toxic metal ion with most chelating agents yields a water-soluble complex which is excreted in the urine.

Another change in the in vivo behavior of the toxic metal ion which can occur after chelation is a considerable alteration in its bodily distribution. In most instances, the toxic metal complex which is formed is an ionic species which is rapidly excreted in the urine, such as the complex formed by mercury and D-penicillamine (SHAMBLY and SACK 1989). More rarely, as with BAL, a complex is formed which is excreted into the bile and then in the feces. However, adverse effects may also arise in this manner. Thus, cadmium ion gains access to the brain quite slowly in chronic cadmium intoxication in rodents. If sodium diethyldithiocarbamate is administered to such animals, this chelating agent forms an electrically neutral, lipid-soluble complex which can readily penetrate the blood-brain barrier and cause a significant increase in the total cadmium content of the brain (JONES et al. 1982). It is significant that such complexed cadmium has no obvious effects on the central nervous system (O'CALLAGHAN and MILLER 1986).

E. Stability Constants of Clinical Chelating Agents with Toxic Metal Ions

The chelating agents used in the clinic (Fig. 5) have been developed over time as a result of: (a) the preparation of specific chelating agent types for a given toxic metal as was the case with arsenic for which BAL, DMSA, and DMPS were developed and (b) trying available compounds as antagonists for toxic metals in animal models. The actual compounds used are those for which clinical experience has provided some justification in terms of availability, effectiveness, and ability to control side effects. The main requirement which must be met by a chelating agent is that the complex which it forms with a toxic metal ion must be more stable than the complex formed by the toxic metal with its in vivo binding site, otherwise it is thermodynamically impossible for the chelating agent to remove the toxic metal from the in vivo binding site. Unfortunately, it is often difficult to determine the stability constant for such a binding site, though various model reactions with enzymes known to be inactivated by the metal ion may be used to approximate this.

I. Conditional or Effective Stability Constants

The typical stability constant is expressed as:

$$M^{x+} + L^{-y} = ML^{x-y}$$

$$K_{stab} = [ML^{x-y}]/[M^{x+}][L^{-y}]$$

where L is derived from H_yL, the actual concentration of L^{-y} at physiological pH (pH = 7.4). The concentration of L^{-y} may be much less than the total concentration of L because there may be appreciable concentrations of other protonated and unprotonated species containing L, as some of the ionization constants may be much less than 10^{-7}. Also, the metal ion may be hydrolyzed at physiological pH, or there may be complexes present which contain hydrolyzed metal ions plus the ligand, or complexes in which additional protons have been lost or gained. The net result of these factors is that the typical measured stability constant does not reflect accurately what happens at physiological pH. The effective stability constant is devised to accommodate to this by providing a stability constant valid at pH 7.4 (HELLER and CATSCH 1959; SCHUBERT 1964; CATSCH and HARMUTH-HOEHNE 1979; MARTIN 1986; CLEVETTE and ORVIG 1990). An additional problem in vivo is the presence of amino acids and other endogenous chelating agents which compete with the exogenous chelating agent for the metal ion and its hydrolysis products. As a result, the a priori sorting out of potential chelating agent antagonists for a toxic metal ion on the basis of their stability constants remains a problem of considerable difficulty (COLE et al. 1985; CLEVETTE and ORVIG 1990).

F. Development of Chelating Agents for Clinical Use

It has been known for about 150 years that reaction with certain compounds can cause a drastic change in the chemical and biological properties of a metal, though the underlying basis for this was not understood until the work of Alfred Werner in the 1890s. Werner demonstrated conclusively that in such metal complexes compounds such as $H_2NCH_2CH_2NH_2$ (ethylenediamine) and $^{-}OOCCOO^{-}$ (oxalate) could, in fact, bond to the metal ion at two sites and that the resultant complexes exhibited behavior which were quite different from that of either the free metal ion or the free species which bound to the metal ion (the ligand). The first use of this difference in a medical setting was the use of tartrate (about 1917) and then tiron (about 1925) to reduce the toxicity of the antimony(III) which was used in the treatment of the parasitical disease schistosomiasis (SCHMIDT 1930). Antimony is toxic to humans, but is much more toxic to the parasite. The administration of the antimony as a complex reduces the toxicity to humans and the parasite, but allows doses of antimony to be administered which are much more toxic to the parasite than to humans. During the 1940s, two chelating agents were developed and became available for clinical use which had a profound effect on the subsequent development of chelation therapy for metal intoxication. These compounds were BAL or 2,3-dimercaptoproan-1-ol, and EDTA or ethylenediaminetetraacetic acid, and they represent the beginning of more a systematic search for toxic metal antidotes.

I. BAL and Its Derivatives

The development of effective antagonists for arsenic followed ultimately from the surmise by Paul Ehrlich that arsenic compounds reacted with sulfhydryl groups on proteins. At the outbreak of the World War II, the possibility of the use of poison gases was recognized by the combatants. In England, a team of scientists under the direction of Sir Rudolf Peters was set the task of developing an antagonist for lewisite, an arsenical compound usable as a poison gas. Its structure is shown in Fig. 7. It was established that sulfhydryl compounds could antagonize the biological activity of this compound and a variety of such compounds were prepared and tested. The success of one of these, BAL, led to the synthesis of DMSA and DMPS. These are shown in Fig. 5. These compounds are all capable of removing arsenic from its linkages to the sulfhydryl groups of proteins to form arsenic complexes which are much less toxic and more readily excreted. BAL is a nonpolar molecule which can gain access to intracellular sites, presumably via passage through the lipid portion of the cellular membrane. It is usually given as a solution in peanut oil which is stabilized by the addition of benzyl benzoate. BAL was soon found to be capable of also acting as an antagonist to other toxic heavy metals, such as lead and mercury, and has been widely used in the clinic for this purpose. DMSA and DMPS bear negative charges at physiological pH values and have a more restricted distribution, but are also much less toxic. They are useful as antagonists for the same types of heavy metals as BAL and have the additional advantage that they can be administered orally. While BAL can enhance the biliary excretion of heavy metals, the complexes formed with DMSA and DMPS are usually ionic and are commonly excreted via the kidneys.

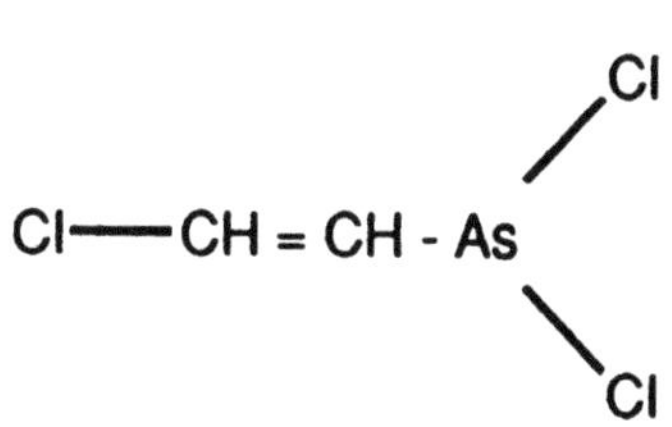

Fig. 7. Structure of lewisite

II. EDTA and Its Analogs

The use of EDTA and its analogs as antidotes for toxic heavy metals was introduced by Rubin and his collaborators, who demonstrated that $Na_2CaEDTA$ was effective in the treatment of lead intoxication (RUBIN et al. 1953). These investigators showed that the use of the calcium complex ($Na_2CaEDTA$) eliminated the danger of tetany, which was found when the parent compound (Na_2EDTA) was administered rapidly. The use of $Na_2CaEDTA$ for the treatment of lead poisoning is now very common, though it has been replaced for some other toxic metals by other compounds. Thus $Na_3ZnDTPA$ is now preferred for the treatment of plutonium intoxication (CATSCH and HARMUTH-HOENE 1979).

III. D-Penicillamine and Triethylenetetramine Dihydrochloride

D-Penicillamine and triethylenetetraamine (Fig. 5) were both developed by WALSHE (1981, 1982) for the treatment of Wilson's disease or hepatolenticular degeneration (STERNLIEB 1990). This is a hereditary disorder in which copper accumulates from the diet to the point where its toxicity is evidenced either as a neurological disorder (DENNING and BERRIOS 1989) or in some cases as an acute liver toxicity. The disease can be controlled by the administration of agents which either enhance the excretion or restrict the absorption of copper to produce a negative copper balance. WALSHE (1981) discovered that patients who had been given penicillin had elevated levels of copper in their urine and went on to characterize the compound responsible for this as D-penicillamine, a metabolic product of penicillin. He subsequently set out to examine the use of this compound to treat individuals with Wilson's disease, which had been discovered several decades earlier, and found that D-penicillamine, given orally, did indeed greatly enchance the copper excretion of such patients and prevented the accumulation of toxic levels of copper. WALSHE (1982) found that some patients developed adverse reactions to D-penicillamine, and to treat such patients he developed triethylenetetramine dihydrochloride, also given orally, which is almost as effective in enhancing the excretion of copper. More recently, WALSHE (1985) has shown that sodium tetrathiomolybdate, Na_2MoS_4, given orally can be used for individuals who have adverse reactions to D-penicillamine or triethylenetetraamine dihydrochloride. This compound appears to operate via a drastic decrease in the uptake of copper from the diet.

IV. Deferoxamine and Hydroxypyridinones

Deferoxamine (Fig. 5) is a compound synthesized by a microorganism to extract iron from its environment (a siderophore) and is effective in increasing the excretion of both iron (EHLERS et al. 1991) and aluminum (ACKRILL et al. 1980). This compound must be administered parenterally and is very expensive. The search for alternative chelating agents for iron has resulted in the finding that 3-hydroxypyrid-4-ones, such as 1,2-dimethyl-3-hydroxypyrid-4-one (Fig. 5), can induce an appreciable increase in the urinary excretion of iron when given orally (KONTOGHIORGHES et al. 1987). The use of therapeutic chelating agents in the treatment of iron overload is treated in detail by Templeton in Chap. 14 of this volume.

V. Sodium Diethyldithiocarbamate

Sodium diethyldithiocarbamate was introduced for the treatment of nickel carbonyl intoxication (SUNDERMAN 1990), for which it is very effective. The use of this compound is not recommended for other toxic metals because of the lipid-soluble complexes which it forms with many of them. Such complexes readily pass into the central nervous system.

G. Toxicity and Adverse Effects of Clinically Used Chelating Agents

Chelating agents, like all chemical compounds, exhibit toxic effects. These can arise from the fact that they increase the excretion of essential trace elements or from more subtle interactions. Thus all compounds which contain sulfhydryl compounds are capable of causing allergic reactions, such as the skin rash reported for DMSA (GRANDJEAN et al. 1991), or the numerous problems which may arise from the continued administration of D-penicillamine such as nephrotic syndrome and anuria (DUBOIS et al. 1990). The administration of EDTA by itself may result in tetany due to the rapid drop in serum calcium which results, this being the reason for its customary administration as the calcium complex.

H. Current Clinical Treatments for Common Metal Intoxications and Their Underlying Chemistry

I. Lead

Lead intoxication is found in adults who have occupational or environmental exposures and produces a variety of symptoms involving the gastrointestinal tract (from oral exposures), the nervous system, the kidneys, and the blood (CULLEN et al. 1983). It is also found in infants living in housing in which lead is present as residues from lead paint, plumbing, or the like (NEEDLEMAN and BELLINGER 1991). The principal sites at which the accumulation of lead gives rise to symptoms of chronic intoxication are the bone, the kidneys, and the brain and nervous system. Lead reacts with and inactivates three of the enzymes involved in the synthesis of heme to produce a characteristic anemia, whose resolution is one sign of the effectiveness of the chelate mobilization of lead. In the kidney, lead reacts with cellular constituents to form slightly soluble materials which can give rise to lead inclusion bodies and there is a continuous decrease in renal function. Lead damages structures in the proximal tubule, which results in a decreased ability to absorb amino acids, glucose, and essential ions from the glomerular filtrate (GOYER and RHYNE 1973; GOYER 1989). In the nervous system, lead produces a variety of symptoms ranging from a decreased conduction velocity in nerves to encephalopathy (WEEDEEN 1984). Lead in the bone is ultimately buried deep in the crystalline matrix, and such lead is metabolized very slowly in the absence of a break in the bone. The removal of lead from these sites results in a partial or total resolution of these symptoms, depending on the extent of exposure.

The coordination preferences of lead include carboxylate oxygens, sulfhydryl sulfurs, and some other donor groups which can be used in combination with these. There are five chelating agents of possible use in the treatment of lead intoxication: BAL, DMSA, DMPS, $Na_2CaEDTA$, and D-penicillamine. The very limited availability of DMPS leaves the other four compounds available for use.

BAL is currently recommended only for the treatment of acute childhood lead encephalopathy in conjunction with $Na_2CaEDTA$ (CHISOLM 1992), as the combination leads to a more rapid decrease in the blood lead levels. BAL, because of its lipid solubility, is expected to be capable of removing lead from sites which are not accessible to $Na_2CaEDTA$, which is confined to the extracellular space. This combination is also more effective in removing lead from children with blood lead levels over 100 μg/dl (4.83 μmol/l), though in children with lower blood levels the combination does not seem to possess any advantage over $Na_2CaEDTA$ alone (O'CONNOR 1992). The use of BAL in children who are glucose-6-phosphate dehydrogenase deficient, however, may lead to hemolysis (JANAKIRAMAN et al. 1978).

$Na_2CaEDTA$ is currently the most commonly used treatment for lead intoxication. The log K_{stab} for PbEDTA is 17.88 (MARTELL and SMITH 1974). $Na_2CaEDTA$ is given parenterally, dissolved in an aqueous medium, and will induce an appreciable increase in the urinary excretion of lead. The urinary lead excretion correlates closely with the blood lead level (TELL et al. 1992) in adults. There is some evidence that under certain circumstances the administration of $Na_2CaEDTA$ to lead-loaded rats may result in an increase in brain levels of lead (CORY-SLECHTA et al. 1987), though in humans such treatments generally lead to a partial or total resolution of neurological symptoms (LINZ et al. 1992; BALESTRA 1991; CULLEN et al. 1983; RUFF et al. 1993).

The development of DMSA has provided an orally administered chelating agent for lead which has several advantages over the traditional $Na_2CaEDTA$ treatment (APOSHIAN 1983; APOSHIAN and APOSHIAN 1990; DING and LIANG 1991; GRAZIANO 1986; GRAZIANO et al. 1992; GLOTZER and

BAUCHNER 1992). DMSA is given orally, is of very modest toxicity, induces a smaller urinary excretion of essential metals and a larger excretion of lead than $Na_2CaEDTA$, and has been found to be superior to $Na_2CaEDTA$ in enhancing the excretion of lead in both animal and clinical studies. The log K_{stab} for the PbDMSA complex is 17.4 (HARRIS et al. 1991), a somewhat smaller value than that reported for the PbEDTA complex, but DMSA can penetrate into some intracellular sites, possibly via transport systems for succinate. There is good reason to believe that it will soon become the preferred compound for the treatment of both acute and chronic lead intoxication.

1. D-Penicillamine

D-Penicillamine (DPA) is used in the treatment of lead intoxication as an oral medication which can be taken in the home (GLOTZER and BAUCHNER 1992). The log K_{stab} value for PbDPA complex is only 12.3 (MARTELL and SMITH 1974), so this compound would be expected to be less effective than either $Na_2CaEDTA$ or DMSA, and this has been found to be the case in both animal and clinical studies.

II. Arsenic

The chemistry involved in both the processes by which arsenic compounds exert much of their toxicity as well as those involved in its treatment using chelating agents which contain vicinal dithiol groups has been investigated in some detail. The types of compounds involved were outlined in early papers (STOCKEN and THOMPSON 1946; WHITTAKER 1947) and these have since been largely confirmed by more recent studies which have used NMR spectra to establish many structural details (DILL et al. 1989, 1991; O'CONNOR et al. 1990). Arsenic compounds of many types inhibit enzymes in both animals and simpler organisms (WEBB 1966). One key process in the intoxication of animals by arsenite and most trivalent arsenic compounds involves the reaction with the lipoic acid moiety (6,8-dithiooctanoic acid), which is an essential part of the pyruvate dehydrogenase complex. The key part of this reaction is shown in Fig. 8. Subsequent to its reaction with arsenious acid [$(HO)_3As$], the pyruvate dehydrogenase is inactive. The six-membered ring formed here has been confirmed by NMR studies (DILL et al. 1991). From observations on the reaction of arsenicals with the protein keratin, it was established that the ratio of As:S was 1:2, which suggested a chelate ring of some sort with one arsenic bound to two sulfhydryl groups. WHITTAKER (1947) has studied the ability of a number of α,ω-dithiols to reverse the effects of arsenic (as lewisite) on pyruvate dehydrogenase. It was found that monothiols are not very effective in reversing the deactivation of the enzyme by arsenic compounds, while dithiols are much more effective. For these dithiols, all compounds of the type $HS\text{-}CH_2(CH_2)_nCH_2\text{-}SH$, the results

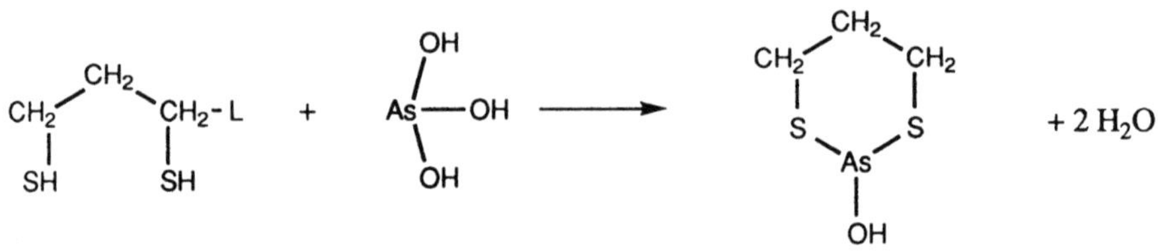

Fig. 8. Reaction of arsenious acid with lipoic acid derivative

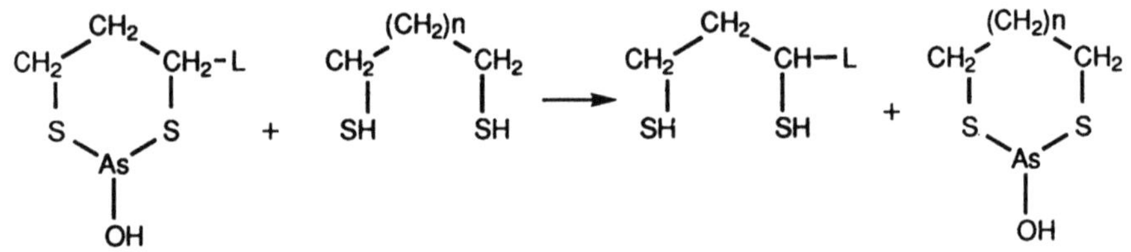

Fig. 9. Removal of arsenic from a lipoic acid derivative by reaction with a dithiol

showed that the effectiveness in reversing the effect of arsenic on the enzyme varied in the order $n = 0 > n = 1 > n = 2 < n = 3 < n = 4 \approx n =$ 5, 6, 7 and 8, with $n = 4$ being superior to $n = 2$, and $n = 0$ superior to compounds with $n = 1$ to 7. These reactions reverse the reaction given above as shown in Fig. 9. The structure of the adduct of lipoic acid and $As(OH)_3$ with a six-membered ring is supported by the study of DILL et al. (1989), who demonstrated the presence of such a ring in the product of the reaction of lipoic acid and C_6H_5-$AsCl_2$. The ability of BAL to act as an antidote for arsenic intoxication is ascribed to the fact that BAL reacts with the lipoic acid-arsenic complex to yield a product with a more stable five-membered chelate ring ($n = 0$ in Fig. 9). Such a five-membered ring was shown to be present in the product of the reaction of C_6H_5-$AsCl_2$ and BAL (DILL et al. 1987a) and to be more stable than the six-membered ring (DILL et al. 1991).

Two compounds closely related to BAL which have been examined as arsenic antidotes for arsenic are the vicinal dithiols meso-2,3-dimercaptosuccinic acid (DMSA) and sodium 2,3-dimercaptopropane-1-sulfonate (DMPS) (APOSHIAN et al. 1984). Both of these compounds form five-membered chelate rings with typical trivalent arsenic compounds (O'CONNOR et al. 1989). These vicinal dithiols do not have identical binding constants for trivalent arsenic. The relative order of binding (O'CONNOR et al. 1990) is BAL $\approx$ DMPS > DMSA, with the binding constants for BAL and DMPS for arsenic being about ten times greater than the constant for DMSA.

In vivo, arsenic compounds also react with glutathione (DILL et al. 1987b) to add one or two glutathione molecules via bonding at their sulfhydryl groups. Reaction with glutathione provides some measure of detoxification of the arsenic.

Rather extensive studies have been carried out on the relative toxicity and antidotal efficacy of vicinal dithiols in animal models, and it has been found that the in vivo behavior also depends very strongly on the nature of the groups other than the vicinal dithiol group (APOSHIAN et al. 1984). While all such compounds form the stable five-membered chelate rings with arsenic, attached groups which modify the polarity have a considerable effect on the distribution of the complex which is formed, as well as its route of excretion. Such compounds are generally capable of removing arsenic from the kidneys, liver, and lungs. BAL is atypical in that it raises the level of arsenic in the brain in arsenic-intoxicated rabbits, guinea pigs, and mice (KREPPEL et al. 1990). In control animals (rabbits) given sodium arsenite, the ratio of urinary to fecal excretion is about 18 to 1. DMSA increases arsenic excretion by both of these routes (APOSHIAN et al. 1983).

III. Mercury

Mercury intoxication usually arises from industrial or occupational exposures. Recently concern has arisen over the possible development of chronic mercury intoxication from mercury amalgam fillings (GOERING et al. 1992; JOKSTAD et al. 1992). The toxic effects of mercury are seen in the kidneys and in the brain and nervous system. There are three forms of mercury of toxicological interest: metallic mercury, inorganic mercury salts, and organomercury compounds. All are toxic, though the effects vary with each type and the nature of the exposure. The removal of metallic mercury from nerve cells or the brain can be an extremely slow process with metallic mercury present as long as 16 years after removal from exposure (HARGREAVES et al. 1988). Mercury may also accumulate by retrograde axonal transport in motor neurons (ARVIDSON 1992). Organomercury compounds are capable of moving through lipid barriers and attaining sites in the brain and nervous system at which they exert their toxic effects and from which they are difficult to remove. Mercury may be transformed from one of these forms to another, but usually at a slow rate. The result of such behavior is to make the treatment of mercury intoxication a time-consuming process.

The removal of mercury from the mammalian body is most readily effected by chelating agents which contain sulfhydryl groups: D-penicillamine (DPA), *N*-acetyl-D,L-penicillamine, *N*-acetyl cysteine, meso-2,3-dimercaptosuccinic acid (DMSA), and sodium 2,3-dimercaptopropane-1-sulfonate (DMPS), all of which can be given orally. Of these DMSA and DMPS are found to be the most effective in animal models and also to be effective in the clinic. DMPS is reported to be somewhat more effective than other compounds in removing mercury from the kidneys in animal models (GABARD 1976; PLANAS-BOHNE 1981b; DING and LIANG 1991; NIELSEN and ANDERSEN

1991), but an equivalent mobilization of mercury can be achieved by a longer treatment period with the more readily available DMSA (Buchet and Lauwerys 1989). In animal studies neither DMSA nor DMPS was found to be notably effective in reducing the accumulation of mercury in the brain of rodents (Buchet and Lauwerys 1989). Both of these compounds are very hydrophilic and neither would be expected to be capable of penetrating the Schwann cells around nerves nor of readily entering those lipid-rich parts of the brain in which the mercury appears to accumulate. Hemodialysis can be used to remove mercury from the blood. In testing a series of chelating agents it was found that *N*-acetylcysteine was the most effective in transferring mercury out of the plasma (Ferguson and Cantilena 1992). Such treatment does not directly remove mercury from intracellular sites, but can speed up removal of the circulating mercury.

Clinical reports indicate that mercury excretion can be enhanced by the oral administration of *N*-acetyl-D,L-penicillamine (Kark et al. 1971), DMPS (Campbell et al. 1986; Hruby and Donner 1987), or DMSA (Roels et al. 1991; Fournier et al. 1988; Bluhm et al. 1992), but the disappearance of symptoms may be very slow or only partial. Since DMSA is superior to *N*-acetyl-D,L-penicillamine in animal experiments and is more readily available, it would appear to be the most useful compound for current clinical treatments.

IV. Copper

The use of chelating agents in enhancing the excretion of copper has been examined because of its relationship to therapy for Wilson's disease (hepatolenticular degeneration), a hereditary disorder in which the normal excretory processes for copper (via the bile) are defective and in which copper is accumulated from the diet in the liver and other organs, to toxic and ultimately lethal levels (Woods and Colón 1989; Marsden 1987). In such individuals the urinary excretion of copper is usually several hundred micrograms per day (but in extreme cases may reach 5000 μg/day) in comparison with normal urinary copper excretion of 80–100 μg/day. The usual treatment is D-penicillamine, 1 g/day, taken in four doses of 0.25 g, though higher doses may be used on occasion to attain a state of negative copper balance. In some individuals D-penicillamine produces severe adverse reactions and for these the usual alternative is triethylenetetramine dihydrochloride, given at 0.5–2.0 g/day, which is almost as effective as D-penicillamine in facilitating the excretion of copper (Dubois et al. 1990).

V. Other Toxic Metals

Of the remaining toxic metals of interest, plutonium has been, perhaps, the most thoroughly studied. The currently recommended treatment for residual plutonium (after surgical removal of any pieces of the element or its compounds) is $Na_3ZnDTPA$ (VOLF 1978), a compound which is also effective in accelerating the removal of americium. Volf's book also contains a discussion of chelate treatment for other intoxication by transuranium elements. Excellent reviews of the biological chemistry of many of the elements of toxicological interest may be found in the corresponding volumes of the *Gmelin Handbook of Inorganic Chemistry*, for example that on polonium (SEIDEL 1990), where data on the use of DMPS to accelerate polonium excretion is presented.

I. Unsolved Problems and Future Prospects

From the summary which has been presented above, it is apparent that there are many situations in which superior therapeutic chelating agents would be very useful. The relative inability of many chelating agents to react either directly or effectively with toxic metal ions in physiologically inaccessible sites is one of the most obvious problems. This is responsible for much of the difficulty which attends attempts to treat heavy lead exposures where a considerable amount of the lead has reached sites in the bone from which its removal is a slow and tedious process at best. Intoxication by any of the radioisotopes of metal ions which deposit readily in the bone is very difficult to treat at present because of the lack of agents which can readily mobilize such deposits and enhance to rate of excretion of the radioactive metal. The problems which are found in low-level chronic mercury intoxication, which is now suspected to be fairly widespread, are imperfectly understood in terms of the most effective procedures for clinical treatment. Cadmium intoxication is another situation in which human cases are difficult to treat, and the possible use of compounds found to be effective in laboratory animals has not yet begun. While the deleterious effects of the exposure of infants and young children to lead on their neurological development are now widely accepted (though the levels at which these occur is presently a matter of dispute), possible treatments of such individuals are presently imperfect at best. The problem of removing toxic metals such as lead from the brain is another unsolved, but significant problem, related to the fact that current chelate treatments for the reduction of blood lead levels often have only a temporary effect as these levels rebound as lead is subsequently mobilized from chelate-inaccessible stores in the bone (CHISOLM 1990). The search for new chelating agents which may ultimately find clinical use in the solution of some of these problems is an active field (JONES 1991).

References

Ackrill P, Ralston AJ, Day JP, Hodge KC (1980) Successful removal of aluminum from patients with dialysis encephalopathy. Lancet 2:692–693
Aposhian HV (1983) DMSA and DMPS – water soluble antidotes for heavy metal poisoning. Annu Rev Pharmacol Toxicol 23:193–215

Aposhian HV, Aposhian MM (1990) Meso-2,3-dimercaptosuccinic acid: chemical, pharmacological and toxicological properties of an orally effective chelating agent. Annu Rev Pharmacol Toxicol 30:279–306

Aposhian HV, Hsu C-A, Hoover TD (1983) DL- and meso-dimercaptosuccinic acid. In vitro and in vivo studies with sodium arsenite. Toxicol Appl Pharmacol 69:206–213

Aposhian HV, Carter DE, Hoover TD, Hsu C-A, Maiorino RM, Stine E (1984) DMSA, DMPS and DMPA – as arsenic antidotes. Fundam Appl Toxicol 4:S58–S70

Arvidson B (1992) Inorganic mercury is transported from muscular nerve terminals to spinal and brainstem motoneurons. Muscle Nerve 15:1089–1094

Balestra DJ (1991) Adult chronic lead intoxication. Arch Intern Med 151:1718–1720

Barregård L, Sällsten G, Schütz A, Attewell R, Skerfving S, Järvholm B (1992) Kinetics of mercury in blood and urine after brief occupational exposure. Arch Environ Health 47:176–184

Bergeron RJ, Liu ZR, McManis JS, Wiegand J (1992) Structural alterations in desferrioxamine compatible with iron clearance in animals. J Med Chem 35: 4739–4744

Bhattacharyya MH, Peterson DP (1979) Action of DTPA on hepatic plutonium. III. Evidence for a direct chelation mechanism for DTPA-induced excretion of monomeric plutonium into rat bile. Radiat Res 80:108–115

Bluhm RE, Bobbit RG, Welch LW, Wood AJJ, Bonfiglio JF, Sarzen C, Heath AJ, Branch RA (1992) Elemental mercury vapour toxicity, treatment, and prognosis after acute, intensive exposure in chloralkali plant workers. I. History, neuropsychological findings and chelator effects. Hum Exp Toxicol 11:201–210

Buchet JP, Lauwerys RR (1989) Influence of 2,3 dimercaptopropane-1-sulfonate and dimercaptosuccinic acid on the mobilization of mercury from tissues of rats pretreated with mercuric chloride, phenylmercury acetate or mercury vapors. Toxicology 54:323–333

Bulman RA (1987) The chemistry of chelating agents in medical sciences. Structure Bonding 67:91–141

Campbell JR, Clarkson TW, Omar MD (1986) The therapeutic use of 2,3-dimercaptopropane-1-sulfonate in two cases of inorganic mercury poisoning. JAMA 256:3127–3130

Catsch A (1968) Dekorporierung radioaktiver und stabiler Metallionen. Therapeutische Grundlagen. Thiemig, Munich

Catsch A, Harmuth-Hoene A-E (1979) Pharmacology and therapeutic applications of agents used in heavy metal poisoning. In: Levine WG (ed) The chelation of heavy metals. Pergamon, Oxford, p 107

Cavanagh JB (1988) Long term persistence of mercury in the brain. Br J Ind Med 45:649–651

Chisolm JJ (1990) Evaluation of the potential role of chelation therapy in the treatment of low to moderate lead exposures. Eviron Health Perspect 89:67–74

Chisolm JJ Jr (1992) BAL, EDTA, DMSA and DMPS in the treatment of lead poisoning in children. J Toxicol Clin Toxicol 30:493–504

Clevette DJ, Orvig C (1990) Comparison of ligands of differing denticity and basicity for the in vivo chelation of aluminum and gallium. Polyhedron 9:151–161

Cole A, Furnival C, Huang ZX, Jones DC, May PM, Smith GL, Whittaker J, Williams DR (1985) Computer simulation models for the low molecular weight complex distribution of cadmium(II) and nickel(II) in human blood. Inorg Chim Acta 108:165–171

Cory-Slechta DA, Weiss B, Cox C (1987) Mobilization and redistribution of lead over the course of calcium disodium ethylenediamine tetraacetate chelation therapy. J Pharmacol Exp Ther 243:804–813

Cullen MR, Robins JM, Eskenazi B (1983) Adult inorganic lead intoxication. Presentation of 31 new cases and a review of recent advances in the literature. Medicine (Baltimore) 62:67–74

Denning TR, Berrios GE (1989) Wilson's disease: psychiatric symptoms in 195 cases. Arch Gen Psychiatry 46:1126–1134
Dill KR, Adams ER, O'Connor RJ, McGown EL (1987a) 2D NMR studies on the phenyldichloroarsine-British anti-lewisite adduct. Magn Reson Chem 25:1074–1077
Dill KR, O'Connor RJ, McGown EL (1987b) Spin-echo NMR investigations of the interaction of phenyldichloroarsine with glutathione in intact erythrocytes. Inorg Chim Acta 138:95–97
Dill KR, Adams ER, O'Connor RJ, McGown EL (1989) Structure and dynamics of a lipoic acid-arsenical adduct. Chem Res Toxicol 2:181–185
Dill KR, Huang L, Bearden DW, McGown EL, O'Connor RJ (1991) Activation energies and formation rate constants for organic arsenical-antidote adducts as determined by dynamic NMR spectroscopy. Chem Res Toxicol 4:295–299
Ding GS, Liang YY (1991) Antidotal effects of dimercaptosuccinic acid. J Appl Toxicol 11:7–14
Dubois RS, Rodgerson DO, Hambridge KM (1990) Treatment of Wilson's disease with triethylene tetramine hydrochloride (trientine). J Pediatr Gastroenterol Nutr 10:77–81
Ehlers KH, Gardina PJ, Lesser ML, Engle MA, Hilgartner MW (1991) Prolonged survival in patients with beta-thalassemia treated with deferoxamine. J Pediatr 118:540–545
Eybl V, Sýkora J (1966) Die Schutzwirkung von Chelatbildern bei der akuten Kadmiumvergiftung. Acta Biol Med Ger 16:61–64
Ferguson CL, Cantilena LR (1992) Mercury clearance from human plasma during in vitro dialysis: screening systems for chelating agents. J Toxicol Clin Toxicol 30:423–441
Foreman H, Trujillo TT (1954) Metabolism of carbon^{14}-labelled ethylenediaminetetraacetate in human beings. J Lab Clin Med 43:566–571
Foulkes EC (1993) Metallothionein and glutathione as determinants of cellular retention and extrusion of cadmium and mercury. Life Sci 52:1617–1620
Friedman JA, Weinberger HL (1990) Six children with lead poisoning. Am J Dis Child 144:1039–1040
Fournier L, Thomas G, Garnier R, Buisine A, Houze P, Pradier F, Dally S (1988) 2,3-Dimercaptosuccinic acid treatment of heavy metal poisoning in humans. Medical Toxicol 3:499–504
Gabard B (1976) Improvement of oral chelation treatment of methyl mercury poisoning in rats. Acta Pharmacol Toxicol (Copenh) 39:250–255
Gale GR, Smith AB, Atkins LM, Jones MM (1985) Effects of diethyldithiocarbamate and *N*-methyl-*N*-dithiocarboxyglucamine on murine hepatic cadmium metallothionein in vitro. Res Commun Chem Pathol Pharmacol 49:423–434
Gale GR, Smith AB, Jones MM, Singh PK (1992) Evidence of active transport of cadmium complexing dithiocarbamates into renal and hepatic cells in vivo. Pharmacol Toxicol 71:452–456
Gibbs KR, Walshe JM (1985) The metabolism of trientine: animal studies. In: Scheinberg IH, Walshe JM (eds) Orphan diseases and orphan drugs. Manchester University Press, Manchester, p 33
Glotzer DE, Bauchner H (1992) Management of childhood lead poisoning: a survey. Pediatrics 89:614–618
Goering PL, Galloway WD, Clarkson TW, Lorscheider FL, Berlin M, Rowland AS (1992) Toxicity assessment of mercury vapor from dental amalgams. Fundam Appl Toxicol 19:319–329
Goyer RA (1989) Mechanisms of lead and cadmium nephrotoxicity. Toxicol Lett 46:153–162
Goyer RA, Rhyne B (1973) Pathological effects of lead. Int Rev Exp Pathol 12:1–77
Grandjean P, Jacobsen IA, Jørgensen PJ (1991) Chronic lead poisoning treated with dimercaptosuccinic acid. Pharmacol Toxicol 68:266–269

Graziano JH (1986) Role of 2,3-dimercaptosuccinic acid in the treatment of heavy metal poisoning. Med Toxicol 1:155–162

Graziano JH, LoIacono NJ, Moulton T, Mitchell ME, Slakovich V, Zarate C (1992) Controlled study of meso-2,3-dimercaptosuccinic acid for the management of childhood lead intoxication. J Pediatr 120:133–139

Hancock RD, Martell AE (1988) The chelate, cryptate and macrocyclic effects. Comments Inorg Chem 6:237–284

Hancock RD, Martell AE (1989) Ligand design for selective complexation of metal ions in aqueous solution. Chem Rev 89:1875–1914

Hargreaves RJ, Evans JG, Janota I, Magos L, Cavanagh JB (1988) Persistent mercury in nerve cells 16 years after metallic mercury poisoning. Neuropathol Appl Neurobiol 14:443–452

Harris WR, Chen Y, Stenback J, Shah B (1991) Stability constants for dimercaptosuccinic acid with bismuth(III), zinc(II) and lead(II). J Coord Chem 23:173–186

Heller HJ, Catsch A (1959) Einige physikalisch-chemische Überlegungen zur Dekorporation radioaktiver Metalle durch Komplexbildner. Strahlentherapie 109:464–482

Hruby K, Donner A (1987) 2,3-Dimercapto-1-propanesulfonate in heavy metal poisoning. Hum Exp Toxicol 2:317–323

Janakiraman N, Seeler RA, Royal JE, Chen MF (1978) Hemolysis during BAL chelation therapy for high blood lead levels in two G6PD deficient children. Clin Pediatr 17:485–487

Jokstad A, Thomassen Y, Bye E, Clench-Aas J, Aaseth J (1992) Dental amalgam and mercury. Pharmacol Toxicol 70:308–313

Jones MM (1991) New developments in therapeutic chelating agents as antidotes for metal poisoning. CRC Crit Rev Toxicol 21:209–233

Jones SG, Basinger MA, Jones MM, Gibbs SG (1982) A comparison of diethyldithiocarbamate and EDTA as antidotes for acute cadmium intoxication. Res Commun Chem Pathol Pharmacol 38:271–278

Joyce DA (1989) D-Penicillamine pharmacokinetics and pharmacodynamics in man. Pharmacol Ther 42:405–427

Kark RA, Poskanzer DC, Bullock JD, Boylen G (1971) Mercury poisoning and its treatment with *N*-acetyl-D,L-penicillamine. N Engl J Med 285:10–16

Kontoghiorghes GJ, Aldouri MA, Sheppard L, Hoffbrand AV (1987) 1,2-Dimethyl-3-hydroxypyrid-4-one, an orally active chelator for treatment of iron overload. Lancet 1:1294–1295

Kreppel H, Reichl FX, Szinicz L, Fichtl B, Forth W (1990) Efficacy of various dithiol compounds in acute As_2O_3 poisoning in mice. Arch Toxicol 64:387–392

Linz DH, Barrett ET, Pflaumer JE, Keith RE (1992) Neuropsychologic and postural sway improvement after Ca^{++}-EDTA chelation for mild lead intoxication. J Occup Med 34:638–641

Margerum DW, Cayley GR, Weatherburn DC, Pagenkopf GK (1978) Kinetics and mechanisms of complex formation and ligand exchange. In: Martell AE (ed) Coordination chemistry, vol 2. American Chemical Society, Washington, p 1

Marsden CD (1987) Wilson's disease. Q J Med 65:959–966

Martell AE, Smith RM (1974) Critical stability constants. Plenum, New York

Martin RB (1986) The chemistry of aluminum as related to biology and medicine. Clin Chem 32:1797–1806

May P, Bulman RA (1983) The present status of chelating agents in medicine. Prog Med Chem 20:225–336

Meier PJ (1988) Transport polarity of hepatocytes. Semin Liver Dis 8:293–307

Møller JV, Sheikh MI (1983) Renal organic anion transport system: pharmacological, physiological, and biochemical aspects. Pharmacol Rev 34:315–357

Moore MR, Meredith PA, Goldberg A (1980) Lead and heme biosynthesis. In: Singhal RL, Thomas JA (eds) Lead toxicity. Urban and Schwarzenberg, Baltimore, p 79

Needleman HL, Bellinger D (1991) The health effects of low level exposure to lead. Annu Rev Public Health 12:111–140
Nielsen JB, Andersen O (1991) Effect of four thiol-containing chelators on disposition of orally administered mercuric chloride. Hum Exp Toxicol 10:423–430
O'Callaghan JP, Miller DB (1986) Diethyldithiocarbamate increases distribution of cadmium to the brain but prevents cadmium-induced neurotoxicity. Brain Res 370:354–358
O'Connor ME (1992) CaEDTA vs CaEDTA plus BAL to treat children with elevated blood lead levels. Clin Pediatr 31:386–390
O'Connor RJ, McGown EL, Dill K, Hallowell SF (1989) Two dimensional NMR studies of arsenical-sulfhydryl adducts. Magn Reson Chem 27: 669–675
O'Connor RJ, McGown EL, Dill, Hallowell SF (1990) Relative binding constants of arsenical-antidote adducts determined by NMR spectroscopy. Res Commun Chem Pathol Pharmacol 69:365–368
Planas-Bohne F (1981a) Metabolism and pharmacokinetics of D-penicillamine in rats – an overview. J Rheumatol 8 [Suppl] 7:35–40
Planas-Bohne F (1981b) The influence of chelating agents on the distribution and biotransformation of methylmercuric chloride in rats. J Pharmacol Exp Ther 217:500–504
Pritchard JB, Miller DS (1992) Proximal tubular transport of organic anions and cations. In: Seldin DW, Giebisch G (eds) The kidney: physiology and pathophysiology, vol 2, 2nd edn. Raven, New York, p 2921
Rabinowitz MB, Kopple JD, Wetherill GW (1976) Kinetic analysis of lead metabolism in healthy humans. Clin Invest 90:700–706
Roels HA, Boeckx M, Ceulemans E, Lauwerys RR (1991) Urinary excretion of mercury after occupational exposure to mercury vapor and influence of the chelating agent meso-2,3-dimercaptosuccinic acid (DMSA). Br J Ind Med 48: 247–253
Rubin M, Gignac S, Bessman SP, Belknap EL (1953) Enhancement of lead excretion in humans by disodium calcium ethylenediaminetetraacetate. Science 117:659–660
Ruff HA, Bijur PE, Markowitz M, Ma YC, Rosen JF (1993) Declining blood levels and cognitive changes in moderately lead-poisoned children. JAMA 269:1641–1646
Schmidt H (1930) Antimon in der Arzeimittelsynthese. Z Angew Chem 43:963–970
Schubert J (1964) The chemical basis of chelation. In: Gross F (ed) Iron metabolism. Springer, Berlin Heidelburg New York, p 466
Seidel A (1990) Metabolism and toxicology of polonium and its removal from the body. In: Buschbeck K-C, Keller C (eds) Polonium supplement, vol 1. Springer, Berlin Heidelburg New York, p 251 (Gmelin handbook of inorganic and organometallic chemistry, 8th edn)
Shambly DJ, Sack JS (1989) Mercury poisoning: a case report and comment on 6 other cases. S Afr Med J 76:114–116
Sternlieb I (1990) Perspectives on Wilson's disease. Hepatology 12:1234–1238
Stewart JR, Diamond G (1987) Renal tubular secretion of the alkanesulfonate 2,3-dimercapto-1-propane sulfonate. Am J Physiol 252:F800–F810
Stocken LA, Thompson RHS (1946) British anti-lewisite. II. Dithiol compounds as antidotes for arsenic. Biochem J 40:535–548
Sunderman FW Sr (1990) Use of sodium diethyldithiocarbamate in the treatment of nickel carbonyl poisoning. Ann Clin Lab Sci 20:12–21
Tell I, Somervaille LJ, Nilsson U, Bensryd I, Schütz A, Chettle DR, Scott MC, Skerfving S (1992) Chelated lead and bone lead. Scand J Work Environ Health 18:113–119
Volf V (1978) Treatment of incorporated transuranium elements. IAEA Tech Rep Ser 184

Walshe JM (1981) The discovery and therapeutic use of D-penicillamine. J Rheumatol 8 [Suppl] 7:3–8
Walshe JM (1982) Treatment of Wilson's disease with trientine (triethylenetetramine) dihydrochloride. Lancet 1:643–647
Walshe JM (1985) Tetrathiomolybdate (MoS_4) as an "anti-copper" agent in man. In: Scheinberg IH, Walshe JM (eds) Orphan diseases and orphan drugs. Manchester University Press, Manchester, p 76
Webb JL (1966) Enzyme and metabolic inhibitors, vol 3. Academic, New York
Wedeen RP (1984) Poison in the pot. The legacy of lead. Southern Illinois University Press, Carbondale
Whittaker VP (1947) An experimental investigation of the "ring hypothesis" of arsenical toxicity. Biochem J 41:56–62
Williams DR, Halstead BW (1982/1983) Chelating agents in medicine. J Toxicol Clin Toxicol 19:1081–1115
Woods SE, Colón VF (1989) Wilson's disease. Am Family Phys 40:171–178

CHAPTER 14

Therapeutic Use of Chelating Agents in Iron Overload

D.M. TEMPLETON

A. Transport, Storage, and Toxicity of Iron

Elemental iron, like molecular oxygen, is a highly toxic substance that is nevertheless essential for life. Therefore, like oxygen, complex biochemical systems have evolved for its safe delivery, transport, and utilization. Iron absorption in the human gut is a highly regulated process that depends in part on total body iron stores (FLANAGAN 1990; McLAREN et al. 1981). Loss of regulation, or stimulated transport in anemic states, contributes to iron overload and in some cases a need for chelation. Very few details of the processes involved in absorption are known. Safe transport of newly absorbed iron through the circulation is achieved by binding to the transport protein, transferrin. Transferrin has two iron-binding sites with association constants ($\log K$) of 22.1 and 22.7 (MARTIN et al. 1987), so iron is bound tightly in a non-redox-active form and is effectively nonexchangeable with other ligands normally present in plasma. Human transferrin is typically about one-third saturated with iron (McLAREN et al. 1981) so there is a large reserve capacity to accommodate additional iron. Controlled delivery of iron to tissues such as the liver is generally achieved by receptor-mediated endocytosis of a transferrin – transferrin-receptor complex, which dissociates in an acidified endosomal compartment (AISEN 1992; THEIL and AISEN 1987), ultimately releasing iron for incorporation into its storage form, ferritin. The completed ferritin molecule comprises 24 protein subunits that surround an iron core with a basic ferric oxohydroxide structure as well as phosphate ligands. Up to 4500 atoms of iron can be accommodated in this core, which is coated by the peptides (THEIL 1987). Utilization of ferritin iron, for example for the synthesis of cytochromes and other hemoproteins, may require reductive enzymatic release of iron from the relatively inert and sequestered core (THEIL 1987).

Under physiological conditions, i.e., in neutral aqueous solution with a. significant pO_2, ferric ion (Fe^{3+}) is the predominant species and ferrous ion (Fe^{2+}) can only exist if protected from oxidation by specific coordination (SCHNEIDER 1988). Transient, enzymatic reduction of Fe^{3+} serves as a control point for the exchange of iron, for example in the mobilization of iron from ferritin stores and transport across lipid membranes. This is important in that most chelators presently in use or under investigation display specificity based on Fe^{3+} chemistry but may depend on reductive mobilization of iron to release it in an accessible form. Reductants like superoxide anion (BIEMOND et al. 1984) and reduced flavins (FUNK et al. 1985) effect the reduction and release of iron from ferritin, and Fe^{3+} is mobilized from endocytic vesicles by reduction (NÚÑEZ et al. 1990). It has been proposed that reductive release of ferritin iron contributes to the free radical-mediated toxicity of a number of chemicals (AUST et al. 1993). Unlike Fe^{3+} – which begins to hydrolyze at pH 1 – millimolar solutions of Fe^{2+} are free from significant hydrolysis at pH 7 (BAES Jr. and MESMER 1986), making the ion availble for diffusion and transport. The redox potential of the Fe^{3+}/Fe^{2+} couple ($E^{\circ\prime}$ = +0.77 V) is sensitive to the ligand set in iron complexes and can easily decrease to biologically accessible values. For example, the $E^{\circ\prime}$ of iron decreases to +0.22 V in cytochrome C and even −0.17 V in horseradish peroxidase (HUGHES 1972). Thus, when Fe^{3+} is not sequestered in stable structures like transferin and ferritin, it may be available to accept an electron from donors like ascorbate or superoxide anion and participate in the generation of harmful hydroxyl radicals through the Fenton reaction (Scheme 1). Such considerations are also relevant to the biochemistry of siderophores (see below), where stabilization of the Fe^{3+} complex lowers the redox potential even further. The iron complex of the hexadentate hydroxamate desferrioxamine B (DFO, [I]) has a formal potential of −0.45 V, accessible to biological reductants which may play a role in iron release and siderophore recycling. In contrast, the catecholate-based enterobactin [II] has an even greater stability and consequently a potential estimated to be −0.75 V at pH 7. This is out of the range of biological reductants and iron release is achieved at the expense of hydrolysis of the siderophore (COOPER et al. 1978). Although at first sight a metabolically extravagant approach, the very great thermodynamic stability of the Fe(III)-enterobactin complex confers a tremendous advantage on an organism competing for the nutrient in an environment where already limited amounts of Fe are rendered insignificant by hydrolysis of ferric ion.

Scheme 1. The iron-catalyzed Fenton reaction

$$\begin{array}{l} Fe^{+3} + O_2^{-\bullet} \rightarrow Fe^{+2} + O_2 \\ \underline{Fe^{+2} + H_2O_2 \rightarrow Fe^{+3} + \bullet OH + OH^-} \\ O_2^{-\bullet} + H_2O_2 \rightarrow O_2 + \bullet OH + OH^- \end{array}$$

The iron-catalyzed generation of hydroxyl radicals accounts for much of the cytotoxicity of the element. The resultant lipid peroxidation causes damage to the plasma membrane and to intracellular membranes of the mitochondria and lysosomes (HERSHKO and WEATHERALL 1988). Free radicals have been implicated in iron-induced injury to many tissues, including lung (ADAMSON et al. 1993), kidney (ALFREY et al. 1989), endothelium (BRIELAND et al. 1992), and heart (HERSHKO et al. 1987; LINK et al. 1993; SCOTT et al. 1985). Iron-catalyzed lipid peroxidation in cell culture systems and its suppression by iron chelators have been used as means of evaluating the potential usefulness of the chelator (HERSHKO et al. 1987). It must be noted that the concomitance of membrane-lipid peroxidation and iron toxicity does not prove that the toxicity arises from physical damage to the membranes. Lipid peroxidation products such as 4-hydroxynonenal may themselves be toxic (ESTERBAUER 1993). Furthermore, other potential effects of iron are poorly understood. For example, Fe^{3+} gradients may interfere with the electrical activity of myocardial cells (LINK et al. 1989) and saturation of transferrin with iron may affect the availability of apotransferrin to function as a growth factor for pituitary cells (SIRBASKU et al. 1992). Nevertheless, the toxicity of excess iron is generally understood primarily as a process of oxygen-radical formation, and the goal of iron-chelation therapy is mainly to prevent the occurrence of redox-active iron complexes.

This chapter discusses circumstances under which the normal pathways for the safe handling of iron become saturated or otherwise circumvented, and describes strategies for iron chelation that then become necessary to prevent radical-mediated tissue damage.

B. Chronic Iron Overload and Clinical Need for Iron Chelators

I. Intake

Chronic iron overload mainly occurs from excessive dietary intake, increased intestinal absorption, or multiple blood tranfusions. Excessive intake is not common; worldwide dietary iron deficiency is a far greater problem. Epidemiologically significant iron overload is rare, occurring for example in African blacks who consume beer brewed in iron pots. Sometimes called Bantu siderosis, it probably affects those with a genetic predisposition (GORDEUK et al. 1992a). However, a recent study has suggested that increased dietary iron intake and serum ferritin concentrations encountered in the general adult male population of Finland may be linked to an increased risk of myocardial infarction (SALONEN et al. 1992). If this is borne out in future investigations, our idea of what comprises dietary iron "overload" may have to be drastically revised.

II. Absorption

The mechanism and regulation of intestinal iron absorption remains elusive. A recent intriguing proposal is that integrins on the luminal surface of gut epithelia mediate the safe delivery to a cytosolic acceptor, mobilferrin, with some sequence similarity to the calcium-binding protein calreticulin (CONRAD and UMBREIT 1993; CONRAD et al. 1993). Hereditary hemochromatosis is an inherited disorder in which hyperabsorption leads to iron deposition in the liver, heart, pancreas, and other endocrine organs, generally over a period of several decades. Because of the variable expression of the underlying genetic disorder, the long time-course of development of symptoms, and exacerbation by alcohol intake, the true incidence of the disease is not known. For a discussion of the natural history of hemochromatosis the reader is referred to the review by McLaren et al. (1981) Reduction of iron stores in hemochromatosis is usually achieved by phlebotomy. Removal of 500 ml blood eliminates about 200 mg iron as hemoglobin, which is ultimately replaced by mobilization of body iron stores. In this way, stores can be decreased and then maintained in balance without the need for chelation therapy. Therefore, long-term chelation for chronic iron overload is almost completely restricted to managing that resulting from transfusion.

III. Transfusion

Patients with absent or ineffective erythropoiesis must have their anemia corrected by blood transfusions. The iron delivered as hemoglobin from the transfused red cells is not excreted and ultimately enters body iron stores. Because the underlying disorder is transfusion-dependent anemia, phlebotomy is not an available option and chelation must be used to manage the iron burden. In the most commonly severe form of thalassemia – β-thalassemia major – there is a complete absence of the β-globin chain. Fetal hemoglobin is a tetramer of two α-globin and two γ-globin chains ($\alpha_2\gamma_2$). At a few months of age, γ-globin gene transcription is switched off and β-globin is switched on, leading to the production of adult $\alpha_2\beta_2$-hemoglobin. Thus, a child with β-thalassemia major survives fetal life but after a few months is incapable of making adult hemoglobin and must be maintained on lifelong transfusions. After receiving about 100 units blood, sufficient tissue iron has accumulated to cause organ damage (McLaren et al. 1981). Presently a large number of the many thousands of children born each year with thalassemia major die untransfused in infancy. As safe and effective transfusion programs become available in developing countries the need for chelation will follow. The molecular bases of the thalassemias (Weatherall and Clegg 1981) and globin gene switching (Stamatoyannopoulos and Nienhuis 1990) have been thoroughly studied, and gene replacement therapy has appeal as a future cure for the disease. However, the complexity of regulatory elements in the globin genes and the requirements for coordinated expression of the α- and β-globin genes to produce functional adult hemoglobin (Behringer et al. 1990; Enver et al. 1990; Hanscombe et al. 1991) combine to raise a number of technical barriers to gene replacement strategies. Modulation of globin gene expression with agents such as hydroxyurea, azacytidine (Lowery and Nienhuis 1993), and arginine butyrate (Perrine et al. 1993), though very promising, are yet in a very early experimental stage. Marrow transplantation is only available to those patients with compatible donors and still carries about a 10% risk of death (LUCARELLI et al. 1993b). Moderate to severe iron loading persisted in a number of patients 7 years after successful marrow transplantation (LUCARELLI et al. 1993a), indicating that chelation should continue following cure of the genetic disorder. Therefore, it is likely that transfusion and the obligatory iron chelation that follows will be the treatment of choice for at least another generation of patients.

C. Other Applications of Iron Chelation

Although the major use of intensive iron chelation is for transfusional overload, the removal or sequestration of iron has been considered in a number of other instances. With the exception of acute iron poisoning – usually an accidental occurrence in the pediatric population – these are generally experimental. Intravenous infusion of DFO is standard treatment for acute iron intoxication and can prevent death from cardiovascular and gastrointestinal injury (LOVEJOY Jr 1982).

Therapeutic applications of iron chelators other than for acute or chronic iron overload have been based on either the desirable suppression iron-dependent free radical chemistry or interference with the actions of iron-dependent enzymes. In the former category are attempts to limit postischemic reperfusion injury (BABBS 1985; HEDLUND and HALLAWAY 1993). Iron released from ischemic tissue can cause damage upon re-oxygenation. It has been suggested that the acidosis that accompanies ischemia may facilitate the reductive release of iron from ferritin stores (VOOGD et al. 1992). The effectiveness of DFO bound to a hydroxyethyl-starch polymer in improving postischemic cardiac function (HALLAWAY et al. 1989) argues for a role of iron-dependent damage to the microvasculature, as this agent obligately acts in the extracellular space. Rheumatoid arthritis may involve a component of oxygen radical-induced damage, and chronic benefits of chelation with DFO were observed in a rat model of the disease (BLAKE et al. 1983) although human trials have been disappointing (POLSON et al. 1986). There have also been attempts to modify the course of immune-mediated inflammatory processes and graft rejection with DFO, and to limit the tissue damage from free radicals in animal models of bleomycin and paraquat toxicity (HERSHKO and WEATHERALL 1988; PIPPARD 1989).

An important enzyme target for iron chelators is ribonucleotide reductase, an enzyme responsible for providing the nucleotide precursors for DNA synthesis (REICHARD and EHRENBERG 1983). The enzyme contains a redox-active iron at its active center, probably in the form of an accessible μ-oxo-bridged iron dimer (SJÖBERG et al. 1987). Its removal by agents such as DFO has been proposed to underlie the antiproliferative effects of these agents, for example by blocking DNA synthesis in lymphocytes (HOFFBRAND et al. 1976). Presumably for this reason, DFO was temporarily effective in controlling drug-resistant acute leukemia (ESTROV et al. 1987). Acquired porphyria cutanea tarda responds both to phlebotomy and DFO infusion (ROCCHI et al. 1986). In this case the disease can result from inhibition of hepatic uroporphyrinogen decarboxylase by iron. Removal of the allosteric inhibitor by chelation restores enzyme activity. *Plasmodium falciparum* and other *Plasmodium* species are strongly iron dependent, and DFO increases clearance of the parasite in adults. In childhood cerebral malaria, a common and frequently fatal complication, DFO infusion enhances the loss of the parasite and increases the rate of recovery from deep coma (GORDEUK et al. 1992b). Melanin pigments bind iron, and the consequent increase in iron in the substantia nigra of patients with Parkinson's disease has been proposed to contribute to the neurodegeneration (BEN-SHACHAR et al. 1992). However, deferiprone (1,2-dimethyl-3-hydroxypyridin-4-one, [III]), but not DFO, was found to inhibit the iron-dependent tyrosine and tryptophan hydroxylases (WALDMEIER et al. 1993) and therefore interfere with dopamine and serotonin metabolism, a potential complication in contemplating chelation therapy for parkinsonism. With the growing awareness of the potential involvement of iron in cardiovascular disease (SALONEN et al. 1992) and other catastrophic illness (CONRAD 1993), experimentation with iron chelation will undoubtedly reach other medical disciplines.

D. Structural Considerations for Iron-Specific Chelators

Under oxidizing conditions and physiological pH, Fe^{3+} is favored over Fe^{2+} and the problem of iron chelation is one of stabilizing ferric ion. In the "hard and soft" classification of ions, Fe^{3+} is hard [its Klopman hardness parameter is 2.22 eV, comparable to Sr^{2+} and Ca^{2+} (KLOPMAN 1968)] and it therefore forms more stable complexes with harder ligands such as oxygen. The d^5 electronic configuration of Fe^{3+} favors octahedral coordination, almost invariably in the high-spin state with poly-oxo chelates (RAYMOND and CARRANO 1979). The oxygen molecular orbitals and those on the metal ion best overlap in octahedral coordination when they are separated by a three-bond spacing. Hider and coworkers have called this the optimum ligand bite size (e.g., BARLOW et al. 1988). It is therefore not surprising that the most stable ferric ion chelates are based on functionalities like catechol, hydroxamate, and α-hydroxy ketone (HIDER and HALL 1991; MARTELL 1981; PITT 1981; PORTER et al. 1989; RAYMOND and CARRANO 1979). The structures of some natural siderophores (Fig. 1) and synthetic iron chelators (Fig. 2) discussed here show the ubiquity of these structures in stable iron complexes. Most natural siderophores contain three such structures, thereby achieving a strong chelate effect as hexadentate ligands. Because three bidentate chelators are required per atom of Fe^{3+}, there is a negative entropic contribution to the binding that generally lowers the stability constants. However, bidentate chelators can relax into optimal geometry around the iron center, and a flexible hexadentate like DFO also demon-

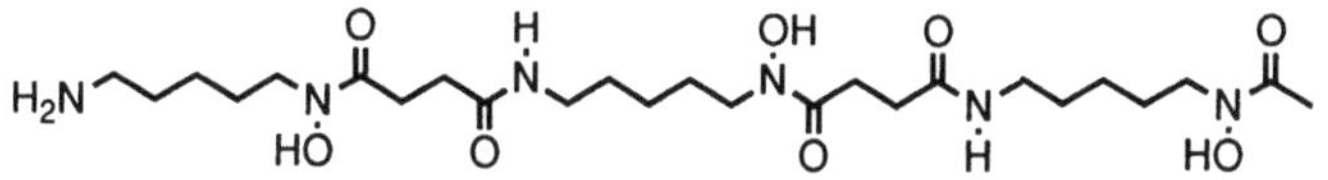

[I]

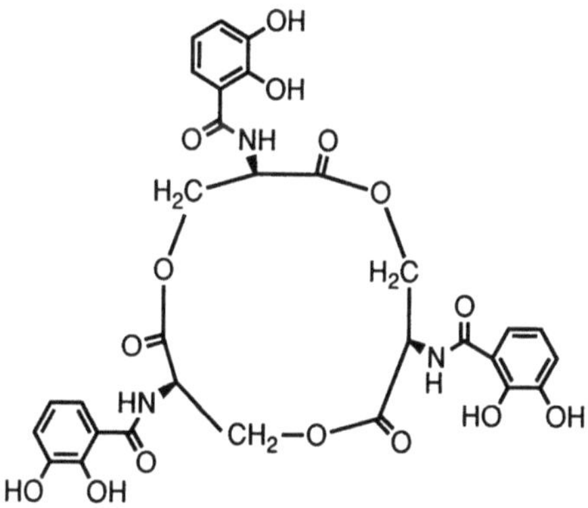

[II]

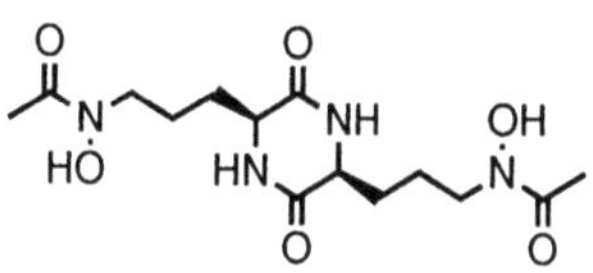

[VI]

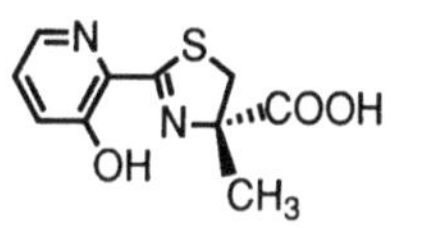

[VIII]

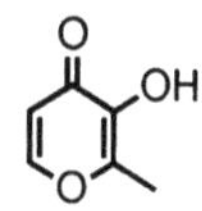

[IX]

Fig. 1. Some naturally occurring iron chelators. Structures are numbered for reference in the text. **I**, desferrioxamine B; **II**, enterobactin; **VI**, rhodotorulic acid; **VIII**, desferrithiocin; **IX**, maltol

strates an entropic effect on binding iron. When the coordinating oxygens in a hexadentate chelator are held in a more rigid cage in the proper geometry, their Fe^{3+} chelates are among the most stable metal complexes known. These effects are seen in the series deferiprone, DFO, and enterobactin, whose log stability constants with Fe^{3+} are 35.9, 31.0 (MOTEKAITIS and MARTELL 1991), and 52 (COOPER et al. 1978), respectively. While such extreme stability constants are unimportant for therapeutic iron chelation – there being no practical advantage of a stability of 10^{50} over one of 10^{30}, for example – they do confer an advantage to the microorganisms synthesizing the siderophore in that they allow the accumulation of Fe^{3+} from solutions

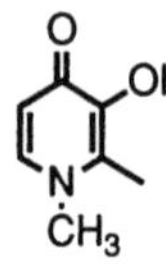

[III]

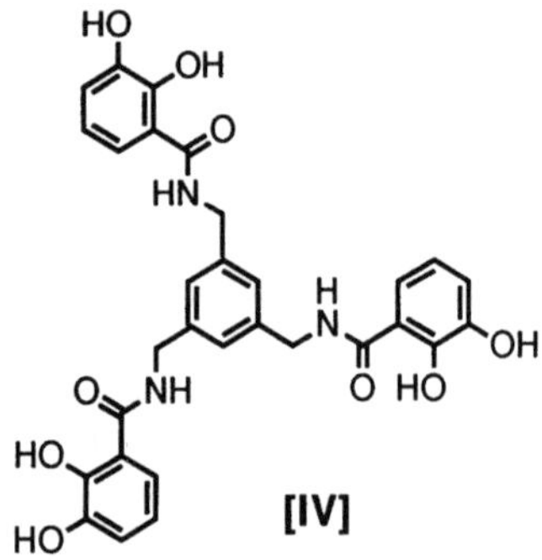

[IV]

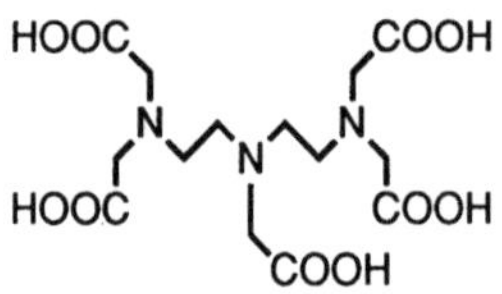

[V]

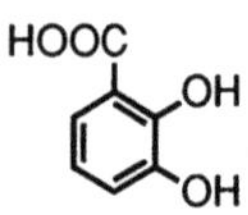

[VII]

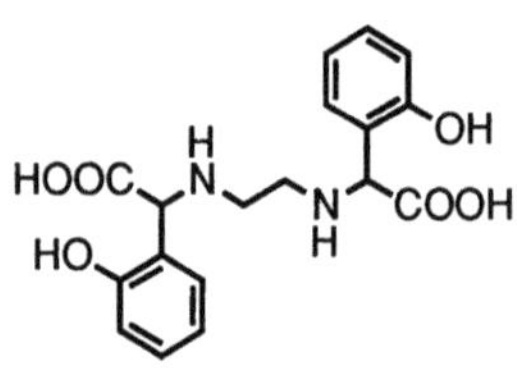

[X]

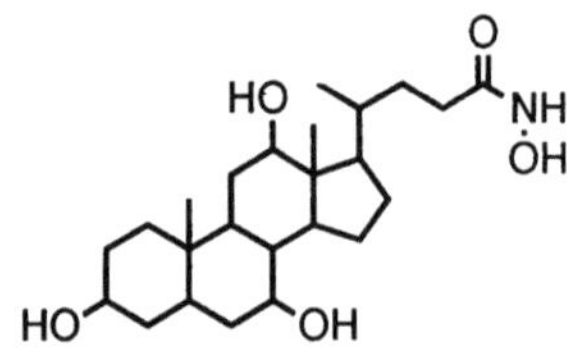

[XI]

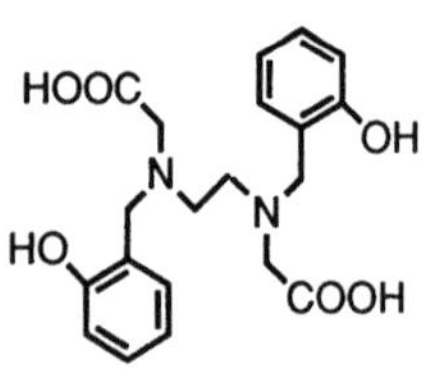

[XII]

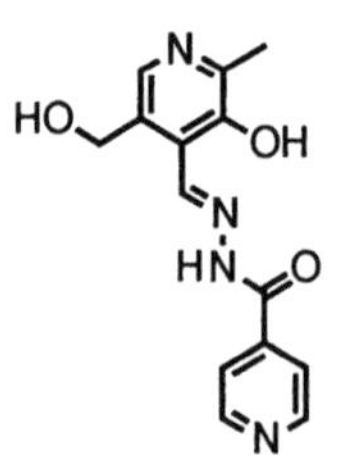

[XIII]

Fig.2. Some synthetic iron chelators. Structures are numbered for reference in the text. **III**, deferiprone; **IV**, MECAM; **V**, DTPA; **VII**, 2,3-DHB; **X**, EHPG; **XI**, cholylhydroxamic acid; **XII**, HBED; **XIII**, PIH

with extremely low concentrations. The contributions of various enthalpic and entropic effects to the stabilities of chelates in solution have been reviewed by MARTELL (1981).

The stability constant of a complex $M_\mu L_\lambda$, where M is metal, L ligand, and the subscripts are the corresponding stoichiometric coefficients is defined as:

$$\beta_{\mu\lambda} = [M_\mu L_\lambda]/[M]^\mu[L]^\lambda.$$

A comparison of β_n values for bi- ($n = 3$) and hexadentate ($n = 1$) chelators can be misleading. For example, log β_3 of deferiprone is 35.9 but the log of the third stepwise formation constant given by $\log(\beta_3/\beta_2)$ is only 9.7 (MOTEKAITIS and MARTELL 1991). Also, this definition of stability constant does not take into account the different acidities of the ligands and the ability of iron to compete for them with proton. Protonation of the ligand and hydrolysis of the metal, as well as competition with other metals and ligands in biological systems, complicate the interpretation of stability constants. Therefore, in comparing the stability of iron chelates it is useful to introduce the additional terms K_{eff} and pM. Martell has defined an effective stability constant (K_{eff}) for Fe^{3+} complexes based on competition for the ligand with Ca^{2+} (MARTELL 1981). Other definitions are possible. Comparison of $\log K_{eff}$ values [e.g., 25 and 16.3 for enterobactin and DFO, respectively (MARTELL 1981)] often illustrates that differences between chelators are not as great as they might appear from comparison of the β values. It is also important to remember that these equilibrium constants reflect the thermodynamic stability of the complexes and are useful for predicting competition between different iron-binding sites, but do not reflect the exchangeability of the chelated Fe^{3+}. Raymond and Caarrano have pointed out that the absence of crystal-field stabilization energy in the high-spin d^5 configuration makes the complexes kinetically labile with respect to ligand exchange in aqueous solution (RAYMOND and CARRANO 1979). This has clear implications for exchange of mobilized iron between chelates and endogenous acceptors that could lead to redistribution rather than removal of iron.

Species distributions of iron chelates are also concentration dependent. Dissociation to free ligand is more favorable at low concentration and bidentate chelators, where three molecules are involved, show this effect more strongly. To compare various chelators under identical conditions and from a practical point of view, Raymond and coworkers (RAYMOND et al. 1981; RAYMOND and XU 1994) have used $pM = -\log[M]_{free}$, the concentration of uncomplexed Fe^{3+} persisting in the presence of ligand under fixed conditions (typically pH 7.4, T = 25°C and a tenfold excess of ligand to metal). Values of pM with enterobactin and DFO are invariant between the millimolar and micromolar concentration range, at 37.6 and 26.3, respectively, while the bidentate deferiprone shows the dilution effect: its value decreases from 24.3 to 18.3 on going from 1 mM Fe^{3+} to 1 μM Fe^{3+} when ligand : metal = 3 : 1 (MOTEKAITIS and MARTELL 1991). Importantly, this change in the stability of the complex reflects a change in the speciation profile, so that in the millimolar concentration range ML_3 accounts for nearly 100% of the iron but in the micromolar range ML_3 and ML_2^+ exist in nearly equal comcentrations. Because the latter is a likely candidate for a Fenton catalyst (see below), the safety of bidentate chelators may be concentration dependent.

Values of pM for most iron chelators are high enough that uncomplexed Fe^{3+} is present in insignificant amounts, and even in the absence of chelators the solubility product of $Fe(OH)_3$ [$K_{sp} \approx 10^{-38}$ (LATIMER 1952)] is so low that the equilibrium concentration of Fe^{3+} at neutarlity is $\approx 10^{-18}\,M$. Nevertheless, the iron chelates themselves are potentially Fenton-active. GRAF et al. (1984) have discussed the importance for Fenton activity of H_2O or an easily displaceable ligand in the first coordination sphere in iron complexes that allows access of $O_2^{-\bullet}$ to Fe^{3+} and H_2O_2 to the Fe^{2+} center. This will obviously be favored in bidentate chelates, and especially under circumstances where significant proportions of the mono- and bis-complexes exist. The bidentate 3-hydroxypyrid-4-ones have Fenton activity comparable to some endogenous ligands [e.g., between that of citrate and ATP (SINGH and HIDER 1988)]. In contrast, a hexadentate pyridin-2-one has low Fenton activity (SINGH and HIDER 1988) and Fe^{3+}-ferrioxamine displays little or no free radical generating capacity (BURKITT et al. 1993; GRAF et al. 1984; SINGH and HIDER 1988); although Fe^{2+}-ferrioxamine can reduce paraquat, it is not re-reduced and does not redox cycle (BURKITT et al. 1993). Therefore, a hexadentate structure is not sufficient to suppress the Fenton reaction. In fact, Fe^{3+}-EDTA catalyzes the generation of HO• at a rate many times that of most lower order chelators (GRAF et al. 1984; HALLIWELL and GUTTERIDGE 1990; SINGH and HIDER 1988).

E. Biological Considerations for Iron Removal

The molecular pattern of iron deposition in tissues determines its accessibility to chelation, and depends upon the source of iron. Excessive intestinal absorption increases circulating levels of diferric transferrin, which deposits iron in hepatocytes and the parenchymal cells of other organs. Deposition in the reticuloendothelial system (RES) occurs only in advanced disease. On the other hand, transfused red cells are turned over in the RES and iron accumulates initially in the bone marrow, spleen, and Kupffer cells of the liver. In patients with aplastic anemia, loading is purely transfusional. However, in thalassemics with ineffective erythropoiesis there is a compensatory increase in intestinal iron absorption, and iron overload can occur even in the absence of transfusion (ELLIS et al. 1954; OLIVIERI et al. 1992b). Therefore, even initial iron deposition in transfused thalassemics affects both RES and parenchymal cells, for example both the Kupffer cells and hepatocytes of the liver.

Ferritin synthesis is stimulated by available iron. In the absence of Fe^{3+}, the apo- form of an iron-binding protein binds to ferritin mRNA and blocks its translation (CAUGHMAN et al. 1988; DIX et al. 1992; LEIBOLD and GUO 1992). However, with excessive iron the maximal rate of translation is unable to accommodate all the metal, and additional iron enters the hemosiderin fraction. Hemosiderin is an insoluble ferric oxo-hydroxide deposit that appears histologically as siderophilic aggregates. It may arise as a degradation product of ferritin (THEIL 1987). If this is taken as part of the definition of hemosiderin, then other siderophilic deposits may also occur. For example, iron delivered rapidly to cells in a non-transferrin-dependent manner may bypass ferritin and enter lysosomal compartments called siderosomes (IANCU et al. 1978; JACOBS et al. 1978) or simply precipitate as hydrolysis products. In the livers of iron-loaded patients up to 90% of the iron accumulates in hemosiderin-like deposits (SELDEN et al. 1980; STUHNE-SEKALEC et al. 1992) that may be a means of protecting the cell from more reactive forms of iron (JACOBS et al. 1978). The normal human liver contains less than 1 mg iron/g dry weight but in highly loaded thalassemics with a high hemosiderin content this may increase 30-fold or more (BRITTENHAM et al. 1993; SELDEN et al. 1980; STUHNE-SEKALEC et al. 1992). The hemosiderin pool can be mobilized under conditions of negative iron balance, achieved by phlebotomy (FAIRBANKS et al. 1971) or chelation (OLIVIERI et al. 1992b). However, it is relatively inert and may limit the rate at which iron can be removed from the body.

In plasma, transferrin iron is tightly bound, but non-transferrin-bound iron should be readily accessible to chelators. Likewise, in tissues iron in transit to or from the ferritin/hemosiderin pool may serve as the source of chelatable iron. Much of what is known about the kinetics of iron in this transit pool has been derived from studies of its accessibility to DFO. Iron delivered to the endosome as diferric transferrin is rapidly available for chelation (PIPPARD et al. 1982), but ferritin iron is not readily removed. When ferritin is given to hepatocytes via transferrin it is available to DFO only several hours later after some lysosomal degradation has occurred (PIPPARD et al. 1982). Citrate, ATP, and ADP are likely candidates for binding the released iron; citrate has been proposed to be the major form of non-transferrin-bound iron in plasma (GROOTVELD et al. 1989). It has also been proposed that approximately $2\,\mu M$ iron in normal rat liver exists in a trimetallic oxoiron cluster with free glutamate and aspartate (DEIGHTON and HIDER 1989). These low molecular mass complexes are the likely source of the chelatable transit pool to and from ferritin. Whatever the nature of the chelatable pool, it must be in equilibrium with the storage pools as they too can be removed. By equilibrium considerations, expansion of the ferritin add/or hemosiderin-like fractions will then also cause an expansion of this hypothetical chelatable pool of iron. It also seems likely that this exchangeable iron is available to initiate Fenton chemistry. It is perhaps fortunate that the same source of iron that initiates radical-mediated tissue damage is most amenable to chelation.

Recent Mössbauer studies by SHILOH et al. (1992) have provided new insight into the time course of iron deposition at the supramolecular, subcellular level. When cultured myocardial cells were loaded with iron it initially entered particles with a diameter less than 3 nm, with transfer to larger ferritin particles occurring over several days. This is in keeping with the idea that in rapid experimental loading iron is at first accumulated in hydrolyzed precipitates, but then finds its way into the more physiologically relevant ferritin core. The increasing inaccessibility of iron to chelation by DFO roughly paralleled its transfer into the ferritin pool, and a "last in–first out" principle (PARKES et al. 1993; SHILOH et al. 1992) probably governs chelation in these experimental culture systems. However, it is not yet clear whether such considerations are relevant to clinical iron overload where iron delivery at the cellular level may be a much more subtle process, and removal may depend on shifting rather one-sided equilibria. The nature of the chelatable iron pool in clinical overload is one of the major questions in iron metabolism.

Although sequestration of circulating non-transferrin-bound iron in a Fenton-inactive form is itself of potential benefit, the depletion of body iron stores requires removal of cellular iron. Therefore, an ideal chelator should enter the cell in its iron-free form and leave upon binding iron. A complex that becomes charged on complexation is at risk of becoming trapped in the cell by iron. The hydroxypyridones have an advantage in this regard; with phenolic pK_as of 9–10 they are neutral at physiological pH and readily absorbed after oral administration. On binding Fe^{3+} loss of one proton per ligand creates a neutral complex that should also easily pass cell membranes to remove iron down its concentration gradient. This is in contrast to catechol, which forms a charged complex with iron that can become trapped in cells.

The principle of optimal hydrophobicity is illustrated by comparing the octanol: water partition coefficients of a series of N-substituted 2-methyl-3-hydroxypyrid-4-ones with their ability to remove iron from hepatocyte cultures (PORTER et al. 1988; PORTER et al. 1986). Maximal removal of iron from the cells was achieved with derivatives having K_{part} values between 0.5 and 1.5 for both the free form and the Fe(III) chelate. Less lipophilic compounds do not enter the cell as readily. Too great a degree of lipophilicity may caues the chelator to partition into the membrane and not access cytosolic iron pools. Molecular size is also an important feature determining the efficacy of iron chelators. Levin has studied the effects of size and lipophilicity on permeability of the blood-brain barrier, assuming a model of diffusion from bulk aqueous phase to a lipid phase (LEVIN 1980). Permeability was proportional to $(K_{part})/(M_r)^{1/2}$ up to about 400 daltons. At higher mass, permeability decreases as for example membrane deformation becomes necessary for penetration. Similar size constraints apply to transcellular passage of molecules undergoing intestinal absorption (HIDER et al. 1994), limiting the oral availability of agents such as DFO. Although cyclic hexadentate chelators are desirable for their ability to suppress Fenton chemistry and their high specificity for iron, these physical considerations may limit their absorption and availability to cells. A minimal moiety, e.g., hydroxamate (HO-NHCOCH$_2$-; $3 \times M_r = 222$) with sufficient methylene bridging to achieve an octahedral disposition of ligands [$(CH_2)_5$; $3 \times M_r = 210$], already approaches the critical mass for membrane exclusion.

While chemical principles can be used to design chelators that form stable and specific Fe^{3+} chelates, uncertainty about the nature and location of the chelatable iron pool and constraints on delivery of suitable chelators to the site of action have determined that iron chelator design is still dominated by empirical testing of structure-function relationships. This in itself adds a new challenge in that there is no perfect animal model for the human iron overload syndromes. PITT (1981) has pointed out that rodents Fe-loaded with heat-damaged erythrocytes have been used most frequently to assess the ability of chelators to remove iron by the fecal (biliary) and urinary routes and lower parenchymal (liver) and RES (splenic) stores. He reviews LD_{50} and iron-removal data on many natural and synthetic hydroxamates, phenols, catechols, tropolones, salicylates, benzoates, azines, and carboxylates. No clear picture emerges and the search for the ideal iron chelator continues.

F. Criteria for the Safe Chelation of Iron

We are dependent on the storage of iron in safe forms, and even in iron-overloaded individuals there are protective mechanisms that minimize the damage potentially arising from the iron load. For instance, iron is stored as the relatively inert hemosiderin precipitate, ferritin is induced, and the transferrin receptor is downregulated to minimize transferrin-mediated uptake. On the other hand, uptake of non-transferrin-bound iron is increased in iron-loaded cells (PARKES et al. 1993), a possible mechanism for its rapid clearance from circulation and diversion into inert storage pools. Therefore, attempts to mobilize and remove iron from the body necessarily interfere with intimately balanced protective mechanisms and are potentially dangerous. An ideal chelator should itself be nontoxic, specific for iron, thoroughly suppress its Fenton activity, remove it from cellular pools, and lead to its rapid excretion without redistribution. Predictably, such an agent has not yet been realized.

Assessment of the toxicity of a chelator should in principle be done in an adequate model of the target population of iron-loaded patients. Observations with subjects that lack excessive iron stores may be based on unrealistic concentrations of the uncomplexed chelator, or alternatively show effects of removal of needed baseline iron stores. For example, it is not yet clear whether toxic effects of the hydroxypyridones are an inherent property of the agents themselves, a result of chelation of iron or other trace elements at strategic cellular locations, or a consequence of the mobilization of iron in toxic form. Toxicity of these compounds in animal studies has not been observed to the same extent in significant experience with patients (see below), and the lack of excessive iron in the animals is a plausible explanation (Hershko 1993). The neurotoxicity of DFO is inversely correlated with body iron stores (Polson et al. 1985), suggesting that the free form of the drug is toxic and binding to iron decreases its toxicity.

In considering the redistribution of iron by chelators, the intermolecular, intracellular, and interorgan levels must all be addressed. Much of what is known about the molecular speciation of iron in overloaded humans pertains to liver (Selden et al. 1980; Stuhne-Sekalec et al. 1992) and has been derived from biopsies, although a similar pattern occurs in the heart (Parkes et al. 1993) and is probably general in the soft tissues. Removal of iron by DFO (Selden et al. 1980) or deferiprone (Olivieri et al. 1992b) does not appear to alter the molecular speciation of hepatic iron, suggesting that iron removal is slow with respect to (and probably depends upon) reequilibration of these pools. Because even these highly avid iron chelators probably access the low molecular mass "chelatable iron pool" (see above) but do not prevent reequilibration of the major iron species, it is unlikely that redistribution of iron among its forms within the cell will become an issue in chelator design. Likewise, redistribution of iron among subcellular compartments, although a theoretical consideration in predicting toxicity, is determined by the molecular speciation and therefore not likely to occur during chronic chelation where the pools remain in equilibrium. A further point is that agents that form lipophilic iron complexes could concentrate iron in membranes and enhance membrane lipid peroxidation. However, such lipophilicity would generally trap iron in the cell rather than remove it effectively, and such situations should be identified early in the design of the chelator by poor activity in cell-culture or animal studies.

It should be noted that there is little information on redistribution phenomena in acute iron poisoning and the possibility of harmful redistribution of newly delivered iron within the cell deserves attention. Tenebein et al. reported deaths from adult respiratory distress syndrome in patients receiving DFO for acute iron poisoning (Tenenbein et al. 1992) and reproduced the phenomenon in mice, showing the lung damage to be dependent on iron, DFO, and O_2 (Adamson et al. 1993). It appears that newly delivered iron is redistributed to the lung by the drug. Such occurrences are uncommon in patients chronically iron-loaded and maintained on DFO therapy, but not unknown (Freedman et al. 1990).

More significant is the risk of interorgan redistribution. Chelates forming in the hepatocyte may of course be transported into the bile and excreted in the feces. This is generally considered a positive attribute, but may simply reflect a limited ability to exit the cell. For instance, because ferrioxamine enters and exits cells with difficulty, almost all of the iron this agent removes from the body by biliary excretion (30%–50% of the total removed) originates in the hepatocyte (Hershko et al. 1978). Ferrioxamine that does make its way into the circulation from other sources is rapidly cleared by the kidney. On the one hand, this means that removal of iron from liver is more efficient than from other, and perhaps more sensitive, organs such as the heart. On the other hand, it means that redistribution of hepatic or marrow stores to these sensitive organs is limited. In contrast, the smaller neutral (deferiprone)$_3$Fe complex exits cells readily, but is also an effective means of delivering iron to cardiomyocytes and hepatocytes in culture (J.G. Parkes and D.M. Templeton, unpublished observations). If harmful deposition of mobilized stores is to be avoided in vivo, biokinetics must intervene to ensure that clearance of the complex by glomerular filtration competes successfully with cellular uptake.

It must also be remembered that the natural siderophores are produced by microorganisms, many of them pathogenic, in order to meet their nutritional requirements. Therefore, administration of such an agent potentially enhances the growth of pathogens with the appropriate receptors. A clear example is the occurrence of opportunistic infection with *Yersinia enterocolitica*, a siderophore-dependent organism with ferrioxamine receptors, in patients receiving DFO (GALLANT et al. 1986). A synthetic compound designed to mimic enterobactin, *N*,*N*′,*N*″-tris (2,3-dihydroxybenzoyl)-1,3,5-triaminomethylbenzene (MECAM [IV]) has been abandoned after producing systemic infections with *Escherichia coli* and *Pseudomonas* sp. in experimental animals (GRADY and JACOBS 1981). In principle, chemical modifications remote from the coordination site could be used to inhibit receptor binding of siderophores and analogs without affecting iron chelation, but the siderophores are of sufficient chemical complexity that their large-scale synthetic or semisynthetic production is not in general commercially attractive.

G. Clinically Useful Iron Chelators

Presently there are two major foci in clinical iron chelator research. On the one hand, there are 3 decades of clinical experience with the natural siderophore DFO. On the other hand, several clinical trials of the synthetic deferiprone are underway. DFO is a generally safe and very effective way of removing iron and has made the successful lifelong management of the thalassemias a reality. However, practical problems of cost and bioavailability have encouraged the ongoing pursuit of other agents. Many patients are noncompliant with the regimen of nightly, continuous subcutaneous infusion of the drug, and costs of the drug alone are in excess of $10 000/year. Several other natural siderophores and orally available synthetic compounds have been considered; none has come as far in clinical trials as deferiprone.

I. Natural Siderophores

1. Desferrioxamine B

The ferrioxamines are a group of structurally related polyhydroxamate siderophores produced by genera of the Actinomycetales. Desferrioxamine B (desferrioxamine, deferoxamine, Desferal, DFO; [I]) is secreted in significant quantities by *Streptomyces pilosus* under iron-limiting conditions (BICKEL et al. 1960) and forms a hexadentate iron chelate with $\log\beta = 35.9$. Although two genes involved in the biosynthesis of DFO B have been cloned (SCHUPP et al. 1988), commercial production still depends on the fermentation of *S. pilosus*. This compound was introduced in 1960 (BICKEL et al. 1960) and by 1974 it had been shown that its i.m. administration, accompanied by i.v. diethylenetriaminepentaacetate (DTPA, [V]), was capable of decreasing hepatic iron stores in patients with thalassemia major and transfusional iron overload (BICKEL et al. 1960). A subsequent improvement in iron excretion was achieved with continuous i.v. infusion of DFO alone (PROPPER et al. 1976) and the introduction of portable pumps allowing ambulatory, continuous, subcutaneous infusion of approximately 50 mg/kg overnight (PROPPER et al. 1977) heralded the contemporary mode of therapy. Ascorbate further enhances DFO-induced iron excretion (HUSSAIN et al. 1977), and – with the rationale that it may increase the chelatable pool of Fe^{2+} or assist in removal of ferritin iron by a reductive mechanism – it is now frequently added to chelation programs. The parenteral mode of DFO administration is necessitated by the poor intestinal bioavailability of the drug and its rapid turnover; its initial $T_{1/2}$ in blood is often quoted to be 5–10 min (PIPPARD 1989; SUMMERS et al. 1979), but its clearance from the circulation is now thought to be biphasic with initial and terminal rates of approximately 0.5 and 3 h (LEE et al. 1993). Metabolism, hepatic uptake, enzymatic hydrolysis, and renal clearance all appear to be important, but experimental details have been hampered by the absence of analytical techniques. A major metabolite arises from deaminative oxidation of a terminal amino group (SINGH et al. 1990). Chemical modifications that would improve its bioavailability and stability, and lower its toxicity, continue to be of interest (BERGERON et al. 1992).

There are now patients who have been living for more than 2 decades with DFO, providing a wealth of clinical experience (AKSOY and BIRDWOOD 1985; BRITTENHAM 1992; COHEN 1990; KATTAMIS 1989; WEATHERALL and CLEGG 1981). Regular chelation with DFO has been shown to decrease body iron stores as assessed by hepatic iron (COHEN et al. 1984) and serum ferritin (COHEN 1990), and to improve cardiac (WOLFE et al. 1985) and endocrine (BRONSPIGEL-WEINTROB et al. 1990; DE SANCTIS et al. 1988) function as well as long-term survival (MODELL et al. 1982; ZURLO et al. 1989). Occasional anaphylactic reactions and local reactions at the infusion site occur. In patients poorly compliant with subcutaneous DFO, very high iron stores, or rapidly deteriorating cardiac status, aggressive i.v. administration of 6–12 g/

day has proven beneficial, though very costly (COHEN 1990). Such aggressive therapy increases the risk of more serious side effects, which include ocular and auditory toxicity (OLIVIERI et al. 1986), and CNS toxicity particularly in patients with compromised blood-brain barrier function or low iron stores (HERSHKO and WEATHERALL 1988). Reports of DFO-associated delays in linear growth (DE VIRGILIIS et al. 1988) are difficult to separate from the effects of the disease process (OLIVIERI et al. 1992a), including endocrine dysfunction. The mechanisms of DFO toxicity are unknown (BRITTENHAM 1992; HERSHKO and WEATHERALL 1988; PIPPARD 1989).

2. Other Siderophores

Rhodotorulic acid [VI] is a hydroxamate produced by *Rhodotorula pilimanae* under iron-deficient conditions. It forms 2:1 complexes with Fe^{3+}. Despite its poor solubility it was effective in animal studies, removing iron mainly from hepatocytes. Preliminary trials in humans were stopped when it was found that s.c. or i.m. injections produced a severe and painful local inflammatory response. Enterobactin [II], produced by *E. coli* and *Salmonella typhimurium*, has not been isolated in sufficient quantities for screening and trials as a chelator (HERSHKO and WEATHERALL 1988) but has suggested several synthetic analogs including MECAM (see above) and 2,3-dihydroxybenzoate (2,3-DHB, [VII]), discussed below. Presently the most promising of the siderophores is desferrithiocin [VIII], produced by *Streptomyces antibioticus* and well absorbed after oral administration. It has been shown to be very effective in removing iron from rodents and *Cebus* monkeys loaded with iron dextran (WOLFE 1990) but is generally considered too toxic for human trials (HERSHKO and WEATHERALL 1988). However, the point has again been made that toxicity is decreased in iron-loaded animals (WOLFE 1990). Recently Bergeron et al. have studied several desferrithiocin analogues in iron-loaded rats and monkeys and found the 2-desmethyl derivative to be free of significant toxicity while retaining its iron-removing properties when given orally (BERGERON et al. 1993).

II. Synthetic Chelators

1. Deferiprone

The 3-hydroxypyridin-4-one derivatives were introduced by Hider and colleagues (HIDER et al. 1982; PORTER et al. 1989) a decade ago on the rationale that the vicinal hydroxyketone should have similar ligand spacing to catechol but acquire a single negative charge on deprotonation, thereby forming a neutral tris-complex with iron. The compounds are generally orally available and easy to synthesize. For example, the 2-methyl and 2-ethyl series are prepared by refluxing maltol [IX] or 2-ethyl maltol with a desired alkyl amine. Of many derivatives now prepared and screened, most studies have focused on 1,2-dimethyl-3-hydroxypyridin-4-one (deferiprone, L1, or CP20; [III]) and the corresponding 1,2-diethyl analog (CP94) (FREDENBURG et al. 1993; GYPARAKI et al. 1986; HUEHNS et al. 1988; PORTER et al. 1988; PORTER et al. 1990), both of which remove hepatocellular iron via the bile and cause urinary excretion of RES-iron in experimental animals (ZEVIN et al. 1992). As noted above, these bidentate chelators form stable, neutral 3:1 complexes with Fe^{3+} under physiological pH and ionic strength, although a significant amount of the $(deferiprone)_2Fe^+$ complex occurs at neutral pH on dilution to the micromolar concentration range (MOTEKAITIS and MARTELL 1991). After the demonstration of good chelating properties and a relatively high LD_{50}, deferiprone was chosen for clinical trials in several centers, and there is now over 500 patient-years of experience with the drug (KONTOGHIORGHES et al. 1993).

In 20 transfusionally iron-loaded patients who received 50 mg/kg daily of deferiprone, net iron excretion was less than that achieved with a comparable dose of DFO (OLIVIERI et al. 1990). However, increasing the dose to 75 mg/kg was well tolerated and increased excretion to values comparable to those achieved with 50 mg/kg DFO, sufficient to maintain negative iron balance. Fecal excretion was variable. Similar results have been reported at doses of 55–80 mg/kg (TONDURY et al. 1990). One thalassemia intermedia patient who was nontransfusionally dependent but nevertheless had a liver iron concentration of more than 15 times normal was found to revert to near normal values after 9 months of chelation with deferiprone (OLIVIERI et al. 1992b). Significant decreases in serum ferritin have not been found consistently (KONTOGHIORGHES et al. 1990; TONDURY et al. 1990). However, at a daily dose of about 100 mg/kg, in the longest running clinical trial to date, it has now been shown that clinically useful iron excretion, depletion of iron stores, and statistically significant lowering of serum ferritin can be achieved with deferiprone (AL-REFAIE et al. 1992).

Trials of deferiprone were undertaken in several countries prior to the usual toxicological evaluation and based on preliminary studies in rodents. Brittenham has concluded that the animal data suggest a delicate balance between safety and efficacy (BRITTENHAM 1992). Ciba-Geigy has now completed toxicity testing in rats, rabbits, and Cynomolgus monkeys, necessary for coordinated worldwide trials. At $100\,mg\,kg^{-1}\,day^{-1}$, rats experienced organ – particularly thymic – atrophy and marrow suppression. Teratogenicity was reported even at 25 mg/kg. No therapeutic safety margin was found. They conclude that further development of the drug is not justified (BERDOUKAS et al. 1993). A different view has been put forward by Hershko on behalf of the International Study Group on Oral Iron Chelators (HERSHKO 1993). He notes that, in contrast to the serious toxicity seen in the animal studies, limited clinical toxicity has been observed in iron-loaded patients. The latter includes several cases of agranulocytosis – with full recovery – and no drug-related fatalities in the major British, Canadian, and Swiss trials. Arthralgia, a common side effect in the patients, has no correlate in the animal studies. It is the view of the International Study Group that more prospective trials of deferiprone are needed, and it is likely that other 3-hydroxypyridin-4-one derivatives will come to clinical trial in future.

2. Other Synthetic Chelators

One of the first synthetic chelators used effectively in patients was the hexadentate *N,N'*-ethylene-bis(2-hydroxyphenylglycine) (EHPG,[X]) (CLETON et al. 1963). It has now been rendered obsolete by less toxic agents. DTPA was also used in the earlier years of treatment of transfusional iron overload (MCLAREN et al. 1981) and, as described above, was used in combination with DFO in early trials of the latter. Unlike EHPG, it is still an alternative to DFO in patients with drug allergy but has several drawbacks (HERSHKO and WEATHERALL 1988). It does not enter the hepatocyte, deriving iron from the RES and clearing it exclusively in the liver. It is not absorbed orally and causes painful local reactions after i.m. injection. Its specificity for iron is poor, potentially causing deficiency of trace elements; it is administered i.v. as the Ca^{2+} salt and often requires zinc supplementation.

Other agents that have come to clinical trial with poor results include 2,3-DHB, a bidentate catechol designed to mimic the Fe^{3+}-binding moieties of enetrobactin. Extremely nontoxic and effective orally in iron removal from experimental animals, it was given to eight thalassemic patients for 1 year (25 mg 2,3-DHB/kg, p.o., q.i.d.) (PETERSON et al. 1979). Disappointingly, there was no difference in the iron status between the patients and untreated controls. Cholylhydroxamic acid [XI] was designed to be absorbable in the gut and remove iron by enterohepatic circulation (GRADY and JACOBS 1981). Because it is nonpolar it is not lost by renal clearance but its more hydrophilic iron complex is poorly reabsorbed in the intestine and is therefore excreted in the feces. Despite ambiguous effectiveness in animal studies (HERSHKO and WEATHERALL 1988), it underwent preliminary clinical trials (25 mg/kg, p.o., q.i.d.) that were abandoned because of problematic diarrhea (GRADY and JACOBS 1981).

N,N'-bis(2-Hydroxybenzoyl)ethylenediamine-*N,N'*-diacetate (HBED, [XII]) is an analog of EDTA in which two of the carboxylic acid groups have been replaced by phenolic groups, improving its lipophilicity. The compound is absorbed in the gut, has a greatly improved stability (log $\beta = 40$) and specificity for Fe^{3+}, is effective in animal models of iron overload, and is quite nontoxic (GRADY and HERSHKO 1990). It has now been administered to four patients in limited trials at 40 and 80 mg/kg, p.o., t.i.d. (GRADY et al. 1994). At these doses, negative iron balance was not achieved. The dimethyl ester of HBED is more readily absorbed than HBED itself and is converted to HBED by esterase activity in the body. GRADY and coworkers are now considering a limited trial with this derivative.

PONKA and coworkers (PONKA et al. 1979) discovered that condensation of pyridoxal with isonicotinoyl hydrazide produced a potent Fe^{3+} chelator, pyridoxal isonicotinoyl hydrazone (PIH, [XIII]). It is orally available, removes iron mainly from the hepatocyte, does not undergo enterohepatic circulation, and has an LD_{50} in rats greater than 800 mg/kg when given i.p. (BRITTENHAM 1990). In early clinical trials, 600 mg PIH given p.o. every 8 h was well tolerated but lacked the expected efficacy, probably due to poor absorption. Further trials await improved formulations or chemical modifications to improve bioavailability.

H. Summary

The major clinical use for chronic iron chelation is in iron overload resulting from increased absorption and multiple transfusions in patients with ineffective erythropoiesis. For this, as well as acute iron poisoning, parenteral administration of DFO is the accepted therapy. As Wolfe has pointed out (WOLFE 1990), the safety and efficacy of this drug have set high standards for the introduction of new agents. Nevertheless, the search for such agents continues actively due to the cumbersome mode of DFO administration, as well as its occasional toxicity and high cost. Technical difficulties in gene replacement or gene switching approaches to the underlying disorders indicate that chelation will be lifesaving for at least another generation of thalassemic patients. As our appreciation increases of the roles iron chemistry and enzymology play in pathology, specific iron chelators should find novel uses. The challenge remains to design an orally active hexadentate chelator capable of accessing parenchymal iron stores, fully suppressing Fenton chemistry, and effectively removing iron without redistribution. The final solution may not be found among today's leading candidates – modified desferrioxamines, hydroxypyridones, desferrithiocins, ethylenediamines, and isonicotinoyl hydrazones – though each may have something to offer along the way.

Acknowledgements. Work on cellular iron uptake and removal in the author's laboratory is supported by grants from the Medical Research Council of Canada and the Heart and Stroke Foundation of Ontario.

References

Adamson IYR, Sienko A, Tenenbein M (1993) Pulmonary toxicity of deferoxamine in iron-poisoned mice. Toxicol Appl Pharmacol 120:13–19
Aisen P (1992) Entry of iron into cells: a new role for the transferrin receptor in modulating iron release from transferrin. Ann Neurol 32:S62–S68
Aksoy M, Birdwood GFB (eds) (1985) Hypertransfusion and iron chelation in thalassemia. Huber, Berne
Al-Refaie FN, Wickens DG, Wonke B, Kontoghiorghes GJ, Hoffbrand AV (1992) Serum non-transferrin-bound iron in beta-thalassaemia major patients treated with desferrioxamine and L1. Br J Haematol 82:431–436
Alfrey AC, Froment DH, Hammond WS (1989) Role of iron in the tubulo-interstitial injury in nephrotoxic serum nephritis. Kidney Int 36:753–759

Aust SD, Chignell CF, Bray TM, Kalyanaraman B, Mason RP (1993) Free radicals in toxicology. Toxicol Appl Pharmacol 120:168–178
Babbs CF (1985) Role of iron ions in the genesis of reperfusion injury following successful cardiopulmonary resuscitation: preliminary data and a biochemical hypothesis. Ann Emerg Med 14:777
Baes CF Jr, Mesmer RE (1986) The hydrolysis of cations. Krieger, Malabar
Barlow DJ, Hider RC, Singh S, Wibley KS (1988) Computer-aided modelling of the structures and metal-ion affinities of chelating agents. Biochem Soc Trans 16: 835–836
Behringer RR, Ryan TM, Palmiter RD, Brinster RL, Townes TM (1990) Human gamma- to beta-globin gene switching in transgenic mice. Genes Dev 4:380–389
Ben-Shachar D, Eshel G, Riederer P, Youdim MBH (1992) Role of iron and iron chelation in dopaminergic-induced neurodegeneration: implication for Parkinson's disease. Ann Neurol 32 [Suppl]:S105–S110
Berdoukas V, Bentley P, Frost H, Schnebli HP (1993) Toxicity of oral iron chelator L1. Lancet 341:1088
Bergeron RJ, Liu Z-R, McManis JS, Wiegand J (1992) Structural alterations in desferrioxamine compatible with iron clearance in animals. J Med Chem 35: 4739–4744
Bergeron RJ, Streiff RR, Creary EA, Daniels RD Jr, King W, Luchetta G, Wiegand J, Moerker T, Peter HH (1993) A comparative study of the iron-clearing properties of desferrithiocin analogues with desferrioxamine B in a Cebus monkey model. Blood 81:2166–2173
Bickel H, Gäumann E, Keller-Schierlein W, Prelog V, Vischer E, Wettstein A, Zähner H (1960) Iron-containing growth factors, the sideramines, and their antagonists, the iron-containing antibiotics, sideromycins. Experientia 16:129–133
Biemond P, van Eijk HG, Swaak AJG, Koster JF (1984) Iron mobilization from ferritin by superoxide derived from stimulated polymorphonuclear leukocytes. J Clin Invest 73:1576–1579
Blake DR, Hall ND, Bacon PA, Dieppe PA, Halliwell B, Gutteridge JMC (1983) Effect of a specific iron chelating agent on animal models of inflammation. Ann Rheum Dis 42:89–93
Brieland JK, Clarke SJ, Karmiol S, Phan SH, Fantone JC (1992) Transferrin: a potential source of iron for oxygen free radical-mediated endothelial cell injury. Arch Biochem Biophys 294:265–270
Brittenham GM (1990) Pyridoxal isonicotinoyl hydrazone: an effective iron-chelator after oral administration. Semin Hematol 27:112–116
Brittenham GM (1992) Development of iron-chelating agents for clinical use. Blood 80:569–574
Brittenham GM, Cohen AR, McLaren CE, Martin MB, Griffith PM, Nienhuis AW, Young NS, Allen CJ, Farrell DE, Harris JW (1993) Hepatic iron stores and plasma ferritin concentration in patients with sickle cell anemia and thalassemia major. Am J Hematol 42:81–85
Bronspigel-Weintrob N, Olivieri NF, Tyler B, Andrews DF, Freedman MH, Holland JF (1990) Effect of age at the start of iron chelation therapy on gonadal function in β-thalassemia major. N Engl J Med 323:713–719
Burkitt MJ, Kadiiska MB, Hanna PM, Jordan SJ, Mason RP (1993) Electron spin resonance spin-trapping investigation into the effects of paraquat and desferrioxamine on hydroxyl radical generation during acute iron poisoning. Mol Pharmacol 43:257–263
Caughman SW, Hentze MW, Rouault TA, Harford JB, Klausner RD (1988) The iron-responsive element is the single element responsible for iron-dependent translational regulation of ferritin biosynthesis. Evidence for function as the binding site for a translational repressor. J Biol Chem 263:19048–19052

Cleton F, Turnbull A, Finch CA (1963) Synthetic chelating agents in iron metabolism. J Clin Invest 42:327–337
Cohen A (1990) Current status of iron chelation therapy with deferoxamine. Semin Hematol 27:86–90
Cohen A, Martin M, Schwartz E (1984) Depletion of excessive liver iron stores with desferrioxamine. Br J Haematol 58:369–373
Conrad ME (1993) Excess iron and catastrophic illness. Am J Hematol 43:234–236
Conrad ME, Umbreit JN (1993) Iron absorption – the mucin-mobilferrin-integrin pathway. A competitive pathway for metal absorption. Am J Hematol 42:67–73
Conrad ME, Umbreit JN, Peterson RDA, Moore EG, Harper KP (1993) Function of integrin in duodenal mucosal uptake of iron. Blood 81:517–521
Cooper SR, McArdle JV, Raymond KN (1978) Siderophore electrochemistry: relation to intracellular iron release mechanism. Proc Natl Acad Sci USA 75:3551–3554
De Sanctis V, Zurlo MG, Senesi E, Boffa C, Callao L, Di Gregorio F (1988) Insulin-dependent diabetes in thalassemia. Arch Dis Child 63:58–62
De Virgiliis S, Congia M, Frau F, Argiolu F, Diana G, Cucca F, Varsi A, Sanna G, Podda G, Fodde M, Pirastu GF, Cao A (1988) Deferoxamine-induced growth retardation in patients with thalassemia major. J Pediatr 113:661–669
Deighton N, Hider RC (1989) Intracellular low molecular weight iron. Biochem Soc Trams 17:490
Dix DJ, Lin P-N, Kimata Y, Theil EC (1992) The iron regulatory region of ferritin mRNA is also a positive control element for iron-independent translation. Biochemistry 31:2818–2822
Ellis JT, Schulman I, Smith CH (1954) Generalized siderosis with fibrosis of the liver and pancreas in Cooley's (Mediterranean) anemia with observations on pathogenesis of siderosis and fibrosis. Am J Pathol 30:287–309
Enver T, Raich N, Ebens AJ, Papayannopoulou T, Constantini F, Stamatoyannopoulos G (1990) Developmental regulation of human fetal-to-adult globin gene switching in transgenic mice. Nature 344:309–313
Esterbauer H (1993) Cytotoxicity and genotoxicity of lipid-peroxidation products. Am J Clin Nutr 57 [Suppl]:779S–786S
Estrov Z, Tawa A, Wang XH, Dube ID, Sulh H, Cohen A, Gelfand EW, Freedman MH (1987) In vitro and in vivo effects of deferoxamine in neonatal acute leukemia. Blood 69:757–761
Fairbanks VF, Fahey JL, Beutler E (1971) Clinical disorders of iron metabolism, 2nd edn. Grune and Stratton, New York
Flanagan PR (1990) Intestinal iron absorption and metabolism. In: Ponka P, Schulman HM, Woodworth RC (eds) Iron transport and storage. CRC, Boca Raton, pp 247–261
Fredenburg AM, Wedlund PJ, Skinner TL, Damani LA, Hider RC, Yokel RA (1993) Pharmacokinetics of representative 3-hydroxypyridin-4-ones in rabbits: CP20 and CP94. Drug Metab Dispos 21:255–258
Freedman MH, Grisaru D, Olivieri N, MacLusky I, Thorner P (1990) Pulmonary syndrome in patients with thalassemia major receiving deferoxamine infusions. Am J Dis Child 144:565–569
Funk F, Lenders J-P, Crichton RR, Schneider W (1985) Reductive mobilisation of ferritin iron. Eur J Biochem 152:167–172
Gallant T, Freedman MH, Vellend H, Francombe WH (1986) Yersinia sepsis in patients with iron overload treated with deferoxamine. N Engl J Med 314:1643
Gordeuk V, Mukiibi J, Hasstedt SJ, Samowitz W, Edwards CQ, West G, Ndambire S, Emmanual J, Nkanza N, Chapanduka Z, Randall M, Boone P, Romano P, Martell RW, Yamashita T, Effler P, Brittenham G (1992a) Iron overload in Africa – interaction between a gene and dietary iron content. N Engl J Med 326:95–100
Gordeuk V, Thuma P, Brittenham G, McLaren C, Parry D, Backenstose A, Biemba G, Msiska R, Holmes L, McKinley E, Vargas L, Gilkeson R, Poltera AA

(1992b) Effect of iron chelation on recovery from deep coma in children with cerebral malaria. N Engl J Med 327:1473–1477

Grady RW, Hershko C (1990) An evaluation of the potential of HBED as an orally effective iron-chelating drug. Semin Hematol 27:105–111

Grady RW, Jacobs A (1981) The screening of potential iron chelating drugs. In: Martell AE, Anderson WF, Badman DG (eds) Development of iron chelators for clinical use. Elsevier/North-Holland, Amsterdam, pp 133–164

Grady RW, Salbe AD, Hilgartner MW, Giardina PJ (1994) Preliminary results from a phase I clinnical trial of HBED. In: Bergeron RJ, Brittenham GM (eds) The development of iron chelators for clinical use. CRC Press, Boca Raton, pp 395–406

Graf E, Mahoney JR, Bryant RG, Eaton JW (1984) Iron-catalyzed hydroxyl radical formation: stringent requirement for free iron coordination site. J Biol Chem 259:3620–3624

Grootveld M, Bell JD, Halliwell B, Aruoma OI, Bomford A, Sadler PJ (1989) Non-transferrin-bound iron in plasma or serum from patients with idiopathic hemochromatosis. Characterization by high performance liquid chromatography and nuclear magnetic resonance spectroscopy. J Biol Chem 264:4417–4422

Gyparaki M, Porter JB, Huehns ER, Hider RC (1986) Evaluation in vivo of hydroxypyrid-4-one iron chelators intended for the treatment of iron overload by the oral route. Biochem Soc Trans 14:1181–1181

Hallaway PE, Eaton JW, Panter SS, Hedlund BE (1989) Modulation of deferoxamine toxicity and clearance by covalent attachment to biocompatible polymers. Proc Natl Acad Sci USA 86:10108–10112

Halliwell B, Gutteridge JMC (1990) Role of free radicals and catalytic metal ions in human disease: an overview. Methods Enzymol 186:1–88

Hanscombe O, Whyatt D, Fraser P, Yannoutsos N, Greaves D, Dillon N, Grosveld S (1991) Importance of globin gene order for correct developmental expression. Genes Dev 5:1387–1394

Hedlund BE, Hallaway PE (1993) High-dose systemic iron chelation attenuates reperfusion injury. Biochem Soc Trans 21:340–343

Hershko C (1993) Development of oral iron chelator L1. Lancet 341:1088–1089

Hershko C, Weatherall DJ (1988) Iron-chelating therapy. CRC Crit Rev Clin Lab Sci 26:303–346

Hershko C, Grady RW, Cerami A (1978) Mechanism of iron chelation in the hypertransfused rat: definition of two alternative pathways of iron mobilization. J Clin Lab Sci 92:144–151

Hershko C, Link G, Pinson A (1987) Modification of iron uptake and lipid peroxidation by hypoxia, ascorbic acid, and α-tocopherol in iron-loaded rat myocardial cell cultures. J Lab Clin Med 110:355–361

Hider RC, Hall AD (1991) Clinically useful chelators of tripositive elements. Prog Med Chem 28:43–173

Hider RC, Kontoghiorges G, Silver J (1982) Pharmaceutical compositions. UK Patent GB 2118176A

Hider RC, Porter JB, Singh S (1994) The design of therapeutically useful iron chelators. In: Bergeron RJ, Brittenham GM (eds) The development of iron chelators for clinical use. CRC Press, Boca Raton, pp 353–371

Hoffbrand AV, Ganeshaguru K, Hooton JWL, Tattersall MHN (1976) Effect of iron deficiency and desferrioxamine on DNA synthesis in human cells. Br J Haematol 33:517–526

Huehns ER, Porter JB, Hider RC (1988) Selection of hydroxypyridin-4-ones for the treatment of iron overload using in vitro and in vivo models. Hemoglobin 12:593–600

Hughes MN (1972) The inorganic chemistry of biological processes. Wiley, London

Hussain MAM, Green N, Flynn DM, Hoffbrand AV (1977) Effect of dose, time, and ascorbate on iron excretion after subcutaneous desferrioxamine. Lancet 1:977–979

Iancu TC, Shiloh H, Link G, Bauminger ER, Pinson A, Hershko C (1987) Ultrastructural pathology of iron-loaded rat myocardial cells in culture. Br J Exp Pathol 68:53–65
Jacobs A, Hoy T, Humphrys J, Perera P (1978) Iron overload in Chang cell cultures: biochemical and morphological studies. Br J Exp Pathol 59:489–498
Kattamis C (ed) (1989) Iron overload and chelation in thalassemia. Huber Toronto
Klopman G (1968) Chemical reactivity and the concept of charge- and frontier-controlled reactions. J Am Chem Soc 90:223–234
Kontoghiorghes GJ, Bartlett AN, Hoffbrand AV, Goddard JG, Sheppard L, Barr J, Nortey P (1990) Long-term trial with the oral iron chelator 1,2-dimethyl-3-hydroxypyrid-4-one (L1). I. Iron chelation and metabolic studies. Br J Haematol 76:295–300
Kontoghiorghes GJ, Agarwal MB, Tondury P, Kersten MJ, Jaeger M, Vreugdenhil G, Vania A, Rahman YE (1993) Future of oral iron chelator deferiprone (L1). Lancet 341:1479–1480
Latimer WM (1952) Oxidation potentials. Prentice-Hall, Englewood Cliffs
Lee P, Mohammed N, Marshall L, Abeysinghe RD, Hider RC, Porter JB, Singh S (1993) Intravenous infusion pharmacokinetics of desferrioxamine in thalassemic patients. Drug Metab Dispos 21:640–644
Leibold EA, Guo B (1992) Iron-dependent regulation of ferritin and transferrin receptor expression by the iron-responsive element binding protein. Annu Rev Nutr 12:345–368
Levin VA (1980) Relationship of octanol/water partition coefficient and molecular weight to rat brain capillary permeability. J Med Chem 23:682–684
Link G, Athias P, Grynberg A, Pinson A, Hershko C (1989) Effect of iron loading on transmembrane potential, contraction, and automaticity of rat ventricular muscle cells in culture. J Lab Clin Med 113:103–111
Link G, Pinson A, Hershko C (1993) Iron loading of cultured cardiac myocytes modifies sarcolemmal structure and increases lysosomal fragility. J Lab Clin Med 121:127–134
Lovejoy FH Jr (1982) Chelation therapy in iron poisoning. J Toxicol Clin Toxicol 19:871–874
Lowery CH, Nienhuis AW (1993) Brief report: treatment with azacitidine of patients with end-stage β-thalassemia. N Engl J Med 329:845–848
Lucarelli G, Angelucci E, Giardini C, Baroncianin D, Galimberti M, Polchi P, Bartolucci M, Muretto P, Albertini F (1993a) Fate of iron stores in thalassemia after bone-marrow transplantation. Lancet 342:1388–1391
Lucarelli G, Galimberti M, Polchi P, Angelucci E, Baronciani D, Giardini C, Andreani M, Agostinelli F, Albertini F, Clift RA (1993b) Marrow transplantation in patients with thalassemia responsive to iron chelation therapy. N Engl J Med 329:840–844
Martell AE (1981) The design and synthesis of chelating agents. In: Martell AE, Anderson WF, Badman DG (eds) Development of iron chelators for clinical use. Elsevier/North-Holland, Amsterdam, pp 67–104
Martin RB, Savory J, Brown S, Bertholf RL, Wills MR (1987) Transferrin binding of Al^{3+} and Fe^{3+}. Clin Chem 33:405–407
McLaren GD, Muir WA, Kellermeyer RW (1981) Iron overload disorders: natural history, pathogenesis, diagnosis, and therapy. CRC Crit Rev Clin Lab Sci 19:205–266
Modell B, Letsky EA, Flynn DM, Peto R, Weatherall DJ (1982) Survival and desferrioxamine in thalassaemia major. Br Med J 284:1081–1084
Motekaitis RJ, Martell AE (1991) Stabilities of the iron(III) chelates of 1,2-diemethyl-3-hydroxy-4-pyridinone and related ligands. Inorg Chim Acta 183: 71–80
Núñez M-T, Gaete V, Watkins JA, Glass J (1990) Mobilization of iron from endocytic vesicles. The effects of acidification and reduction. J Biol Chem 265:6688–6692

Olivieri NF, Buncic JR, Chew E, Gallant T, Harrison RV, Keenan N, Logan W, Mitchell D, Ricci G, Skarf B, Taylor M, Freedman MH (1986) Visual and auditory neurotoxicity in patients receiving subcutaneous deferoxamine infusions. N Engl J Med 314:869–873

Olivieri NF, Koren G, Hermann C, Bentur Y, Chung D, Klein J, St Louis P, Freedman MH, McClelland RA, Templeton DM (1990) Comparison of oral iron chelator L1 and desferrioxamine in iron-loaded patients. Lancet 336:1275–1279

Olivieri NF, Koren G, Harris J, Khattak S, Bailey JD, Poon AO, Templeton DM, Reilly BJ (1992a) Growth failure and bony changes induced by deferoxamine. Am J Pediatr Hematol Oncol 14:48–56

Olivieri NF, Koren G, Matsui D, Liu PP, Blendis L, Cameron R, McClelland RA, Templeton DM (1992b) Reduction of tissue iron stores and normalization of serum ferritin during treatment with the oral iron chelator L1 in thalassemia intermedia. Blood 79:2471–2748

Parkes JG, Hussain RA, Olivieri NF, Templeton DM (1993) Effects of iron loading on uptake, speciation and chelation of iron in cultured myocardial cells. J Lab Clin Med 122:36–47

Perrine SP, Ginder GD, Faller DV, Dover GH, Ikuta T, Witkowska HE, Cai S-P, Vichinsky EP, Olivieri NF (1993) A short-term trial of butyrate to stimulate fetal-globin-gene expression in the β-globin disorders. N Engl J Med 328:81–86

Peterson CM, Graziano JH, Grady RW, Jones RL, Markenson A, Lavi U, Canale V, Gray GF, Cerami A, Miller DR (1979) Chelation therapy in β-thalassemia major: a one-year double blind study of 2,3-dihydroxybenzoic acid. Exp Hematol 7:74–80

Pippard MJ (1989) Clinical use of iron chelation. In: de Sousa M, Brock JH (eds) Iron in immunity, cancer and inflammation. Wiley, Chichester, pp 361–392

Pippard MJ, Johnson DK, Finch CA (1982) Hepatocyte iron kinetics in the rat explored with an iron chelator. Br J Haematol 52:211–224

Pitt CG (1981) Structure and activity relationships of iron chelating drugs. In: Martell AE, Anderson WF, Badman DG (eds) Development of iron chelators for clinical use. Elsevier/North-Holland, Amsterdam, pp 104–131

Polson RJ, Jawed A, Bomford A, Berry H, Williams R (1985) Treatment of rheumatoid arthritis with desferrioxamine: relation between stores of iron before treatment and side effects. Br Med J 291:448

Polson RJ, Jawed ASM, Bomford A, Berry H, Williams R (1986) Treatment of rheumatoid arthritis with desferrioxamine. Q J Med 61:1153–1158

Ponka P, Borova J, Neuwirt J, Fuchs O (1979) Mobilisation of iron from reticulocytes. FEBS Lett 97:317–321

Porter JB, Gyparaki M, Huehns ER, Hider RC (1986) The relationship between lipophilicity of hydroxypyrid-4-one iron chelators and cellular iron mobilization using an hepatocyte culture model. Biochem Soc Trans 14:1180

Porter JB, Gyparaki M, Burke LC, Huehns ER, Sarpong P, Saez V, Hider RC (1988) Iron mobilization from hepatocyte monolayer cultures by chelators: the importance of membrane permeability and the iron-binding constant. Blood 72:1497–1503

Porter JB, Huehns ER, Hider RC (1989) The development of iron chelating drugs. Baillieres Clin Haematol 2:257–292

Porter JB, Hider RC, Huehns ER (1990) Update on the hydroxypyridinone oral iron-chelating agents. Semin Hematol 27(2):95–100

Propper RD, Shurin SB, Nathan DG (1976) Reassessment of the use of desferrioxamine B in patients with iron overload. N Engl J Med 294:1421–1423

Propper RD, Cooper B, Rufo RR, Nienhuis AW, Anderson WF, Bunn HF, Rosenthal A, Nathan DG (1977) Continuous subcutaneous administration of desferrioxamine in patients with iron overload. N Engl J Med 297:418–423

Raymond KN, Carrano CJ (1979) Coordination chemistry and microbial iron transport. Accts Chem Res 12:183–190

Raymond KN, Pecoraro VL, Weitl FL (1981) Design of new chelating agents. In: Martell AE, Anderson WF, Badman DG (eds) Development of iron chelators for clinical use. Elsevier/North-Holland, Amsterdam, pp 165–187

Raymond KN, Xu J (1994) Siderophore-based hydroxypyridonate sequestering agents. In: Bergeron RJ, Brittenham GM (eds) The development of iron chelators for clinical use. CRC Press, Boca Raton, pp 307–327

Reichard P, Ehrenberg A (1983) Ribonucleotide reductase – a radical enzyme. Science 221:514–520

Rocchi E, Gilbertinin P, Cassanelli M, Pietrangelo A, Borghi A, Pantaleoni M, Jensen J, Ventura E (1986) Iron removal therapy in porphyria cutanea tarda: phlebotomy versus slow subcutaneous desferrioxamine infusion. Br J Dermatol 114:621–629

Salonen JT, Nyyssönen K, Korpela H, Tuomilehto J, Seppänen R, Salonen R (1992) High stored iron levels are associated with excess risk of myocardial infarction in Eastern Finnish men. Circulation 86:803–811

Schneider W (1988) Iron hydrolysis and the biochemistry of iron – the interplay of hydroxide and biogenic ligands. Chimia 42:9–20

Schupp T, Toupet C, Divers M (1988) Cloning and expression of two genes of Streptomyces pilosus involved in the biosynthesis of the siderophore desferrioxamine B. Gene 64:179–188

Scott JA, An Khaw B, Locke E, Haber E, Homcy C (1985) The role of free radical-mediated processes in oxygen-related damage in cultured murine myocardial cells. Circ Res 56:72–77

Selden C, Owen M, Hopkins JMP, Peters TJ (1980) Studies on the concentration and intracellular localization of iron proteins in liver biopsy specimens from patients with iron overload with special reference to their role in lysosomal disruption. Br J Haematol 44:593–603

Shiloh H, Iancu TC, Bauminger E, Link G, Pinson A, Hershko C (1992) Deferoxamine-induced iron mobilization and redistribution of myocardial iron in cultured rat heart cells: studies of the chelatable iron pool by electron microscopy and Mössbauer spectroscopy. J Lab Clin Med 119:429–437

Singh S, Hider RC (1988) Colorimetric detection of the hydroxyl radical: comparison of the hydroxyl-radical-generating ability of various iron complexes. Anal Biochem 171:47–54

Singh S, Hider RC, Porter JB (1990) Separation and identification of desferrioxamine and its iron chelating metabolites by high-performance liquid chromatography and fast atom bombardment mass spectrometry: choice of complexing agent and application to biological fluids. Anal Biochem 187:212–219

Sirbasku DA, Pakala R, Sato H, Eby JE (1992) Thyroid hormone and apotransferrin regulation of growth hormone secretion by GH_1 rat pituitary tumor cells in iron restricted serum-free defined medium. In Vitro Cell Dev Biol 28A:67–71

Sjöberg BM, Sanders-Loehr J, Loehr TM (1987) Identification of a hydroxide ligand at the iron center of ribonucleotide reductase by resonance Raman spectroscopy. Biochemistry 26:4242–4247

Stamatoyannopoulos G, Nienhuis AW (eds) (1990) The regulation of hemoglobin switching. Johns Hopkins University Press, Baltimore

Stuhne-Sekalec L, Xu SX, Parkes JG, Olivieri NF, Templeton DM (1992) Speciation of tissue and cellular iron with on-line detection by inductively coupled plasma-mass spectrometry. Anal Biochem 205:278–284

Summers MR, Jacobs A, Tudway D, Perera P, Ricketts C (1979) Studies in desferrioxamine and ferrioxamine metabolism in normal and iron-loaded subjects. Br J Haematol 42:547–555

Tenenbein M, Kowalski S, Sienko A, Bowden DH, Adamsom IYR (1992) Pulmonary toxic effects of continuous desferrioxamine administration in acute iron poisoning. Lancet 339:699–701

Theil EC (1987) Ferritin: structure, gene regulation, and cellular function in animals, plants and microorganisms. Annu Rev Biochem 56:289–315

Theil EC, Aisen P (1987) The storage and transport of iron in animal cells. In: Winkelmann G, van der Helm G, Neilands JB (eds) Iron transport in microbes, plants and animals. VCH, Weinheim, pp 491–520

Tondury P, Kontoghiorghes GJ, Ridolfi-Luthy A, Hirt A, Hoffbrand AV, Lottenbach AM, Sonderegger T, Wagner HP (1990) L1 (1,2-dimethyl-3-hydroxypyrid-4-one) for oral iron chelation in patients with beta-thalassaemia major. Br J Haematol 76:550–553

Voogd A, Sluiter W, van Eijk HG, Koster JF (1992) Low molecular weight iron and the oxygen paradox in isolated heart cells. J Clin Invest 90:2025–2055

Waldmeier PC, Buchle A-M, Steulet A-F (1993) Inhibition of catechol-O-methyl-transferase (COMT) as well as tyrosine and tryptophan hydroxylase by the orally active iron chelator, 1,2-dimethyl-3-hydroxypyridin-4-one (L1, CP20), in rat brain in vivo. Biochem Pharmacol 45:2417–2424

Weatherall DJ, Clegg JB (1981) The thalassemia syndromes. Blackwell, Oxford

Wolfe LC, Olivieri NF, Sallan D, Colan S, Rose V, Propper R, Freedman MH, Nathan DG (1985) Prevention of cardiac disease by subcutaneous deferoxamine in patients with thalassemia major. N Engl J Med 312:1600–1603

Wolfe LC (1990) Desferrithiocin. Semin Hematol 27:117–120

Zevin S, Link G, Grady RW, Hider RC, Peter RR, Hershko C (1992) Origin and fate of iron mobilized by the 3-hydroxypyridin-4-one oral chelators: studies in hypertransfused rats by selective radioiron probes of reticuloendothlial and hepatocellular iron stores. Blood 79:248–253

Zurlo MG, De Stefano P, Borgna-Pignatti C, Di Palma A, Piga A, Melevendi C, Di Gregorio F, Burattini MG, Terzoli S (1989) Survival and causes of death in thalassemia major. Lancet 2:27–30

CHAPTER 15

Zinc Fingers and Metallothionein in Gene Expression

J. ZENG and J.H.R. KÄGI

A. Introduction

Zinc, the second most abundant transition metal in eukaryotic cells, is one of the major regulatory ions for cell growth and differentiation. Besides serving as an essential component in the many zinc metalloenzymes, this metal has recently been found to be present and to play an indispensable role in a steadily increasing number of gene regulatory proteins. These so-called zinc finger proteins are now believed to constitute the majority of the eukaryotic nucleic acid-binding regulatory proteins and are involved in the control of the transcription of a large number of genes. Numerous in vitro studies have shown that zinc activates the nucleic acid-binding domains of these proteins and that its removal abrogates binding to DNA and thereby their transcription-controlling competence. Thus, it seems reasonable that there is a link between intracellular disposability of zinc and the activities of zinc finger proteins.

While the detailed molecular processes of zinc uptake by and zinc release from eukaryotic cells remain to be elucidated, there is now increasing evidence that the intracellular zinc supply is controlled, at least in part, by the powerful chelator thionein. The synthesis of this ubiquitous metal-scavenging, cysteine-rich protein of low molecular weight is positively modulated by many exogenous and endogenous inducers including metals, certain hormones, and growth factors. The couple thionein/Zn-thionein displays the qualities of an adjustable metal-buffering system permitting the regulation of the free zinc concentration in the subnanomolar range. As binding of zinc in zinc-thionein is stronger than in zinc finger proteins, it stands to reason that changes in the supply of thionein might exert a regulating effect on their action. Such actions of thionein on zinc finger proteins have now been observed experimentally in a number of systems, both in vitro and in vivo, and are subject to this review.

B. Zinc Finger Proteins in Gene Expression

The *Xenopus* 5S RNA gene-specific transcription factor IIIA (*TFIIIA*) was the first zinc finger protein characterized one decade ago (KLUG and RHODES 1987; BERG 1990; RHODES and KLUG 1993). This protein specifically binds to both the 5S RNA gene in its approximately 50-base-pair internal control region and to 5S RNA. It forms with 5S RNA the 7S ribonucleoprotein particles that are stored in immature oocytes (SHASTRY 1991). *TFIIIA* was reported to contain 2–3 or 9–11 zinc ions per molecule, as first analyzed by HANAS et al. (1983) and later by MILLER et al. (1985), respectively. The omission of chelating agents in the purification procedure was reported to allow the recovery of more zinc in the *TFIIIA* preparation (MILLER et al. 1985). Interestingly, however, the latest study by SHANG et al. (1989) showed that *TFIIIA* requires only two to three firmly bound zinc ions for its functioning in vitro. *TFIIIA* requires zinc for activating the transcription of the 5S RNA gene by RNA polymerase III, an effect which is lost upon exposure to chelating agents. In its primary structure derived from the cDNA sequence, the protein contains nine tandem imperfectly repeated segments of approximately 30 amino acids which share the consensus sequence motif (Tyr, Phe)-X-**Cys**-X_{2-4}-**Cys**-X_3-Phe-X_5-Leu-X_2-**His**-X_3-**His**-X_{2-5}, where X stands for relatively invariable amino acids. MILLER et al. (1985) hypothesized that each sequence segment represents a structural motif organized by the coordination of zinc to tetrahedrally arranged sites of cysteine and histidine residues and introduced for them the designation "zinc finger."

Evidence supporting the presence of several zinc-stabilized polypeptide loops in *TFIIIA* came from the results of limited proteolytic digestion (MILLER et al. 1985). This was corroborated by EXAFS (extended X-ray absorption fine structure) measurements on 7S RNP particles (DIAKUN et al. 1986) and complemented by footprinting studies on the interaction of *TFIIIA* mutants with the internal control region of the 5S RNA gene (TULLIUS and DOMBROSKI 1986; VRANA et al. 1988). The ultimate confirmation of the zinc finger hypothesis came from the X-ray crystallographic determination of a complex between a domain of the mouse immediate early zinc finger protein *zif268* and its cognate DNA (PAVLETICH and PABO 1991). This DNA-binding domain contains three zinc fingers in a semicircular arrangement. Important components of this structure are the three tetrahedrally coordinated zinc ions. They are essential to the protein-DNA interaction since the metal ion holds an α-helix and an antiparallel β-sheet together so that the protein fits specifically into the major groove of the 9-bp DNA recognition site.

The zinc-finger-encoding genes are estimated to make up as much as 0.01% of the human genome (PELLEGRINO and BERG 1991). They have widely different biological functions. Thus, the nuclear transcription factor *Sp1*, which has three tandem zinc fingers interacting with GC boxes in the promoter region of several genes, functions as a general positive transcription factor of RNA polymerase II common to all vertebrates (DYNAN and TJIAN 1983, 1985). Another one, yeast *SW15*, is a cell-type-specific transcription factor involved in the control of mating type switching (STILLMAN et al. 1988), and the *MAZ* (*Myc*-associated zinc finger protein) gene cloned

Table 1. Average apparent stability constants for zinc in *TFIIIA*, the synthetic zinc finger peptides CP-1 and NC_p10, and Zn_7-thionein, at pH 7.0

	TFIIIA[a]	CP-1[b]	NC_p10[c]	Zn_7-MT[a]
Competitive chelation	$5 \times 10^{11}\,M^{-1}$	n.d.	n.d.	$7 \times 10^{12}\,M^{-1}$
Titration studies	n.d.	$5 \times 10^{11}\,M^{-1}$	$5 \times 10^{11}\,M^{-1}$	$2 \times 10^{12}\,M^{-1}$

n.d., not determined.
[a] From BIENZ (1992) and VAŠÁK and KÄGI (1983).
[b] From KRIZEK et al. (1991).
[c] From MELY et al. (1991); value interpolated to pH 7.0.

recently by BOSSONE et al. (1992) encodes a 447-amino-acid regulatory factor with dual roles in transcription initiation and termination of the protooncogene c-*MYC*. Furthermore, many of the zinc finger proteins are involved in the control of embryogenesis and developmental processes (see reviews by EL-BARADI and PIELER 1991; MELLERICK et al. 1992).

The metal-coordinating properties of the Cys_2His_2 zinc fingers thus far studied reveal closely comparable affinities for the metal. Thus, recent measurements of the binding constants of zinc for CP-1, a synthetic single finger model peptide (KRIZEK et al. 1991), and for *Moloney* murine leukemia virus nucleocapside protein *NCp10* (MELY et al. 1991), as well as from *Xenopus* oocyte TFIIIA (BIENZ 1992), are all close to $5 \times 10^{11}\,M^{-1}$, when measured at pH 7.0 (Table 1).

In addition to the classical *TFIIIA*-like Cys_2His_2 class, there are also other types of zinc finger proteins. Thus, steroid/thyroid hormone receptors and *GAL4* superfamilies are representative zinc finger proteins containing solely Cys residues as ligands. They share the consensus sequence $\mathbf{CysX_2CysX_{13}CysX_2CysX_{15-17}CysX_5CysX_9CysX_2CysX_4Cys}$ and yet are structurally quite different (see review in EVANS 1988; BERG 1989). The first convincing evidence for the presence of a metal ion in steroid receptors came from X-ray absorption spectra which revealed that the glucocorticoid receptor binds two zinc or cadmium ions which are required for its specific binding to DNA (FREEDMAN et al. 1988). The two-dimensional NMR solution structures of the DNA-binding domains of the glucocorticoid and estrogen receptors have been solved. In these receptors each of the two zinc ions is coordinated to 4 Cys and separated from each other by a helical region which determines the site-specific interaction with DNA (HÄRD et al. 1990; SCHWABE et al. 1990). In contrast, for *GAL4* in which zinc was substituted by cadmium, ^{113}Cd NMR studies showed that this yeast transcription factor contains a metal-thiolate cluster formed by two metal ions and six Cys reminiscent of those in metallothionein (PAN and COLEMAN 1990). This result has been confirmed also by the determination of the two-dimensional NMR structure of the DNA-binding domain of *GAL4* and of the crystal structure of the *GAL4*/DNA complex (KRAULIS et al. 1992; BALEJA et al. 1992; MARMORSTEIN et al. 1992).

Besides these well-established zinc finger proteins, a number of other eukaryotic regulatory proteins possess a Cys-rich sequence motif represented as $\mathbf{CysX_2CysX_{9-27}CysXHisX_2CysX_2CysX_{6-17}CysX_2Cys}$, which is also suggestive of a metal-binding domain (FREEMONT et al. 1991; REDDY et al. 1992). Among these so-called ring finger proteins, *Xenopus* nuclear factor 7 (*Xnf7*) is a maternally expressed transcription factor, *RET* is a human transforming gene, and *Rpt*-1 is a downregulator of the interleukin 2 receptor gene. Recently, the properties of metal and DNA binding of the ring finger motif have been examined on a 55-amino-acid synthetic peptide (LOVERING et al. 1993). The optical absorption spectra of the synthetic peptide-cobalt complex revealed that the ring finger preferentially binds zinc in tetrahedral coordination with a dissociation constant estimated to be in the nanomolar range. Evidence from bandshift assays indicated zinc-dependent DNA binding by the peptide. Furthermore, zinc-dependent in vivo transcription by a ring finger protein was recently demonstrated in a study on the bovine herpesvirus immediately-early transactivator *BICPO* (see below).

C. Modulation of Zinc Finger-Dependent Gene Expression by pZn

The control of the expression of genes by dissociable zinc finger protein complexes allows in principle for a regulatory role by the metal. Thus, it is of interest to see if zinc finger-dependent functions are influenced by the availability of zinc, i.e., by fluctuations in intracellular pZn ($= -\log Zn^{2+}$). Although there is little information available as yet on intracellular pZn and on its maintenance, it is obvious that this quantity must be set by the supply of zinc-binding ligands and their affinities for the metal, among them most prominently glutathione, clusters of metal-binding ligands in proteins, and in particular the potent intracellular chelator thionein (=apometallothionein).

Thionein is a unique sulfur-rich 6-kDa polypeptide with a high affinity for d^{10} metal ions including zinc. It has long been known to be responsible for the intracellular sequestration of such metals and is usually present in parenchymatous mammalian tissues as a complex with seven zinc ions. Most mammalian forms contain 61 amino acid residues, among them 20 cysteines, which serve as metal ligands in 2 metal-thiolate clusters, i.e., Zn_3Cys_9 and Zn_4Cys_{11}, located within the amino- and carboxyl-terminal domains, respectively (cited in KÄGI and SCHÄFFER 1988). The protein occurs in all animal phyla as well as in some plants and microorganisms. Its formation is characteristically stimulated both by environmental factors and by many regulators of cellular activity, e.g., metal salts, steroid hormones, cytokines, and various second messengers (cited in KARIN 1985; HAMER 1986; KÄGI 1991). Its synthesis is also enhanced in tissue regeneration and in certain stages of embryogenesis and of tumor promotion (OHTAKE et al. 1978; NEMER et al. 1984; HASHIBA et al. 1989), suggesting functions in growth processes.

Since thionein is usually found in the metal-saturated form Me_7-thionein, the search for its biological function has been focused in the past primarily on roles as a metal repository in cells and tissues exposed to elevated concentrations of Zn^{2+}, Cu^+, Cd^{2+}, and others (WEBB 1987), or as an in vitro donor of metals (Zn, Cu) to newly synthesized apoforms of metalloenzymes, such as carbonic anhydrase, alkaline phosphatase, and superoxide dismutase (BREMNER 1991). However, as biosynthesis provides for a steady supply of the metal-free and metal-seeking apoprotein thionein, it is equally important to view the role of this form as a powerful intracellular acceptor and competitor for metal ions, among them in particular for zinc. In fact, it is chemically most reasonable to consider both Zn_7-thionein and thionein as conjugates of a zinc donor/zinc acceptor couple whose concentrations fix pZn and thereby regulate the availability of zinc within the cell. Values for the average apparent stability constant, $K_{Zn,\ pH\ 7.0}$, determined by two independent methods for pH 7.0, are listed in Table 1.

The binding isotherm calculated from this stability constant assuming cooperative cluster formation of Zn_7-thionein as a function of pZn is shown in Fig. 1. The curve, centered about $\log K_{Zn,\ pH\ 7.0} = 12.8$, divides the continuum of affinities of zinc-binding systems into the weaker ones which are thermodynamically accessible to competition by thionein, on the left, and into the stronger ones to which Zn_7-thionein may donate the metal, on the right.

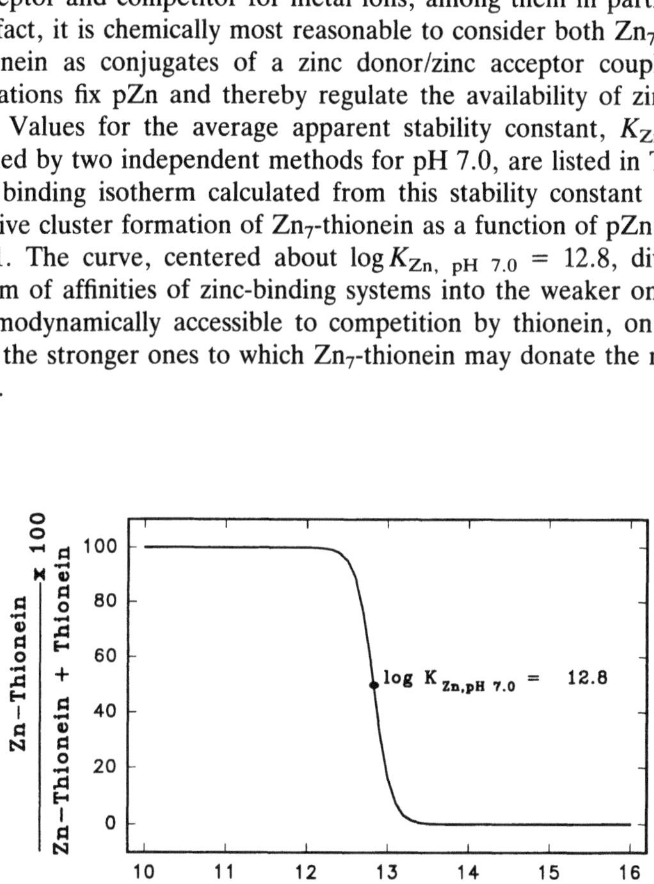

Fig. 1 Isotherm for cooperative binding of Zn^{2+} to thionein at pH 7.0, 25°C, in 1 *M* HEPES calculated from the average apparent association constant for Zn^{2+} in human Zn_7-thionein, $K_{Zn,\ pH\ 7.0} = 7 \times 10^{12}\ M^{-1}$, and an assumed cooperativity coefficient of 4. $pZn = -\log[Zn^{2+}]$

D. Effects of Thionein on Zinc Finger-Dependent Gene Expression

The data given in Table 1 reveal that at pH 7.0 thionein is a better chelator for Zn^{2+} than *Xenopus TFIIIA* and model peptides possessing the same Cys_2His_2 zinc finger arrangement. Thus, one might predict that thionein can compete for the metal of zinc finger proteins and inhibit their binding to DNA, thereby interfering with their functions. Indeed, we have now observed such effects in a number of systems.

The first example pertains to the effects of thionein on the interaction of the classical *Xenopus* oocyte transcription factor *TFIIIA* with the internal control region (*ICR*) of the 5S RNA gene and with its transcription product 5S RNA (ZENG et al. 1991b). Thus, it was found by band retardation assay that preincubation of purified *TFIIIA* with a slight excess of thionein is sufficient to abolish binding of the zinc finger protein to the labeled cognate DNA probe (Fig. 2A, lane 7) and that the ability to bind the probe and, hence, to retard its movement in the gel is restored by readdition of appropriate amounts of zinc (Fig. 2B, lane 4). In the same manner, preexposure

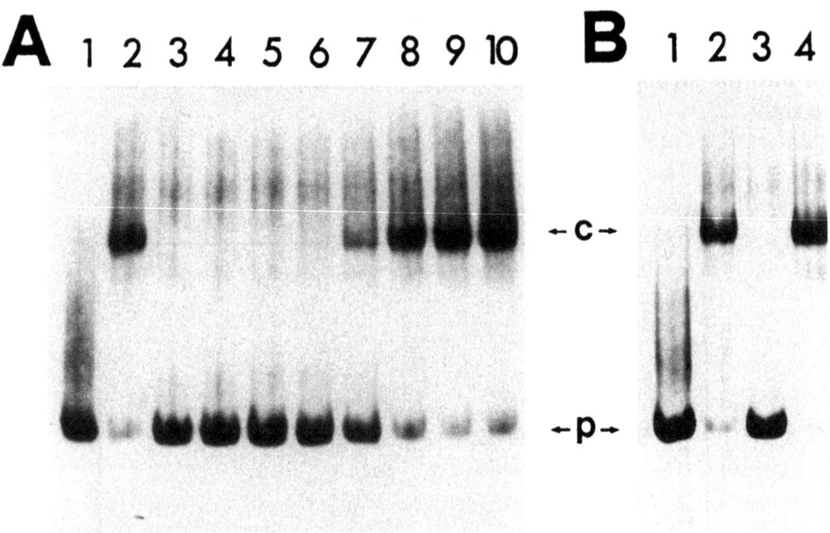

Fig. 2 Gel retardation assay for *Xenopus TFIIIA* binding to the internal control region (*ICR*) of the 5S RNA gene using a 6% native polyacrylamide gel. *A*, Thionein inhibition of *TFIIIA* binding to *ICR*. *Lane 1*, no *TFIIIA*; *lane 2*, with *TFIIIA*; *lanes 3–9, addition of 64, 32, 16, 8, 4, 2, and 1 μM* thionein, respectively; *lane 10*, addition of 64 μM Zn_7-thionein. *B*, Reversal of thionein-induced blocking of *TFIIIA* binding to the *ICR* by addition of zinc. *Lane 1*, no *TFIIIA*; *lane 2*, with *TFIIIA* (300 n*M*); *lane 3*, addition of 32 μM thionein; *lane 4*, addition of 32 μM thionein and then 130 μM zinc. *c*, *TFIIIA-ICR* complex; *p*, ^{32}P-end-labeled *ICR* probe. (From ZENG et al. 1991b)

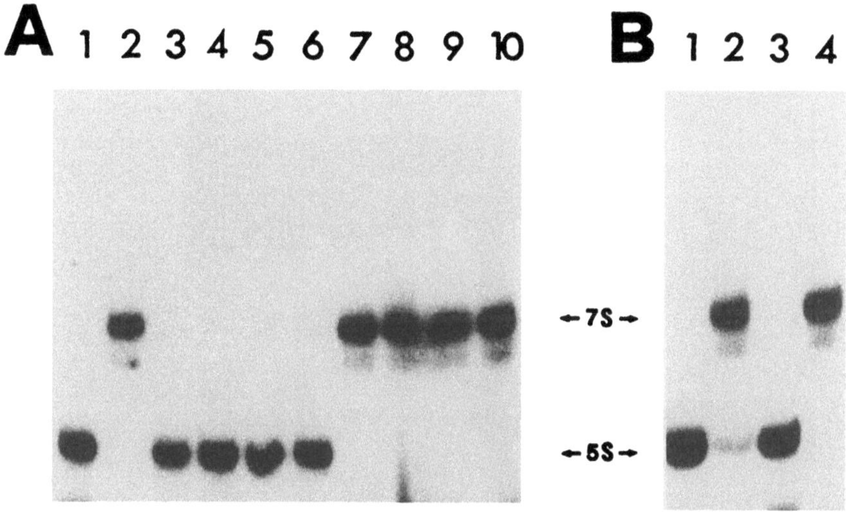

Fig. 3 Gel retardation assay for *Xenopus TFIIIA* binding to 5S, ^{32}P-labeled RNA using 6% native polyacrylamide gel. *A*, Thionein inhibition of *TFIIIA* binding to 5S RNA. *Lane 1*, no *TFIIIA*; *lane 2*, with *TFIIIA* (final concentration, 300 n*M*); *lanes* 3–9, addition of 32, 16, 8, 4, 2, 1, and 0.5 *M* thionein, respectively; *lane 10*, 32 μM Zn_7-thionein. *B*, Restoration of *TFIIIA* binding to 5S RNA in the presence of thionein by addition of zinc. *Lane 1*, no *TFIIIA*; *lane 2*, with *TFIIIA*; *lane 3*, addition of micromolar thionein; *lane 4*, addition of 32 μM thionein and then 130 μM zinc. 5S, 5S RNA; 7S, *TFIIIA*-5S RNA complex (7S RNP). (From Zeng et al. 1991b)

of *TFIIIA* to thionein also resulted in a total loss of its ability to combine with 5S RNA, an effect which is also counteracted by zinc (Fig. 3). Neither binding of *TFIIIA* to DNA nor to RNA is interfered with when the transcription factor is preincubated with Zn_7-thionein instead of thionein (Fig. 2, lane 10; Fig. 3, lane 10). Thus, one can infer that it is the zinc-sequestering power of thionein which is responsible for the observed effects.

The expected functional consequence of the suppression of *TFIIIA* binding to DNA by thionein is the loss of the expression of *TFIIIA*-dependent genes by RNA polymerase III. This effect, first demonstrated in an in vitro transcription system composed of plasmid-encoded 5S RNA gene and RNA polymerase III from HeLa cell extract (Zeng et al. 1991b), has now also been verified under in vivo conditions in mature *Xenopus* oocytes, a system possessing the advantage of highly efficient RNA polymerase III-catalyzed transcription (Brown and Gurdon 1977; Grosjean and Kubli 1986). In such experiments, we introduced by microinjections, at first into the cytoplasm, a predetermined amount of thionein and, 4 h later into the nucleus, a mixture of plasmids carrying either the *TFIIIA*-dependent 5S RNA gene or the *TFIIIA*-independent $tRNA^{arg}$ gene, and 10^{-4} mCi α-^{32}P GTP, and assessed transcription by monitoring the formation of ^{32}P-labeled RNA products. The results shown in Fig. 4 demonstrate that a dosage of

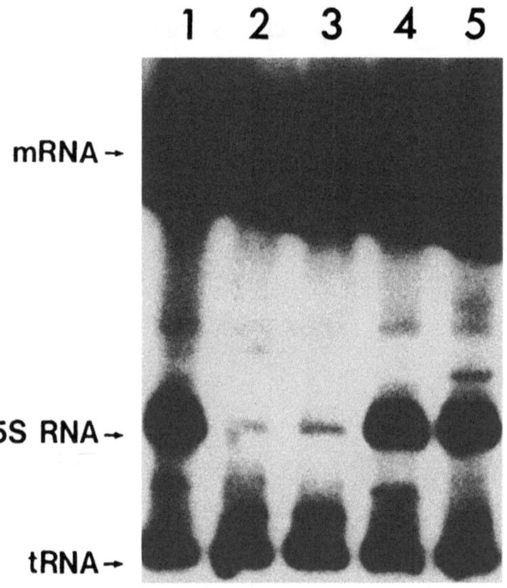

Fig. 4 In vivo transcription of 5S RNA and $tRNA^{Arg}$ genes in *Xenopus* oocytes and the inhibition of 5S RNA gene transcription by preinjection of thionein into oocyte cytoplasm. *Lane 1*, preinjection of 50 m*M* Tris-HCl buffer (control), pH 7.4; *lanes 2–4*, preinjection of 70, 35, and 1.2 μM thionein, respectively; *lane 5*, preinjection of 70 μM Zn_7-thionein (for further details see text). (Adapted from Fig. 4 of ZENG 1994)

32 pmol thionein/oocyte nearly completely abolished the *TFIIIA*-dependent 5S RNA production while the *TFIIIA*-independent formation of $tRNA^{arg}$ and of endogenous messenger RNA remained unaffected. The experiment thus established that thionein introduced into the cytosol is capable of modulating the action of *TFIIIA* in the nucleus. The lack of any effect of the holoprotein Zn_7-thionein reemphasizes that it is the metal-chelating power of thionein which is responsible for the inhibition. The fact that a nearly 20-fold excess of thionein over the estimated amount of *TFIIIA* in mature oocytes (2 pmol) was needed for the in vivo inhibition was accounted for by the considerable stock of extra zinc accessible to the injected chelator in mature oocytes.

The absence of any noticeable inhibitory effect of thionein on the *TFIIIA*-independent transcription at the plasmid-encoded $tRNA^{arg}$ gene by RNA polymerase III and on the overall production of endogenous *Xenopus* oocyte mRNA by RNA-polymerase II (Fig. 4) indicates that, at the intracellular concentrations attained in these studies, thionein affected the transcription process exclusively at the level of the regulatory zinc finger protein and did not interfere with the transcriptional machineries. As both RNA polymerase II and III are also zinc metalloproteins and inhibitable by high concentrations of the chelator 1,10-phenantroline (cited in COLEMAN

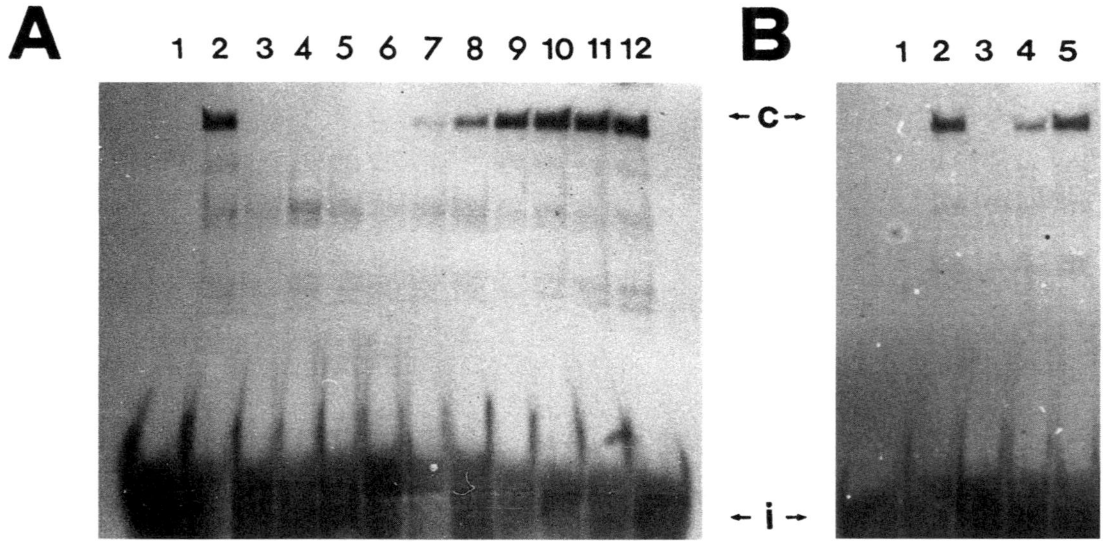

Fig. 5 Gel retardation assay with HeLa cell nuclear extract. *A*, Thionein (*apoMT*) inhibition of binding of *Sp1* to ^{32}P-labeled cognate DNA probe. *Lane 1*, no extract; *lane 2*, with extract; *lane 3*, competition with 1000-fold excess unlabeled *Sp1* binding sequence; *lanes 4–10*, addition of 13.6, 6.8, 3.4, 1.7, 0.85, 0.43, 0.2 μM thionein, respectively; *lanes 11, 12*, addition of 6.8 and 13.6 μM Zn_7-thionein, respectively. *i*, input fragment; *c*, major complex. *B*, Restoration of *Sp1* binding to DNA probe in the presence of thionein by addition of zinc. *Lane 1*, no extract; *lane 2*, with extract; *lane 3*, addition of 11.7 μM thionein; *lanes 4, 5*, addition of 11.7 μM thionein, and then 54, 100 μM zinc, respectively. (From ZENG et al. 1991a)

1983), their refractoriness to microinjected thionein probably reflects their relatively stronger binding of zinc and may thus illustrate the thermodynamic limits of the zinc-sequestering power of thionein (Fig. 1).

The second example of a classical His_2Cys_2 zinc finger protein shown to be inhibitable by thionein is the vertebrate transcription factor *Sp1* (ZENG et al. 1991a). Previous studies have shown that binding of *Sp1* from nuclear HeLa cell extract to a labeled *Sp1*-responsive DNA segment was specifically abolished in the presence of the chelator EDTA or 1,10-phenantroline (WESTIN and SCHAFFNER 1988). Using the same band retardation assay, we have now found that thionein has much the same effect albeit at much lower concentration (Fig. 5A). No inhibition was noticed when native Zn_7-thionein was used instead of thionein and, again, the inhibition of *Sp1* binding to the labeled DNA probe was reversed by the addition of zinc (Fig. 5B).

The functional consequences of the inhibition of *Sp1* binding by thionein were assessed in an in vitro transcription experiment using an *OVEC* reporter gene with an upstream-located, duplicated strong *Sp1*-binding sequence and the components of HeLa cell nuclear extract. As indicated in Fig. 6, RNA formation monitored by S1-nuclease mapping was reduced to a low level when 50 m*M* thionein was present in the reaction mixture, whereas the same concentration of Zn_7-thionein was without inhibitory effect. The transcription of the same *OVEC* template with an "octamer" enhancer sequence (*OTF-OVEC*) instead of the *Sp1* element, which does not require a zinc finger transcription factor, was insensitive to the presence of thionein (ZENG et al. 1991a). Thus, one may conclude that thionein could also

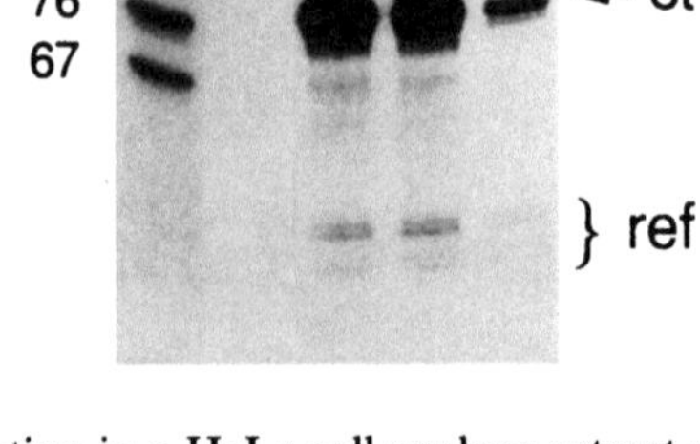

Fig. 6 In vitro transcription in a HeLa cell nuclear extract and S1 nuclease mapping of RNA products. *OVEC Sp1* template (*lane 1*, control; *lane 2*, with 50 m*M* Zn_7-thionein; *lane 3*, with 50 m*M* thionein); *S1* probe, ^{32}P-end-labeled 93 nucleotide probe extending between positions −18 and +75 of the noncoding strand of the *β*-globin gene. *MHap*II-digested pBR 322 marker DNA. (Adapted from Fig. 2 of ZENG et al. 1991a)

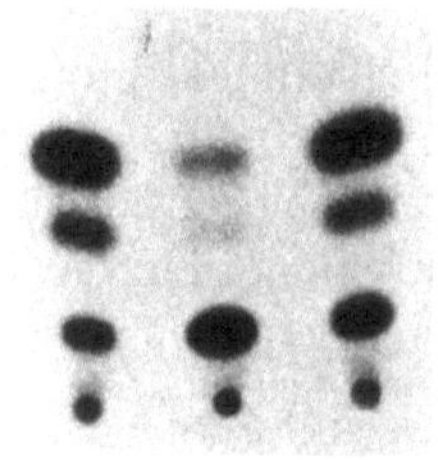

Fig. 7 Partial suppression of transactivation of the IE 4.2/2.9 bovine herpesvirus 1 (*MHV-1*) promoter by the viral "ring finger" protein *BICPO* in *Xenopus* oocytes by thionein as monitored by CAT activity expression assay. Three groups of 30 oocytes each were injected into the cytosol with either 0.5 μg thionein (*lane 2*) or 0.5 μg Zn_7-thionein (*lane 3*) or a buffer blank (*lane 1*) prior to coinjection of 50 ng each of pCE26X (effector plasmid) and pCE29CAT (reporter plasmid). (Adapted from Fig. 4 of FRAEFEL et al. 1994)

function as a selective modulator of zinc finger-controlled RNA polymerase II catalyzed transcription.

That inhibition of transcription activation by thionein also extends to transcription factors possessing the ring finger motif of metal-binding residues was recently demonstrated for the bovine herpesvirus 1 (BHV-1) immediate early gene product *BICPO*, which has properties of a potent transactivator in viral reproduction (WIRTH et al. 1992). Its transactivator effect has now been made observable in *Xenopus* oocytes into which both an effector plasmid (pCE26X) containing the *BICPO*-coding sequence with its own E2.6 viral promoter and a polyadenylation signal, and a reporter plasmid (pCE29CAT) containing a *BICPO*-sensitive viral IE4.2/2.9 promoter-*CAT* construct, were introduced by microinjection and in which *CAT* activity was monitored (Fig. 7). The results revealed that in oocytes containing both plasmids the expression of *CAT* activity was increased 72-fold on average. This effect was reduced to about one-fourth when oocytes were preinjected with 80 pmol thionein, thus providing strong evidence that the transactivating function of *BICPO* requires zinc and that thionein is capable of removing zinc from this member of the cystein-rich ring finger proteins (FRAEFEL et al. 1994).

E. Implications and Speculations

The ability to bind zinc stoichiometrically at subnanomolar concentrations (Fig. 1) suggests that thionein may play a significant role in limiting the intracellular concentration of Zn^{2+} and, hence, in modulating the activity of zinc-dependent processes. The measured binding constant of $7 \times 10^{12} M^{-1}$ allows the inference that all systems with more moderate affinity for zinc, i.e., $<10^{12} M^{-1}$, are potential targets for competition by thionein, among them, as shown above, some zinc finger proteins and, probably, also zinc-activated protein components of signal transduction pathways (AHMED et al. 1991; FORBES et al. 1991). From such considerations one is led to expect that relationships exist between the extent of intracellular production of thionein, intracellular movements of zinc, and regulation effects on zinc-dependent systems.

The proposition that the endogenous thionein production serves an important function in metalloregulation would be in accord with the well-established facts that the synthesis of thionein is regulated by a number of biological messengers such as hormones and growth factors (cited in KÄGI 1991) and that the emergence of thionein mRNA and of metallothionein is cell specific and programmed for different stages of spermatogenesis (DE et al. 1991), embryogenesis (NEMER et al. 1984), and fetal (ANDREWS et al. 1984) and perinatal (NISHIMURA et al. 1989) development. A connection between the emergence of thionein mRNA and tissue differentiation was also noticed in mouse epidermis tumor promotion by phorbol-12-myristate-13-acetate (BOHM et al. 1990). The same agent had been shown in thymocytes to induce a movement of zinc from the nucleus to the cytosol (CSERMELY et al. 1988). Strongly increased production of thionein mRNA and Zn-thionein accompanied by a redistribution of zinc was also reported to occur in rainbow trout liver following estradiol induction of vitellogenin synthesis (OLSSON et al. 1989), and there is experimental evidence that a comparable programmed transfer of zinc from high molecular weight protein including *TFIIIA* to a newly synthesized smaller protein takes place in the late stages of *Xenopus* oogenesis (Zeng, unpublished observation), where the concentration of *TFIIIA* is known to drop sharply to low levels (SHASTRY et al. 1984). Progesterone, an inducer of vertebrate development (SLATER et al. 1988) and regulator of oocyte maturation (SCHÜTZ 1985), could be instrumental in the induction of this metal-seeking protein which, by removing zine from *TFIIIA*, may inactivate the protein and promote its degradation by proteolysis.

F. Summary

During the past decade zinc finger proteins have emerged as the largest class of transcription factors involved in the control of nearly every aspect of life. That these factors depend on zinc to bind to nucleic acids and activate transcription suggests that their activities are subject to regulation by this metal. Our studies have now shown that the zinc-chelating protein thionein suppresses binding of several zinc finger transcription factors to their cognate nucleic acids and thereby abrogates their ability to promote RNA synthesis. Since zinc and zinc-dependent proteins are known to play pivotal regulatory roles in embryogenesis and cell division, it is conceivable that a programmed competition for zinc between thionein and the zinc-requiring proteins serves a modulatory purpose in cellular differentiation and proliferation.

Acknowledgements. This work was supported by grants to J.H.R. Kägi from the Swiss National Science Foundation (grant #3100-31012.91), the Kanton Zürich, and the Bonizzi-Theler Stiftung.

References

Ahmed S, Kozma R, Lee J, Monfries C, Harden N, Lim L (1991) The cysteine-rich domain of human proteins, neuronal chimaerin, protein kinase C and diacylglycerol kinase binds zinc. Biochem J 280:233–241

Andrews GK, Adamson ED, Gedamu L (1984) The ontogeny of expression of murine metallothionein: comparison with the α-fetoprotein gene. Dev Biol 103:294–303

Baleja JD, Marmorstein R, Harrison SC, Wagner G (1992) Solution structure of the DNA-binding domain of CD_2-GAL4 from S. cerevisiae. Nature 356:450–453

Berg JM (1989) DNA binding specificity of steroid receptors. Cell 57:1065–1068

Berg JM (1990) Zinc fingers and other metal-binding domains. J Biol Chem 265: 6513–6516

Bienz A (1992) Diploma thesis, University of Zürich

Bohm S, Berghard A, Pereswetoff-Morath C, Toftgard R (1990) Isolation and characterization of complementary DNA clones corresponding to genes induced in mouse epidermis in vivo by tumor promoters. Cancer Res 50:1626–1633

Bossone SA, Asselin C, Patel AJ, Marcu KB (1992) MAZ, a zinc finger protein, binds to c-MYC and C2 gene sequences regulating transcriptional initiation and termination. Proc Natl Acad Sci USA 89:7452–7456

Bremner I (1991) Nutritional and physiologic significance of metallothionein. Methods Enzymol 205:25–35

Brown DD, Gurdon JB (1977) High-fidelity transcription of 5S DNA injected into *Xenopus* oocytes. Proc Natl Acad Sci USA 74:2064–2068

Coleman JE (1983) The role of Zn(II) in RNA and DNA polymerases. In: Spiro TG (ed) Metal ions in biology, vol 5: Zinc enzymes. Wiley, New York, p 219

Csermely P, Szamel M, Resch K, Somogyi J (1988) Zinc can increase the activity of protein kinase C and contributes to its binding to plasma membranes in T lymphocytes. J Biol Chem 263:6487–6490

De SK, Enders GC, Andrews GK (1991) High levels of metallothionein messenger RNAs in male germ cells of the adult mouse. Mol Endocrinol 5:628–636

Diakun GP, Fairall L, Klug A (1986) EXAFS study of the zinc-binding sites in the protein transcription factor IIIA. Nature 324:698–699

Dynan WS, Tjian R (1983) The promoter-specific transcription factor Sp1 binds to upstream sequences in the SV40 early promoter. Cell 35:79–87

Dynan WS, Tjian R (1985) Control of eukaryotic messenger RNA synthesis by sequence-specific DNA-binding proteins. Nature 316:774–778

El-Baradi T, Pieler T (1991) Zinc finger proteins: what we know and what we would like to know. Mech Dev 35:155–169

Evans RM (1988) The steroid and thyroid hormone receptor superfamily. Science 240:889–895

Forbes IJ, Zalewski PD, Giannakis C (1991) Role for zinc in a cellular response mediated by protein kinase C in human B lymphocytes. Exp Cell Res 195: 224–229

Fraefel C, Zeng J, Choffat Y, Engels M, Schwyzer M, Ackerman M (1994) Identification and zinc dependence of the bovine herpesvirus 1 transactivator protein BICPO. J Virol 68:3154–3162

Freedman LP, Luisi BF, Korszun ZR, Basavappa R, Sigler PB, Yamamoto KR (1988) The function and structure of the metal coordination sites within the glucocorticoid receptor DNA binding domain. Nature 334:543–546

Freemont PS, Hanson IM, Trowsdale J (1991) A novel cysteine-rich sequence motif. Cell 64:483–484
Grosjean H, Kubli E (1986) Functional aspects of tRNAs microinjected into *Xenopus laevis* oocytes: results and perspectives. In: Celis JE, Graessmann A, Loyter A (eds) Microinjection and organelle transplantation techniques. Academic, London, p 301
Hamer DH (1986) Metallothionein. Annu Rev Biochem 55:913–951
Hanas JS, Hazuda DF, Bogenhagen DF, Wu FY-H, Wu C-W (1983) *Xenopus* transcription factor A requires zinc for binding to the 5S RNA gene. J Biol Chem 258:14120–14125
Härd T, Kellenbach E, Boelens R, Maler BA, Dahlman K, Freedman LP, Carlstedt-Duke J, Yamamoto KR, Gustafsson JÅ, Kaptein R (1990) Solution structure of the glucocorticoid receptor DNA-binding domain. Science 249:157–160
Hashiba H, Hosoi J, Karasawa M, Yamada S, Nose K, Kuroki T (1989) Induction of metallothionein mRNA by tumor promoters in mouse skin and its constitutive expression in papillomas. Mol Carcinogenesis 2:95–100
Kägi JHR (1991) Overview of metallothionein. Methods Enzymol 205:613–626
Kägi JHR, Schäffer A (1988) Biochemistry of metallothionein. Biochemistry 27: 8509–8515
Karin M (1985) Metallothioneins: proteins in search of function. Cell 41:9–10
Klug A, Rhodes D (1987) Zinc finger: a novel protein motif for nucleic acid recognition. TIBS 12:464–469
Kraulis PJ, Raine ARC, Gadhavi PL, Laue ED (1992) Structure of the DNA-binding domain of zinc GAL4. Nature 356:448–450
Krizek BA, Amann BT, Kilfoil VJ, Merkle DL, Berg JM (1991) A consensus zinc finger peptide: design, high-affinity metal binding, a pH-dependent structure, and a His to Cys sequence variant. J Am Chem Soc 113:4518–4523
Lovering R, Hanson IM, Borden KLB, Martin S, O'Reilly NJ, Evan GI, Rahman D, Pappin DJC, Trowsdale J, Freemont PS (1993) Identification and preliminary characterization of a protein motif related to the zinc finger. Proc Natl Acad Sci USA 90:2112–2116
Marmorstein R, Carey M, Ptashne M, Harrison SC (1992) DNA recognition by GAL4: structure of a protein-DNA complex. Nature 356:408–414
Mellerick DM, Kassis JA, Zhang SD, Odenwald WF (1992) Castor encodes a novel zinc finger protein required for the development of a subset of CNS neurons in *Drosophila*. Neuron 9:789–803
Mely Y, Cornille F, Fournié-Zaluski M-C, Darlix J-L, Roques BP, Gérard D (1991) Investigation of zinc-binding affinities of *Moloney* murine leukemia virus nucleocapsid protein and its related zinc finger and modified peptides. Biopolymers 31:899–906
Miller J, McLachlan AD, Klug A (1985) Repetitive zinc-binding domains in the protein transcription factor IIIA from *Xenopus* oocytes. EMBO J 4:1609–1614
Nemer M, Travaglini EC, Rondinello E, D'Alonzo J (1984) Developmental regulation, induction and embryonic tissue specificity of sea urchin metallothionein gene expression. Dev Biol 102:471–482
Nishimura H, Nishimura N, Tohyama C (1989) Immunohistochemical localization of metallothionein in developing rat tissues. J Histochem Cytochem 37:715–722
Ohtake H, Hasegawa K, Koga M (1978) Zinc-binding protein in the livers of neonatal, normal and partially hepatectomized rats. Biochem J 174:999–1005
Olsson P-E, Zafarullah M, Gedamu L (1989) A role of metallothionein in zinc regulation after oestradiol induction of vitellogenin synthesis in rainbow trout, *Salmo gairdneri*. Biochem J 257:555–559
Pan T, Coleman JE (1990) GAL4 transcription factor is not a "zinc finger" but forms a $Zn(II)_2Cys_6$ binuclear cluster. Proc Natl Acad Sci USA 87:2077–2081
Pavletich NP, Pabo CO (1991) Zinc finger-DNA recognition: crystal structure of a Zif268-DNA complex at 2.1 Å. Science 252:809–817

Pellegrino GR, Berg JM (1991) Identification and characterization of "zinc-finger" domains by the polymerase chain reaction. Proc Natl Acad Sci USA 88:671–675
Reddy BA, Etkin LD, Freemont PS (1992) A neurofilament-specific sequence motif. TIBS 17:344–345
Rhodes D, Klug A (1993) Zinc fingers. Sci Am 2:56–65
Schütz AW (1985) Oogenesis. In: Browder LW (ed) Developmental biology, comprehensive synthesis, vol 2. Plenum, New York, p 3
Schwabe JWR, Neuhaus D, Rhodes D (1990) Solution structure of the DNA-binding domain of the oestrogen receptor. Nature 348:458–461
Shang Z, Liao Y-D, Wu F Y-H, Wu C-W (1989) Zinc release from *Xenopus* transcription factor IIIA induced by chemical modifications. Biochemistry 28: 9790–9795
Shastry BS (1991) *Xenopus* transcription factor IIIA (XTFIIIA): after a decade of research. Prog Biophys Mol Biol 56:135–144
Shastry BS, Honda BM, Roeder RG (1984) Altered levels of a 5S gene-specific transcription factor (TFIIIA) during oogenesis and embryonic development of *Xenopus laevis*. J Biol Chem 259:11373–11382
Slater EP, Catto AC, Karin M, Baxter JD, Beato M (1988) Progesterone induction of metallothionein IIA gene expression. Mol Endocrinol 2:485–491
Stillman DJ, Bankier AT, Seddon A, Groenhout EG, Nasmyth KA (1988) Characterization of a transcription factor involved in mother cell specific transcription of the yeast HO gene. EMBO J 7:485–494
Tullius TD, Dombroski BA (1986) Hydroxyl radical "footprinting": high-resolution information about DNA-protein contacts and application to λ repressor and Cro protein. Proc Natl Acad Sci USA 83:5469–5473
Vašák M, Kägi JHR (1983) Spectroscopic properties of metallothionein. Met Ions Biol Syst 15:213–273
Vrana KE, Churchill MEA, Tullius TD, Brown DD (1988) Mapping function regions of transcription factor IIIA. Mol Cell Biol 8:1684–1696
Webb M (1987) Toxicological significance of metallothionein. Experientia Suppl 52:109–134
Westin G, Schaffner W (1988) Heavy metal ions in transcription factors from HeLa cells: Sp1, but not octamer transcription factor requires zinc for DNA binding and for activator function. Nucleic Acids Res 16:5771–5781
Wirth UV, Fraefel C, Vogt B, Vlček Č, Pačes V, Schwyzer M (1992) Immediate-early RNA 2.9 and early RNA 2.6 of bovine herpesvirus 1 are 3′ coterminal and encode a putative zinc finger transactivator protein. J Virol 66:2763–2772
Zeng J (1994) Regulation of gene expression by metals. In: Berthon G (ed) Handbook of metal-ligand interactions in biological fluids. Dekker, New York (in press)
Zeng J, Heuchel R, Schaffner W, Kägi JHR (1991a) Thionein (apometallothionein) can modulate DNA binding and transcription activation by zinc finger containing factor Sp1. FEBS Lett 279:310–312
Zeng J, Vallee B, Kägi JHR (1991b) Zinc transfer from transcription factor IIIA fingers to thionein clusters. Proc Natl Acad Sci USA 88:9984–9988

CHAPTER 16

Role of Active Oxygen Species in Metal-Induced DNA Damage

S. KAWANISHI

A. Introduction

The carcinogenic risks of metal compounds to humans have been evaluated by the International Agency for Research on Cancer (IARC) (Table 1). Arsenic, chromium(VI), and nickel(II) compounds are confirmed human carcinogens. Beryllium, cadmium, and cisplatin are probably carcinogenic to humans. Cobalt, lead, and certain iron complexes have shown carcinogenic effects in animal studies and are probably carcinogenic to animals. However, the mechanism of metal carcinogenesis has not been well understood, although recent evidence suggests that carcinogenic metals induce genotoxicity in a multiplicity of ways (SNOW 1992).

DNA damage is a critical event not only in the initiation but also in the promotion phase of carcinogenesis. It is well accepted that many organic carcinogens can be readily converted to reactive intermediates by drug-metabolizing enzymes. A variety of DNA-carcinogen adducts have been identified in cultured cells or intact organisms treated with carcinogens. On the other hand, carcinogenic metal compounds cannot be enzymatically metabolized to reactive intermediates forming DNA-carcinogen adducts, although Cr(VI), Co(II), and Ni(II) have been shown to cause DNA damage in animals and cultured cells. Since these metals do not cause damage to isolated DNA, it is speculated that these metals react with endogenous compounds to produce active species causing DNA damage. Hydrogen peroxide (H_2O_2) can be one such compound. H_2O_2 is directly produced by several enzymes such as urate oxidase, glycolate oxidase, and flavoprotein dehydrogenases involved in the β-oxidation of fatty acids (Fig. 1). Superoxide (O_2^-)-generating systems in mitochondria and microsomes produce H_2O_2 by nonenzymatic or superoxide dismutase(SOD)-catalyzed dismutation (HALLIWELL and GUTTERIDGE 1990).

With regards to mechanism of metal carcinogenesis, KAWANISHI et al. (1986a) reported that carcinogenic chromate(VI) reacts with H_2O_2 to produce •OH and singlet oxygen ($^1O_2$), which cause damage to isolated DNA. Further studies were performed by the DNA-sequencing technique (MAXAM and GILBERT 1980) using ^{32}P 5′-end-labeled DNA fragments obtained from human c-Ha-*ras*-1 protooncogene (CAPON et al. 1983), and electron spin resonance (ESR) spin-trapping techniques. Carcinogenic Fe(III)

Table 1. Carcinogenicity of metal compounds

Exposure	Target organ		IARC evaluation[a]
	Human	Animal	
Arsenic	Skin, lung (liver, hematopoietic system gastrointestinal tract, kidney)[b]	Mouse, hamster (lung, respiratory tract)	1
Asbestos	Lung, pleura, peritoneum gastrointestinal tract, larynx	Rat: lung, pleura, peritoneum Mouse: peritoneum	1
Chromium(VI)	Lung (gastrointestinal tract)	Mouse: local Rat: lung	1
Nickel(II)	Nasal sinus, lung (larynx)	Rat: lung	1
Beryllium	(Lung)	Rat, monkey: lung Rabbit: osteosarcoma	2A
Cadmium	(Prostate, lung)	Mouse, rat: testis Rat: lung	2A
Cisplatin		Mouse: lung, skin, leukemia	2A
Cobalt		Rat: local, (lung)	2B
Iron-dextran	(Local)	Mouse, rat, rabbit: local	2B
Iron-NTA		Rat: kidney	2B
Lead	(Stomach, kideny, bladder)	Rat, mouse: kidney	2B
Metallic nickel		Rat: lung	2B
Chromium(III)			3
Metallic chromium			3

[a] Group 1, the agent is carcinogenic to humans; group 2A, the agent is probably carcinogenic to humans; group 2B, the agent is possibly carcinogenic to humans; group 3, the agent is not classifiable as to its carcinogenicity to humans.
[b] Suspected target organs in parentheses.

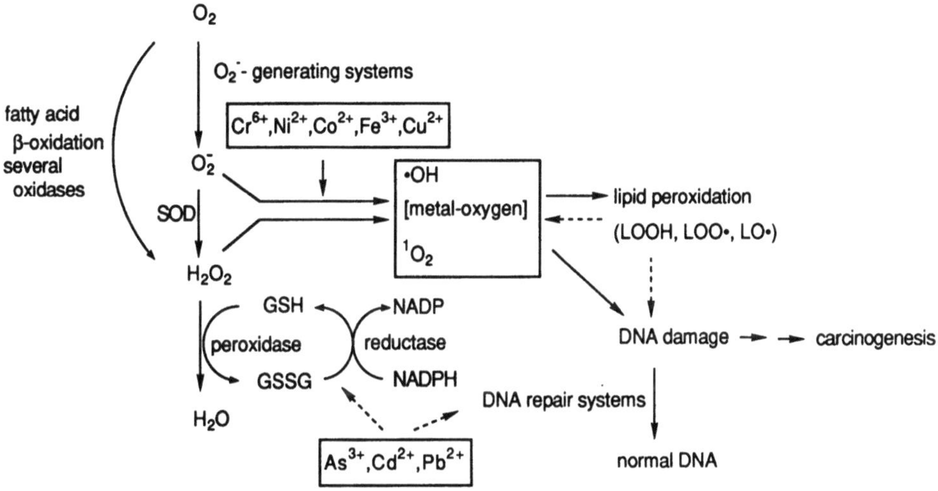

Fig. 1. Possible mechanisms of metal-mediated oxidative DNA damage

nitrilotriacetate, cobalt(II), and nickel(II) were shown to react with H_2O_2 to produce •OH, $^1O_2$, and metal-oxygen complex, which cause site-specific DNA damage (Table 2). On the basis of these findings, KAWANISHI et al. (1989d) have emphasized the role of active oxygen species in metal carcinogenesis. Cell culture experiments suggested that chromate(VI) induced cellular DNA damage via •OH formation (SUGIYAMA et al. 1989). KASPRZAK et al. (1990) reported 8-hydroxyl-2′-deoxyguanosine (8-OH-dG) formation in the kidney of rats treated with nickel acetate. Oxidative damage to cellular DNA is reflected in the formation of 8-OH-dG (FLOYD 1990; KASAI et al. 1991). These studies have demonstrated that some carcinogenic metal compounds induce DNA damage via active oxygen species formation in vitro and in vivo. In this chapter, the role of active oxygen species in metal-induced DNA damage is reviewed with relation to the metal carcinogenesis mechanism. In addition, the participation of copper and manganese in oxidative DNA damage by some carcinogens is discussed.

B. Chromium

Chromium(VI) compounds, widely used in various forms in industry, have serious toxic and carcinogenic effects on humans. There is sufficient evidence in humans for the carcinogenicity of chromium(VI) compounds as encountered in the chromate production, chromate pigment production, and chromate plating industries (IARC Working Group 1990b). Extensive epidemiological evidence has been published on the high incidence of respiratory tract cancer in men occupationally exposed to chromium(VI) compounds. The incidence of cancer in the gastrointestinal tract and pan-

Table 2. Active oxygen species formation and site-specific DNA damage induced by metal compounds in the presence of hydrogen peroxide

Metal	Cr(VI)	Ni(II)	Fe(III)-NTA	Co(II)	Cu(II)
Carcinogenicity	++	++	+	+	?
Possible active species	$\bullet OH$, $^1O_2$	$[Ni^{IV}\text{-}O]^{2+}$	$\bullet OH$	$^1O_2$	$[Cu^{II}O_2^{2}\text{-}Cu^{II}]^{2+}$
Site specificity of DNA damage	G>T~C~A	C~T~G>A	G~T~C~A	G>T~C>A	G~T~C>A
Reference	KAWANISHI et al. (1986a)	KAWANISHI et al. (1989a)	INOUE and KAWANISHI (1987)	YAMAMOTO et al. (1989)	YAMAMOTO/ KAWANISHI (1989)

++, sufficient evidence of carcinogenicity in humans and animal experiments.
+, evidence of carcinogenicity in animal experiments.

creas may also have increased. Chromium(VI) compounds have been shown to be potent carcinogens in animal experiments. Bronchial carcinomas were produced in rats after intrabronchial implantation of strontium chromate and zinc chromate. Calcium chromate induced lung tumors by intratracheal administration in a single study and by intrabronchial administration in three studies in rats, and increased the incidence of lung adenoma by inhalation in mice. In contrast, chromium(III) compounds did not produce tumors (IARC Working Group 1990b).

De Flora and Wetterhahn (1989) have reviewed the mechanism of chromium metabolism and genotoxicity. The anionic hexavalent form of chromium can easily cross the cell membranes, and it is reduced inside the cell to its trivalent form, and this may react with DNA, resulting in DNA single-strand break and cross-linking of DNA-nuclear proteins. Since chromium(VI) compound itself does not cause DNA damage, the metabolism of chromium(VI) appears to be important in chromium(VI)-induced DNA alterations. Reduction of Cr(VI) by glutathione (GSH) and ascorbate has been suggested to occur both outside and inside the cells after the administration of chromate(VI). Similarly, hydrogen peroxide (H_2O_2) may be an endogenous reductant of chromate(VI).

Kawanishi et al. (1986a) investigated reactivities of chromium compounds with DNA by the DNA-sequencing technique using ^{32}P 5′-end-labeled DNA fragments. When double-stranded DNA was incubated with sodium chromate(VI) plus H_2O_2, cleavage occurred at positions of every base residue, but the cleavage at the position of guanine was more dominant than that of the other three bases (Table 2). ESR studies using 5,5-dimethylpyrroline-*N*-oxide (DMPO) and (4-pyridyl 1-oxide)-*N*-*tert*-butylnitrone (4-POBN) as •OH traps have demonstrated that •OH is generated during the reaction of chromate(VI) with H_2O_2. The •OH adduct spectrum with a 1:2:2:1 pattern of four lines increased almost linearly with time when DMPO was added to the reaction mixture of sodium chromate(VI) and H_2O_2. The signals of DMPO-OH were efficiently decreased by •OH scavengers. The addition of ethanol, a •OH scavenger, inhibited the production of DMPO-OH, resulting in the appearance of a new signal due to trapping of the α-hydroxyethyl radical, indicating the •OH formation. An ESR method was also used for detecting singlet oxygen production. It is based on the reaction of singlet oxygen with sterically hindered 2,2,6,6-tetramethyl-4-piperidone and 2,2,6,6-tetramethylpiperidine, leading to stable free radical nitroxides. This method has demonstrated that $^1O_2$ is also generated during the reaction of chromate(VI) with H_2O_2, and reacts specifically with dGMP. Similarly, photochemically generated $^1O_2$ oxidized specifically the guanine residues of DNA (Kawanishi et al. 1986b; Ito et al. 1993). These results indicate that sodium chromate(VI) reacts with H_2O_2 to produce •OH and $^1O_2$; •OH causes every base alteration and deoxyribose-phosphate backbone breakage, and $^1O_2$ oxidizes the guanine residues resulting in the formation of alkali-labile sites.

Both the •OH formation and DNA damage induced by Cr(VI) plus H_2O_2 were confirmed by several researchers. AIYAR et al. (1990) found the enhancing effect of GSH on the •OH formation and DNA damage. That is, incubation of chromium(VI) with GSH prior to addition of H_2O_2 led to formation of peroxochromium(V) species and a marked increase in •OH production over that detected in the reaction of chromium(VI) with H_2O_2 alone. Reaction of chromium(VI) with calf thymus DNA in the presence of a preincubated mixture of GSH and H_2O_2 led to detection of the 8-OH-dG, whose formation correlated with that of •OH production. SHI et al. (1992) showed that •OH generated by the Cr(VI)/flavoenzyme/NAD(P)H, enzymatic system reacts with 2′-deoxyguanosine to form 8-OH-dG.

SUGIYAMA et al. (1989, 1992) have studied whether chromate-induced DNA damages through oxygen radicals occur in cells. Incubation of Chinese hamster V-79 cells with Na_2CrO_4 plus vitamin B_2 resulted in an increase of DNA single-strand breaks, chromosomal aberrations, and mutations. On the other hand, pretreatment with α-tocopherol succinate (vitamin E) for 24h prior to chromate exposure resulted in a decrease of metal-induced chromosomal aberrations (SUGIYAMA et al. 1991). In addition, the production of DNA single-strand breaks in H_2O_2-resistant cells treated with Na_2CrO_4 was reduced by about 50% as compared with that in parent cells. Concomitantly, ESR studies revealed that the level of chromium(V) in H_2O_2-resistant cells during treatment with Na_2CrO_4 was about 50% that in parent cells (SUGIYAMA et al. 1993). On the basis of these results, it has been proposed that the formation of active oxygen species and/or chromium(V) during reduction of chromium(VI) inside cells might be associated with the induction of the DNA strand breaks and the clastogenic and mutagenic action caused by the metal.

C. Iron Complex

High body stores of iron may increase the risk of cancer in humans (STEVENS et al. 1988). Iron overload can occur through diet, through the parenteral administration of iron dextran complex, or occupationally. Some studies of metal workers exposed to ferric oxide dusts have shown an increased incidence of lung cancer, but the influence of factors other than ferric oxide in the workplace cannot be discounted. No carcinogenic effect of ferric oxide has been observed in animal experiments, whereas ferric oxide enhanced lung and nasal cavity carcinogenesis induced by *N*-nitrosodiethylamine and *N*-nitrosodimethylamine, respectively, and repeated intratracheal instillation to hamsters of benzo[a]pyrene bound to ferric oxide particles induced carcinomas (IARC Working Group 1987). Several woman had developed sarcoma following multiple injections of iron-dextran complex. Iron-dextran complex has been tested in mice, rabbits, and rats by repeated subcutaneous or intramuscular injections, producing local tumors at the injection site (IARC Working Group 1987).

Nitrilotriacetate (NTA) is a chelating agent for metals and a potential substitute for phosphates in household detergents. NTA has been shown to be a potent carcinogen and promoter (IARC Working Group 1990a). It has been reported that long-term administration of NTA at high doses to rats induced tumors of the urinary bladder and kidney (GOYER et al. 1981). Some papers also reported that NTA showed promoting effects on renal tubular cell tumors and urinary bladder carcinogenesis initiated by nitrosamine derivatives. Furthermore, renal cell carcinoma was observed in Fe(III)-NTA-treated rats, while no tumors were formed in NTA- or Al(III)-NTA-treated rats (EBINA et al. 1986). Thus, it can be presumed that NTA binds Fe(III) to form the complex participating in carcinogenesis.

INOUE and KAWANISHI (1987) examined Fe(III)-NTA-induced DNA damage in the presence of H_2O_2 by using ^{32}P 5′-end-labeled DNA fragments of defined sequence obtained from human c-Ha-*ras*-1 protooncogene, in comparison with the ferric complex of other aminopolycarboxylic acids. The chemical basis for the different behavior of these iron chelates with respect to radical production was investigated using ESR spin-trapping techniques. Fe(III)-NTA catalyzed the decomposition of H_2O_2 to produce •OH, which subsequently caused the base alterations and backbone breakages. The DNA damage showed almost no marked site specificity. In contrast, Fe(III) chelates of other aminopolycarboxylic acids did not cause DNA damage under the conditions used. The Fe(III)-HEDTA/H_2O_2 system did not cause DNA damage although it produced as much •OH as the Fe(III)-NTA/H_2O_2 system. It may be interpreted by the structural factor. Fe(III)-NTA is supposed to approach the groove of the DNA double helix more readily than Fe(III)-HEDTA because the latter is bulkier than the former. Since •OH is short lived, it damages DNA only when produced in the vicinity of the DNA.

The mechanism of •OH production during the reaction of Fe(III)-NTA with H_2O_2 seems to be more complex than the Fenton reaction. The process shown in Eqs. 1–4 can be envisioned as accounting for most of the observations:

$$Fe^{3+}\text{-NTA} + H_2O_2 \rightarrow Fe^{3+}\text{-NTA}(HO_2^-) + H^+ \quad (1)$$

$$Fe^{3+}\text{-NTA}(HO_2{}^-) \rightarrow Fe^{2+}\text{-NTA} + H^+ + O_2^- \quad (2)$$

$$Fe^{3+}\text{-NTA} + O_2{}^- \rightarrow Fe^{2+}\text{-NTA} + O_2 \quad (3)$$

$$Fe^{2+}\text{-NTA} + H_2O_2 \rightarrow Fe^{3+}\text{-NTA} + \bullet OH + OH^- \quad (4)$$

$$Fe^{2+}\text{-NTA} + H_2O_2 \rightarrow FeO^{2+}\text{-NTA} + H_2O \quad (5)$$

$$FeO^{2+}\text{-NTA} + H_2O \rightarrow Fe^{3+}\text{-NTA} + \bullet OH + OH^- \quad (6)$$

Fe(III)-NTA is first reduced by H_2O_2 to form Fe(II)-NTA and O_2^- via an intermediate complex (Eqs. 1, 2). Fe(III)-NTA is readily reduced by superoxide to form Fe(II)-NTA (Eq. 3). The formed Fe(II)-NTA reacts with H_2O_2 to give rise to •OH (Eq. 4). Although it is believed that •OH is formed in the Fenton system employing an excess of H_2O_2, YAMAZAKI and PIETTE (1991) stated that •OH formed in the Fenton reaction is not totally free in solution. The extent to which it is not free but bound in some kind of complex depends upon the type of chelator used. RUSH and KOPPENOL (1986) have also proposed a ferryl (FeO^{2+}) complex as an intermediate of the reaction between Fe(II)-EDTA and H_2O_2 as shown in Eq. 5. The ferryl (FeO^{2+}) complex is found to be significantly less reactive than the •OH toward benzoate and *tert*-butyl alcohol. These results showed the inhibitory effects of benzoate and *tert*-butyl alcohol on the DNA damage induced by Fe(III)-NTA plus H_2O_2, suggesting that, although the ferryl (FeO^{2+})-NTA complex may be involved as the intermediate, the species causing the DNA damage is a state close to •OH.

DIZDAROGLU et al. (1991) reported that Fe(III)-NTA plus H_2O_2 caused 8-OH-dG formation in isolated chromatin. UMEMURA et al. (1990) reported a significant increase of 8-OH-dG in the kidney DNA of rats given Fe(III)-NTA by single i.p. injection. By contrast, non- or weakly carcinogenic compounds, the Al(III)-NTA complex, non complexed NTA, and ferric chloride, had no effect on 8-OH-dG production in the kidney DNA. Therefore, it is considered that Fe-NTA induces DNA damage via oxygen radicals not only in vitro but also in vivo.

GODDARD et al. (1986) described the marked stimulation of lipid peroxidation by Fe(III)-NTA administered to mice, suggesting that DNA damage in vivo is induced indirectly via lipid peroxidation products by Fe(III)-NTA (OKADA et al. 1987). Although lipid peroxidation products mediate the formation of 8-OH-dG in DNA, the reaction rate is relatively slow (PARK and FLOYD 1992). In addition, UMEMURA et al. (1991) reported that administration of exogenous GSH an Cys suppressed 8-OH-dG formation, an indicator of oxidative DNA damage following Fe(III)-NTA treatment, whereas peroxidation was found to be inhibited only by GSH and not Cys treatment. It is considered that the lipid peroxidation and the DNA damage might be mediated by different pathways. The fact that the amount of H_2O_2 is particularly high in rat kidney (RONDONI and CUDKOWICZ 1953; SZATROWSKI and NATHAN 1991) is of interest in connection with the observation that NTA and Fe(III)-NTA induced renal tumors. Thus, •OH may play an important role in the initiation and/or promotion phase of carcinogenesis by Fe(III)-NTA.

D. Nickel

Extensive epidemiological studies showed that nickel sulfides, nickel oxides, and soluble nickel were associated with increased risk of respiratory cancer in the production industry (IARC Working Group 1990b). SUNDERMAN and MAENZA (1976) compared the carcinogenicities of nickel compounds in rats and showed that Ni_3S_2 was highly carcinogenic. POTT et al. (1987) also reported that Ni_3S_2, NiO, and nickel powder caused lung tumors in rats after intratracheal instillation and that the effect was stronger for Ni_3S_2 than for nickel powder, while NiO was clearly the least effective. In mammalian cells, nickel compounds are genotoxic, including the induction of sister chromatid exchanges (SCEs) and chromosomal aberrations. COSTA and MOLLENHAUER (1980) reported that carcinogenic activity of particulate nickel compounds was proportional to their cellular uptake. The intracellular fate of a phagocytized nickel particle is to produce nickel ion potentially capable of interaction with DNA (COSTA and HECK 1984). In addition, SUNDERMAN (1989) reviewed possible mechanisms of nickel carcinogenesis and speculated that Ni(II) enters nuclei and might replace Zn(II) in finger-loop domains of transforming proteins to induce morphological transformation of cells.

Even if Ni(II) binds DNA or the protein close to DNA, Ni(II) itself causes no or little damage to DNA. Therefore, it is speculated that Ni(II) reacts with endogenous compounds such as H_2O_2 to produce active species causing DNA damage. Iron and copper are candidates for catalysts of •OH formation by the Fenton reaction, but nickel-dependent generation of activated oxygen species has not been demonstrated so far. INOUE and KAWANISHI (1989) using ESR spectroscopy examined whether activated oxygen species are produced by the reaction of Ni(II) oligopeptides with H_2O_2. The •OH adduct of DMPO was formed by the decomposition of H_2O_2 in the presence of Ni(II) GlyGlyHis. It is known that free •OH reacts with ethanol and formate to produce α-hydroxyethyl radical and CO_2^- radical, respectively. However, in the case of Ni(II) oligopeptides and H_2O_2, the spin adducts of α-hydroxyethyl radical and CO_2^- radical were scarcely observed although ethanol and formate showed inhibitory effects. This result led us to the idea that the •OH adduct is formed in the reaction of the nickel-oxygen complex and DMPO. Recently, it was reported that the oxidizing species produced by some ferrous iron chelates and H_2O_2 is not •OH, but an iron-oxo species such as the ferryl ion (YAMAZAKI and PIETTE 1991). Similarly, an oxo-Ni(IV) complex or Ni(III)-peroxide complex may be able to release •OH in concerted reactions.

The experiments with isolated DNA showed that Ni(II) induced strong DNA cleavage in the presence of H_2O_2 even without piperidine treatment (KAWANISHI et al. 1989a). Piperidine-labile sites were induced frequently at cytosine, thymine, and guanine residues, and rarely at the adenine residue (Table 2). The characteristic cleavage site was the 5′-CC-3′ sequence. Tandem double CC → TT mutations are known to occur via UV damage to DNA and are thought to be a specific indicator of UV exposure. However, REID and LOEB (1993) reported that tandem double CC → TT mutations were produced by reactive oxygen species. With Ni(II), the active oxygen species damaging DNA may be bound hydroxyl radicals (e.g., $[Ni^{IV}\text{-}O]^{2+}$) which have been suggested by analysis of the effects of •OH scavengers on the DNA damage (KAWANISHI et al. 1989a) and that of ESR data (INOUE and KAWANISHI 1989). KASPRZAK and HERNANDEZ (1989) reported that addition of Ni(II) doubled the 8-OH-dG formation in double-stranded DNA by H_2O_2 plus ascorbic acid.

The possible roles of active oxygen species in nickel-induced DNA damage are supported by cell cultures and animal experiments. Vitamin E (α-tocopherol succinate) pretreatment significantly inhibited the chromosomal aberrations induced by crystalline NiS particles (LIN et al. 1991). MISRA et al. (1993) investigated the formation of oxidatively damaged DNA bases with the use of gas chromatography/mass spectrometry techniques in the nuclei from kidneys of rats after a single i.v. injection of the $Ni(II)(His)_2$ complex. Administration of $20\,\mu$mol of the Ni(II) complex/kg body wt. significantly increased levels of oxidatively damaged DNA bases, including 2,6-diamino-4-hydroxy-5-formamidopyrimidine (FapyGua; 3.5-fold vs. the control value) 3 h postinjection and 8-OH-dG (2.6-fold), cytosine glycol (2.5-fold), 8-oxoadenine (2-fold), and FapyGua (1.9-fold) 18 h postinjection. The DNA base derivatives found were typical products of oxygen radical attack on DNA. Furthermore, HIGINBOTHAM et al. (1992) reported that nickel subsulfide induced rat renal sarcomas containing GGT to GTT transversions in codon 12 of the K-*ras* oncogene. Iron enhanced the frequency of these transforming mutations. These mutations are consistent with the known ability of nickel, in the presence of an oxidizing agent, to catalyze formation of 8-OH-dG. The 8-OH-dG promotes misincorporation of dATP opposite the oxidized guanine residue and thus may be important in carcinogenesis (SHIBUTANI et al. 1991).

E. Cobalt

The major use of cobalt has been in the production of metal alloys, and about one-third of the cobalt used is in the production of cobalt chemicals, which are used primarily as catalysts and pigments. The IARC Working Group (1992) estimated that cobalt and cobalt compounds are possibly carcinogenic to humans (group 2B). In a study in Sweden of hard metal workers with exposure to cobalt-containing dusts, a significant increase in lung cancer risk was seen in people exposed for more than 10 years. In a French electrochemical plant, there was a significant increase in the risk of lung cancer among workers in cobalt production. Confounding by nickel and arsenic is a major problem, and the size of most of the investigated populations has been rather small, giving inadequate evidence for the carcinogenicity of cobalt and cobalt compounds in humans. There is sufficient evidence for the carcinogenicity of cobalt metal powder and cobalt(II) oxide in experimental animals. Injections or implantation of cobalt metal, cobalt alloys, and cobalt compounds induced local and sometimes metastasizing sarcomas in rats, rabbits, and mice. Intratracheal administration of a high dose of one type of cobalt oxide induces lung tumors in rats (STEINHOFF and

MOHR 1991). Cobalt chloride was found to be clastogenic to bone marrow cells of mice when administered orally in vivo (PALIT et al. 1991). The clastogenic effects, mainly chromosomal breaks, increased significantly with increasing concentration. In cultured mammalian cells in vitro, predominantly positive results were obtained, with induction of DNA strand breakage and SCE. Cobalt sulfide induced morphological transformation in Syrian hamster embryo cells (IARC Working Group 1992).

Although Co(II) itself binds DNA, it causes no or little DNA damage in vitro. With isolated DNA, YAMAMOTO et al. (1989) demonstrated that cobalt(II) ion caused extensive site-specific damage (G>T~C>A) in the presence of H_2O_2. Guanine residue was the most alkali-labile site and the extent of cleavages at the positions of thymine and cytosine was dependent on the sequence. Adenine residue was relatively resistive. Various $^1O_2$ scavengers (dimethylfuran, sodium azide, 1,4-diazobicyclo[2,2,2]octane, dGMP), sulfur compounds (methional, methionine), and SOD inhibited DNA cleavage completely. Scavengers of •OH were not as effective as $^1O_2$ scavengers. ESR studies using 2,2,6,6-tetramethyl-4-piperidone as an $^1O_2$ trap suggested that Co(II) reacts with H_2O_2 to produce $^1O_2$ or its equialent cobalt-oxygen complex. We measured fluorescein-dependent chemchemiluminescence induced by Co(II) and H_2O_2. The Co(II)-induced chemiluminescence increased with increasing concentration of H_2O_2. The intensity was enhanced about threefold in D_2O, in which the lifetime of $^1O_2$ is ten or more times longer than in H_2O. These results suggest that $^1O_2$ generated during the reaction of Co(II) with H_2O_2 participates in the DNA damage.

NACKERDIEN et al. (1991) have investigated the ability of Co(II) ions in the presence of H_2O_2 to cause chemical changes in DNA bases in chromatin extracted from cultured cells of human origin. Treatment of chromatin with Co(II)/H_2O_2 caused formation of significant amounts of products. Eleven modified DNA bases in chromatin were identified and quantitated by the use of gas chromatography-mass spectrometry. Since the products identified were typical •OH-induced products of DNA bases, NACKERDIEN et al. (1991) thought that the •OH was involved in their formation and that partial inhibition of product formation indicated a possible "site-specific" formation of •OH by unchelated Co(II) ions bound to chromatin. Alternatively, partial inhibition of product formation by typical •OH scavengers suggests that active species are $^1O_2$ and/or cobalt peroxide complexes with similar reactivity to $^1O_2$ and •OH.

Sulfite (SO_3^{2-}) is an endogenous and exogenous reductant as is H_2O_2. Sulfite is formed in vivo, as a normal product of the degradation of sulfur-containing amino acids. It is used as a preservative in food and drugs, and is formed in the lung by the hydration of sulfur dioxide, a major air pollutant. An epidemiological study has shown that sulfur dioxide is associated with mortality from malignancies of the respiratory tract of males. Although sulfite has not been shown to be a carcinogen in animal experiments, there is a possibility that it is a cocarcinogen or promoter (SHAPIRO 1977). Incubation of DNA with sulfite plus Co(II) followed by the piperidine treatment led to predominant cleavage at the positions of guanine especially those located 5′ to guanine (KAWANISHI et al. 1989c). The photolysis of peroxydisulfate ($S_2O_8^{2-}$), which is known to produce SO_4^- radicals, gave a DNA cleavage pattern similar to that induced by sulfite plus Co(II). The ESR spin trapping method provided evidence for the formation of sulfate radical ($SO_4^{\bullet -}$) during Co(II)-catalyzed autoxidation of sulfite (ITO and KAWANISHI 1991). These results suggest that sulfite is rapidly autoxidized in the presence of Co(II) to produce SO_4^- radical, causing site-specific DNA damage.

Thus, it is considered that carcinogenic metal compounds react with various endogenous compounds such as H_2O_2, SO_3^{2-}, and GSH to produce active oxygen species causing DNA damage and that oxidative damage to DNA is mutagenic and thus plays a role in carcinogenesis.

F. Copper

The biological significance of copper has recently attracted much interest in connection with carcinogenicity and mutagenicity (ARGARWAL et al. 1989). Copper is an essential component of chromatin (DIJKWEL and WENINK 1986; SAUCIER et al. 1991) and is known to accumulate preferentially in the heterochromatic regions (BRYAN et al. 1976). LI et al. (1991) reported that copper accumulated in liver tissues of LEC rats which had spontaneously developed hepatocellular carcinomas, suggesting that the abnormal copper metabolism is involved in hepatic carcinogenesis in LEC rats.

A recent case-cohort study showed a U-shaped relationship between premorbid plasma copper levels and the risk of developing breast cancer (OVERVAD et al. 1993). The adjusted odds ratios for breast cancer were 1.8, 1.0, 1.6, and 3.2, respectively, in the four quartiles of the copper distribution. Although epidemiological studies showed significantly increased lung cancer mortality in the group of copper smelter workers, the mortality was generally considered to be due to arsenic exposure. The IARC has not evaluated the carcinogenicity of copper and copper compounds in humans. The intratesticular injection of $CuSO_4$ or $CuCl_2$ induced testicular tumors in mice and chicken (MAGOS 1991). Copper sulfate ($CuSO_4$) showed clastogenic effects on the bone marrow chromosomes of mice in vivo (ARGARWAL et al. 1990). A low concentration of copper complex with neocuproine ($\sim$0.1 mmol/10^5 cells/ml) resulted in DNA single-strand breakage during 1-h treatments as measured by DNA alkaline elution (BYRNES et al. 1992).

Hydroxyl free radicals are generated from H_2O_2 by means of the Fenton reaction with reduced iron, but whether copper acts like iron or not remains to be clarified. The observation that DNA cleavage is efficiently induced by treatment with Cu(II) ion, H_2O_2, and various reducing agents has stimulated interest in the mechanism of the reaction of Cu(I) with H_2O_2 (GOLDSTEIN and CZAPSKI 1986). It has been suggested that Cu(I) complex binds DNA, and subsequent oxidation by H_2O_2 causes the damage due to the •OH at the binding site (QUE et al. 1980). However, recent studies on the reaction of H_2O_2 with the Cu(I) ion or Cu(I) complex have indicated that under some conditions •OH is not formed (JOHNSON et al. 1988; MASARWA et al. 1988).

YAMAMOTO and KAWANISHI (1989) showed that Cu(II) plus H_2O_2 induced strong DNA cleavage even without piperidine treatment. Piperidine-labile sites were induced frequently at thymine and guanine residues, and rarely at the adenine residue (Table 2). A Cu(I)-specific chelating agent, bathocuproine, inhibited the DNA damage. Neither ethanol nor mannitol inhibited it. Of alcohols, *tert*-butyl alcohol, having relatively low reactivity to •OH, inhibited the DNA damage most strongly. ESR studies using DMPO showed that the •OH adduct of DMPO was formed during the reaction of Cu(II) with H_2O_2, and that the addition of sodium formate produced the CO_2^- radical adduct of DMPO more efficiently than expected. ESR studies showed that the nitroxide radical was formed from 2,2,6,6-tetramethyl-4-piperidone in the presence of Cu(II) plus H_2O_2, indicating the formation of $^1O_2$ or its equivalent. The effects of scavengers on DNA damage have a high degree of correlation with the effects of scavengers on nitroxide radical production and DMPO-OH formation. The results suggest that the main active species causing DNA damage are more likely copper-peroxide complexes, with a similar reactivity to $^1O_2$ and/or •OH, rather than •OH. A mechanism for Cu(II) plus peroxide-induced DNA damage has been proposed. A unique peroxide bridge between the two copper ions facilitates the initial reduction of the metal centers and subsequent reoxidation to generate the active oxygen species causing DNA damage. The possible mechanism has been supported by KAGAWA et al. (1991) and GEIERSTEANGER et al. (1991), who have solved the single crystal structure of oligonucleotides soaked with Cu(II). TKESHELASHVILI et al. (1991) reported the mutation spectrum of copper-induced DNA damage. The predominant mutagenic sequence changes are single-base substitutions, the most frequent being replacement of a template C by a T. Copper-induced mutations are found predominantly in clusters, suggesting direct interaction of copper ions with specific nucleotide sequences in DNA.

Our previous works raised the possibility of copper involvement in the DNA damage by some carcinogens in vivo. In the presence of Cu(II), 1,2,4-benzenetriol (a benzene metabolite) (KAWANISHI et al. 1989b), 2,5-dihydroxybiphenyl (an *o*-phenylphenol metabolite) (INOUE et al. 1990), and caffeic acid (INOUE et al. 1992) caused damage to isolated DNA through H_2O_2 formation. These carcinogens have not been proved to be mutagenic in bacterial systems. It is of interest that the "nonmutagenic" carcinogens or their metabolites cause oxidative DNA damage in the presence of transition metals.

Most hydrazine derivatives have been shown to be carcinogenic. Recently, methyl radical generated from hydrazine derivatives has been shown to have the ability to alkylate the 8-position of guanine residues (AUGUSTO et al. 1990). YAMAMOTO and KAWANISHI (1992) subsequently investigated carbon-centered radicals or oxygen-derived active species, which are more important in hydrazine-induced DNA damage. With Cu(II), DNA cleavage was caused by every hydrazine derivative tested, and the predominant cleavage site was thymine residue, especially of the 5′-GTC-3′ sequence. The cleavage was not inhibited by •OH scavengers. Since the cleavage pattern was similar to that induced by Cu(I) plus H_2O_2 but not to that induced by Cu(II) plus H_2O_2, it is speculated that the copper-oxygen complex derived from the reaction of H_2O_2 with Cu(I) participates in DNA damage by hydrazines plus Cu(II). During the oxidation of a high concentration of phenylhydrazine by ferricyanide, phenyl radical seemed to cause DNA damage, especially the backbone breakage. These results suggest the possibility that active oxygen species (copper-oxygen complex, •OH) are more important in hydrazine-induced DNA damage than carbon-centered radicals. Similarly, the active oxygen species was shown to participate in Cu(II)-dependent DNA damage by hydroxylamine and its derivative (e.g., a metabolite of 4-hydroxyaminoquinoline 1-oxide) (YAMAMOTO et al. 1993).

G. Manganese

There have been no reports indicating that manganese is carcinogenic to humans (MAGOS 1991). In rats, manganese acetylacetone induced injection-site tumors. The prolonged intraperitoneal administration of manganese sulfate ($MnSO_4$) increased the incidence of lung adenoma in strain A mice. In vivo, Mn(II) was clastogenic in mouse bone marrow (JOARDAR and SHARMA 1990). When mice were administered different doses of $MnSO_4$ orally over a period of 3 weeks, the frequencies of both chromosomal aberrations in bone marrow cells and micronuclei were increased significantly. The clastogenic effects were directly related to the concentrations used. Mn(II) produced an increase in the incidence of SCE in human lymphocytes and induced chromosomal aberrations in cultured mammalian cells (UMEDA and NISHIMURA 1979).

Manganese(II) has the ability to mediate oxidative damage of cellular and isolated DNA by certain carcinogens (ITO et al. 1992; INOUE et al. 1992). Hydrazine and its derivatives induced DNA single- and double-strand breaks in cells pretreated with Mn(II). Concomitantly, hydrazine plus Mn(II) caused cleavage of isolated DNA at every nucleotide with a little weaker cleavage at positions of adenine. The result suggests •OH involvement, since it is known that •OH causes DNA cleavage at every nucleotide with no marked site specificity (KAWANISHI et al. 1986a; CELANDER and CECH 1990). The cleavage was inhibited by •OH scavengers and SOD but not by catalase. In the ESR spin trapping experiment the •OH was consistently trapped by DMPO upon the reaction of hydrazine with Mn(II). The DMPO-OH formation was inhibited by •OH scavengers and SOD, while catalase had no effect. Similar results were obtained with Mn(III). Based on all the above findings, it is considered that Mn(II) readily reacts with O_2 in the presence of hydrazine to produce Mn(III) and O_2^-, and subsequently •OH is generated to cause DNA damage. A similar mechanism could be considered with 1,2-dimethylhydrazine and isoniazid (KAWANISHI and YAMAMOTO 1991; YAMAMOTO and KAWANISHI 1991a).

The process which leads to the formation of •OH and H_2O_2 during the Mn(II)-mediated autoxidation of hydrazine is likely to involve the following reactions.

$$Mn^{3+} + NH_2NH_2 \rightarrow Mn^{2+}(\bullet NHNH_2) + H^+ \quad (1)$$

$$Mn^{2+}(\bullet NHNH_2) + O_2 \rightarrow Mn^{2+} + NH{=}NH + O_2^- + H^+ \quad (2)$$

$$Mn^{2+} + O_2^- \rightarrow MnO_2^+ \quad (3)$$

$$MnO_2^+ + NH_2NH_2 \rightarrow [MnO_2H^+] + \bullet NHNH_2 \quad (4)$$

$$[MnO_2H^+] + H^+ \rightarrow Mn^{3+} + \bullet OH(0.5H_2O_2) + OH^- \quad (5)$$

$$\bullet NHNH_2 \rightarrow 0.5NH_2NH_2 + 0.5NH{=}NH \quad (6)$$

$$1.5NH{=}NH + 1.5O_2 \rightarrow 1.5N_2 + 1.5H_2O_2 \quad (7)$$

$$1.5NH_2NH_2 + 2.5O_2 \rightarrow 1.5N_2 + 2H_2O_2 + H_2O \quad \text{(overall reaction)}$$

The amounts of production of H_2O_2 and consumption of NH_2NH_2 and O_2 agreed stoichiometrically with the overall reaction. The formation of MnO_2^+ appears to play an important role in the process of •OH formation not via H_2O_2.

H. Arsenic, Lead, and Cadmium

Carcinogenic arsenic, beryllium, cadmium, and lead do not have the capability to produce highly reactive species such as •OH from O_2^- or H_2O_2. Therefore, active oxygen species cannot directly participate in the carcinogenesis pathway, although these metal compounds may induce oxidative stress by binding GSH or by inhibiting antioxidant enzymes such as GSH peroxidase (Fig. 1). The inhibition of the DNA repair system may play a more important role in the carcinogenesis pathway of these metal compounds, since SNYDER and LACHMANN (1989) have demonstrated that a number of metal salts bind the SH group and inhibit the DNA repair process in human cells.

Arsenic shows positive clastogenic and negative mutagenic effects, suggesting cocarcinogenicity mediated through inhibition of DNA repair. After treatment with As(V), most of the arsenic in the muclei of cells is the formation of As(III). One possible consequence of the reaction of As(III) with SH groups is the inhibition of DNA repair enzymes (MAGOS 1991).

Relevantly, LEE-CHEN et al. (1992) reported that arsenite inhibited postreplication repair in UV-irradiated cells. On the other hand, because arsenic can be methylated in vivo, the genotoxic role of dimethylarsenic has attracted some attention. YAMANAKA et al. (1989a,b) reported that methylated arsenics induced pulmonary DNA strand breaks via the production of active oxygen species in mice. However, as the dose given to mice was very high, the question has arisen as to the significance of this finding in arsenic carcinogenesis.

The genotoxicity of lead is also due to indirect effects such as interference with DNA-repair processes (HARTWIG et al. 1990). As possible mechanisms of repair inhibition, either the interaction with repair enzymes or the interaction with calcium-regulated processes has been suggested. Calcium acetate prevented the carcinogenic effect of lead compound. Therefore, lead may substitute for Ca(II) in the calcium-specific calmodulin, stimulating calcium-dependent protein kinase (MAGOS 1991). Protein kinase C, which is stimulated by picomolar concentrations of lead (MARKOVAC and GOLDSTEIN 1988), is a receptor for tumor promoters such as phorbol esters. Phorbol esters stimulate the formation of active oxygen species and induce DNA strand breakage (BIRNBOIM 1982; SHACTER et al. 1990; WITZ 1991).

Cadmium and cadmium compounds are carcinogenic both by inhalation and by injection, but have only limited genotoxicity. KOIZUMI et al. (1992) reported that H_2O_2 induced strong DNA single-strand breakage in Leydig cells, a target cell for cadmium carcinogenesis in the rat testis, whereas cadmium did not enhance the H_2O_2-induced DNA damage. Although there remains the possibility that the initiation of carcinogenesis in the rat testis by cadmium is triggered by H_2O_2, which is generated by the metal exposure (ZHONG et al. 1990), cadmium cannot participate in producing highly reactive species causing DNA damage from O_2^- or H_2O_2. Cadmium accumulates mainly in the liver and kidney, where most cadmium is bound to metallothionein. The mechanism of cadmium-induced testicular and prostatic carcinogenesis may be due to a deficiency of metallothionein in testes and prostate of rats (WAALKES et al. 1989). Metallothionein is a potent •OH scavenger, and its antioxidant properties in vivo have been suggested by the observation that metallothionein protected cellular H_2O_2-induced DNA damage (CHUBATSU and MENEGHINI 1993). On the other hand, cadmium/zinc metallothionein was shown to induce strand breaks to isolated DNA. The DNA damage might be caused by a radical species formed by the cysteine residues of metallothionein charged with the heavy metal ions (MULLER et al. 1991). At present, it is not clear whether or not the role of metallothionein is to protect from or enhance cadmium carcinogenesis (CHERIAN et al. 1993).

I. Role of Active Oxygen Species in Carcinogenesis

DNA damage by active oxygen species has attracted much interest in relation to carcinogenesis. Active oxygen species may be involved in initiation, promotion, and conversion of multistage carcinogenesis. These highly reactive species not only cause DNA damage, but may also alter the cellular antioxidant defense system. The defense systems consist of: (a) low molecular weight antioxidants (e.g., vitamin E, GSH, vitamin C, β-carotene, uric acid); (b) antioxidant enzymes (e.g., GSH peroxidase, catalase, SOD, DT diaphorase [i.e., NAD(P)H dehydrogenase], GSSG reductase); and (c) DNA degradation repair enzymes (exonuclease III, endonucleases III and IV, glycosylases, polymerases) (SARAN and BORS 1990). New repair enzymes (MutY glycosylase and MutM glycosylase) prevent mutations by an oxidatively damaged form of guanine in DNA (TCHOU and GROLLMAN 1993). Antioxidants, the free radical scavengers, are shown to be anticarcinogens. Similarly, agents that suppress oxygen radical formation can prevent cancer (TROLL 1991). When compared to their appropriate normal cell counterparts, tumor cells are always low in manganese SOD activity, usually low in copper and zinc SOD activity and almost always low in catalase activity. These findings led to the concept that active oxygen species play important roles in carcinogenesis (SUN 1990).

Free radicals and other reactive oxygen species are constantly formed in the human body. Many of them serve useful physiological functions, but they can be toxic when generated in excess and this toxicity is often aggravated by the presence of ions of such transition metals (HALLIWELL and GUTTERIDGE 1990). Metal ions react with O_2^- and H_2O_2 to produce highly reactive species such as •OH and metal-oxygen complexes in biological systems. The Fenton reaction of Fe(II) with H_2O_2 is the well-known mechanism for the generation of •OH and/or the ferryl ion (IMLAY et al. 1988). Carcinogenic metal compounds such as chromium(VI), Fe(III) NTA, nickel(II), and cobalt (II) produced various types of active oxygen species from H_2O_2 (Fig. 1). These active oxygen species were suggested to give different kinds of site-specific DNA damage (Table 2). Since H_2O_2 can reach the nucleus if it survives in significant concentrations, and may be produced even in the nucleus (PESKIN and SHLYAHOVA 1986; PUNTARULO and CEDERBAUM 1992), this DNA damage is supposed to occur in cells. DIZDAROGLU (1992) also observed oxidative DNA modifications in chromatin of cultured mammalian cells treated with H_2O_2 and in chromatin of organs of animals treated with carcinogenic metal salts. On the other hand, copper(II) and manganese(II) have the ability to mediate both H_2O_2 formation and oxidative DNA damage by certain carcinogens which have no or weak mutagenicity. Thus, the active oxygen species in metal-induced DNA damage play important roles in the metal carcinogenesis and the unknown carcinogenic mechanism of some organic carcinogens.

References

Agarwal K, Sharma A, Talukder G (1989) Effects of copper on mammalian cell components. Chem Biol Interact 69:1–16

Agarwal K, Sharma A, Talukder G (1990) Clastogenic effects of copper sulphate on the bone marrow chromosomes of mice in vivo. Mutat Res 243:1–6

Aiyal J, Berkovits HJ, Floyd RA, Wetterhahn KE (1990) Reaction of chromium(VI) with hydrogen peroxide in the presence of glutathione: reactive intermediates and resulting DNA damage. Chem Res Toxicol 3:595–603

Augusto O, Cavalieri EL, Rogan EG, RamaKrishna NVS, Kolar C (1990) Formation of 8-methylguanine as a result of DNA alkylation by methyl radicals generated during horseradish peroxidase-catalyzed oxidation of methylhydrazine. J Biol Chem 265:22093–22096

Birnboim HC (1982) DNA strand breakage in human leukocytes exposed to a tumor promotor, phorbol myristate acetate. Science 215:1247–1249

Bryan SE, Simon SJ, Vizard DL, Hardy KJ (1976) Interaction of mercury and copper with constitutive heterochromatin and euchromatin in vivo and in vitro. Biochemistry 15:1667–1676

Byrnes RW, Antholine WE, Petering DH (1992) Oxidation-reduction reactions in Ehrlich cells treated with copper-neocuproine. Free Radic Biol Med 13:469–478

Capon DJ, Chen EY, Levinson AD, Seeburg PH, Goeddel DV (1983) Complete nucleotide sequences of the T24 human bladder carcinoma oncogene and its normal homologue. Nature 302:33–37

Celander DW, Cech TR (1990) Iron(II)-ethylenediaminetetraacetic acid catalyzed cleavage of RNA and DNA oligonucleotides: similar reactivity toward single- and double-stranded forms. Biochemistry 29:1355–1361

Cherian MG, Huang PC, Klaassen CD, Liu YP, Longfellow DG, Waalkes MP (1993) National Cancer Institute workshop on the possible roles of metallothionein in carcinogenesis. Cancer Res 53:922–925

Chubatsu LS, Meneghini R (1993) Metallothionein protects DNA from oxidative damage. Biochem J 291:193–198

Costa M, Heck JD (1984) Perspectives on the mechanism of nickel carcinogenesis. Adv Inorg Biochem 6:285–309

Costa M, Mollenhauer HH (1980) Carcinogenic activity of particulate nickel compounds proportional to their cellular uptake. Science 209:515–517

De Flora S, Wetterhahn KE (1989) Mechanism of chromium metabolism and genotoxicity. Life Chem Rep 7:169–244

Dijkwel PA, Wenink PW (1986) Structural integrity of the nuclear matrix: differential effects of thiol agents and metal chelators. J Cell Sci 84:53–67

Dizdaroglu M (1992) Oxidative damage to DNA in mammalian chromatin. Mutat Res 275:331–342

Dizdaroglu M, Rao G, Halliwell B, Gajewski E (1991) Damage to the DNA bases in mammalian chromatin by hydrogen peroxide in the presence of ferric and cupric ions. Arch Biochem Biophy 285:317–324

Ebina Y, Okada S, Hamazaki S, Ogino F, Li FJ, Midorikawa O (1986) Nephrotoxicity and renal cell carcinoma after use of iron- and aluminum-nitrilotriacetate complexes in rats. JNCI 76:107–113

Floyd RA (1990) The role of 8-hydroxyguanine in carcinogenesis. Carcinogenesis 11:1447–1450

Geierstanger BH, Kagawa TF, Chen S-L, Quigley GJ, Ho PS (1991) Base-specific binding of copper(II) to Z-DNA: the 1.3 A single crystal structure of d(m5CGUAm5CG) in the presence of $CuCl_2$. J Biol Chem 266:20185–20191

Goddard JG, Basford D, Sweeney GD (1986) Lipid peroxidation stimulated by iron nitrilotriacetate in rat liver. Biochem Pharmacol 35:2381–2387

Goldstein S, Czapski G (1986) Mechanisms of the reaction of some copper complexes in the presence of DNA with O_2^-, H_2O_2, and molecular oxygen. J Am Chem Soc 108:2244–2250

Goyer RA, Falk HL, Hogan M, Feldman DD, Richter W (1981) Renal tumors in rats given trisodium nitrilotriacetic acid in drinking water for 2 years. JNCI 66:869–880

Halliwell B, Gutteridge JMC (1990) Role of free radicals and catalytic metal ions in human disease: an overview. Methods Enzymol 186:1–85

Hartwig A, Schlepegrell R, Beyersmann D (1990) Indirect mechanism of lead-induced genotoxicity in cultured mammalian cells. Mutat Res 241:75–82.
Higinbotham KG, Rice JM, Diwan BA, Kasprzak KS, Reed CD, Perantoni AO (1992) GGT to GTT transversions in codon 12 of the K-ras oncogene in rat renal sarcomas induced with nickel subsulfide or nickel subsulfide/iron are consistent with oxidative damage to DNA. Cancer Res 52:4747–4751
IARC Working Group (1987) IARC Monographs on the evaluation of carcinogenic risk of chemicals to humans, overall evaluations of carcinogenicity: an updating of IARC Monographs vol 1 to 42. IARC Monogr Eval Carcinog Risk Chem Hum [Suppl 7]
IARC Working Group (1990a) Nitrilotriacetate and its salts. IARC Monogr Eval Carcinog Risk Chem Hum 48:181–214
IARC Working Group (1990b) Chromium and chromium compounds. IARC Monogr Eval Carcinog Risk Chem Hum 49:49–256
IARC Working Group (1990c) Nickel and nickel compounds. IARC Monogr Eval Carcinog Risk Chem Hum 49:257–446
IARC Working Group (1992) Cobalt and cobalt compounds. IARC Monogr Eval Carcinog Risk Chem Hum 52:363–473
Imlay JA, Chin SM, Linn S (1988) Toxic DNA damage by hydrogen peroxide through the Fenton reaction in vivo and in vitro. Science 240:640–642
Inoue S, Kawanishi S (1987) Hydroxyl radical production and human DNA damage induced by ferric nitrilotriacetate and hydrogen peroxide. Cancer Res 47:6522–6527
Inoue S, Kawanishi S (1989) ESR evidence for superoxide, hydroxyl radicals and singlet oxygen produced from hydrogen peroxide and nickel(II) complex of glycylglycyl-L-histidine. Biochem Biophys Res Commum 159:445–451
Inoue S, Yamamoto K, Kawanishi S (1990) DNA damage induced by metabolites of *o*-phenylphenol in the presence of copper(II) ion. Chem Res Toxicol 3:144–149
Inoue S, Ito K, Yamamoto K, Kawanishi S (1992) Caffeic acid causes metal-dependent damage to cellular and isolated DNA through H_2O_2 formation. Carcinogenesis 13:1497–1502
Ito K, Kawanishi S (1991) Site-specific fragmentation and modification of albumin by sulfite in the presence of metal ions or peroxidase/H_2O_2: role of sulfate radical. Biochem Biophys Res Commun 176:1306–1312
Ito K, Yamamoto K, Kawanishi S (1992) Manganese-mediated oxidative damage of cellular and isolated DNA by isoniazid and related hydrazines: non-Fenton-type hydroxyl radical formation. Biochemistry 31:11606–11613
Ito K, Inoue S, Yamamoto K, Kawanishi S (1993) 8-Hydroxydeoxyguanosine formation at the 5′ site of 5′-GG-3′ sequences in double-stranded DNA by UV radiation with riboflavin. J Biol Chem 286:13221–13227
Joardar M, Sharma A (1990) Comparison of clastogenicity of inorganic Mn adminstered in cationic and anionic forms in vivo. Mutat Res 240:159–163
Johnson GRA, Nazhat NB, Saadalla-Nazhat RA (1988) Reaction of the aquacopper(I) ion with hydrogen peroxide: evidence for a Cu(III) (cupryl) intermediate. J Chem Soc Faraday Trans 1 84:501–510
Kagawa TF, Geierstanger BH, Wang AH-J, Ho PS (1991) Covalent modification of guanine bases in double-stranded DNA: the 1.2 A Z-DNA structure of d(CGCGCG) in the presence of $CuCl_2$. J Biol Chem 266:20175–20184
Kasai H, Chung MH, Jones DS, Inoue H, Ishikawa H, Kamiya H, Ohtsuka E, Nishimura S (1991) 8-Hydroxyguanine, a DNA adduct formed by oxygen radicals: its implication on oxygen radical-involved mutagenesis/carcinogenesis. J Toxicol Sci 16 [Suppl 1]:95–105
Kasprzak KS, Hernandez L (1989) Enhancement of hydroxylation and deglycosylation of 2′-deoxyguanosine by carcinogenic nickel compounds. Cancer Res 49: 5964–5968

Kasprzak KS, Diwan BA, Konishi N, Mirsa M, Rice JM (1990) Initiation by nickel acetate and promotion by sodium barbital of renal cortical epithelial tumors in male F344 rats. Carcinogenesis 11:647–652
Kawanishi S, Yamamoto K (1991) Mechanism of site-specific DNA damage induced by methylhydrazines in the presence of copper(II) or manganese(III). Biochemistry 30:3069–3075
Kawanishi S, Inoue S, Sano S (1986a) Mechanism of DNA cleavage induced by sodium chromate(VI) in the presence of hydrogen peroxide. J Biol Chem 261:5952–5958
Kawanishi S, Inoue S, Sano S, Aiba H (1986b) Photodynamic guanine modification by hematoporphyrin is specific for single-stranded DNA with singlet oxygen as a mediator. J Biol Chem 261:6090–6095
Kawanishi S, Inoue S, Yamamoto K (1989a) Site-specific DNA damage induced by nickel(II) ion in the presence of hydrogen peroxide. Carcinogenesis 10:2231–2235
Kawanishi S, Inoue S, Kawanishi M (1989b) Human DNA damage induced by 1,2,4-benzenetriol, a benzene metabolite. Cancer Res 49:164–168
Kawanishi S, Yamamoto K, Inoue S (1989c) Site-specific DNA damage induced by sulfite in the presence of cobalt(II) ion: role of sulfate radical. Biochem Pharmacol 38:3491–3496
Kawanishi S, Inoue S, Yamamoto K (1989d) Hydroxyl radical and singlet oxygen production and DNA damage induced by carcinogenic metal compounds and hydrogen peroxide. Biol Trace Elem Res 21:367–372
Koizumi T, Li ZG, Tatsumoto H (1992) DNA damaging activity of cadmium in Leydig cells, a target cell population for cadmium carcinogenesis in the rat testis. Toxicol Lett 63:211–220
Lee-Chen SF, Yu CT, Jan KY (1992) Effect of arsenite on the DNA repair of UV-irradiated Chinese hamster ovary cells. Mutagenesis 7:51–55
Li Y, Togashi Y, Sato S, Emoto T, Kang J-H, Takeichi N, Kobayshi H, Kojima Y, Une Y, Uchino J (1991) Abnormal copper accumulation in non-cancerous and cancerous liver tissues of LEC rats developing hereditary hepatitis and spontaneous hepatoma. Jpn J Cancer Res 82:490–492
Lin XH, Sugiyama M, Costa M (1991) Differences in the effect of vitamin E on nickel sulfide or nickel chloride-induced chromosomal aberrations in mammalian cells. Mutat Res 260:159–164
Magos L (1991) Epidemiological and experimental aspects of metal carcinogenesis: physicochemical properties, kinetics, and the active species. Environ Health Perspect 95:157–189
Masarwa M, Cohen H, Meyerstein D, Hickman DL, Bakac A, Espenson JH (1988) Reactions of low-valent transition-metal complexes with hydrogen peroxide: are they "Fenton-like" or not? I. The case of Cu^{+}aq and Cr^{2+}aq. J Am Chem Soc 110:4293–4297
Markovac J, Goldstein GW (1988) Picomolar concentration of lead stimulate brain protein kinase C. Nature 334:71–73
Maxan AM, Gilbert W (1980) Sequencing end-labeled DNA with base-specific chemical cleavages. Methods Enzymol 65:499–560
Misra M, Olinski R, Dizdaroglu M, Kasprzak KS (1993) Enhancement by L-histidine of nickel(II)-induced DNA-protein cross-linking and oxidative DNA base damage in the rat kidney. Chem Res Toxicol 6:33–37
Nackerdien Z, Kasprzak KS, Rao G, Halliwell B, Dizdaroglu M (1991) Nickel(II)- and cobalt(II)-dependent damage by hydrogen peroxide to the DNA bases in isolated human chromatin. Cancer Res 51:5837–5842
Muller T, Schuckelt R, Jaenicke L (1991) Cadmium/zinc-metallothionein induces DNA strand breaks in vitro. Arch Toxicol 65:20–26
Okada S, Hamazaki S, Ebina Y, Li F-J, Midorikawa O (1987) Nephrotoxicity and its prevention by vitamin E in ferric nitrilotriacetate-promoted lipid peroxidation. Biochim Biophys Acta 922:28–33

Overvad K, Wang DY, Olsen J, Allen DS, Thorling EB, Bulbrook RD, Hayward JL (1993) Copper in human mammary carcinogenesis: a case-cohort study. Am J Epidemiol 137:409–414
Palit S, Sharma A, Talukder G (1991) Cytotoxic effects of cobalt chloride on mouse bone marrow cells in vivo. Cytobios 68:85–89
Park L-W, Floyd RA (1992) Lipid peroxidation products mediate the formation of 8-hydroxydeoxyguanosine in DNA. Free Radic Biol Med 12:245–250
Peskin AV, Shlyahova L (1986) Cell nuclei generate DNA-nicking superoxide radicals. FEBS Lett 194:317–321
Pott F, Ziem U, Reiffer F-J, Huth F, Ernst H, Mohr U (1987) Carcinogenicity studies on fibers, metal compounds, and some other dusts in rats. Exp Pathol 32:129–152
Puntarulo S, Cederbaum AI (1992) Effect of phenobarbital and 3-methylcholanthrene treatment on NADPH- and NADH-dependent production of reactive oxygen intermediates by rat liver nuclei. Biochim Biophys Acta 1116:17–23
Que BG, Downey KM, So AG (1980) Degradation of deoxyribonucleic acid by a 1,10-phenanthroline-copper complex: the role of hydroxyl radicals. Biochemistry 19:5987–5991
Reid TM, Loeb LA (1993) Tandem double CC→TT mutations are produced by reactive oxygen species. Proc Natl Acad Sci USA 90:3904–3907
Rondoni P, Cudkowicz G (1953) Hydrogen peroxide in tumours. Experientia 9:348–349
Rush JD, Koppenol WH (1986) Oxidizing intermediates in the reaction of ferrous EDTA with hydrogen peroxide. Reactions with organic molecules and ferrocytochrome *c*. J Biol Chem 261:6730–6733
Saran M, Bors W (1990) Radical reactions in vivo – an overview. Radiat Environ Biophys 29:249–262
Saucier MA, Wang X, Re RN, Brown J, Bryan SE (1991) Effects of ionic strength on endogenous nuclease activity in chelated and nonchelated chromatin. j Inorg Biochem 41:117–124
Shacter E, Lopecz RL, Beecham EJ, Janz S (1990) DNA damage induced by phorbol ester-stimulated neutrophils is augmented by extracellular cofactors. Role of histidine and metals. J Biol Chem 265:6693–6699
Shapiro R (1977) Genetic effects of bisulfite (sulfur dioxide). Mutat Res 39:149–176
Shi X, Sun X, Gannett PM, Dalal NS (1992) Deferoxamine inhibition of Cr(VI)-mediated radical generation and deoxyguanine hydroxylation: ESR and HPLC evidence. Arch Biochem Biophys 293:281–286
Shibutani S, Takeshita M, Grollman AP (1991) Insertion of specific bases during DNA synthesis past the oxidation-damaged base 8-oxodG. Nature 349:431–434
Snow ET (1992) Metal carcinogenesis: mechanistic implications. Pharmacol Ther 53:31–65
Snyder RD, Lachmann PJ (1989) Thiol involvement in the inhibition of DNA repair by metals in mammalian cells. Mol Toxicol 2:117–128
Steinhoff D, Mohr D (1991) On the question of a carcinogenic action of cobalt-containing compounds. Exp Pahtol 41:169–174
Stevens RG, Jones DY, Micozzi MS, Taylor PR (1988) Body iron stores and the risk of cancer. N Engl J Med 319:1047–1052
Sugiyama M, Ando A, Ogura R (1989) Effects of vitamin B_2-enhancement of sodium chromate(VI)-induced DNA single strand breaks: ESR study of the action of vitamin B_2. Biochem Biophys Res Commun 159:1080–1085
Sugiyama M, Lin X, Costa M (1991) Protective effect of vitamin E against chromosomal aberrations and mutation iduced by sodium chromate in Chinese hamster V79 cells. Mutat Res 260:19–23
Sugiyama M, Tsuzuki K, Lin X, Costa M (1992) Potentiation of sodium chromate(VI)-induced chromosomal aberrations and mutation by vitamin B_2 in Chinese hamster V79 cells. Mutat Res 283:211–214

Sugiyama M, Tsuzuki K, Haramaki N (1993) DNA single-strand breaks and cytotoxicity induced by sodium chromate(VI) in hydrogen peroxide-resistant cell lines. Mutat Res 299:95–102

Sun Y (1990) Free radicals, antioxidant enzymes, and carcinogenesis. Free Radic Biol Med 8:583–599

Sunderman FW J (1989) Mechanisms of nickel carcinogenesis. Scand J Work Environ Health 15:1–12

Sunderman FW Jr, Maenza RM (1976) Comparisons of carcinogenicities of nickel compounds in rats. Res Commun Chem Pathol Pharmacol 14:319–330

Szatrowski TP, Nathan CF (1991) Production of large amounts of hydrogen perpoxide by human tumor cells. Cancer Res 51:794–798

Tchou J, Grollman AP (1993) Repair of DNA containing the oxidatively-damaged base, 8-oxoguanine. Mutat Res 299:277–287

Tkeshelashvili LK, McBride T, Spence K, Loeb LA (1991) Mutation spectrum of copper-induced DNA damage. J Biol Chem 266:6401–6406

Troll W (1991) Prevention of cancer by agents that suppress oxygen radical formation. Free Radic Res Commun 12/13(2):751–757

Umeda M, Nishimura M (1979) Inducibility of chromosomal aberrations by metal compounds in cultured mammalian cells. Mutat Res 57:221–229

Umemura T, Sai K, Takagi A, Hasegawa R, Kurokawa Y (1990) Formation of 8-hydroxydeoxyguanosine (8-OH-dG) in rat kidney DNA after intraperitoneal administration of ferric nitrilotriacetate (Fe-NTA). Carcinogenesis 11:345–347

Umemura T, Sai K, Takagi A, Hasegawa R, Kurokawa Y (1991) The effects of exogenous glutathione and cysteine on oxidative stress induced by ferric nitrilotriacetate. Cancer Lett 58:49–56

Waalkes MP, Perantoni A, Rehm S (1989) Tissue susceptibility factors in cadmium carcinogenesis. Correlation between cadmium-induction of prostatic tumors in rats and an apparent deficiency of metallthionein. Biol Trace Elem Res 21:483–490

Wang DY, Olsen J, Allen DS, Thorling EB, Bulbrook RD, Hayward JL (1993) Copper in human mammary carcinogenesis: a case-cohort study. Am J Epidemiol 137:409–414

Witz G (1991) Active oxygen species as factors in multistage carcinogenesis. Proc Soc Exp Biol Med 198:675–682

Yamamoto K, Kawanishi S (1989) Hydroxyl free radical is not the main active species in site-specific DNA damage induced by copper(II) ion and hydrogen peroxide. J Biol Chem 264:15435–15440

Yamamoto K, Kawanishi S (1991a) Site-specific DNA damage induced by hydrazine in the presence of manganese and copper ions: the role of hydroxyl radical and hydrogen atom. J Biol Chem 266:1509–1515

Yamamoto K, Kawanishi S (1991b) Free radical production and site-specific DNA damage induced by hydralazine in the presence of metal ions or peroxidase/hydrogen peroxide. Biochem Pharmacol 41:905–914

Yamamoto K, Kawanishi S (1992) Site-specific DNA damage induced by phenylhydrazine and phenelzine in the presence of Cu(II) ion or Fe(III) complexes: role of active oxygen species and carbon radicals. Chem Res Toxicol 5:440–446

Yamamoto K, Inoue S, Yamazaki A, Yoshinaga T, Kawanishi S (1989) Site-specific DNA damage induced by cobalt(II) ion and hydrogen peroxide: role of singlet oxygen. Chem Res Toxicol 2:234–239

Yamamoto K, Inoue S, Kawanishi S (1993) Site-specific DNA damage and 8-hydroxydeoxyguanosine formation by hydroxylamine and 4-hydroxyaminoquinoline 1-oxide in the presence of Cu(II): role of active oxygen species. Carcinogenesis 14:1397–1401

Yamanaka K, Hasegawa A, Sawamura R, Okada S (1989a) Dimethylated arsenics induce DNA strand breaks in lung via the production of active oxygen in mice. Biochem Biophys Res Commun 165:43–50

Yamanaka K, Ohiba H, Hasegawa A, Swamura R, Okada S (1989b) Mutagenicity of dimethylated metabolites of inorgainc arsenics. Chem Pharma Bull (Tokyo) 37:2753–2756

Yamazaki I, Piette LH (1991) EPR spin-trapping study on the oxidizing species formed in the reaction of the ferrous ion with hydrogen peroxide. J Am Chem Soc 113:7588–7593

Zhong ZJ, Troll W, Koenig KL, Frenkel K (1990) Carcinogenic sulfide salts of nickel and cadmium induce H_2O_2 formation by human polymorphonuclear leukocytes. Cancer Res 50:7564–7570

CHAPTER 17

Metal Mutagenesis

T.G. Rossman

A. Introduction

Interest in the mutagenicity of metal compounds has its origin in the search for a reasonable mechanism for the carcinogenicity of some metal compounds. Epidemiological studies have provided evidence that occupational and environmental exposures to some metal compounds (arsenic, beryllium, cadmium, hexavalent chromium, nickel, and possibly lead and mercury) are associated with human cancers (IARC 1973, 1980, 1987, 1993; Magos 1991). In addition, there are a number of other metal compounds which induce tumors in animals, but for which the human data is not available (IARC 1980). The recent identification and cloning of mutant oncogenes and tumor suppressor genes provides convincing evidence that mutational events in somatic cells are involved in the conversion of normal cells to malignancy. Thus, understanding the types of genetic changes induced by metal compounds should yield important clues for understanding their carcinogenic effects as well as other toxic effects which have a genotoxic component.

The focus of this review will be on mechanisms of gene mutations induced by metal compounds, especially in mammalian cells. No attempt will be made to provide a complete survey of all genotoxicity results. Clastogenesis (induction of chromosome aberrations) and other genotoxic effects, such as the induction of sister chromatid exchanges, morphological transformation, and DNA damage and repair, will be discussed insofar as these effects help to elucidate the molecular mechanisms of mutagenesis. Special attention will be given to "comutagenic" (enhancing) effects of metal compounds, since these often occur at much lower concentrations than the primary effects (mutagenesis, etc.), and might actually be more physiologically relevant. A number of excellent reviews have appeared in recent years dealing with aspects of metal genotoxicity not covered in this review (Beyersmann and Hartwig 1992; Bianchi et al. 1983; Bianchi and Levis 1987; Christie and Costa 1983; Christie and Katsifis 1990; Coogan et al. 1988; Costa et al. 1991; DeFlora and Wetterhahn 1989; DeFlora et al. 1990; Gebhart and Rossman 1991; Hansen and Stern 1984; Kazantzis 1987; Kazantzis and Lilly 1986; Klein et al. 1991a; Leonard 1984; Leonard and Lauwerys 1980, 1987; Leonard et al. 1983, 1986; Magos 1991; Snow 1992; Sunderman 1984).

With the exception of chromate, bacterial mutagenicity systems are notoriously insensitive at detecting metal compounds, although new improved methods show promise (PAGANO and ZEIGER 1992). Although compounds of nickel and arsenic are considered to be human carcinogens, neither cause reversions in bacterial systems (reviewed in GEBHART and ROSSMAN 1991). Carcinostatic platinum complexes represent a special case, in that they cause cross-links in DNA and are usually positive in short-term assays.

The lack of mutagenicity by metal compounds in bacterial assays may be due to a number of factors (ROSSMAN et al. 1984, 1987). Some metal ions may be excluded from bacterial cells. In that case, no toxicity is seen, even at very high concentrations (ROSSMAN et al. 1984). Essential bacterial enzymes may be more sensitive than their mammalian counterparts to metal ion inhibition, resulting in toxicity which masks the mutagenicity. In some

Table 1. Response of metals in bacterial assays

Metal	Induction of SOS response[a]	Ames test[b]	Modified Ames test[c]
As(III)	−	−	
Ag(II)	−		
Ba(II)	−		
Be(II)	−	−	
Ca(II)	−		
Cd(II)	−	−	+
Co(II)	−	−	+
Cr(III)	+	−	
Cr(VI)	+++	+	
Cu(II)	−	−	
Fe(II)	+	+	+
Fe(III)	+		
Hg(II)	−	−	
Mn(II)	+++	−	+
Mn(VII)	+	−	
Mo(VI)	+		
Ni(II)	+	−	
Pb(II)	++	−	
Se(IV)	i	+	
Se(VI)	++	+	
Sn(II)	++	−	
Wo(VI)	+		
Zn(II)	+	+	+

[a] −, no response; +, <5 times background; ++, 5–10 times background; +++, >10 times background; i, inconsistent response. Data is from ROSSMAN et al. (1984, 1991).

[b] +, positive response; −, negative response. References given in ROSSMAN et al. (1991).

[c] Data is from PAGANO and ZEIGER (1992).

cases, insoluble metal compounds are the genotoxic species. Phagocytosis of insoluble (precipitated) compounds of Pb(II), Ba(II), Be(II), Ni(II), and Mn(II) may be an important route of entry into mammalian cells (ROSSMAN et al. 1987). Bacteria are unable to phagocytize.

In the Ames test, the genetic endpoint being measured (reversion) may be inappropriate for compounds which cause mainly deletions, rearrangements, aneuploidy, or gene amplification. When a bacterial system using a broader endpoint (induction of λ prophage, resulting from induction of the SOS system) was utilized, a number of metal compounds which were not mutagenic in the Ames test were able to induce the SOS system (Table 1, data taken from ROSSMAN et al. 1991).

B. Mutagenesis by Oxidative Reactions

Participation of oxygen free radicals in mutagenesis and carcinogenesis has been proposed by numerous investigators in the past 10 years (AMES 1989; CERUTTI 1985; SIMIC 1989; WEI and FRENKEL 1991). The involvement of free radicals is well established in ionizing radiation damage. Recent findings suggest that oxidative processes may also play an important role in metal mutagenesis and carcinogenesis (reviewed in KLEIN et al. 1991a). Carcinogenic metals may act by altering the oxidative status of cells either directly (e.g., via Fenton-type reactions with endogenous hydrogen peroxide) or indirectly by affecting cellular antioxidative defenses such as glutathione (GSH). The reader is referred to Chap. 12 in this volume for details on the DNA damage caused by oxygen radicals.

Neither $\bullet O_2^-$ nor H_2O_2 reacts with DNA in the absence of metal ions (DIZDAROGLU et al. 1991). DNA damage can be induced in isolated DNA by H_2O_2 in the presence of Cu(II), Fe(II), Cr(VI), Pb(II), and Ni(II), but not Cd(II) (KAWANISHI et al. 1986; AIYAR et al. 1991; ROY and ROSSMAN 1992 and unpublished). In addition to causing frank DNA strand breaks, which are believed to contribute to clastogenesis and deletion mutations, active oxygen species also cause oxidative base damage which is likely to play an important role in the generation of point mutations. The highly reactive species $^\bullet OH$ induces a pattern of DNA base damage that is highly characteristic (DIZDAROGLU 1991). The oxidation product 8-oxo-7,8 dihydro-2′ deoxyguanosine (8-oxo dG), produced at guanine residues in DNA, causes G·C → T·A transversions (WOOD et al. 1990; MORIYA et al. 1991). Another oxidation product, hydroxymethyl deoxyuridine (HMdU), can mediate DNA-protein cross-links with histones and nonhistone proteins (MEE and ADELSTEIN 1981). HMdU can apparently cause various types of base pair substitutions and frameshift mutations (SHIRNAME-MORE et al. 1987). A third oxidation product, thymine glycol, mispairs with dG causing T·A → C·G transitions (BASU et al. 1989).

Metal compounds can also interact with thiols. The interaction of metals with GSH has been reviewed (CHRISTIE and COSTA 1984). Metals which have been reported to promote oxidation of GSH include Cu(II), Co(II), Mn(II), Fe(II), and Cr(VI), whereas nonoxidizing complexes are formed by Zn(II), Cd(II), Hg(II), Pb(II), and Ni(II). Metal compounds also cause oxidation of cysteine (HARMON et al. 1984). The free radical intermediates (e.g., thyl radical, •OH) can then react with a number of other molecules, and are probably responsible for the mutagenicity of GSH and cysteine in the presence of kidney extracts, a process that requires metal ions (GLATT et al. 1983; STARK et al. 1988).

C. Confounding Factors in Metal Mutagenesis

Studies on mutagenesis by metal compounds may be complicated by inducible tolerance mechanisms in some cells. Metallothioneins (MT) are small cysteine-rich proteins which bind a number of metals with high affinity. The reader is referred to Chap. 5 in this volume for more detail. Besides cadmium, zinc, and copper salts, a number of other metal salts have also been shown to induce MT synthesis and/or to bind to the MT protein. Hg(II), Co(II), and Ni(II), but not Pb(II), induce MT synthesis in primary cultures of rat hepatocytes (BRACKEN and KLAASSEN 1987), and may do so in the cultured cells used for mammalian mutagenesis experiments. Thus, under some experimental protocols, the metal salt being assayed may itself induce metallothionein, or the results of mutagenicity assays may be confounded by components of the medium or serum which affect the levels of metallothionein in the cell (Rossman and Goncharova, unpublished).

The same argument can be made for other inducible proteins which may cause tolerance. Heat shock proteins are induced in human fibroblasts by Cd(II), Cu(II), As(III), Zn(II), and Hg(II), but not by Co(II), Ni(II), Fe(II), Fe(III), Mn(II), Pt(II), or Pb(II) (LEVINSON et al. 1980). Like metallothionein, heat shock proteins have been postulated to play a role in the acquisition of tolerance to agents which induce its synthesis. Cd(II) and arsenite also induce heme oxygenase, a protein associated with oxidative stress (KEYSE and TYRRELL 1989).

Rat kidney cells adapted to growth in 5 and $10\,\mu M$ $Pb(NO_3)_2$ become resistant to challenge with otherwise toxic Pb(II) concentrations. This resistance is associated with de novo protein synthesis. A large number of new proteins are synthesized within 6h after the cells are exposed to Pb(II) (HITZFELD et al. 1989). The C6 rat glioma cell line also shows an adaptive response to lead (LAKE et al. 1980) and a similar response has been reported for human cells (SKREB et al. 1981). We have recently identified an inducible mechanism for arsenite resistance, with cross-resistance to arsenate and antimonite, in Chinese hamster V79 cells (WANG and ROSSMAN 1993). This inducible response is different from the heat shock response (WANG et al. 1994). The reader is referred to Chaps. 10, 14, 15, and 17 in this volume for more information on metal-induced responses.

D. Mutagenic Effects of Human Carcinogens

I. Arsenic

Epidemiological studies indicate that inorganic arsenic compounds are human carcinogens. However, no reliable animal model has been established for arsenic carcinogenicity (IARC 1980; LEONARD 1984). Humans are exposed to both arsenite [As(III)] and arsenate [As(V)], the former being the more toxic and presumably more carcinogenic (LEONARD 1984). Metabolism is toward the less toxic forms: As(III) → As(V) → methylated forms (APOSHIAN 1989; VAHTER and MARAFANTE 1989). Unlike many carcinogens, arsenic compounds do not induce mutations in either bacteria or mammalian cells, at least under conditions of high survival (ROSSMAN et al. 1980). Since the nonmutagenicity of arsenite in bacteria or in Chinese hamster cells could not rule out the possibility of mutations caused by large deletions (which are either unselectable or lethal events in these systems), the mutagenicity of arsenite was also assayed in a transgenic line, G12, a pSV2*gpt*-transformed *hprt*$^-$ V79 cell which is able to detect deletions in addition to point mutations (KLEIN and ROSSMAN 1990). Arsenite was not significantly mutagenic at the *gpt* locus in these cells (LI and ROSSMAN 1991). The arsenic metabolite dimethylarsine, however, was mutagenic in bacteria (YAMANAKA et al. 1990), but it is unlikely that this metabolite would be present in high enough concentration in mammalian cells to cause mutagenesis (VAHTER and MARAFANTE 1989). The inability of arsenite to induce the SOS system in *Escherichia coli* is consistent with its lack of direct genotoxicity (ROSSMAN et al. 1984).

Arsenic compounds may act as cocarcinogens or comutagens (enhancing agents) rather than primary carcinogens or mutagens. Arsenite is comutagenic with UV in *E. coli* (ROSSMAN 1981) and with UV, methyl methanesulfonate (MMS), and methyl nitrosourea (MNU) in Chinese hamster cells (LEE et al. 1985; LI and ROSSMAN 1989a; YANG et al. 1992a). In addition to comutagenic effects, arsenic compounds have been shown to induce chromosomal aberrations, sister chromatid exchanges (SCE), and cell transformation (PATON and ALLISON 1972; LARRAMENDY et al. 1981; DIPAOLO and CASTO 1979). It is also an inducer of gene amplification in hamster cells (LEE et al. 1989) and in immortalized human keratinocytes (ROSSMAN and WOLOSIN 1992). Arsenite also enhances the clastogenicity of diepoxybutane (WIENCKE and YAGER 1992). Since there is little evidence that arsenite, the presumed carcinogenic species, interacts with DNA, most of these genetic effects probably occur by mechanisms involving reactions with proteins. Inorganic trivalent arsenicals are especially good inhibitors of enzymes containing vicinal sulfhydryl groups (APOSHIAN 1989).

The comutagenicity of arsenite and its lack of DNA-damaging ability suggests that it may modify the repair of DNA lesions induced by other agents. This theory is supported by the finding that arsenic trioxide inhibits the removal of pyrimidine dimers from the DNA of human SF34 cells after UV irradiation and potentiates the lethal effects of UV in normal and excision-proficient xeroderma pigmentosum variant cells, but not in excision-defective xeroderma pigmentosum cells (OKUI and FUJIWARA 1986). Repair of both UV- and X-ray-induced DNA damage is blocked by arsenite (SNYDER et al. 1989). Arsenite also potentiated X-ray and UV-induced chromosomal damage in peripheral human lymphocytes and fibroblasts (JHA et al. 1992). There is some evidence that arsenite also inhibits postreplication repair of UV-induced damage (LEE-CHEN et al. 1992). Arsenite alters the mutational spectrum (but not the strand bias) of UV-irradiated Chinese hamster ovary cells, suggesting that it may interfere with the processing of mutation fixation at thymine-thymine (TT) and cytosine-thymine (CT) dimers (YANG et al. 1992a).

The mechanism of arsenite's comutagenesis with alkylating agents such as MMS cannot be explained in the same way, since different damage and repair pathways are involved. Alkylating agents react with DNA to form a variety of adducts in which MNU favors primarily the base oxygens while MMS favors the base nitrogens (SINGER and GRUNBERGER 1983). With the exception of O^6-methylguanine, which is removed by O^6-alkylguanine-DNA alkyltransferase (AGT), these DNA adducts are excised by specific DNA glycosylases (LINDAHL 1982). Since Chinese hamster CHO snd V79 cell lines are deficient in AGT activity and are incapable of removing O^6-methylguanine from DNA (WARREN et al. 1979; GOTH-GOLDSTEIN 1980), the comutagenicity of methylating agents with arsenite might therefore involve inhibition of the base excision repair of the other lesions.

The repair of DNA lesions induced by MNU can be monitored by measuring endonuclease-induced single-strand breaks (SSB). A decrease of SSB with time after MNU treatment would indicate completion of DNA repair. The nick translation assay (SNYDER and MATHESON 1985), which measures DNA strand breaks by incorporating radioactive deoxyribonucleoside triphosphate at the free 3′OH ends by endogenous polymerase(s) or with addition of *E. coli* DNA polymerase I in permeabilized cells, was used to demonstrate that the repair of MNU-induced strand breaks is inhibited by arsenite (LI and ROSSMAN 1989a). This result is consistent with the finding that the DNA repair enzyme DNA ligase II is a target of arsenite's action in cells (LI and ROSSMAN 1989b). DNA ligases contain essential sulfhydryl groups (SÖDERHALL and LINDAHL 1987), and are therefore likely targets of other sulfhydryl-binding metal ions as well.

II. Beryllium

In the environment of tissues, beryllium probably exists as colloidal beryllium phosphate/beryllium hydroxide complexes and is accumulated by macrophage-like cells (SKILLETER 1984). Cells in culture also seem to accumulate beryllium by endocytosis of its insoluble form. Beryllium is toxic to mammalian cells only at concentrations where a visible precipitate was seen in the culture medium (ROSSMAN et al. 1987). The lack of toxicity by beryllium compounds in many of the studies (especially in bacteria) suggests possible lack of uptake. Intracellular transfer is from lysozyme to nucleus (reviewed by SKILLETER 1984). Beryllium salts cause blockage of $G_1 \rightarrow S$ transition, in contrast to salts of a number of other metals such as Cd, Mg, Co, Cu, Ni, Zn, and Pb, which produce S-specific blockage (COSTA et al. 1982). The binding of purified DNA to beryllium ion is a very weak interaction ($K_a = 7 \times 10^3$) (TRUHAUT et al. 1968). In the primary hepatocyte culture/DNA repair test, which measures unscheduled DNA synthesis by autoradiographic light nuclear labeling, beryllium sulfate tetrahydrate (0.1–10 mg/ml) was inactive, suggesting a lack of repairable DNA damage by Be(II) (WILLIAMS et al. 1982).

Beryllium salts are inactive as mutagens in most bacterial mutagenesis assays (SIMMON 1979; DUNKEL et al. 1984; KURODA et al. 1991; OGAWA et al. 1987; ASHBY et al. 1990). ARLAUSKAS et al. (1985) reported positive results in fluctuation tests with both *E. coli* and *S. typhimurium* TA 100 (but not other strains), suggesting possible base pair substitution at GC base pairs. A very modest increase in mutations in the *lacI* gene of *E. coli* grown in $BeCl_2$ was reported with no clear dose-response relationship and a mutant spectrum similar to that of the spontaneous spectrum (ZAKOUR and GLICKMAN 1984). $BeCl_2$ (up to 2 m*M*) did not cause Trp^+ reversion or enhance the mutagenicity of UV in *E. coli* (ROSSMAN and MOLINA 1986). In study of comutagenesis by $BeCl_2$ with the frameshift mutagen 9-aminoacridine in *S. typhimurium* TA1537 and TA2637, enhancement was found (OGAWA et al. 1987). In most of these studies, no toxic levels of Be(II) could be established, and it is doubtful that much Be(II) actually entered the cells.

In the only report on gene mutagenicity in mammalian cells, toxic levels of $BeCl_2$ increased 8-azaguanine-resistant mutants in Chinese hamster V79 cells by a factor of about 6 (MIYAKI et al. 1979). LARRAMENDY et al. (1981) found that $BeSO_4$ caused a tenfold increase in chromatid aberrations in human lymphocytes. Even stronger effects were seen in SHE cells. In contrast, ASHBY et al. (1990) report that up to 2.5 mg/ml $BeSO_4$ is nonclastogenic to Chinese hamster lung cells (toxicity was seen only at 2.5 mg/ml), nor were chromosome aberrations noted with the same compound in human diploid fibroblasts and human leukocytes in vitro (PATON and ALLISON 1972). BROOKS et al. (1989) also find little effect of $BeSO_4$ on chromosome damage in CHO cells, but did note an enhancement by fairly high concentrations of $BeSO_4$ on X-ray-induced chromatid-type exchanges.

Nontoxic concentrations of $BeSO_4$ increased the level of sister chromotid exchanges (SCE) in Syrian hamster embryo cells and in human lymphocytes (LARRAMENDY et al. 1981). KURODA et al. (1991) found that $BeCl_2$ and $Be(NO_3)_2$ (but not BeO) induced SCE in V79 cells at concentrations of 63 μg/ml and higher.

Beryllium chloride (1–10 mM) increased the misincorporation of nucleoside triphosphates during polymerization of poly-d(A-T) by *Micrococcus luteus* DNA polymerase, and strongly inhibited the $3' \rightarrow 5'$ exonuclease activity of this enzyme (LUKE et al. 1975). In a similar system, beryllium chloride reduced the fidelity of DNA synthesis in an in vitro assay using avian myeloblastosis virus DNA polymerase. This effect was dose related (2–10 mM) and was ascribed to the noncovalent binding of ionic divalent beryllium to the DNA polymerase rather than to DNA (SIROVER and LOEB 1976). It is not known whether these effects of Be(II) are relevant to concentrations likely to be found within cells.

III. Cadmium

Cadmium had been known to be a rodent carcinogen for some time, but only recently has IARC concluded that it is a human carcinogen (IARC 1993). There is nothing known about the spectrum of mutations induced by Cd(II). At treatments which cause little toxicity, Cd(II) is a weak mutagen at most genetic loci. At a dose which resulted in only 12% survival, $CdSO_4$ induced a tenfold increase in mutations at the thymidine kinase (TK) locus in mouse lymphoma line L51784/$TK^{+/-}$ but only a fourfold increase over background is seen at a dose resulting in 55% survival (OBERLY et al. 1982). $CdCl_2$ also produced about a fourfold increase in mutations at the *hprt* locus in Chinese hamster V79 cells at 70% survival (OCHI and OHSAWA 1983). Both of these loci are able to detect a wide range of mutations, but the TK locus is thought to be superior at detecting clastogens, which tend to cause large deletions (DEMARINI et al. 1989).

The G12 cell line contains a single copy of the *E. coli* gpt gene integrated into the end of chromosome 1 of Chinese hamster V79 cells (KLEIN and ROSSMAN 1990; KLEIN and SNOW 1993). G12 cells show the same mutability toward base pair substitution mutagens as their V79 parent, but give an enhanced mutagenic response toward radiation and oxidative mutagens. $CdCl_2$ can induce gpt^- mutants in these cells at concentrations that are not toxic (2 μM for 24 h) with a maximum response at 4 μM (80% survival) (ROSSMAN et al. unpublished data). However, the mutagenic potency is highly variable from experiment to experiment. Preliminary data on the mutagenic effects of insoluble CdS in G12 cells have so far also been inconsistent. Chromosome aberrations (CA) were induced in Chinese hamster V79 cells treated with 20–50 μM concentrations of Cd(II) for 2 h (OCHI and OHSAWA 1985). However, no aberrations were induced in human lymphocytes treated with up to 6.2 μM for 72 h (BASSENDOWSKA-KARSKA and ZAWADZKAK-KOS 1987). The differences between the treatments (as well as the cells) might explain the different results. A 2-h pulse with a high dose apparently produced DNA damage that is converted to CA, in contrast to a long-term exposure to a lower concentration which may induce the synthesis of metallothionein to protect the cell from the genotoxic effects of Cd(II) (see Sect. C).

One of the paradoxes of cadmium's genotoxic activity is the lack of direct evidence for DNA damage by Cd(II) itself. Treatment of both bacterial and mammalian cells with Cd(II) results in apparent DNA strand breaks as measured by reduced size of DNA in alkaline elution or sucrose gradient (MITRA et al. 1975; ROBISON et al. 1982). Treatment with the insoluble CdS (10 μg/ml for 24 h) also caused a reduction in the DNA size. In V79 cells, as little as 20 μM $CdCl_2$ (2-h exposure) caused decreased DNA size, measured by alkaline elution. However, this effect could only be seen after proteinase K digestion, indicating the formation of DNA-protein cross-links (OCHI and OHSAWA 1983). DNA repair synthesis was stimulated in Syrian hamster embryo cells after a 1-h treatment with 10 μM $CdCl_2$ (ROBISON et al. 1984).

Cd(II) can bind to both the bases and the phosphate groups of DNA, and tends to destabilize the DNA helix (JACOBSON and TURNER 1980). However, these effects do not constitute "DNA damage" in the usual sense of the term (i.e., strand breaks, cross-links, adducts). Incubation of DNA with Cd(II) alone or with H_2O_2 did not result in strand breaks (Roy and Rossman, unpublished data). Cadmium-metallothionein complex is able to induce DNA strand breaks (MÜLLER et al. 1991). However, the production of DNA strand breaks in cells by a cadmium-metallothionein complex is unlikely. Pretreatment of cells with Zn(II) to induce metallothionein resulted in decreased toxicity and DNA strand breaks by Cd(II) (COOGAN et al. 1992). Transfection of a metallothionein overproducing plasmid into G12 cells abolished the mutagenicity of Cd(II) (Goncharova and Rossman, in preparation).

There is evidence that the genotoxic effects of Cd(II) may be mediated by oxidative damage to DNA. Various scavengers of active oxygen species were assayed for their abilities to block CA induced by $CdCl_2$. No protection was see with superoxide dismutase (SOD) or dimethylfuran (a scavenger of singlet oxygen) but catalase, D-mannitol (a scavenger of hydroxyl radicals), and the antioxidant butylated hydroxytoluene (BHT) were protective (OCHI et al. 1987). These results suggest that $CdCl_2$ causes genotoxic effects via H_2O_2, which can form hydroxyl radical in the presence of endogenous iron or copper ions. In support of this is the finding that Cd(II)-induced DNA strand breaks can be blocked by catalase, KI, and mannitol, but not by SOD (SNYDER 1988).

$CdCl_2$ treatment also reduced the cellular glutathione (GSH) level (OCHI et al. 1987). GSH, along with additional antioxidants such as vitamin A, C, and E and antioxidant enzymes, help to reduce cellular damage by hydroxyl radicals. In the absence of such protection one might expect direct DNA damage by endogenous oxygen free radicals as well as lipid peroxidation, whose products can also cause DNA damage and mutations (MUKAI and Goldstein 1976). Cd(II) has been shown to induce lipid peroxidation (Stacey et al. 1980).

In addition to the effects of Cd(II) discussed, insoluble particles of CdS may indirectly cause oxidative damage in vivo via inflammation. CdS, when phagocytized by human polymorphonuclear leukocytes (PMNs), causes an oxidative burst, releasing H_2O_2, which can damage DNA in neighboring cells (Zhong et al. 1990). This response does not occur with soluble $CdSO_4$.

$CdCl_2$ was found to exert a strong comutagenic (synergistic) effect on mutagenicity with the methylating agents methyl nitrosourea (MNU) and *N*-methyl-*N'*-nitro-*N*-nitroguanidine (MNNG) is *S. tymphimurium* (Mandel and Ryser 1984). There was no synergism for toxicity. Its comutagenic effect with MNU was confirmed by Takahashi et al. (1988) in *E. coli* suggesting that Cd(II) can interfere with the repair of a mutagenic, but nontoxic, lesion on DNA caused by methylating agents. One candidate is O^6-methylguanine, which causes G:C to A:T transitions during DNA replication (Snow et al. 1984), and has been proposed as a critical lesion in mutagenesis by methylating agents.

O^6-Methylguanine (as well as O^4-methylthymine and methylphosphotriester) residues on DNA are repaired by O^6-alkylguanine DNA alkyltransferase (AGT). Cd(II) was found to be a strong inhibitor of both mammalian and bacterial AGT (Scicchitano and Pegg 1987; Bhattacharyya et al. 1988). In *E. coli*, Cd(II) inhibits induction of the "adaptive response" to methylating and ethylating agents which involves induction of at least four genes, one of which encodes AGT. The methylated form of AGT is thought to serve as the transcriptional regulator, so that inhibition of the transmethylase reaction by Cd(II) would also affect regulation of the adaptive response. Cd(II) potentiates mutagenesis in *E. coli* by methylating agents but not by UV irradiation or N^4-aminocytidine (Rossman and Molina 1986; Takahashi et al. 1991), results which are consistent with Cd(II) being a specific inhibitor of AGT activity and induction in *E. coli*. The synergism between Cd^{2+} and methylating agents is also reflected in one tumorigenicity study (Wade et al. 1987).

Cd(II) ($4\,\mu M$) blocked UV-induced unscheduled DNA synthesis (i.e., repair replication) in human cells, and this inhibition was blocked by a higher molarity of Zn(II) (Nocentini 1987). The same concentration of Cd(II) caused the accumulation of DNA strand breaks, and this was also blocked by a tenfold molar excess of Zn(II) (Nocentini 1987). Because the initial decrease in size of DNA from UV-irradiated cells was not affected by Cd(II), it is unlikely that the initial incision steps of excision repair were inhibited. However, a more toxic treatment of cells with $CdCl_2$ did inhibit pyrimidine dimer removal (Snyder et al. 1989). UV-induced cytotoxicity was increased by Cd(II), an effect also blocked by Zn(II). Taken together, these results suggest that Cd(II) and Zn(II) compete for a common binding site on enzymes involved in DNA replication and repair. Cd(II) inhibits repair of X-ray-induced strand breaks (DNA ligase?) (Snyder et al. 1989).

Cd(II) has been shown to inhibit human DNA polymerase β, but this inhibition is not blocked by Zn(II) (POPENOE and SCHMAELER 1979). However, since Zn(II) is present in other DNA polymerases as well as in other enzymes involved in DNA metabolism, additional targets for Cd(II) would be expected (JACOBSON and TURNER 1980).

Interference of repair of UV-induced DNA damage would be expected to result in enhancement of UV mutagenesis. Treatment with Cd(II) caused a dose-dependent enhancement (up to $3\,\mu M$) of mutagenesis at the *hprt* locus in UV-irradiated V79 cells. Cd(II) itself was not mutagenic except for a twofold increase at $2\,\mu M$ (HARTWIG and BEYERSMANN 1989a). Thus, at relatively low concentrations, Cd(II) acts as a comutagen with UV in mammalian cells (although not in *E. coli*; compare ROSSMAN and MOLINA 1986). Recently, Cd(II) was also shown to act as a coclastogen with mitomycin C and 4-nitroquinoline 1-oxide in CHO K1 cells, but was not coclastogenic with 4-nitroquinoline 1-oxide in the excision repair deficient human line XP2055V (YAMADA et al. 1993). This data supports the hypothesis that some step in the excision repair pathway is a target for Cd(II).

IV. Chromium

More is known about the mutagenic effects of chromium compounds than any other metal because soluble chromates are easily detected as mutagens in a large number of cell types and loci (reviewed by DEFLORA et al. 1990). Yet even in this case, the molecular mechanisms for chromate's mutagenicity remain elusive. In general, compounds of Cr(VI), but not Cr(III), are mutagenic (BIANCHI et al. 1983). However, the "nonmutagenicity" of Cr(III) may be a function of its limited uptake by cells (BEYERSMANN et al. 1984). When Cr(III) is complexed with an organic ligand such as 2,2′-bipyridyl or 1,10-phenanthroline, which facilitates uptake of Cr(III) into cells (KORTENKAMP et al. 1987), some Cr(III) compounds are mutagenic in *Salmonella* (WARREN et al. 1981; SUGDEN et al. 1990). In addition, when insoluble Cr(III) is phagocytized by mammalian cells, it is mutagenic (BIEDERMAN and LANDOLPH 1990). In *E. coli*, a small amount of λ prophage was induced by Cr(III) (Table 1). In bacteria, chromate causes mutations in *S. typhimurium*, especially in strain TA102, which reverts at an A·T site and is susceptible to oxidative mutagens (BENNICELLI et al. 1983). However, it also causes base pair substitutions and frameshift mutations in other strains (PETRILLI and DEFLORA 1977).

In mammalian cells, Cr(VI) induces mutations at the *hprt* locus in Chinese hamster CHO and V79 cells, but the response is weak compared with the strong mutagenicity in bacteria and tends to occur over a narrow dose range (SUGIYAMA et al. 1991; COHEN et al. 1992). A stronger response is seen at the TK locus in L51784 mouse lymphoma cells (OBERLY et al. 1982). The insoluble lead chromate is an anomaly in that it does not appear to induce mutations (unless solubilized first), but it does act as a clastogen and transforming agent in $10T^{1/2}$ cells (BIEDERMAN and LANDOLPH 1990). Chromate also induces mutations in V79 cells at the Na/K ATPase locus, which detects only base pair substitutions (RAINALDI et al. 1982). The mutation spectrum of chromate has not yet been elucidated, but when five chromate-induced *hprt* mutants were sequenced, five out of the five were at A·T sites (four A·T → T·A transversions and one transition). Of 24 CrO_3^- induced mutants, 22 were at A·T sites (18 transversions and 4 transitions) (YANG et al. 1992b). Since 75% of the mutated T's were located on the nontranscribed strand (which is expected to be repaired more slowly than the transcribed strand), the authors argue that T was the damaged base. In addition, approximately 20% of the mutants did not yield a PCR product from their mRNA, suggesting the possibility of deletions. These results are consistent with oxidative mutagenesis. Chromate was not a strong mutagen at the transgenic *gpt* locus in G12 cells, which usually shows greater mutability with oxidative mutagens compared to *hprt* (KLEIN et al. 1993b), but it gave a stronger response in another transgenic line, G10, which usually has the strongest response to oxidative mutagens.

Cr(VI) (chromate), the carcinogenic species of chromium, is actively transported across the plasma membrane by the anion transport system (CONNETT and WETTERHAHN 1983). Chromate is reduced (ultimately) to Cr(III) by a complex interplay of different cellular reducing agents including cytochrome P_{450}, NADH, glutathione (GSH), ascorbate, DT diaphorase, and H_2O_2 (reviewed in DEFLORA and WETTERHAHN 1989). Exposure of cells or organisms to chromate results in many different types of DNA damage, including DNA interstrand cross-links, DNA-protein cross-links, oxidative base damage, and DNA strand breaks (CUPO and WETTERHAHN 1985; TSAPAKOS and WETTERHAHN 1983; AIYAR et al. 1991; MILLER and COSTA 1989). Damage to DNA occurs only when chromate undergoes intracellular reduction (TSAPAKOS and WETTERHAHN 1983) and is strongly dependent on Cr(V) intermediates (CUPO and WETTERHAHN 1985) as well as active oxygen species (see below). Different pathways of Cr(VI) reduction can lead to different intermediates resulting in different types of DNA damage, but the relative importance of these lesions for mutagenesis has only begun to be evaluated.

Because pretreatment by ascorbate oxidase, which eliminates ascorbate, severely blocked Cr(VI) reduction by rat liver, kidney, and lung ultrafiltrates, it has been argued that ascorbate is the principal reductant of Cr(VI) in these cells (STANDEVEN and WETTERHAHN 1991, 1992). Using lung ultrafiltrate to reduce Cr(VI) in the presence of salmon sperm nuclei, Cr-DNA binding was detected, and could be completely blocked by ascorbate oxidase. Significant inhibition of binding was also seen with *N*-ethylmaleimide, suggesting that sulfhydryls such as GSH may play a significant role in the binding of Cr to DNA (STANDEVEN and WETTERHAHN 1992). Pretreatment of Chinese hamster V-79 cells with ascorbate decreased Cr(VI)-induced alkali-labile sites but increased the cytotoxicity and number of DNA-protein cross-links (SUGIYAMA et al. 1991).

The second most important route of chromate reduction may be via GSH and other thiols. In vitro incubation of plasmid DNA with chromate and GSH induces DNA strand breaks, a process which may require Fe(III) (KORTENKAMP et al. 1989, 1990). In the absense of other metal ions, the reaction of isolated DNA with chromate and GSH resulted in DNA damage that was not caused by active oxygen species, since the GSH thiyl radical was produced, but •OH was not (AIYAR et al. 1991) and DNA-Cr(V)-GSH adducts, but not 8-oxo-dG or DNA strand breaks, were seen (BORGES and WETTERHAHN 1989; BORGES et al. 1991). Increasing the GSH level in cells results in increased Cr(VI)-induced DNA strand breaks, while GSH depletion results in decreased breaks (CUPO and WETTERHAHN 1985; SNYDER 1988). Since the amount of DNA strand breaks in cells treated with Cr(VI) was proportional to the GSH content of the cells, then either Fe(III) or some other metal ions are available in cells, or some of the strand breaks are due to repair of the DNA-Cr(V)-GSH adducts.

Treatment of cells with chromate also results in DNA-protein cross-links. These are repaired much more slowly than are DNA strand breaks or DNA-DNA cross-links (SUGIYAMA et al. 1986). The proteins which are cross-linked include actin and other proteins found in the nuclear matrix, and Cr(III) is present in at least some of these cross-links (WILLER et al. 1991; LIN et al. 1992). Amino acids, especially tyrosine, cysteine, and histidine, can be complexed to DNA by Cr(III), and proteins appear to bind to DNA via these amino acids as well (LIN et al. 1992; SALNIKOW et al. 1992). The binding to DNA is on the phosphates, and is not base specific.

Reduction of Cr(VI) by H_2O_2 produces highly genotoxic species such as singlet oxygen ($^1O_2$), hydroxyl radical (•OH), and superoxide anion (O_2^-) (KAWANISHI et al. 1986; AIYAR et al. 1991). In the presence of H_2O_2, but not in its absence, levels of 8-oxo-dG in calf thymus DNA increased with the concentration of Cr(VI) (AIYAR et al. 1991). Incubation of plasmid DNA with Cr(VI) and H_2O_2 also causes DNA strand breaks (KAWANISHI et al. 1986). Vitamin E, a free radical scavenger, decreases chromium-mediated strand breaks in cells whereas riboflavin (vitamin B2), which reduces Cr(VI) to Cr(V), enhances DNA strand breaks (SUGIYAMA et al. 1991). Neither vitamin affects Cr(III) formation or DNA-protein cross-linking. Isolated Cr(V) intermediates can break plasmid DNA and can cause mutations in bacteria (RODNEY et al. 1989). DNA strand breaks induced in Cr(VI)-treated human fibroblasts can be blocked by SOD and catalase, but not by the hydroxyl radical scavengers mannitol and KI (SNYDER 1988).

Although a great deal of attention has been paid in recent years to the genotoxic consequences of intermediates in the various pathways of chromate reduction, this does not rule out possible effects by the end product, Cr(III). As discussed above, if Cr(III) compounds can enter cells, mutagenesis and clastogenesis can occur. When single-stranded M13mp with a *lacZ* gene insert is treated with $CrCl_3$ and transfected into *E. coli*, a dose-dependent increase in *lacZ* mutations is seen (SNOW 1991). Cr(III) also effects DNA replication. At low concentrations ($0.5-5.0\,\mu M$) the replication rate and processivity is increased while the fidelity is decreased (SNOW and XU 1991), and there is increased bypass of DNA lesions (SNOW 1993). The latter effect may contribute to the synergism seen when cells are treated with Cr(VI) and another mutagen (LAVELLE 1986). At a higher concentration ($10\,\mu M$), replication was inhibited possibly due to DNA-DNA cross-linking. Another consequence of Cr(III) is the inhibition of enzymes involved in the biosynthesis of nucleotide pools (BIANCHI et al. 1982). Unbalanced pools can affect the fidelity of DNA replication and cause mutations in cells (reviewed in KUNZ and KOHALMI 1991).

V. Nickel

Nickel compounds are carcinogenic to humans, causing cancers of the nasosinus, lung, larynx, and bladder (IARC 1987). In animals, the most potent carcinogenic effects are found with particulate water-insoluble nickel compounds, especially crystalline αNi_3S_2, which is among the strongest carcinogens known (SUNDERMAN 1984). Water-soluble nickel compounds are taken up poorly by cells (KASPRZAK et al. 1983). Among the insoluble compounds, the carcinogenic potency varies with the ease with which cells can phagocytize the particles (COSTA and MOLLENHAUER 1980). Particles with a negative surface charge are more easily phagocytized and are more active in transformation in vitro (ABBRACCHIO et al. 1982). Cellular uptake by phagocytosis and solubilization of particulate nickel compounds are necessary for nickel-induced carcinogenesis. Once inside the cell, αNi_3S_2 particles are continuously solubilized (facilitated by the acidic pH of phagocytic vacuoles) to a form, probably hydrated Ni(II), capable of entering the nucleus and interacting with nuclear macromolecules (COSTA et al. 1981).

Despite its potent carcinogenicity, until recently nickel compounds were not thought to be highly mutagenic (COOGAN et al. 1988). Ni(II) compounds are almost always negative in bacterial mutagenicity assays, including *S. typhimurium* strain TA102, which detects oxidative mutagens (BIGGART and COSTA 1986). Since bacteria are not phagocytic, most of these assays have of necessity been of soluble compounds. The only bacterial assay which detects Ni(II)-induced genotoxicity is λ prophage induction (Table 1) (ROSSMAN et al. 1984, 1991).

Earlier work in mammalian systems also tended to use soluble Ni(II) compounds, such as $NiCl_2$. These compounds are, at most, weakly mutagenic at toxic concentrations at the *hprt* locus of Chinese hamster V79 or mouse FM3A cells (MIYAKI et al. 1979; HARTWIG and BEYERSMANN 1989b; MORITA et al. 1991) and at the TK locus in mouse lymphoma cells (AMACHER and PAILLET 1980). Nontoxic concentrations of Ni(II) were able to alter the expression of a temperature-sensitive v-*mos* gene in transformed rat kidney cells (BIGGART et al. 1987), via a 70-base pair duplication (CHIOCCA et al. 1991).

In contrast to its limited mutagenicity, both soluble and insoluble Ni(II) compounds have easily detectable chromosomal effects. Ni(II) ions induce SCEs (HARTWIG and BEYERSMANN 1989b). Chromosomal gaps, deletions, and rearrangements are seen in many cell types after exposure to Ni(II) (reviewed in COOGAN et al. 1988; CHRISTIE and KATSIFIS 1990). Chromosomal aberrations induced by $NiCl_2$ occurred in all chromosomes, but with preference to the heterochromatic centromeric regions. These are protein- and Mg-rich regions containing densely packed DNA which are thought to be transcriptionally silent. Crystalline NiS particles caused similar types of aberrations, but also induced fragmentation of the heterochromatic long arm of the X chromosome (Xp) (SEN and COSTA 1985). A large proportion of Chinese hamster embryo cells transformed by Ni(II) also had partial or complete deletions of Xp (CONWAY and COSTA 1989). Transfer of a normal Chinese hamster X chromosome into these cells caused senescence of the previously immortal cells, suggesting that the Chinese hamster X chromosome contains one or more senescence genes which are targeted by Ni(II) (KLEIN et al. 1991b). It is hypothesized that the dissolution of NiS from phagocytic vacuoles which surround the nucleus delivers a higher dose of Ni(II) to the nucleus, where it immediately encounters heterchromatin situated on the inner nuclear membrane (COSTA and MOLLENHAUER 1980).

Recent studies in the field of nickel mutagenesis have focused more on the particulate carcinogenic forms. The potent carcinogen αNi_3S_2 is (not surprisingly) not mutagenic in *S. typhimurium*, but also causes no significant mutations at the *hprt* locus in V79 cells (ARROUIJAL et al. 1990) or at the Na/K ATPase locus in C3H/10T$^{1/2}$ cells (MIURA et al. 1989). However, in the transgenic G12 cell line, crystalline NiS was a powerful mutagen inducing mutant frequencies >60 times background (CHRISTIE et al. 1992). This result was not seen with soluble $NiSO_4$, which enhanced mutagenesis only twofold. The *gpt*$^-$ mutants induced by NiS were not caused by gene deletion, as full-length PCR products were produced (LEE et al. 1993). G12 cells showed a substantial (>20-fold) mutagenic response to other insoluble nickel compounds, i.e., NiO (black), NiO (green), and Ni_3S_2 (KARGACIN et al. 1993). In contrast, the parental V79 cell line (at the *hprt* locus) did not respond to NiS or NiO (black). The *hprt* gene is located on the short arm of the Chinese hamster X chromosome (FENWICK 1980). As described above, the heterochromatic long arm of the X chromosome is a target for damage by Ni(II) (SEN and COSTA 1985; CONWAY and COSTA 1989). It is possible that much of the damage on the X chromosome is lethal, and thus potential Chinese hamster *hprt*$^-$ mutants cannot be recovered.

Interestingly, another transgenic line, G10 (KLEIN and ROSSMAN 1990), showed a much lower response to the same group of compounds compared with G12 (KARGACIN et al. 1993). These results are surprising because other clastogens (e.g., X-rays, bleomycin) are more mutagenic in G10 (KLEIN et al. 1993b). The two transgenic lines differ in that, although each contains a single copy of the *E. coli gpt* gene, the integration sites differ, as demonstrated by Southern blot (KLEIN and ROSSMAN 1990). Fluorescent in situ hybridization localized the *gpt* gene in G12 to the subtelomeric region of the largest chromosome 1 (KLEIN and SNOW 1993). This is a heterochromatic region which may be particularly sensitive to Ni(II)-induced damage. In G10, the *gpt* gene appears to be on a medium-sized chromosome (Klein, unpublished data). Preliminary mapping data suggests that G12 cells contain a partial plasmid insertion, whereas G10 cells have a partial tandem insertion of plasmid sequences flanking the *gpt* gene (KLEIN et al. 1993a). This repeated region may be responsible for the greater frequency of deletion mutations seen in G10, and its greater sensitivity to most clastogens (KLEIN et al. 1993b). The lower mutagenic response of G10 to nickel compounds indicates that deletion mutagenesis is not the predominant mode of nickel mutagenesis. Some preliminary evidence suggests that at least some of the "mutants" induced in G12 may have arisen via epigenetic mechanisms involving changes in gene expression (LEE and CHRISTIE 1993).

Human kidney epithelial cells can be immortalized by treatment with Ni(II). One such immortalized line had abnormal p53 expression and was found to contain a T → C transition in codon 238 of the p53 gene (MAEHLE et al. 1992). A high proportion of Ni(II)-induced renal carcinomas in F344 rats contain G → T mutations in codon 12 of the K-*ras* oncogene (HIGINBOTHAM et al. 1992). Although these mutations represent a small sample size, the changes are consistent with (but do not prove) oxidative DNA damage.

A number of different nickel-induced DNA lesions might be expected to give rise to mutations. In mammalian cells in vitro, nickel compounds produce both DNA single-strand breaks, which are repaired rapidly, and DNA-protein cross-links, which are not (ROBISON et al. 1982; PATIERNO and COSTA 1985; PATIERNO et al. 1985). Ni(II) alone does not damage DNA, but in the presence of H_2O_2 it produces DNA damage characteristic of •OH. This effect is enhanced in the presence of peptides (NACKERDIEN et al. 1991). The failure of free radical scavengers to prevent DNA damage is attributed to the "site-specific" formation of •OH. There is some evidence that Ni(III) can be formed in cells, and may be involved in redox cycling to give rise to the oxidative damage seen after treatment of cells and animals with nickel compounds (reviewed in KLEIN et al. 1991a). A number of particulate Ni(II) compounds caused large increases in H_2O_2 production in human polymorphonuclear leukocytes (ZHONG et al. 1990).

Ni(II) has a much higher affinity for proteins than for DNA (CICCARELLI and WETTERHAHN 1985), which might help to explain its propensity to react with heterchromatin. Ni_3S_2 was also reported to interfere with spindle proteins, which could lead to aneuploidy (SWIERENGA and MCLEAN 1984). Oxidized bases resulted when chromatin was exposed to Ni(II) alone, and it was suggested that some Ni(II)-protein complexes are able to generate free radicals in the presence of oxygen. The Ni(II)-induced oxidative DNA damage and DNA-protein cross-links seen in rat kidney after injection can be enhanced by injecting a Ni(II)-(histidine)$_2$ complex (MISRA et al. 1993). When the DNA from Ni(II)-treated cells is isolated by a procedure utilizing proteinase K, the major residual amino acids attached to the DNA were cysteine > histidine. In contrast to the DNA-protein cross-links formed with chromium, most of the cross-links induced by Ni(II) do not contain Ni(II), and appear to result from Ni(II)/Ni(III) redox cycling (LIN et al. 1992).

Ni(II) also effects DNA polymerases. It can substitute inefficiently for Mg(II) in some polymerases, resulting in a slower rate of activity along with reduced fidelity (SIROVER and LOEB 1976; MIYAKI et al. 1977; ZAKOUR et al. 1981; CHRISTIE et al. 1991). Recently, it was found that the effects of Ni(II) on different polymerases are quite variable with regard to enzyme inhibition, fidelity, and processivity (SNOW et al. 1993).

Ni(II) has been shown to act as a comutagen with MMS, but not with UVC, in bacteria (DUBINS and LAVELLE 1986; ROSSMAN and MOLINA 1986). In HeLa cells, Ni(II) inhibits the repair of UV-induced DNA damage (SNYDER et al. 1989) and in V79 cells it enhances UV mutagenesis and SCEs (HARTWIG and BEYERSMANN 1989a,b). The comutagenic effect of Ni(II) with UV is greater under conditions which allow more time to repair lesions on DNA, suggesting that the comutagenesis occurs via inhibition of DNA repair by Ni(II). No comutagenic effect is seen with Ni(II) and MMS, leading to the hypothesis that nucleotide excision repair (which repairs UV damage) but not base excision repair (which repairs much of the methylation damage) is inhibited by Ni(II) in mammalian cells (HARTWIG and BEYERSMANN 1989b). The repair replication step in these two pathways may use different polymerases (PERRINO and LOEB 1990), and since, as pointed out above, Ni(II) can affect different polymerases differently, one may speculate that the polymerase acting in the nucleotide excision repair pathway may be most susceptible to Ni(II).

E. Mutagenic Effects of Other Metals

I. Lead

Lead compounds induce tumors in rodents but have not been established conclusively as human carcinogens (IARC 1980). They are considered as possible human carcinogens based on the animal data. Synergistic tumorigenic effects with some, but not all, rodent carcinogens have been reported for lead compounds (reviewed in MAGOS 1991). Pb(II) can also induce cell transformation (DIPAOLO et al. 1978), and can enhance viral transformation in vitro (CASTO et al. 1979). In the few studies carried out on mutagenicity in bacterial systems, the results were usually negative (NISHIOKA 1975;

KANEMATSU et al. 1980) except for the induction of lambda prophage in *E. coli* (ROSSMAN et al. 1984).

Both soluble and insoluble Pb(II) compounds were weakly mutagenic at the *hprt* locus in Chinese hamster V79 cells (ZELIKOFF et al. 1988). When G12 cells were exposed to lead compounds, a small but significant mutagenic response at the transgenic *gpt* locus was seen only at toxic (20% survival) doses of Pb(II) (ROY and ROSSMAN 1992). The mechanism of mutagenesis at high doses is unclear. No direct DNA damage such as strand breaks or DNA-protein cross-links were detected by alkaline elution (ZELIKOFF et al. 1988) or nucleoid sedimentation (HARTWIG et al. 1990). However, nick translation to detect DNA strand breaks in permeabilized cells treated with Pb(II) revealed increased nucleotide incorporation by *E. coli* DNA polymerase I, but only a small increase by endogenous polymerases (ROY and ROSSMAN 1992). The nick translation results suggest that high concentrations of lead(II) might introduce nicks into chromosomal DNA, inhibit the ligation of preexisting nicks, or both. These nicks may be caused by hydroxyl radicals. When cells were treated simultaneously with lead acetate and the hydroxyl radical scavenger potassium iodide, the introduction of nick-translatable sites was partially blocked (Roy and Rossman, unpublished data). It is not clear why no DNA strand breaks were seen using the alkaline elution or nucleoid sedimentation technique. Lead salts at extremely high doses have also been reported to decrease the fidelity of DNA replication (ZAKOUR et al. 1981; SIROVER and LOEB 1976). Pb(II) in the presence of H_2O_2, but not alone, causes nicks in supercoiled plasmid DNA (ROY and ROSSMAN 1992).

It was noted earlier that lead salts are only cytotoxic to mammalian cells at concentrations where an insoluble precipitate is formed, suggesting that phagocytosis may be a more efficient route of entry than ion transport (ROSSMAN et al. 1987). Phagocytosis of some insoluble metal compounds by human polymorphonuclear leukocytes causes the production of H_2O_2 (ZHONG et al. 1990). There is also some indication that the process of phagocytosis itself can generate H_2O_2 inside cells (K. Frenkel, personal communication). Perhaps H_2O_2 generated as a result of the phagocytic process is responsible for the greater mutagenicity of lead acetate, which precipitates in the culture medium, compared with lead nitrate which does not.

Although very high concentrations of Pb(II) were required to detect mutagenesis, enhancement of mutagenicity by UV (HARTWIG et al. 1990; ROY and ROSSMAN 1992) and MNNG (ROY and ROSSMAN 1992) is seen at a lower (nontoxic) Pb(II) doses. This enhancement is thought to be due to the inhibition of DNA repair. Nucleoid sedimentation analysis shows that, in the presence of Pb(II), DNA strand breaks generated during UV repair persist (HARTWIG et al. 1990). This suggests that either the polymerase or the ligase step of repair is inhibited. Pb(II) inhibits the activity of DNA polymerase β (POPENOE and SCHMAELER 1979), an enzyme thought to be involved in DNA repair, and has been reported to inhibit repair replication in X-irradiated cells (SKREB and HABAZIN-NOVAK 1977). Studies using the nick translation assay would seem to favor an effect on DNA ligase, since nick-translatable sites were increased and endogenous polymerase activity was not inhibited by Pb(II) (ROY and ROSSMAN 1992). However, nick-translatable sites after a combination of UV (or MNNG) and lead treatments have not yet been studied. Pb(II) also inhibits O^6-alkyguanine DNA alkyltransferase (BHATTACHARYYA et al. 1988). If this occurs in vivo, Pb(II) would contribute to mutagenesis and perhaps carcinogenesis of small alkylating agents.

II. Mercury

Although there is some evidence that both inorganic Hg(II) and methylmercury chloride (Me-Hg-Cl) cause renal tumors in rodents, there is inadequate evidence for the carcinogenicity of mercury compounds in humans (IARC 1993). The mutagenic and genotoxic effects of mercury compounds have been reviewed (RAMEL 1972; LEONARD et al. 1983; COSTA et al. 1991). In general, inorganic Hg(II) was less effective than organomercury compounds in inducing genetic effects in vitro.

Mercury compounds are highly toxic to both prokaryotic and eukaryotic cells (ROSSMAN et al. 1987). There are no reports on mutagenicity by Hg(II) in bacteria, and it also failed to induce λ prophage (Table 1). There is one report that Hg(II) induced mutations at the *tk* locus in mouse lymphoma L5178Y cells, but, strangely, only in the presence of a rat liver S9 system (OBERLY et al. 1982). Organic Hg compounds have yielded varying responses in mammalian cells. In one study, both Me-Hg-Cl and methoxyethylmercury chloride induced mutations at the Na/K ATPase and *hprt* loci in Chinese hamster V79 cells, whereas another study with the same cells did not detect mutations at the *hprt* locus after treatment with methylmercury hydroxide (FISKESJO 1979; ONFELT and JENSSEN 1982). Me-Hg-Cl did not induce mutations at the *gpt* locus in G10 or G12 cells, nor at the *hprt* locus of the parental V79 cell line (KLEIN et al. 1993a).

Hg(II) enhanced cell transformation by SA7 virus in Syrian hamster embryo cells, but did not induce transformation of human fibroblasts (CASTO et al. 1979; BIEDERMAN and LANDOLPH 1987). It also inhibited O^6-alkylguanine-DNA-alkyltransferase, suggesting the possibility of synergism with small alkylating agents (SCICCHITANO and PEGG 1987; BHATTACHARYYA et al. 1988).

Me-Hg-Cl is more toxic than Hg(II), and both forms can induce DNA strand breaks in cells (reviewed in COSTA et al. 1991). Hg(II)-induced strand breaks are not repaired, and, in fact, Hg(II) may itself inhibit repair of its own lesions since it has been found to inhibit X-ray-induced strand breaks (CHRISTIE et al. 1986; SNYDER et al. 1989) as well as UV-induced damage (SNYDER et al. 1989). Hg(II) is capable of generating oxygen radicals in cells, but the amounts do not appear to be sufficient to account for the high levels of DNA strand breaks seen (CANTONI et al. 1984). Addition of $^{203}Hg(II)$ to cells results in its binding to DNA, and this binding could not be dissociated with EDTA or high NaCl. However, label was released when the DNA was hydrolyzed, suggesting possible binding to phosphodiester groups (COSTA et al. 1991).

Hg(II) induced SCEs only slightly in CHO cells but a greater response was seen in human lymphocytes (HOWARD et al. 1991; MORIMOTO et al. 1982). Chromosomal aberrations have been induced by Hg(II) and by organomercury compounds in a variety of cells (LEONARD et al. 1983; HOWARD et al. 1991). In addition to its clastogenic effects, mercury compounds cause a disturbance of mitosis due to their reactions with the sulfhydryl groups in the spindle fiber proteins. Deleterious effects on the spindle apparatus were seen with Me-Hg-Cl in a number of studies (WATANABE et al. 1982; CURLE et al. 1987). Effects on the spindle apparatus are likely to lead to errors of chromosomal segregation (i.e., aneuploidy) (RAMEL and MAGNUSSON 1979; VERSCHAEVE et al. 1984).

III. Other Metals

Compounds of *antimony* have not been assayed for mutagenicity in mammalian cells. Sb(III) was shown to enhance transformation by SA7 virus (CASTO et al. 1979). Because antimonite resembles arsenite, and can induce arsenite tolerance (WANG and ROSSMAN 1993), one might speculate that antimonite could have a comutagenic effect similar to that of arsenite.

The genetic toxicology of *cobalt* has been recently reviewed (BEYERSMANN and HARTWIG 1992). Co(II) is not mutagenic in bacteria, nor does it induce the SOS system (Table 1). In fact, Co(II) tends to be antimutagenic in bacteria. It is not mutagenic in mouse lymphoma cells (AMACHER and PAILLET 1980) but is very weakly mutagenic at toxic doses in V79 cells (MIYAKI et al. 1979). More pronounced effects are seen in modulation of mutagenicity in mammalian cells. It acts as a comutagen with UV, but not with X-rays, and inhibits removal of UV-induced pyrimidine dimers (SNYDER et al. 1989; BEYERSMANN and HARTWIG 1992). High concentrations of Co(II) can substitute for Mg(II) in some DNA polymerases, resulting in reduced fidelity (SIROVER and LOEB 1976). Co(II) treatment induced DNA strand breaks in mammalian cells (reviewed in BEYERSMANN and HARTWIG 1992). In the presence of H_2O_2, Co(II) causes DNA strand breaks and alkali-labile sites in isolated DNA (YAMAMOTO et al. 1989; NACKERDIEN et al. 1991). Crystalline CoS particles can be phagocytized by cells and cause formation of H_2O_2 in human polymorphonuclear leukocytes (ZHONG et al. 1990).

Although there is little evidence that *copper* compounds alone are mutagenic to cells, Cu(II) can react with some compounds (e.g., ascorbate) to produce genotoxic free radicals (STICH et al. 1979). Since Cu(II) takes part in Fenton reactions, a cell which is in a prooxidant state might be mutagenized by Cu(II) if it were available. In bacteria, Cu(II) is not an SOS inducer (Table 1), but it can enhance UVC mutagenesis (ROSSMAN 1989). This comutagenic effect results at least in part from DNA strand breakage in the presence of Cu(II) + UVC. Cu(II) is also a photosensitizer with UVB (LLOYD et al. 1993) and would be expected to enhance UVB mutagenesis. The repair of UV-induced prymidine dimers was blocked by Cu(II) (SNYDER et al. 1989). Cu(II) causes increased misincorporation during DNA replication (SIROVER and LOEB 1979), and increased depurination of DNA (SCHAAPER et al. 1987). Treatment of plasmid DNA with Cu(II) or Cu(I) causes C → T transitions (TKESHELASHVILI et al. 1991).

There is little evidence for mutagenicity of *iron* compounds in mammalian cells, although there is in bacteria (Table 1). Like copper, iron is also a well-known Fenton reagent participant. In both cases, mammalian cells are protected from potential Fenton reactions by these essential metals by sequestering them. Like Cu(II), Fe(III) also acts as a photosensitizer with UV (LARSON et al. 1992). Fe(II) was reported to enhance SA7 virus transformation (CASTO et al. 1979). Treatment of plasmid DNA with Fe(II) caused G → C transversions most frequently, followed by C → T transitions and G → T transversions MCBRIDE et al. 1991). It is of interest that the mutagenic specificity of Fe(II) was different from that of Cu(II) (see above).

Manganese compounds are mutagenic in bacteria, but only under some conditions and in certain strains, especially *S. typhimuium* TA102 (ZAKOUR and GLICKMAN 1984; ARLAUSKAS et al. 1985; ROSSMAN and MOLINA 1986; DEMÉO et al. 1991). Mn(II) and to a lesser degree Mn(VII), are SOS inducers (Table 1). The genotoxic effects of MnO_4^- are attributed to its conversion to Mn(II) in acidic solution (DEMÉO et al. 1991). Both forms are also clastogenic (JOARDAR and SHARMA 1990). Mn(II) enhanced the clastogenic action of ascorbate (STICH et al. 1979). It was reported to be weakly mutagenic at the *hprt* locus in V79 cells at toxic concentrations (MIYAKI et al. 1970). Treatment of human fibroblasts with high ($\geqslant 10\,mM$) Mn(II) induced DNA strand breaks which were rapidly repaired. Strand breakage could be blocked by the hydroxyl radical scavengers mannitol and KI, but not by SOD or catalase (SNYDER 1988).

Substitution of Mn(II) for Mg(II) in DNA polymerases results in decreased fidelity (GOODMAN et al. 1983; BECKMAN et al. 1985). Under conditions where Mn(II) is not mutagenic, it enhanced UV-induced mutagenesis in *E. coli* (ROSSMAN and MOLINA 1986). A likely explanation for the comutagenic effect is that Mn(II) affects DNA replication of a damaged template. This was shown in the case of alkylation damage, where Mn(II) facilitated postlesion synthesis which was accompanied by misincorporation opposite the alkylated bases (BHANOT and SOLOMON 1994).

There is no evidence that *molybdenum* compounds are mutagenic in mammalian cells. In *E. coli*, Mo(VI) is an SOS inducer (Table 1) and enhances UV mutagenesis (ROSSMAN and MOLINA 1986). Mo(VI) is reported to inhibit X-ray-induced DNA strand breaks (SNYDER et al. 1989).

A great deal of attention has been paid to *selenium* as a chemopreventive agent against carcinogenesis. Yet, compounds of selenium can have either genotoxic or antigenotoxic properties (reviewed by SHAMBERGER 1985). Both selenate [Se(VI)] and selenite [Se(IV)] are mutagenic to *Salmonella* and induce the SOS system in *E. coli* (Table 1). In mammalian cells, both induce chromosome aberrations, with Se(IV) > Se(VI). The same order of reactivity is seen with regard to DNA strand breaks, repair replication, and SCEs. In general, selenocysteine is much less effective than the inorganic forms. Selenite-induced DNA damage and chromosome aberrations depend upon formation of a selenite-GSH conjugant and are not affected by oxygen radical scavengers (WHITING et al. 1980; SNYDER 1988). Stepwise reduction of Se compounds by GSH + GSH reductase is thought to yield HSe^- (selenide) and $GS\text{-}Se^-$ (selenopersulfide), whose free radicals may be the ultimate genotoxic species.

Se(IV) has an antimutagenic effect against several mutagens (reviewed in SHAMBERGER 1985). Se(IV) acts as an antagonist against arsenic and mercury compounds, but Se(VI) may enhance the clastogenic effects of mercury. The antimutagenicity of Se is thought to operate at the physiological levels found in blood, whereas the mutagenic effects occur at higher levels (SHAMBERGER 1985).

Tungsten W(VI) enhanced transformation of SA7 virus (CASTO et al. 1979) and induced the *E. coli* SOS system (Table 1).

Vanadium (V) compounds (NH_4VO_3 and V_2O_5) gave inconclusive results in bacterial mutagenicity assays (HANSEN and STERN 1984). These compounds act as clastogens and induce SCEs in some studies but not in others. Recently, NH_4VO_3 was shown to be a weak mutagen in V79 cells at the *hprt* locus and in G12 at the *gpt* locus (COHEN et al. 1992). DNA-protein cross-links were seen in V(V)-treated cells in the same study.

The mutagenic effects of *zinc* were reviewed in 1986 (LEONARD et al. 1986). At that time it was concluded that most bacterial mutagenesis assays yielded negative results with Zn(II). In the same year, in an evaluation of the *S. typhimurium* assay by the USEPA, Zn(II) was mutagenic in this assay (KIER et al. 1986). Pesticides containing zinc (e.g., zeneb, ziram) are also mutagenic in bacteria (WARREN et al. 1976). However, this has been attributed to the carbamates present. A 3-h exposure to Zn(II) did not induce TK mutants in mouse lymphoma cells (AMACHER and PAILLET 1980). Recently, we have found that long-term growth of cells in some concentrations of Zn(II) results in increased "spontaneous" mutagenesis (GONCHAROVA and ROSSMAN 1994). Treatment of human fibroblasts with high concentrations of Zn(II) caused the appearance of DNA strand breaks which were rapidly repaired (SNYDER 1988). Repair of X-ray-induced strand breaks were inhibited by Zn(II) (SNYDER et al. 1989), which also inhibited the repair enzyme O^6-alkylguanine-DNA-alkyltransferase (SCICCHITANO and PEGG 1987;

Bhattacharyya et al. 1988). Zn(II) also enhanced transformation by SA7 virus (Casto et al. 1979).

Acknowledgements. I thank my colleagues Drs. E. Snow, J. Solomon, M. Costa, E. Goncharova, and C. Klein for their thoughtful discussions and for allowing me the use of their data from manuscripts in press.

References

Abbracchio MP, Heck JD, Costa M (1982) The phagocytosis and transforming activity of crystalline metal sulfide particles are related to their negative surface charge. Carcinogenesis 3:175–180

Aiyar J, Berkovits HJ, Floyd RA, Wetterhahn EK (1991) Reaction of chromium(VI) with glutathione or with hydrogen peroxide: identification of reactive intermediates and their role in chromium(VI)-induced DNA damage. Environ Health Perspect 92:53–62

Amacher DE, Paillet SC (1980) Induction of trifluorothymidine resistant mutants by metal ions in L5178Y/TK$^{+/-}$ cells. Mutat Res 78:279–288

Ames B (1989) Endogenous DNA damage as related to cancer and aging. Mutat Res 214:41–46

Aposhian HV (1989) Biochemical toxicology of arsenic. Rev Biochem Toxicol 10: 265–299

Arlauskas A, Baker SU, Bonin AM, Tandon RK, Crisp PT, Ellis J (1985) Mutagenicity of metal ions in bacteria. Environ Res 36:379–388

Arrouijal FZ, Hildebrand HF, Vophi H, Marzin D (1990) Genotoxic activity of nickel subsulphide α-Ni_3S_2. Mutagenesis 5:583–589

Ashby J, Ishidate JR, Stoner GD, Morgan MA, Ratpan F, Callander RD (1990) Studies on the genotoxicity of beryllium sulfate in vitro and in vivo. Mutat Res 240:217–225

Bassendowska-Karska E, Zawadzkak-Kos M (1987) Cadmium sulfate does not induce sister chromatid exchanges in human lymphocytes in vitro. Toxicol Lett 37:173–174

Basu AK, Loechler EL, Leadon SA, Essigman JM (1989) Genetic effects of thymine glycol: site specific mutagenesis and molecular modeling studies. Proc Natl Acad Sci USA 86:7677–7681

Beckman RA, Mildvan AS, Loeb LA (1985) On the fidelity of DNA replication: manganese mutagenesis in vitro. Biochemistry 24:5810–5817

Bennicelli C, Camoirano A, Petruzzelli S, Zanacchi P, DeFlora S (1983) High sensitivity of Salmonella TA102 in detecting hexavalent chromium mutagenicity and its reversal by liver and lung preparations. Mutat Res 122:1–5

Beyersmann D, Hartwig A (1992) The genetic toxicology of cobalt. Toxicol Appl Pharmacol 115:137–145

Beyersmann D, Köster A, Buttner B (1984) Model reactions of chromium compounds with mammalian and bacterial cells. Toxicol Environ Chem 8:279–286

Bhanot OS, Solomon JJ (1994) The roleof mutagenic metal ions in mediating in vitro mispairing by alkylpyrimidines. Environ Health Perspect (in press)

Bhattacharyya D, Boulden AM, Foote RS, Mitra S (1988) Effect of plyvalent metal ions on the reactivity of human O^6-methylguanine-DNA methyltransferase. Carcinogenesis 9:683–685

Bianchi V, Levis AG (1987) Recent advances in chromium genotoxicity. Toxicol Environ Chem 15:1–24

Bianchi V, Debetto P, Zantedeschi A, Levis AG (1982) Effects of hexavalent chromium on the adenylate pool of hamster fibroblasts. Toxicology 25:19–30

Bianchi V, Celotti L, Lanfranchi G, Majone F, Marin G, Montaldi A, Sponza G, Tamino G, Vernier P, Zantedeschi A, Levis AG (1983) Genetic effects of chromium compounds. Mutat Res 117:279–300
Biederman KA, Landolph JR (1987) Induction of anchorage independence in human diploid foreskin fibroblasts by carcinogenic metal salts. Cancer Res 47:3815–3823
Biederman KA, Landolph JR (1990) Role of valence state and solubility of chromium compounds on induction of cytotoxicity, mutagenesis and anchorage independence in diploid fibroblasts. Cancer Res 50:7835–7842
Biggart NW, Costa M (1986) Assessment of the uptake and mutagenicity of nickel chloride in Salmonella tester strains. Mutat Res 175:209–215
Biggart NW, Gallick GE, Murphy EC (1987) Nickel-induced heritable alterations in retroviral transforming gene expression. J Virol 61:2378–2388
Borges KM, Wetterhahn KE (1989) Chromium cross-links glutathione and cysteine to DNA. Carcinogenesis 10:2165–2168
Borges KM, Boswell JS, Liebross RH, Wetterhahn KE (1991) Activation of chromium (VI) by thiols results in chromium (V) formation, chromium binding to DNA and altered DNA conformation. Carcinogenesis 12:551–561
Bracken WM, Klaassen CD (1987) Induction of metallothionein in rat primary hepatocyte cultures; evidence for direct and indirect induction. J Toxicol Environ Health 22:163–174
Brooks AL, Griffith WC, Johnson NF, Finch GL, Cuddihy RG (1989) The induction of chromosome damage in CHO cells by beryllium and radiation given alone and in combination. Radiat Res 120:494–507
Cantoni O, Christie NT, Swann A, Drath DB, Costa M (1984) Mechanism of $HgCl_2$ cytotoxicity in cultured mammalian cells. Mol Pharmacol 26:360–368
Casto BC, Meyer A, DiPaolo FA (1979) Enhancement of viral transformation for evaluation of the carcinogenic or mutagenic potential of inorganic lead. Cancer Res 39:193–197
Cerutti PH (1985) Pro-oxidant states and tumor promotion. Science 227:375–381
Chiocca SM, Sterner DA, Biggart NW, Murphy EC (1991) Nickel mutagenesis: alteration of the MiSV40110 thermosensitive splicing phenotype by a nickel-induced duplication of the 3′ splice-site. Mol Carcinog 4:61–71
Christie NT, Costa M (1983) In vitro assessment of the toxicity of metal compounds. III. Effects of metals on DNA struture and function in intact cells. Biol Trace Elem Res 5:55
Christie NT, Costa M (1984) In vitro assessment of the toxicity of metal compounds. IV. Disposition of the metals in cells: interactions with membranes glutathione, metallothionein and DNA. Biol Trace Elem Res 6:139–158
Chrisitie NT, Katsifis SP (1990) Nickel carcinogenesis. In: Foulkes EC (ed) Biological effects of heavy metals, vol 2. CRC, Boca Raton, pp 95–128
Christie NT, Cantoni O, Sugiyama M, Cattabeni F, Costa M (1986) Differences in the effcts of Hg(II) on DNA repair induced in Chinese hamster ovary cells by ultraviolet or X-rays. Mol Pharmacol 29:173–178
Christie NT, Chin YE, Snow ET, Cohen MD (1991) Kinetic analysis of Ni^{2+}-effects on DNA replication by polymerase α. J Cell Biochem 15D:114
Christie NT, Tummolo DM, Klein CB, Rossman TG (1992) The role of Ni(II) in mutation. In: Nieboer E, Antio A (eds) Nickel and human health: current perspectives. Wiley, New York, pp 305–317
Ciccarelli RB, Wetterhahn KE (1985) In vitro interaction of 63-nickel(II) with chromatin and DNA from rat kidney and liver nuclei. Chem Biol Interact 52:347–360
Cohen M, Klein C, Costa M (1992) Forward mutations and DNA-protein crosslinks induced by ammonium metavanadate in cultured cells. Mutat Res 269:141–148
Connett PH, Wetterhahn KE (1983) Metabolism of the carcinogenic chromate by cellular constituents. Struct Bond 54:93–124
Conway K, Costa M (1989) Nonrandom chromosomal alterations in nickel-transformed Chinese hamster embryo cells. Cancer Res 49:6032–6038

Coogan TP, Latta DM, Snow ET, Costa M (1988) Toxicity and carcinogenicity of nickel compounds. CRC Crit Rev Toxicol 19:341–394
Coogan TP, Bare RM, Waalkes MP (1992) Cadmium-induced DNA strand damage in cultured liver cells: reduction in cadmium genotoxicity following zinc pretreatment. Toxicol Appl Pharmacol 113:227–233
Costa M, Mollenhauer HH (1980) Carcinogenic activity of particulate nickel compounds is proportional to their cellular uptake. Science 209:515–517
Costa M, Simmons-Hansen J, Bedrossian CWM, Bonura J, Caprioli RM (1981) Phagocytosis, cellular distribution and carcinogenic activity of particulate nickel compounds in cell culture. Cancer Res 41:2868
Costa M, Cantoni O, de Mars M, Swartzendruber DE (1982) Toxic metals produce an S phase-specific cell cycle block. Res Commun Chem Path Pharmacol 38: 405–419
Costa M, Christie NT, Cantoni O, Zelikoff J, Wang XW, Rossman TG (1991) DNA damage by mercury compound. In: Imura N, Clarkson T (eds) Advances in mercury toxicology. Plenum, New York, pp 255–273
Cupo DY, Wetterhahn KE (1985) Modification of chromium (VI)-induced DNA damage by glutathione and cytochromes P-450 in chicken embryo hepatocytes. Proc Natl Acad Sci USA 82:6755–6759
Curle DC, Ray M, Persaud TVN (1987) In vivo evaluation of teratogensis and cytogenetic changes following methylmercuric chloride treatment. Anat Record 2129:286–295
DeFlora S, Wetterhahn KE (1989) Mechanisms of chromium metabolism and genotoxicity. Life Chem Rep 7:169–244
DeFlora S, Bagnasco M, Serra D, Zanucchi P (1990) Genotoxicity of chromium compounds: a review. Mutat Res 238:99–172
DeMéo M, Laget M, Castegnaro M, Duménil G (1991) Genotoxic activity of potassium permanganate in acidic solutions. Mutat Res 260:295–306
DeMarini D, Brockman HE, deSerres FJ, Evans HH, Stankowski LF, Hsie AW (1989) Specific-locus mutations induced in eukaryotes (especially in mammalian cells) by radiation and chemicals: a perspective. Mutat Res 220:11–29
Demple B (1990) Oxidative DNA damage: repair and inducible cellular responses. In: Mendelsohn ML, Albertini RJ (eds) Mutation and the environment, part A. Wiley-Liss, New York, pp 157–167
DiPaolo JA, Casto BLC (1979) Quantitative studies of in vitro morphological transformation of Syrian hamster cells by inorganic metal salts. Cancer Res 39:1008–1313
DiPaolo JA, Nelson RL, Casto BC (1978) In vitro neoplastic transformation of SHE cells by lead acetate and its relevance to environmental carcinogenesis. JNCI 38:452–455
Dizdaroglu M, Rao G, Halliwell B, Tajecwski E (1991) Damage to the bases in mammalian chromation by hydrogen peroxide in the presence of ferric and cupric ions. Arch Biochem Biophys 285:317–324
Dubins JS, LaVelle JM (1986) Nickel(II) genotoxicity: potentiation of mutagenesis of simple alkylating agents. Mutat Res 162:187–199
Dunkel VC, Zeiger E, Brusick D, McÇoy E, McGregor D, Mortelmans K, Rosenkranz HS, Simmon VF (1984) Reproducibility of microbial mutagenicity assays. I. Tests with Salmonella typhimurium and Escherichia coli using a standardized protocol. Environ Mutagen 2(6):1–254
Fenwick RG (1980) Reversion of mutation affecting the molecular weight of HGPRT: intragenic suppression and localization of X-linked genes. Somat Cell Genet 6:477–494
Fiskesjo G (1979) Two organic mercury compounds tested for mutagenicity in mammalian cells by use of the cell line V79-4. Hereditas 90:103–109
Gebhart E, Rossman TG (1991) Mutagenicity, carcinogenicity, teratogenicity. In: Merian E (ed) Metals and their compounds in the environment. VCH, Weinheim, pp 617–641

Glatt H, Protic-Sablijic M, Oesch F (1983) Mutagenicity of glutathione and cysteine in the Ames test. Science 220:961–963

Goncharova EI, Rossman TG (1994) A role for metallothionein and zinc in spontaneous mutagenesis. Cancer Res (in press)

Goodman MF, Keener S, Guidotti S (1983) On the enzymatic basis for mutagenesis by manganese. J Biol Chem 258:3469–3475

Goth-Goldstein R (1980) Inability of Chinese hamster ovary cells to excise O^6-alkylguanine. Cancer Res 40:2623–2624

Hansen K, Stern RM (1984) A surver of metal-induced mutagenicity in vitro and in vivo. J Am Coll Toxicol 3:381–430

Harmon LS, Motley C, Mason RP (1984) Free radical metabolites of L-cysteine oxidation. J Biol Chem 159:5606–5611

Hartwig A, Beyersmann D (1989a) Comutagenicity and inhibition of DNA repair by metal ions in mammalian cells. Biol Trace Elem Res 21:359–365

Hartwig A, Beyersmann D (1989b) Enhancement of UV-induced mutagenesis and sister chromatid exchanges by nickel ions in V79 cells: evidence for the inhibition of DNA repair. Mutat Res 217:65–73

Hartwig A, Schlepegrell R, Beyersmann D (1990) Indirect mechanism of lead-induced genotoxicity in cultured mammalian cells. Mutat Res 241:75–82

Higinbotham KG, Rice JM, Diwan BA, Kasprzak KS, Reed CD, Perantoni AO (1992) GGT to GTT transversions in codon 12 of the K-ras oncogene in rat renal sarcomas induced with nickel subsulfide or nickel subsulfide/iron are consistent with oxidative damage to DNA. Cancer Res 52:4747–4751

Hitzfeld B, Planas-Bohne F, Taylor D (1989) The effect of lead on protein and DNA metabolism of normal and lead-adapted rat kidney cells in culture. Biol Trace Elem Res 21:87–95

Howard V, Leonard B, Moody W, Kochlat TS (1991) Induction of chromosome changes by metal compounds in cultured CHO cells. Toxicol Lett 56:179–186

IARC (1973) Arsenic and inorganic arsenic compounds. IARC Monogr Eval Carcinog Risk Chem Hum 2

IARC (1980) Some metals and metallic compounds. IARC Monogr Eval Carcinog Risk Chem Hum 23

IARC (1987) Overall evaluations of carcinogenicity: an updating of IARC monographs volumes 1–42, supplement 7. International Agency for Research on Cancer, Lyon

IARC (1993) Beryllium, cadmium, mercury and exposures in the glass manufacturing industry. IARC Monogr Eval Carcinog Risk Chem Hum

Jacobson KB, Turner JE (1980) The interaction of cadmium and certain other metal ions with proteins and nucleic acids. Toxicology 16:1–37

Jha AN, Noditi M, Nilsson R, Natarajan AT (1992) Genotoxic effects of sodium arsenite on human cells. Mutat Res 284:215–221

Joardar M, Sharma A (1990) Comparison of clastogenicity of inorganic Mn administered in cationic and anionic forms in vivo. Mutat Res 240:159–163

Kanematsu N, Hara M, Kada T (1980) Rec assay and mutagenicity studies on metal compounds. Mutat Res 77:109–116

Kargacin B, Klein CB, Costa M (1993) Mutagenic response of nickel oxides and nickel sulfides in Chinese hamster V79 cell lines at the xanthine-guanine phosphoribosyl transferase locus. Mutat Res 300:63–72

Kasprzak KS, Gabryel P, Jarczewska K (1983) Carcinogenicity of nickel(II) hydroxides and nickel(II) sulfate in Wistar rats and its relation to the in vitro dissolution rates. Carcinogenesis 4:275–279

Kawanishi S, Inoue S, Sano SJ (1986) Mechanism of DNA cleavage induced by sodium chromate (VI) in the presence of hydrogen peroxide. Biol Chem 261: 5952–5958

Kazantzis G (1987) The mutagenic and carcinogenic effects of cadmium: an update. Toxicol Environ Chem 15:83–100

Kazantzis G, Lilly LJ (1986) Mutagenic and carcinogenic effects of metals. In: Friberg L (ed) Handbook on the toxicology of metals, 2nd edn. Elsevier, Amsterdam

Keyse SM, Tyrrell RM (1989) Heme oxygenase is the major 32Kda stress protein induced in human skin fibroblasts by UVA radiation, hydrogen peroxide and sodium arsenite. Proc Natl Acad Sci USA 86:99–103

Kier LE, Brusick DJ, Auletta AE, von Halle ES, Brown MM, Simmon VF, Dunkel V, McCann J, Mortelmans K, Prival M, Rao TK, Ray V (1986) The *S. typhimurium*/mammalia microsomal assay. A report of the E.S. EPA Gene-Tox Program. Mutat Res 168:69–240

Klein CB, Rossman TG (1990) Transgenic chinese hamster V79 cell lines which exhibit variable levels of gpt mutagenesis. Environ Mol Mutagen 16:1–12

Klein CB, Snow ET (1993) Localization of the gpt sequence in transgenic G12 cells via fluorescent in situ hybridization. Environ Mol Mutagen 21:35a

Klein CB, Frenkel K, Costa M (1991a) The role of oxidative processes in metal carcinogenesis. Chem Res Toxicol 4:592–604

Klein CB, Conway K, Wang XW, Bhamra RK, Lin X, Cohen MD, Annab L, Barrett JC, Costa M (1991b) Senescence of nickel-transformed cells by an X chromosome: possible epigenetic control. Science 251:796–799

Klein CB, Su L, Rossman TG, Snow ET (1993) Transgenic gpt^+ V79 cell lines differ in their mutagenic response to clastogens. Mutat Res 304:217–228

Klein CB, Kargacin B, Su L, Cosentino S, Snow ET, Costa M (1994) Metal mutagenesis in transgenic Chinese hamster cell lines. Environ Health Perspect (in press)

Kortenkamp A, Beyersman D, O'Brien P (1987) Uptake of chromium(III) complexes by erythrocytes. Toxicol Environ Chem 14:23

Kortenkamp A, Ozolins Z, Beyersmann D, O'Brien P (1989) Generation of PM2 DNA breaks in the course of reduction of chromium(VI) by glutathione. Mutation Res 216:19–26

Kortenkamp A, Oetken G, Beyersmann D (1990) The DNA cleavage induced by chromium(V) complex and by chromium and glutathione is mediated by activated oxygen species. Mutation Res 232:155–161

Kunz BA, Kohalmi SE (1991) Modulation of mutagenesis by deoxyribonucleoside levels. Annu Rev Genet 25:339–359

Kuroda K, Endo G, Okamoto A, Yoo YLS, Horiguchi S-I (1991) Genotoxicity of beryllium, gallium and antimony in short-term assays. Mutat Res 264:163–170

Lake LM, O'Cheskey SO, Masuji H, Gerschenson LE (1980) Isolation of lead-resistant cells from an established rat glioma cell line. Chem Biol Interact 30:235–240

LaVelle JM (1986) Chromium(VI) comutagenesis: characterization of the interaction of K_2CrO_4 with azide. Environ Mutagen 8:717–725

Larramendy ML, Popescu NC, DiPaolo JA (1981) Induction by inorganic metal salts of sister chromatid exchanges and chromosome aberrations in human and syrian hamster cell strains. Environ Mutagen 3:597–606

Larson RA, Lloyd RE, Marley KA, Tuveson RW (1992) Ferric ion-photosensitized damage to DNA by hydroxyl and non-hydroxyl radical mechanisms. J Photochem Photobiol Biol 14:345–357

Lee TC, Huang RY, Jan KY (1985) Sodium arsenite enhances the cytotoxicity clastogenicity and 6-thioguanine-resistant mutagenicity of ultraviolet light in Chinese hamster ovary cells. Mutat Res 148:83–89

Lee TC, Tanaka N, Lamb WP, Gilmer TM, Barrett JC (1989) Induction of gene amplification by arsenic. Science 241:79–81

Lee YW, Christie NT (1993) Analysis of 6-TG resistant cells induced by insoluble nickel compounds in hamster G12 cells. Environ Mol Mutatgen 21 [Suppl 22]: 39

Lee YW, Pons C, Tummolo DM, Klein CB, Rossman TG, Christie NT (1993) Mutagenicity of soluble and insoluble nickel compounds at the gpt locus in G12 Chinese hamster cells. Environ Mol Mutagen 21:365–371
Lee-Chen SF, Yu CT, Jan KY (1992) Effect of arsenite on the DNA repair of UV-irradiated Chinese hamster ovary cells. Mutagenesis 7:51–55
Leonard A (1984) Recent advances in arsenic mutagenesis and carcinogenesis. Toxicol Environ Chem 7:241–250
Leonard A, Lauwerys RR (1980) Carcinogenicity and mutagenicity of chromium. Mutat Res 76:227–239
Leonard A, Lauwerys R (1987) Mutagenicity, carcinogenicity and teratogenicity of beryllium. Mutat Res 186:35–42
Leonard A, Jacquet P, Lauwerys RR (1983) Mutagenicity and teratogenicity of mercury compounds. Mutat Res 114:1–18
Leonard A, Gerber GB, Leonard F (1986) Mutagenicity, carcinogenicity and teratogenicity of zinc. Mutat Res 168:343–353
Levinson W, Oppermann H, Jackson J (1980) Transition series metals and sulfhydryl reagents induce the synthesis of four proteins in eukaryotic cells. Biochim Biophys Acta 606:170–180
Li J-H, Rossman TG (1989a) Mechanism of comutagenesis of sodium arsenite with N-methyl-N-nitrosourea. Biol Trace Elem Res 21:373–381
Li J-H, Rossman TG (1989b) Inhibition of DNA ligase activity by arsenite: a possible mechanism of its comutagenesis. Mol Toxicol 2:1–9
Li J-H, Rossman TG (1991) Comutagenesis of sodium arsenite with ultraviolet radiation in Chinese hamster V79 cells. Biol Metals 4:197–200
Lin X, Zhuang Z, Costa M (1992) Analysis of residual amino acid-DNA crosslinks induced in intact cells by nickel and chromium compounds. Carcinogenesis 13:1763–1768
Lindahl T (1982) DNA repair enzymes. Annu Rev Biochem 51:61–87
Lloyd RE, Larson RA, Adair TL, Tuveson RW (1993) Cu(II) sensitizes pBR322 plasmid DNA to inactivation by UV-B (280–315 nm). Photochem Photobiol 57:1011–1017
Luke MZ, Hamilton L, Hollocher TC (1975) Beryllium-induced misincorporation by a DNA-polymerase: a possible factor in beryllium toxicity. Biochem Biophys Res Commun 62:497–501
Maehle L, Metcalf RA, Ryberg D, Bennett WP, Harris CC, Haugen A (1992) Altered p53 gene structure and expression in human epithelial cells after exposure to nickel. Cancer Res 52:218–221
Magos L (1991) Epidemiological and experimental aspects of metal carcinogenesis: physicochemical propertics, kinetics, and the active species. Environ Health Perspect 95:157–189
Mandel R, Ryser HJP (1984) Mutagenicity of cadmium in Salmonella typhimurium and its synergism with two nitrosamines. Mutat Res 18:9–16
McBride TJ, Preston BD, Loeb LA (1991) Mutagenic spectrum resulting from DNA damage by oxygen redicals. Biochemistry 30:207–213
Mee KL, Adelstein J (1981) Predominance of core histones in formation of DNA-protein crosslinks in gamma-irradiated chromatin. Proc Natl Acad Sci USA 78:2194–2198
Miller CA III, Costa M (1989) Immunological detection of DNA-protein complexes induced by chromate. Carcinogenesis 10:667–672
Miller CA, Cohen MD, Costa M (1991) Complexing of actin and other nuclear proteins to DNA by cis-diamminodichloroplatinum (II) and chromium compounds. Carcinogenesis 12:269–276
Misra M, Olinski R, Dizdaroglu M, Kasprzak K (1993) Enhancement by L-histidine of nickel(II)-induced DNA-protein cross-linking and oxidative DNA base damage in the rat kidney. Chem Res Toxicol 6:33–37

Mitra RS, Gray RH, Chin B, Bernstein IA (1975) Molecular mechanisms of accommodation in Escherichia coli to toxic levels of Cd^{2+}. J Bacteriol 121:1180–1188

Miura T, Patiern SR, Sakuramoto T, Landolph JR (1989) Morphological and neoplastic transformation of C3H/10T$_{1/2}$Cl 8 mouse embryo cells by insoluble carcinogenic nickel compounds. Environ Mol Mutagen 14:65–78

Miyaki M, Murata I, Osabe M, Ono T (1977) Effect of metal cations on misincorporation by E. coli DNA polymerases. Biochem Biophys Res Commun 77: 854–860

Miyaki M, Akamatsu N, Ono T, Koyama H (1979) Mutagenicity of metal cations in cultured cells from Chinese hamster. Mutat Res 68:259–263

Morimoto K, Iijima S, Koizumi A (1982) Selenite prevents the induction of sister-chromatid exchanges by methyl mercury and mercuric chloride in human whole-blood cultures. Mutat Res 10:183–192

Morita H, Umeda M, Ogawa HI (1991) Mutagenicity of various chemicals including nickel and cobalt compounds in cultured mouse FM3A cells. Mutat Res 261: 131–137

Moriya M, Ou C, Bodipudi V, Takeshita M, Grollman AA (1991) Site specific mutagenesis using a gapped duplex vector: a study of translesion synthesis past 8-oxodeoxyguanosine in E. coli. Mutat Res 254:281–288

Mukai FH, Goldstein BD (1976) Mutagenicity of malonaldehyde, a decomposition product of peroxidized polyunsaturated fatty acid. Science 191:868–869

Müller T, Schuckelt R, Jaenicke L (1991) Cadmium/zinc-metallothionein induces DNA strand breaks in vitro. Arch Toxicol 65:20–26

Nackerdien Z, Kasprzak KS, Rao G, Halliwell B, Dizdaroglu M (1991) Nickel(II)- and cobalt(II)-dependent damage by hydrogen peroxide to the DNA bases in isolated chromatin. Cancer Res 51:5837–5842

Nishioka H (1975) Mutagenic activities of metal compounds in bacteria. Mutat Res 31:185–189

Nocentini S (1987) Inhibition of DNA replication and repair by cadmium in mammalian cells. Protective interaction by zinc. Nucleic Acids Res 15:4211–4225

Oberly TJ, Piper CE, McDonald DS (1982) Mutagenicity of metal salts in the L5178Y mouse lymphoma assay. J Toxicol Environ Health 9:367–376

Ochi T, Ohsawa M (1983) Induction of 6-thioguanine-resistant mutants and single-strand scission of DNA by cadmium chloride in cultured Chinese hamster cells. Mutat Res 111:69–78

Ochi T, Ohsawa M (1985) Participation of active oxygen species in the induction of chromosomal aberrations by cadmium chloride in cultured Chinese hamster cells. Mutat Res 143:137–142

Ochi T, Takahashi K, Ohsawa M (1987) Indirect evidence for the induction of a prooxidant state by cadmium chloride in cultured mammalian cells and a possible mechanism for the induction. Mutat Res 180:257–266

Ohawa HI, Tsuruta S, Niyitani Y, Mino H, Sakata K, Kato Y (1987) Mutagenicity of metal salts in combination with 9-amino-acridine in Salmonella typhimurium. Jinrui Idengaku Zasshi 62:159–162

Okui T, Fujiwara Y (1986) Inhibition of human excision DNA repair by inorganic arsenic and the comutagenic effect in V79 Chinese hamster cells. Mutat Res 172:69–76

Onfelt A (1983) Spindle disturbances in mammalian cells. I. Changes in the quantity of free sulfhydryl groups in relation to survival and c-mitosis in V79 Chinese hamster cells after treatment with colcemid, diamide, carbaryl and methyl mercury. Chem Biol Interact 46:201–217

Onfelt A, Jenssen D (1982) Enhanced mutagenic response of MNU by posttreatment with methylmercury, caffeine or thymidine in V79 Chinese hamster cells. Mutat Res 106:297–303

Pagano DA, Zeiger E (1992) Conditions for detecting divalent metals as direct-acting mutagens in Salmonella. Environ Mol Mutagen 19:139–146

Patierno SR, Costa M (1985) DNA-protein cross-links induced by nickel compounds in intact cultured mammalian cells. Chem Biol Interact 55:75–91

Patierno SR, Sugiyama M, Basilion JP, Costa M (1985) Preferential DNA-protein crosslinking by $NiCl_2$ in magnesium insoluble regions of fractionated Chinese hamster ovary cell chromatin. Cancer Res 45:5787

Paton GR, Allison AC (1972) Chromosome damage in human cell cultures induced by metal salts. Mutat Res 16:332–336

Perrino FW, Loeb LA (1990) Animal cell DNA polymerases in DNA repair. Mutat Res 236:289–300

Petrilli FL, DeFlora S (1977) Toxicity and mutagenicity of hexavalent chromium on Salmonella typhimurium. Appl Environ Microbiol 33:805–809

Popenoe EA, Schmaeler MA (1979) Interaction of human DNA polymerase β with ions of copper, lead and cadmium. Arch Biochem Biophys 106:190–201

Rainaldi G, Colella CM, Piras A, Marini T (1982) Thioguanine resistance, ouabain resistance and sister chromatid exchanges in V79/AP4 Chinese hamster cells treated with potassium dichromate. Chem Biol Interact 42:45–51

Ramel C (1972) Genetic effects. In: Friberg L, Vosta D (eds) Mercury in the environment: toxicological effects of epidemiological and toxicological appraisal. CRC, Cleveland, pp 169–181

Ramel C, Magnusson J (1979) Chemical induction of nondisjunction in Drosophila. Environ Health Perspect 31:59–66

Robison SH, Cantoni O, Costa M (1982) Strand breakage and decreased molecular weight of DNA induced by specific metal compounds. Carcinogenesis 3:657–662

Robison SH, Cantoni O, Costa M (1984) Analysis of metal-induced DNA lesions and DNA repair replication in mammalian cells. Mutat Res 131:173–181

Rodney PF, Robert JJ, Lay PA, Dixon NE, Raker RSU, Bonin AM (1989) Chromium (V)-induced cleavage of DNA: are chromium (V) complexes the active carcinogens in chromium (VI)-induced cancer? Chem Res Toxicol 2: 227–229

Rossman TG (1981) Enhancement of UV-mutagenesis by low concentrations of arsenite in E. coli. Mutat Res 91:207–211

Rossman TG (1989) On the mechanism of the comutagenic effect of Cu(II) with ultraviolet light. Biol Trace Elem Res 21:383–388

Rossman TG, Molina M (1986) The genetic toxicology of metal compounds. II. Enhancement of ultraviolet light-induced mutagenesis in E. coli WP2. Environ Mutagen 8:263–271

Rossman TG, Wolosin D (1992) Differential susceptibility to carcinogen-induced amplification of SV40 and dhfr sequences in SV40-transformed human keratinocytes. Mol Carcinogen 6:203–213

Rossman TG, Stone D, Molina M, Troll W (1980) Absence of arsenite mutagenicity in E. coli and Chinese hamster cells. Environ Mutagen 2:371–379

Rossman TG, Molina M, Meyer LW (1984) The genetic toxicology of metal compounds. I. Induction of λ prophage in E. coli $WP2_s$ (λ). Environ Mutagen 6:59–69

Rossman TG, Zelikoff JT, Agarwal S, Kneip TJ (1987) Genetic toxicology of metal compounds: an examination of appropriate cellular models. Toxicol Environ Chem 14:251–262

Rossman TG, Molina M, Meyer L, Boon P, Klein CB, Wang Z, Li F, Lin WC, Kinney PL (1991) Performance of 133 compounds in the lambda induction endpoint of the Microscreen assay and a comparison with Salmonella mutagenicity and rodent carcinogenicity bioassays. Mutat Res 260:349–367

Roy NK, Rossman TG (1992) Mutagenesis and comutagenesis by lead compounds. Mutat Res 298:97–103

Salnikow K, Zhitkovich A, Costa M (1992) Analysis of the binding sites of chromium to DNA and protein in vitro and in intact cells. Carcinogenesis 13:2341–2346
Schaaper RM, Koplitz RM, Tkeshelashvili LK, Loeb LA (1987) Metal-induced lethality and mutagenesis: possible role of apurinic intermediates. Mutat Res 177:179–188
Schultz PM, Warren G, Kosso C, Rogers S (1982) Mutagenicity of a series of hexacoordinate cobalt(III) compounds. Mutat Res 102:393–400
Scicchitano DA, Pegg AE (1987) Inhibition of O^6-alkylguanine-DNA-alkyltransferase by metals. Mutat Res 192:207–210
Sen P, Costa M (1985) Induction of chromosomal damage in Chinese hamster ovary cells by soluble and particulate nickel compounds: preferential fragmentation of the heterochromatic long arm of the X-chromosome by carcinogenic crystalline NiS particles. Cancer Res 45:2320–2325
Shamberger RJ (1985) The genotoxicity of selenium. Mutat Res 154:29–48
Shirname-More L, Rossman TG, Troll W, Teebor GW, Frenkel K (1987) Genetic effects of 5-hydroxymethyl-2′-deoxyuridine, a product of ionizing radiation. Mutat Res 178:177–186
Simic MG (1989) Mechanisms of inhibition of free-radical processes in mutagenesis and carcinogenesis. Mutat Res 202:377–386
Simmon VF (1979) In vitro mutagenicity assays of chemical carcinogens and related compounds with Salmonella typhimurium. JNCI 63:893–899
Singer B, Grunberger D (1983) Molecular biology of mutagens and carcinogens. Plenum, New York, pp 55–65
Sirover MA, Loeb LA (1976) Metal-induced infidelity during DNA synthesis. Proc Natl Acad Sci USA 73:2331–2335
Skilleter DN (1984) Biochemical properties of beryllium potentially relevant to its carcinogenicity. Toxicol Environ Chem 7:213–228
Skreb Y, Habazin-Novak V (1977) Lead induced modification of the response to X-rays in human cells in culture. Stud Biophys 63:97–103
Skreb Y, Habazin-Novak V, Hors N (1981) The rate of DNA synthesis in Hela cells during combined long-term and acute exposures to lead. Toxicology 19:1–10
Snow ET (1991) A possible role for chromium (III) in genotoxicity. Environ Health Perspect 92:75–81
Snow ET (1992) Metal carcinogenesis: mechanistic considerations. Pharmacol Ther 53:31–65
Snow ET (1994) Effects on chromium on DNA replication in vitro. Environ Health Perspect (in press)
Snow ET, Xu L-S (1991) Chromium (III) bound to DNA templates enhances DNA polymerase processivity during replication in vitro. Biochemistry 30: 11238–11245
Snow ET, Foote RS, Mitra S (1984) Base-pairing properties of O^6-methylguanine in template DNA during in vitro DNA replication. J Biol Chem 259:8095–8100
Snow ET, Xu L-S, Kinney PL (1993) Effects of nickel ions on polymerase activity and fidelity during DNA replication in vitro. Chem Biol Interact 88:155–173
Snyder RD (1988) Role of active oxygen species in metal-induced DNA strand breakage in human diploid fibroblasts. Mutat Res 193:237–246
Snyder RD, Matheson DW (1985) Nick-translation: a new assay for monitoring DNA damage and repair in cultured human fibroblasts. Environ Mutagen 7: 267–279
Snyder RD, Davis GF, Lachmann P (1989) Inhibition by metals of x-ray and ultraviolet-induced DNA repair in human cells. Biol Trace Elem Res 21: 389–398
Söderhall S, Lindahl T (1987) DNA ligase of eukaryotes. FEBS Lett 67:1–8
Stacey NH, Cantilen LR Jr, Klaassen CD (1980) Cadmium toxicity and lipid peroxidation in isolated rat hepatocytes. Toxicol Appl Pharmacol 53:4700–480

Standeven AM, Wetterhahn KE (1991) Ascorbate is the principal reductant of chromium (VI) in rat liver and kidney ultrafiltrates. Carcinogenesis 12: 1733–1737

Standeven AM, Wetterhahn KE (1992) Ascorbate is the principal reductant of chromium (VI) in rat lung ultrafiltrates and cytosols, and mediates chromium-DNA binding in vitro. Carcinogenesis 13:1319–1324

Stark A-A, Zeiger E, Pagano DA (1988) Glutathione mutagenesis in Salmonella typhimurium is a γ-glutamylpeptidase-enhanced process involving active oxygen species. Carcinogenesis 9:771–777

Stich HF, Wei L, Whiting RF (1979) Enhancement of the chromosome-damaging action of some reducing agents. Cancer Res 39:4145–4151

Sugden K, Burris RB, Rogers SJ (1990) An oxygen dependence in chromium mutagenesis. Mutat Res 244:239–244

Sugiyama M, Wang XW, Costa M (1986) Comparison of DNA lesions and cytotoxicity induced by calcium chromate in human, mouse, and hamster cell lines. Cancer Res 46:4547–4551

Sugiyama M, Lin X, Costa M (1991) Protective effect of vitamin E against chromosomal aberrations and mutations induced by sodium chromate in Chinese hamster V79 cells. Mutat Res 260:19–23

Sunderman FW Jr (1984) Carcinogenicity of nickel compounds in animals. IARC Sci Publ 53:127–142

Swierenga SHH, McLean JR (1984) Further insights into mechanisms of nickel-induced DNA damage: studies with cultured rat liver cells. In: Brown SS, Sunderman FW Jr (eds) Progress in Nickel Toxicology. Blackwell Scientific, London, pp 101–104

Takahashi K, Imaeda T, Kawazoe Y (1988) Effect of metal ions on the adaptive response induced by N-methyl-N-nitrosourea in Escherichia coli. Biochem Biophys Res Commun 157:1124–1130

Takahashi K, Imaeda T, Kawazoe Y (1991) Inhibitory effect of cadmium and mercury ions on the induction of the adaptive response in Escherichia coli. Mutat Res 254:45–53

Tkeshelashvili LK, McBride T, Spence K, Loeb LA (1991) Mutation spectrum of copper-induced DNA damage. J Biol Chem 266:6401–6406

Truhaut R, Festy B, LeTalaer J-Y (1968) Interaction of beryllium with DNA and its incidence with some enzymatic system (Fr.). C R Acad Sci [D] (Paris) 266: 1192–1195

Tsapakos MJ, Wetterhahn KE (1983) The interaction of chromium with nucleic acids. Chem Biol Interact 46:265–277

Vahter M, Marafante E (1989) Intracellular distribution and chemical forms of arsenic in rabbits exposed to arsenate. Biol Trace Elem Res 21:233–239

Verschaeve L, Kirsch-Volders M, Susanne C (1984) Mercury-induced segregational errors of chromosomes in human lymphocytes and Indian muntjac cells. Toxicol Lett 21:247–253

Wade GG, Mandel R, Ryser HJP (1987) Marked synergism of dimethylnitrosamine carcinogenesis in rats exposed to cadmium. Cancer Res 47:6606–6613

Wang Z, Rossman TG (1993) Stable and inducible arsenite resistance in Chinese hamster cells. Toxicol Appl Pharmacol 118:80–86

Wang Z, Hou G, Rossman TG (1994) Induction of arsenite tolerance and thermotolerance occur by different mechanisms. Environ Health Perspect (in press)

Warren G, Skaar PD, Rogers SJ (1976) Genetic activity of dithiocarbamate and thiocarbamoyl disulfide fungicides in Saccharomyces cerevesiae, Salmonella typhimurium and Escherichia coli. Mutat Res 38:391–392

Warren G, Schultz P, Bancroft D, Bennett K, Abbott EH, Rogers S (1981) Mutagenicity of a series of hexacoordinate chromium(III) compounds. Mutat Res 90:111–122

Warren W, Crathorn AR, Shooter KV (1979) The stability of methylated purines and of methylphosphotriesters in the DNA of V79 cells after treatment with N-methyl-N-nitrosourea. Biochim Biophys Acta 563:82–88
Watanabe T, Shimada T, Endo A (1982) Effects of mercury compounds on ovulation and meiotic and mitotic chromosomes in female golden hamsters. Teratology 25:381–384
Wei H, Frenkel K (1991) In vivo formation of oxidized DNA bases in tumor promoter-treated mouse skin. Cancer Res 51:4443–4449
Whiting RF, Wei L, Stich HF (1980) Unscheduled DNA synthesis and chromosomal aberrations induced by inorganic and organic selenium compounds in the presence of glutathione. Mutat Res 78:159–169
Wiencke JK, Yager JW (1992) Specificity of arsenite in potentiating cytogenetic damage induced by the DNA crosslinking agent diepoxybutane. Environ Mol Mutagen 19:195–200
Willams GM, Laspia MF, Dunkel VC (1982) Reliability of the hepatocyte primary culture/DNA repair test in testing of coded carcinogens and noncarcinogens. Mutat Res 97:359–370
Wood ML, Dizdaraglu M, Gajewski E, Essigman JM (1990) Mechanistic studies of ionizing radiation and oxidative mutagenesis: genetic effects of a single 8-hydroxyguanine (7-hydro-8 oxoguanine) residue inserted at a unique site in a viral genome. Biochemistry 29:7024–7033
Yamada H, Miyahara T, Sasaki YF (1993) Inorganic cadmium increases the frequency of chemically induced chromosome aberrations in cultured mammalian cells. Mutat Res 302:137–145
Yamamoto K, Inoue S, Yamazaki A, Yoshinaga T, Kawanishi S (1989) Site-specific DNA damage induced by cobalt(II) ion and hydrogen peroxide: role of singlet oxygen. Chem Res Toxicol 2:234–239
Yamanaka K, Hoshino M, Pkamoto M, Sawamura R, Hasegawa A, Okada S (1990) Induction of DNA damage by dimethylarsine, a metabolite of inorganic arsenics, is for the major part likely due to its peroxyl radical. Biochem Biophys Res Commun 168:58–64
Yang J-L, Chen M-F, Wu C-W, Lee T-C (1992a) Posttreatment with sodium arsenite alters the mutation spectrum induced by ultraviolet light irradiation in Chinese hamster ovary cells. Environ Mol Mutagen 20:156–164
Yang J-L, Hsieh Y-C, Wu C-W, Lee T-C (1992b) Mutational specificity of chromium (VI) compounds in the hprt locus of Chinese hamster ovary-K1 cells. Carcinogenesis 13:2053–2057
Zakour RA, Glickman BW (1984) Metal-induced mutagenesis in the lac I gene of Escherichia coli. Mutat Res 126:9–18
Zakour RA, Kunkle TA, Loeb LA (1981) Metal induced infidelity of DNA synthesis. Environ Health Perspect 40:197–205
Zelikoff JT, Li JH, Hartwig A, Wang XW, Costa M, Rossman TG (1988) Genetic toxicology of lead compounds. Carcinogenesis 9:1727–1732
Zhong Z, Troll W, Koenig KL, Frenkel K (1990) Carcinogenic sulfide salts of nickel and cadmium induce H_2O_2 formation by human polymorphonuclear leukocytes. Cancer Res 50:7564–7570

CHAPTER 18

Biological Mechanisms and Toxicological Consequences of the Methylation of Arsenic*

M. STYBLO, M. DELNOMDEDIEU, and D.J. THOMAS

A. Introduction

Since ancient times, arsenic (As) has been recognized as an agent with potent biological effects (FROST 1967). Both Hippocrates and Galen described the medicinal use of As-containing sulfides and oxides (BUCHANAN 1962). In the era before penicillin, the antibiotic potency of organic As (organoAs) compounds made them an important part of the pharmaceutical armamentarium (GOODMAN and GILMAN 1941). The potent toxicity of As results in its continued use as a component of insecticides, herbicides, and rodenticides and as a wood preservative. As has long been used in glass-making, and As-containing semiconductors will likely play an increasingly large role in electronics manufacture.

In addition to occupational exposure to As, environmental exposure to As occurs mainly by the ingestion of organoarsenicals, which are natural components of a wide variety of foods, especially oceanic animals and plants, and by the ingestion of drinking water that contains inorganic As (iAs) as a natural contaminant. Episodes of exposure to As as a natural contaminant of drinking water have been described repeatedly. Probably the most extensively studied of these is the large population in Taiwan which was exposed to high concentrations of As in drinking water for many years (TSENG et al. 1968). A dose-response relation between the prevalence of Blackfoot disease, a peripheral vascular disorder characterized by severe atherosclerosis, and the duration of intake of As-contaminated drinking water has been observed in Taiwan (TSENG 1989). Studies of the health status of this As-exposed population, examining both carcinogenic and non-carcinogenic effects of As exposure, have served as the basis of several evaluations of the risk associated with chronic exposure to As (CHEN et al. 1985, 1988; US Environmental Protection Agency 1988; WU et al. 1989; SMITH et al. 1992).

---

*The first two authors contributed equally to the preparation of this review. This manuscript has been reviewed in accordance with the policy of the Health Effects Research Laboratory, United States Environmental Protection Agency, and approved for publication. Approval does not signify that the contents necessarily reflect the views and policies of the Agency, nor does mention of trade names or commercial products constitute endorsement or recommendation for use.

Based on data from occupational exposures, As has been classified as a human carcinogen (International Agency for Research on Cancer 1987). Although a widely accepted model for As-induced carcinogenesis in an experimental species has not yet been developed, there are reports of As-induced cancer in some species (RUDNAI and BORSONYI 1980; PERSHAGEN et al. 1984; PERSHAGEN and BJORKLUND 1985). The absence of an animal model has impeded our understanding of the molecular basis of As carcinogenicity and toxicity. However, substantial progress has been made in elucidating the distribution, metabolism, and fate of As in humans and other species. Perhaps the most striking aspect of the metabolism of As is the methylation of iAs to monomethyl (MMA), dimethyl (DMA), and trimethyl (TMA) derivatives which has been reported in species as diverse as fungi and humans. Hence, exposure to iAs or an organoAs compound results in the formation of a variety of methylated metabolites which likely differ in their tissue distribution, binding, toxicity, and carcinogenicity. The existence of multiple metabolites following exposure to As poses unique problems in terms of its risk assessment. For example, which As metabolite causes toxicity or carcinogenicity that is observed in an As-exposed individual? What are the factors which determine the formation of the putative toxic or carcinogenic metabolite? Thus, improved risk assessment for As depends greatly on answering questions about the metabolism and fate of this agent.

In this chapter, the literature on the formation, distribution, excretion, toxicity, and carcinogenicity of As-containing species in humans and other species is reviewed. Recommendations for future research conclude this chapter.

B. Biological Methylation of Arsenic

I. Role of Methylation in Arsenic Metabolism

Methylated As (MetAs) species are less acutely toxic than iAs and methylation of iAs is generally considered as a detoxification pathway. However, neither the enzymatic mechanism(s) for formation of organoAs species nor the consequences of the formation of organoAs species and their disposition in tissues is fully understood. In the following paragraphs, the biological basis and toxicological significance of methylation of As are discussed.

An overview of the role of methylation reactions in the toxicity of As has been presented (MCKINNEY 1992). Two working hypotheses can be offered for the role of methylation in the metabolism of As. In the first, methylation reactions are viewed as a means for the detoxification of iAs. Conversely, the second hypothesis is that the methylation of As is a process yielding organoAs species with unique toxic and/or carcinogenic effects. Thus, in this model, methylation of iAs is viewed as a toxification process.

At least on the basis of acute toxicity, the data are consistent with the first hypothesis: namely, that methylation of iAs is a detoxification mechanism. However, organoAs species exert some unique toxic effects. For example, DMA induces an unusual teratologic alteration (altered palatal rugae) in rats (ROGERS et al. 1981), acute lung injury in mice (YAMANAKA et al. 1991), and renal cortical degeneration, proximal tubule necrosis, and papillary necrosis in rats (MURAI et al. 1993). High doses of DMA cause single-strand DNA breaks by a mechanism that may involve formation of a DMA peroxyl radical (YAMANAKA et al. 1989, 1990; TEZUKA et al. 1993). Treatment with DMA also induces DNA-protein cross-links in mice and human cultured cells (YAMANAKA et al. 1993). Although the data are not conclusive, they suggest that methylation of iAs can be viewed as an activation process. Additional data on the fate and toxicity of organoAs compounds are needed.

II. Determinants of Interindividual Variation in Methylation Capacity

In humans as in other species, the capacity for methylation of iAs is probably determined by four factors. These are: (a) the extent to which individual steps in the methylation pathway are saturable processes, (b) the availability of cofactors and substrates needed for methylation reactions, (c) the range of genetically determined capacity for As methylation, and (d) competition between As and other substrates at rate-limiting steps in the methylation pathway.

1. Saturable Capacity for Arsenic Methylation

The enzymatically catalyzed reactions involved in the methylation of As are likely saturable. Thus, the capacity for metabolism of As could be exceeded at high exposure levels. For example, if the conversion of iAs to MMA is the rate-limiting reaction, then exposure to high levels of iAs would be associated with proportionally greater urinary excretion of iAs and the excretion of proportionally less MMA and DMA. The studies of BUCHET and associates (1981a) in volunteers indicate that the capacity for production of MetAs metabolites as reflected by an increase in the biological half-life of As in urine might be exceeded at a p.o. intake above 125 μg sodium metaarsenite/day. However, studies in a small number of volunteers are unlikely to detect the variation in capacity for methylation of As that may exist in a large population. The existence of interindividual variation in capacity for methylation of As is consistent with the concept that As is a threshold carcinogen (PETITO and BECK 1991; MASS 1992; SMITH et al. 1992). In one model of As as a threshold carcinogen, iAs is assumed to be the carcinogenic form of this metalloid. If the capacity for the conversion of iAs to MMA is saturated, then the body burden of iAs will increase and exceed some critical or threshold concentration needed to trigger carcinogenesis. A test of the threshold hypothesis was performed by HOPENHAYN-RICH et al.

(1993), who examined the data from a number of studies of individuals exposed to various levels of As from occupational and environmental sources in which the As contents of urine were speciated into iAs, MMA, and DMA. The threshold hypothesis would predict that high exposure to As would result in proportionally more iAs in urine of individuals with the highest exposure to As. Analysis of several data sets found that approximately 20%–25% of the As in urine (range, 2%–38%) was present as iAs. The lack of a clear trend suggested that the fraction of urinary As present as iAs was unaffected by the level of exposure to As. This finding is consistent with the absence of threshold. However, replication of this finding in a single large study population would alleviate concerns over possible differences which might exist in the design and performance of the studies analyzed by Hopenhayn-Rich and associates.

2. Influence of Nutritional Status

The composition of the diet can affect As metabolism. The most significant of these factors is probably variation in the size of the methyl donor pool available for the production of methylated As species. A group of nutrients which includes methionine, choline, vitamin B_{12}, and folic acid which are termed lipotropes regulate the synthesis and utilization of *S*-adenosylmethionine (*S*-adomet), the source of labile methyl groups required for enzymatically catalyzed methylation reactions (Newberne and Rogers 1986; Zeisel 1988). If the availability of lipotropes from diet controls the availability of *S*-adomet, then variation in extent of As methylation within and between populations consuming different diets might be expected. Concern over the role of diet as a factor influencing the health effects of As has prompted the recommendation that future epidemiological and experimental studies address the issue of the role of methyl donor availability in As metabolism, toxicity, and carcinogenicity (US Environmental Protection Agency 1988).

Studies in experimental species demonstrate that methyl donor availability affects the methylation of iAs. In mice exposed to arsenite (As^{III}), depletion of the intracellular pool of *S*-adomet by treatment with periodate-oxidized adenosine (PAD), an inhibitor of *S*-adomet synthesis (Hoffman 1980), results in reduced urinary excretion of DMA (Marafante and Vahter 1984). Marafante and Vahter (1986) showed that consumption of choline-deficient diet decreased the urinary excretion of DMA and increased tissue As retention in rabbits given 0.4 mg arsenate (As^{V})/kg i.v. The same effect of a diet deficient in choline, methionine, and protein was observed in rabbits after administration of 0.4 mg As^{III}/kg i.v. (Vahter and Marafante 1987).

In relation to the threshold hypothesis discussed above, Mass (1992) has proposed an interesting hypothesis for the role of methylation in the toxicity and carcinogenicity of As which incorporates elements related to the dietary availability of lipotropes. In this model, iAs competes with other substrates for a limited supply of methyl groups which are available from the *S*-adomet pool. Increased demand for *S*-adomet as a methyl donor for the methylation of As reduces the availability of methyl groups for reactions catalyzed by other methyltransferases. MASS (1992) postulates that the utilization of methyl donors for As methylation can cause carcinogenesis by disrupting the normal pattern of DNA methylation. Because the formation of specific methylated bases in DNA (e.g., 5-methylcytosine) is an important mechanism for regulation of gene transcription (DOERFLER 1983; MICHALOWSKY and JONES 1989), hypomethylation of DNA due to reduced availability of methyl donors could affect the expression of specific genes and consequently the transformation of cells to a carcinogenic phenotype. For example, altered methylation of oncogenes is an early step in the development of some human epithelial cells cancers (FEARON and VOGELSTEIN 1990). This model of As carcinogenesis is also consistent with the concept that As is a threshold carcinogen. If the requirement for methyl donors for the methylation of As does not reduce the availability of donors for the methylation of some critical (but as yet unidentified) target, then As exposure will not result in carcinogenesis. However, if the demand for methyl donors for the methylation of As does exceed this threshold, then methylation of the critical target will be affected and cellular transformation will occur.

3. Genetically Determined Capacity for Arsenic Methylation

The role of genotype in interindividual variation in capacity for the steps in As methylation pathway has not been determined. In humans and in inbred mouse strains, the activities of various enzymes catalyzing the methylation of a number of xenobiotics is under genetic control (for a review, see WEINSHILBOUM 1988). However, there is no direct evidence to indicate whether or not the activity of the enzymes catalyzing As methylation is under genetic control. If humans are polymorphic for activities of the enzymes that catalyze As methylation, then a range of catalytic capacities will be expressed within any population. As described by THOMAS (1994), assessing the importance of genetically determined interindividual variation in capacity for As methylation will require determination of the activities of specific enzymes in a large number of individuals. Depending on the frequency of the alleles for a given enzyme activity, the number of individuals who must be tested for the activity could range into the thousands. Thus, assessing allele frequency is not a project to be attempted in studies of a few volunteers but rather would require screening individuals from a large population with known and reasonably constant exposure to As. A number of populations are known to have reasonably uniform exposure to iAs from drinking water and could be used for such studies. These populations could also provide kindreds for studies of pattern of inheritance of methylation capacity within kindreds.

III. Enzymology of Arsenic Methylation

1. Characteristics of Arsenic Methyltransferases

Although methylation of As by methylcobalamin can occur by strictly chemical processes (BUCHET and LAUWERYS 1985), a significant component of the methylation of As in cells is enzymatically catalyzed. The enzymatic reaction requires *S*-adomet as a methyl group donor, glutathione disulfide, oxidized glutathione (GSH), methylcobalamin, and Mg^{II} (BUCHET and LAUWERYS 1988; HIRATA et al. 1989; TAKAHASHI et al. 1990). Based on differential sensitivity of methyltransferase reactions to inhibition by Hg^{II}, BUCHET and LAUWERYS (1985) postulated that unique enzymes catalyze the mono- and dimethylation reactions. In vitro, the first methylation reaction appears to be rate limiting, can be stimulated by GSH at a high As^{III} concentration, and is also sensitive to inhibition by iAs. Methylation reactions can be prevented by a large excess of thiol-containing molecules, probably by decreasing the amount of unbound (free) As^{III}. In rat and mouse liver, methylation activities are localized in the cytosol (BUCHET and LAUWERYS 1985, 1988). Treatment of mice or rabbits with PAD, which depletes *S*-adomet before injection of As^{III}, resulted in a 25%–70% decrease in the production of DMA, indicating that *S*-adomet is the methyl group donor in vivo. This effect on methylation was first observed in the liver, suggesting that this organ is the main site for As methylation. Inhibition of methyltransferase resulted in a two- to sixfold increase of accumulation of iAs in tissues (MARAFANTE and VAHTER 1984). A similar experiment in rabbits with injection of PAD before As^{V} administration found a 1.5- to 4-fold increase in tissue accumulation of iAs (MARAFANTE et al. 1985).

2. Role of GSH in Arsenic Reduction, Binding, and Methylation

Recent studies have shown that GSH reduces As^{V} to As^{III}, MMA^{V} to MMA^{III}, and DMA^{V} to DMA^{III} and forms the following complexes: $As^{III}(GS)_3$, $MMA^{III}(GS)_2$, and $DMA^{III}(GS)$ (SCOTT et al. 1993; DELNOMDEDIEU et al. 1994). Studies in Chinese hamster ovary (CHO) cells indicated that the cytotoxicity of As^{III} was inversely related to intracellular GSH concentration (HUANG et al. 1993). In addition, As resistance in CHO cells has been associated with a high intracellular GSH concentration, elevated GSH-*S*-transferase activity, and increased efflux of As (LEE et al. 1989; WANG and LEE 1993). The relation between the possible roles of GSH in As-resistant CHO cells and mechanisms of As resistance in prokaryotes is discussed below.

The methylation of As^{III} or MMA requires the presence of a normal intracellular (millimolar) concentration of GSH (BUCHET and LAUWERYS 1985). Methylation of arsenite or MMA did not occur in either rat or mouse liver or kidney homogenates without addition of GSH. Addition of GSH promoted both mono- and dimethylation in these homogenates. MMA formation increased with increasing arsenite concentrations; however, the percentage of total MetAs decreased (HIRATA et al. 1989). Studies in rats (BUCHET and LAUWERYS 1987) and hamsters (HIRATA et al. 1988) indicated that the concentration of GSH in tissues may be the limiting factor in As methylation. Rat liver, kidney, and lung methylated As^{III} to MMA and DMA, with liver having the greatest methylation capacity. DMA can be produced from MMA by rat liver slices and this methylation is stimulated by GSH (GEORIS et al. 1990). Depletion of GSH in rats by phorone pretreatment significantly increased the urinary excretion of iAs during 24 h after injection of 1 mg As^{III}/kg (BUCHET and LAUWERYS 1987). In contrast, treatment of hamsters with buthionine sulfoximine (BSO), an inhibitor of GSH synthesis, resulted in reduced urinary As excretion and oliguric renal failure after p.o. administration of 5 mg As^{III}/kg (HIRATA et al. 1990). These differences in the effects of GSH depletion in rat and hamster may reflect species differences in susceptibility and metabolism. GSH depletion in rat liver appears to reduce preferentially the monomethylation of As^{III}, resulting in the persistent elevation of iAs in the liver (BUCHET and LAUWERYS 1988). Because As^{III} is highly reactive with monothiols and with vicinal thiol groups (DELNOMDEDIEU et al. 1993, 1994; SCOTT et al. 1993), depletion of GSH could not only reduce the rate of methylation of As^{III} but also result in the diversion of As^{III} from its normal binding site on GSH to critical thiol residues in proteins (HIRATA et al. 1988).

The interaction between the metabolism of iAs and GSH utilization has been examined in rat liver. In the anesthetized rat, treatment with As^{III} or As^{V} increased biliary excretion of endogenous non-protein thiols in a dose-dependent fashion up to 24- and 31-fold, respectively. Simultaneously, GSH excretion increased to a similar extent, suggesting that the increment in biliary thiol output originated from enhanced hepatobiliary transport of GSH. The transport of As as a GSH complex was proposed to account for the GSH dependence of biliary As transport (GYURASICS et al. 1991a,b). In the isolated perfused rat liver, exposure to As^{III} increased the excretion of both GSH and GSH disulfide (GSSG) into bile (ANUNDI et al. 1982). The release of GSH and GSSG in the bile following the accumulation of iAs in the liver may reflect alteration of GSH cycling induced by As or alterations in the hepatobiliary transport pathway for GSH.

C. Comparative Metabolism, Kinetics, and Toxicity

I. Methylation in Prokaryotes

The pioneering work of Challenger and associates (CHALLENGER 1945, 1951) demonstrated that microorganisms have the capacity to methylate As to a volatile species. In a series of studies, Gosio gas, a volatile species with a strong garlic odor produced by several species of fungus, was identified by these investigators as trimethylarsine, $(CH_3)_3As$. Identification of MetAs species in pristine lakes suggested that the geochemical cycle of As probably included the biological methylation of iAs (BRAMAN and FOREBACK 1973). COX and ALEXANDER (1973) identified fungal species (*Candida humicola, Gliocladium roseum, Penicillium* species) in sewage which methylate As^V, As^{III}, MMA, and DMA to TMA. The conversion of arsenate to MMA and DMA by a methanobacterium has been reported (MCBRIDE and WOLFE 1971). Notably, this conversion occurred under anaerobic conditions and this species did not produce TMA. As summarized by GADD (1993), the conversion of As to methylated species by bacteria is likely a mechanism of detoxification, yielding less reactive and more volatile species which can be excreted.

An alternative scheme exists for the resistance of some bacterial species to As. The plasmid-borne *ars* operon encodes a group of proteins involved in the reduction and translocation of As out of the bacterial cell (ROSEN et al. 1991; WU and ROSEN 1993; SILVER et al. 1993a). The increased efflux of As^{III} in *ars*-bearing bacteria confers resistance by reducing the intracellular concentration of As^{III}. Because the translocator is specific for As^{III} (and Sb^{III}), the *ars* operon also includes an enzyme that catalyzes reduction of As^V (JI and SILVER 1992). Increased efflux of metals or metalloids is a common strategy to attain resistance in bacteria (MERGEAY 1991). However, little is known about the role of efflux pathways as a component of the cellular response to metals and metalloids in higher organisms. The recently identified candidate gene for Menkes' disease, an inborn error of Cu homeostasis in humans (MERCER et al. 1993; VULPE et al. 1993), may encode a defective P-type ATPase with high homology to a staphylococcal Cd-resistance ATPase (SILVER et al. 1993b). As noted above, increased efflux of As has been reported in an As-resistant CHO cell line which attains high cellular GSH concentrations and high GSH-*S*-transferase activity (LEE et al. 1989; WANG and LEE 1993). Given the important roles for GSH in the reduction and complexation of As which have been described recently (SCOTT et al. 1993; DELNOMDEDIEU et al. 1993, 1994; THOMPSON 1993), it is interesting to speculate that a system analogous to the well-characterized efflux system for As in bacteria exists in eukaryotes.

II. Methylation in Eukaryotes

1. Methylation in Nonhuman Species

a) Domestic Species

The major sources of exposure of domestic animals to arsenicals are ingestion of feed contaminated with As from industrial pollution or with residues of As-containing pesticides. Other potential sources of As exposure are drinking water contaminated with iAs or organoAs preparations used as feed additives. Many studies describing retention, organ distribution, and excretion of As in domestic animals exposed to different As compounds have been summarized (National Research Council 1977); however, only limited data are available on As methylation in these animal species.

In the cow, ingestion of As (As^V) acid, sodium arsenite, DMA, or MMA increased tissue concentrations of As but did not increase As concentration in milk (PEOPLES 1964, 1969), suggesting that neither iAs nor its possible metabolites can cross the blood-mammary barrier. OrganoAs formulations have been used as feed additives for disease control and improvement of weight gain in swine and poultry since the mid-1940s. The most widely used compounds are arsanilic acid, sodium arsinilate, and 3-nitro-4-hydroxyphenylarsonic acid (National Research Council 1977). The extent of metabolic conversion of these arsenicals in vivo remains controversial. Some investigators reported the degradation and reduction of these compounds to iAs^{III} (HARVEY 1970). Other studies, however, indicated that arsanilic acid and acetylarsanilic acid (4-acetylaminophenylarsonic acid) were excreted unchanged by chickens with no evidence of biotransformation (OVERBY and FREDRICKSON 1963, 1965; OVERBY and STRAUBE 1965). Similar results were obtained in chickens treated with 3-nitro-4-hydroxyphenylarsonic acid and 4-nitrophenylarsonic acid (National Research Council 1977). In rats, rabbits, and swine, most of a p.o. dose of phenylarsonic acid was eliminated unchanged in urine, although limited biotransformation occurred (MOODY and WILLIAMS 1964a,b).

Because fungi and bacteria methylate As in anaerobic conditions (see above), ruminants might exhibit unique patterns of As metabolism and excretion. A comparative study of the concentration of iAs and methylated MetAs in urine of cows and dogs fed normal diet showed similar As excretory patterns (LAKSO and PEOPLES 1975). Feeding of either species with a diet containing As^V (2.75 mg/kg for cows, 3.4 mg/kg for dogs) or As^{III} (1.57 mg/kg for cows, 1.94 mg/kg for dogs) for 5 days resulted in the urinary excretion of iAs and MetAs. In the cow, 73%–75% of As in urine was MetAs. In the dog, 38%–49% of As in urine was MetAs. Although cows excreted a significantly larger percentage of administered As as MetAs, the authors of this study expressed doubt that the rumen was the only site of methylation in this species. More recent work in the rat (see below) has shown that gastrointestinal microflora can contribute to the methylation of iAs.

b) Experimental Species

With the exception of rats and marmoset monkeys, most experimental species show similar patterns of As methylation. Typically, DMA was found to be the major methylated metabolite excreted in urine together with a smaller fraction present as MMA. The proportion of methylated metabolites and rate of excretion vary among species, depending on route of administration, dosage level, and chemical form of As administered.

a) Rabbits, Mice, and Hamsters. In mice and rabbits receiving 0.04–0.4 mg As^{V}/kg i.v., both As^{III} and As^{V} appeared in urine during the 1st h post-injection; excretion of DMA became significant only after about 2 h (VAHTER and ENVALL 1983). This pattern suggests that As^{V} was reduced to As^{III} before being methylated. As^{III} was also detected in the plasma of As^{V}-treated animals. Ultrafiltration of urine showed that about 3% of As was associated with macromolecules with an $M_r \geqslant 25$ kDa. In mice administered As^{V} or As^{III} (p.o., 0.04–2.0 mg As/kg, or s.c., 0.4 mg As/kg), DMA was found to be the major urinary metabolite during 48 h after administration. As^{III} was methylated to a greater extent than As^{V} and resulted in higher whole-body retention (VAHTER 1981). Retention increased with increasing dose in parallel with a decrease in the extent of methylation. Subcutaneously administered As^{III} yielded less MetAs in urine during the first 48 h postdosing than did As^{III} given p.o.

Rabbits injected with 1 μg As^{III}/kg i.p. excreted 60% of the dose in urine and 6% with feces during the first posttreatment day (BERTOLERO et al. 1981). iAs was found to be the predominant species in urine during the first 2 h posttreatment. At later time points, DMA was the major metabolite (80% of urinary As at 6 h). MMA was detected in urine as a minor metabolite (<5%). MMA, DMA, and iAs were found in plasma in similar proportions. At 6 h after injection, the ultrafiltrable fraction of As in cytosol of lung, liver, and kidney from injected animals consisted of 3.6%, 30.1%, and 48.5% of iAs, respectively; the portion as DMA reached 95.3%, 68.5%, and 47.6%. MMA ranged between 1.3% and 3.3%. In liver and kidney cytosol, most of the As (70.3%–86.5% of total As) was bound to proteins at 4 and 6 h after injection.

Dose-dependent disposition of As^{V} in mice was studied by HUGHES and coworkers (1994). An acute p.o. dose (0.5–5000 μg/kg) had no effect on the percentage of the dose of As excreted in urine (66%–79%) and in feces (10%–18%). DMA was the predominant urinary metabolite (51%–64% of dose) but no effect of dose on its elimination was detected. However, peak DMA elimination in urine shifted from 4 h postexposure at lower dosage levels to 8 h at the 5000-μg/kg dose. With increasing dose, a significant increase of the MMA portion in urine (0.1%–1.0% of dose) was observed. At the 5000-μg/kg dose, there was an increase in the amount of arsenate and arsenite in the 1- and 2-h urines. These results suggest that an acute dose of As^{V} (5000 μ/kg) can affect the metabolism of arsenicals.

Treatment with dimercaptans was shown to increase the rate of As excretion in urine of rabbits after s.c. injection of 1 mg As^{III}/kg (MAIORINO and APOSHIAN 1985). Injection intramuscularly of 2,3-dimercapto-1-propanesulfonic acid (DMPS), *meso*-2,3-dimercaptosuccinic acid (DMSA), or *N*-(2,3-dimercaptopropyl)-phthalamidic acid (DMPA) before As^{III} treatment increased the rate of urinary excretion of total As during the first 24 h after administration but did not affect the cumulative amount of As excreted between 0 and 48 h. A change in As species in urine (increased iAs and decreased DMA) was observed in dimercaptan-treated animals. Pretreatment with DMPS or DMPA also increased the urinary excretion of MMA. In contrast, DMSA pretreatment did not alter MMA excretion.

In addition to MMA and DMA, a small amount of a TMA compound was found in urine and liver but not in blood, feces, or any other organs of hamsters given a single p.o. dose (4.5 mg/kg) of As^{III} trioxide (YAMAUCHI and YAMAMURA 1985). Forty-nine percent of the As dose was excreted in urine and 11% in feces during 5 days after administration. DMA was the major urinary metabolite. TMA has been reported to be a normal urinary metabolite of As in male Syrian golden hamsters, accounting for 66% of urinary As in hamsters fed a commercial laboratory diet (YAMAUCHI and YAMAMURA 1986). However, neither the total concentration of As nor its chemical speciation in this diet has been published[1].

The influence of carbon tetrachloride-induced liver cirrhosis on the metabolism and excretion of As in male Syrian golden hamster was examined by TAKAHASHI and coworkers (1988). Compared with the metabolic profile found in hamsters receiving 1.5 mg As^{III} trioxide/kg alone, chronic s.c. treatment with 40 μl 5% carbon tetrachloride in olive oil twice weekly for 15, 20, or 30 weeks before p.o. administration of As^{III} trioxide resulted in increased urinary excretion of both iAs and DMA and no change in the urinary excretion of MMA. The increase in the formation and excretion of DMA in hamsters with liver cirrhosis was associated with a decrease in the concentration of *S*-adomet in the liver, suggesting that utilization of *S*-adomet as a methyl donor may be increased in hamsters with altered liver function.

YAMAUCHI and coworkers (1988) reported that MMA was only partly methylated in vivo to DMA when administered p.o. or i.p. to hamsters (50 mg/kg); 33.6% of the p.o. dose and 82.7% of the i.p. dose was eliminated in urine during first 120 h after administration. Fecal excretion represented 60.9% and 1.0%, respectively. Most of the dose was excreted as the parent compound. The amount of DMA excreted in urine within 120 h of dosing accounted for 4.2% of the p.o. dose and 1.2% of the i.p. dose. Less than 0.2% of the p.o. or i.p. dose was excreted as DMA in feces. The highest tissue concentration of DMA was attained in lung 12 h after p.o. administration. TMA excreted in urine after MMA administration accounted for 0.3% of the p.o. dose and 0.1% of the i.p. dose. A higher portion of DMA and TMA was found in urine when a smaller dose of MMA (5 mg/kg) was administered p.o.: 7.5% and 1.9% of the dose, respectively. The portion of DMA in feces was unaffected. No TMA was detected in feces at either dosage level, and in vivo demethylation of MMA was not detected.

[1] In the work of Yamauchi and coworkers cited in this review, all data on the species of As present in tissues and excreta are corrected for background (i.e., pretreatment) concentrations of each As species in tissues and excreta.

Another TMA compound (trimethylarsineoxide, TMAO) is a product of further methylation of DMA. In urine of mice and hamsters collected during 2 days after p.o. administration of DMA (40 mg As/kg), MARAFANTE and associates (1987) found 3.5% and 6.4% of total urinary As in the form of TMAO, respectively. No demethylation of DMA was observed. YAMAUCHI and YAMAMURA (1984a) reported that TMAO represented 32% of urinary As excreted during 24 h following p.o. administration of 50 mg DMA/kg to hamsters. DMA represented 68% of the As in 24-h urine and almost 100% of total As excreted in feces during this interval. In addition to these compounds, increased levels of iAs and MMA were found in urine, feces, and some organs. TMAO was found in urine of hamsters as the sole metabolite of TMA (10 mg As/kg) administered p.o. The calculated whole body half-life for TMA was 3.7 h in comparison with 5.3 h for TMAO, 5.6 h for DMA, 6.1 h for arsenobetaine, 7.4 h for MMA, and 28.6 h for As^{III} trioxide (YAMAUCHI et al. 1990). $(CH_3)_3As$ was detected in air expired by hamsters injected i.p. at the 50 mg As/kg dosage level, suggesting that a part of the dose was eliminated in expired air.

Metabolism of arsenobetaine and arsenocholine, the most common organoAs compounds in marine organisms (EDMONDS et al. 1977; KUROSAWA et al. 1980; CANNON et al. 1981; LUTEN et al. 1982, 1983; NORIN and CHRISTAKOPOULOS 1982; NORIN et al. 1983), has been studied in mice, rabbits, and rats. During 3 days after i.v. administration of arsenobetaine, about 75% of the dose (4–400 mg As/kg) was excreted in urine of rabbits and more than 98% in urine of mice and rats (VAHTER et al. 1983). Chemical speciation of the As in urine and soluble extract of tissues detected no biotransformation of arsenobetaine in either species. Mice, rabbits, and rats given arsenocholine (4 mg As/kg) p.o. or i.v. excreted 70%–80% of the dose in urine within 3 days (MARAFANTE et al. 1984). Unmetabolized arsenocholine was found in urine during the 1st day after administration. At later time points, arsenobetaine was the major or only urinary metabolite. Arsenobetaine was not metabolized to iAs, MMA, DMA, or TMAO and As was present in tissues as arsenobetaine or arsenolipids.

The metabolism and methylation of As from gallium arsenide (GaAs) was studied after its p.o. or i.p. administration (100 mg/kg) to hamsters (YAMAUCHI and YAMAMURA 1986). In contrast to the other As compounds, about 80% of the As from GaAs given p.o. was excreted in feces within 120 h after administration. Due to its low solubility, As was poorly excreted in both feces and urine following i.p. injection of GaAs. MMA, DMA, and iAs were found in urine and tissues of GaAs-injected hamsters. Toxicity and metabolic conversion of methylated As compounds used in the fabrication of arsenide-containing semiconductors have been studied (YAMAMURA 1993). The LD_{50} for s.c. administration of $(CH_3)_3As$, triethylarsine (TEAs), and trisdimethyl-aminoarsine (TDMAAs) in mice was calculated at 8000, 750, and 15 mg/kg, respectively. TDMAAs was dissociated in vivo, yielding iAs which was methylated to DMA. TEAs was deethylated forming diethylAs.

$(CH_3)_3As$ was oxidized to TMAO and excreted in urine with a half-life of 3.7 h.

β) Rats. The rat differs from other species in the whole-body retention, excretion, and distribution of iAs and its metabolites. In an early study, COULSON and associates (1935) reported that rats excreted As derived from ingestion of shrimp faster than iAs which was added to the diet. HUNTER and coworkers (1942) noted that the erythrocyte was the major depot for As following administration of iAs. Subsequent studies have confirmed the high accumulation and retention of As in the rat erythrocyte. In a comparison among species 48 h after intramuscular administration of sodium $[^{74}As^V]$ arsenate, LANZ and coworkers (1950) found that the blood compartment of the rat accounted for nearly 45% of the administered dose of ^{74}As. In contrast, the blood compartment in cat accounted for 5.6% of the administered dose; in dog, 0.1%; in rabbit, 0.27%, in guinea pig, 0.25%; in chicken, 0.19%; and in mouse, 0.07%.

The mechanistic basis of the accumulation and retention of As in the rat erythrocyte remains unresolved. Whether administered as iAs or as DMA, available evidence indicates that DMA is rapidly accumulated in the rat erythrocyte. The avid accumulation of As in the rat erythrocyte has been attributed to the relatively high cysteine content of rat hemoglobin. Notably, the high cysteine content of rat hemoglobin has been postulated to account for the rapid accumulation of methylmercury in the rat erythrocyte (DOI 1991). However, it is not clear whether the accumulation of As in the rat erythrocyte significantly influences the toxicity of As. Comparison of LD_{50} estimates for iAs or DMA in the rat and mouse, species which differ greatly in the accumulation of As in the erythrocyte, shows no association between the magnitude of As accumulation in erythrocytes and the acute lethality of these As species (STEVENS et al. 1979). If accumulation of As in the rat erythrocyte sequesters As and reduces its toxicity, then one would expect the rat to be extremely resistant to the lethal effects of iAs and DMA. Additional studies are required to elucidate the molecular basis for the accumulation and retention of As in the erythrocyte.

More recent studies have provided additional data on the distribution, metabolism, and excretion of As in the rat. A comparative metabolic study after i.v. injection of arsenic (As^V) acid (5 mg/kg) showed that during 48 h about 50% of the dose was excreted in urine (17.2%) and feces (33.0%) of rats. In comparison, hamsters excreted 87.9% of the dose (43.8%, urine; 44.1%, feces) and mice excreted 97.3% of the dose (48.5%, urine; 48.8% feces) over this interval (ODANAKA et al. 1980). In urine and feces of rats, iAs was the predominant form and methylated metabolites represented only 18% and 36%, respectively. In mice and hamsters, about 60% of urinary As and 35%–49% of fecal As was found in the form of DMA and MMA. Significant differences in whole-body retention and excretion of methylated metabolites were also demonstrated when As was given to rats and mice p.o. as As^{V} or As^{III} (VAHTER 1981). The high retention of As in blood accounted for the slower rate of excretion and higher whole-body retention of As in rats than in other species (ODANAKA et al. 1980). Two days after p.o. (5 mg/kg) or i.v. (1 mg/kg) administration of arsenic (As^{V}) acid, about 40% of the dose remained in blood, almost entirely as DMA. In comparison, As was undetectable in the blood of mice and hamsters at 2 days posttreatment with arsenic (As^{V}) acid. Similarly, almost all the As from As trioxide given to the rats in drinking water (4–20 μg/ml) for 2–3 months was accumulated in the blood, with 88% in hemoglobin (FUENTES et al. 1981). About 99% of the As in hemoglobin was found to be bound to the hematin fraction after acid acetone treatment. Accumulation of DMA (75% of the dose) by rat erythrocytes was observed at 4 h after i.v. injection of As^{III} and As^{V} (LERMAN and CLARKSON 1983). Studies of the role of the gastrointesinal microflora in the metabolism of As in the rat have been reported (ROWLAND and DAVIES 1981). The microflora of both the small intestine and cecum was found to reduce As^{V} to As^{III}. Cecal microflora but not small intestinal microflora was shown to metabolize iAs to MMA and DMA. The reductive and methylation capacity of cecal and small intestinal contents was destroyed by antibiotics or autoclaving. The contribution of the microflora of the gastrointestinal tract to As metabolism warrants further study.

γ) *Other Animal Species*. In addition to the laboratory species discussed above, dogs and cats have been used in some of the experiments which examined As methylation in vivo. DMA was detected as the predominant species in plasma of dogs injected i.v. with a submicrogram dose of arsenic (As^{V}) acid. At 3 days after treatment, 99% of As in urine was DMA (TAM et al. 1978). Cats injected i.v. with arsenic (As^{V}) acid (1 mg/kg) excreted more than 60% of the dose in urine within 48 h. DMA accounted for 48%, MMA for 2%, and iAs for 51% of the urinary As in this species (ODANAKA et al. 1980).

c) Marmoset Monkeys

Unlike all other species, marmoset monkeys cannot metabolize iAs to methylated forms. During 4 days after i.p. administration of 0.4 mg As^{III}/kg, marmoset monkeys eliminated only 30% of the dose, mainly via urine (VAHTER et al. 1982). As in urine and tissues of marmosets was iAs. Liver accumulated the largest portion of the dose (20%). Half of the liver As was bound to the rough microsomal membranes. No biotransformation (except of partial reduction of As^{V} to As^{III}) was found after i.v. injection of 0.4 mg As^{V}/kg (VAHTER and MARAFANTE 1985). Within 3 days posttreatment, about 39% of the dose was excreted in urine and only 2% in feces. Urinary As consisted of iAs^{V} and iAs^{III} present at a ratio of 1:1. The basis for the absence of iAs methylation in marmoset monkeys remains unknown. In contrast to the marmoset, other monkey species (*Cynomologus* and *Rhesus*) have been stated to methylate iAs (Vahter and Marafante 1985). It has been hypothesized that methylation of As may be dependent on a specific methyltransferase, which is absent or inactivated in the marmoset.

2. Arsenic Methylation in Humans

Creselius (1977) reported the urinary excretion of MMA and DMA by a volunteer who had ingested wine containing 63 μg iAs. To date, a number of studies have examined the methylation of As in humans during environmental or occupational exposure to different forms of As. In addition, several laboratory experiments have been performed on volunteers administered with single or repeated doses of arsenicals. The main purpose of this work has been to characterize the kinetic behavior of iAs and organoAs compounds and to find reliable markers of exposure of humans to As.

a) Occupational and Environmental Exposure

Exposure to As^{III} trioxide can occur in the smelting and glass-making processes. In workers exposed to As dust in the production of arsenic (As^{V}) acid, Yamamura and Yamauchi (1980) found increased As^{III} and As^{V} concentrations in blood with little increase in the DMA concentration in blood. When workers were classified on the basis of airborne As concentrations into low (202 μg As/m^3) or high (338 μg As/m^3), the output of As^{III} in urine was greater in the high exposure group than in the low exposure group; however, the urinary output of DMA was similar in the two groups. The similarity in ouput of DMA by the two groups could indicate that capacity for methylation was exceeded in the high As exposure group. Analysis of urine of workers in nonferrous smelter in northern Sweden (Vahter 1986) showed that, after the 1st day of the working week, 19% of urinary As was iAs, 20% MMA, and 61% DMA. A similar pattern of urinary As metabolites was observed in a randomly selected general population from Swedish cities. All three metabolites showed a significant correlation ($r = 0.94–0.99$) with the total urinary As. A similar study in glass workers exposed to As^{III} trioxide was performed in Italy (Foa et al. 1984). Urinary excretion of As in exposed individuals consisted mainly of iAs, MMA, and DMA (70%–85% of total urinary As). From 15% to 30% of the As in urine of glass workers was present in other unidentified chemical forms. These unidentified As species accounted for as much as 70% of total urinary As in an Italian reference population with normal environmental exposure to As. In the reference population the 30% of the urinary As which was speciated consisted of equal portions of iAs, MMA, and DMA. It is possible that the unidentified As species in the urine of the Italian reference population is an organoAs species derived from seafood. In workers exposed to As in copper smelting and in GaAs production, Yamauchi and coworkers (1989) found that urinary excretion of iAs was sufficiently sensitive to monitor low-level exposure to As typically found in

GaAs workers. Exposure to higher levels of As in copper smelting resulted in increased urinary excretion of iAs, MMA, and DMA. Urinary excretion of TMA was not increased under high exposure conditions and can be used to monitor the dietary intake of TMA species such as arsenobetaine. The results of this study emphasize again the need for complete speciation in evaluating exposure to As in humans.

The relation between occupational exposure to As^{III} trioxide fumes and dust and urinary excretion of As metabolites has been studied in workers from a sulfuric acid producing plant in Belgium (OFFERGELT et al. 1992). Statistically significant correlations were found between the weighted average exposure concentrations of As in air and the metabolites (iAs, MMA, DMA, and total As) in urine collected at the end of the shift or just before the next shift. The low correlation between airborne As levels and urinary As excretion ($r = 0.4–0.5$ for iAs and DMA, and $r = 0.2–0.3$ for MMA) might be attributed to exposure to As via diet and/or ingestion of dust, both of which were not evaluated in this study.

MAHIEU and coworkers (1981, 1987) have examined the urinary excretion of iAs and its metabolites in individuals acutely poisoned by ingestion of As^{III} trioxide who were treated with 2,3,-dimercaptopropanol (British antilewisite, BAL). The proportion of iAs, MMA, and DMA excreted in urine was found to change during the course of BAL therapy. During the first 2–4 days post-intoxication, iAs was the major urinary metabolite. After the first 4 days, the portion of As excreted as MMA and DMA increased. At later time points, more than 95% of the excreted As was organoAs, mainly DMA. These data suggest that saturation of methylation processes might occur after ingestion of large doses of iAs and that iAs inhibits the methyltransferases that catalyze the production of MMA and DMA, a hypothesis reviewed by THOMPSON (1993). Studies in the mouse following p.o. administration of As^{V} also suggest that high doses of iAs can alter the pattern of production of methylated As metabolites (HUGHES et al. 1994). However, the exact dose-response dependence of the relation between iAs intake and extent of methylation has not been adequately elucidated. Notably, the data of MAHIEU and associates (1981, 1987) on the species of As excreted in urine after As intoxication may be confounded by treatment with BAL to promote decorporation of As. Although the effect of BAL treatment on As methylation in humans has not been examined, MAIORINO and APOSHIAN (1985) have previously reported that dimercaptans reduce DMA production in iAs-treated rabbits. Hence, the reduced output of DMA in As-intoxicated humans who received BAL could reflect the inhibition of methylation by BAL rather than the inhibition of this enzyme by iAs.

The relation between environmental exposure to As and the relative amounts of each As species excreted in urine has been investigated. Among the residents of the Region Laguera in Mexico who ingest drinking water containing up to 400 μg As/l, CEBRIÁN and coworkers (1993) have found

DMA to be the major As metabolite (70% of urinary As). Notably, the significant increase in the amount of MMA was found in urine of As-exposed individuals accompanied by a decrease in the amount of DMA. This alteration in the ratio of MetAs species in urine could reflect dosage-dependent changes in metabolism of iAs. Thus, the urinary ratio of MMA/DMA may be an appropriate indicator of iAs exposure in humans and warrants examination in other As-exposed and control populations.

α) *Variation in Arsenic Consumption.* In populations not occupationally exposed to As, consumption of food containing As is the major route of exposure to As. Arsenicals may appear in the food as residues of pesticides or herbicides, as the result of soil contamination from industrial sources, or as products of bacterial action. Accumulation of As by certain types of plants or animals (National Research Council 1977) may also lead to high exposure to As compounds. Seafood has been repeatedly reported to contain high levels of TMA, mainly arsenobetaine. This compound has been found in urine and blood of individuals who consumed large quantities of seafood (Kurosawa et al. 1980; Cannon et al. 1981; Shibata et al. 1993). It was shown that following the ingestion of a fish meal (10.1 ± 0.03 mg As/person), 50% of the As dose was recovered in the urine in the first day and 78.3% within 8 days after dosing; only 0.33% of the dose was excreted in feces (Tam et al. 1982).

Arsenic intake and excretion was monitored in two adult men and two adult women who had eaten a customary Japanese diet for 7 days (Mohri et al. 1990). Mean daily intake of As was 182 μg (5.7% iAs, 3.6% MMA, 27.4% DMA, and 47.9% TMA). The mean amounts of As excreted daily in urine and feces were 148 and 46 μg, respectively. Urinary As composition was 1.4% iAs, 3.5% MMA, 33.6% DMA, and 61.4% TMA. The daily As intake correlated significantly with the total urinary As and with the amount excreted in feces during the following day. In a more recent study, Yamauchi and coworkers (1992) examined patterns of As intake in 35 volunteers consuming a typical Japanese diet. Mean daily intake of As was 195 ± 235 (SD) μg (range, 15.8–1039 μg) and consisted of 17.3% iAs, 0.8% MMA, 5.8% DMA, and 76% TMA. In another 56 healthy volunteers, these investigators found urinary As to be 6.7% iAs, 2.2% MMA, 26.7% DMA, and 64.6% TMA with a mean urinary total As concentration of $129 \pm 92/dm^3$. Mean blood total As in the 56 volunteers was $0.73\% \pm 0.57$ μg/dl and consisted of 9.6% iAs, 14% DMA, and 73% TMA. Urinary TMA concentrations were significantly correlated ($P < 0.01$) with blood TMA concentrations. A recent Canadian national survey estimated daily total As intake to be 38.1 μg for the average individual (Dabeka et al. 1993). Differences in As intake between Japanese and Canadian subjects reflect the higher intake of As-rich seafood by the Japanese. Among Canadians, the highest daily intake (59.2 μg As) was found in 20- to 39-year-old males. For 1- to 4-year-old children, the average estimated intake was 14.9 μg As/day.

Based on a 10-kg body weight, this yields an intake of 1.5 μg As/kg/day, which is less than the WHO/FAO provisional tolerable weekly intake of iAs (2.1 μg As kg^{-1} day^{-1}) (Joint WHO/FAO 1989). In the absence of data on the species of As present in foods, it is widely assumed that dietary As is primarily organoAs species (e.g., arsenobetaine and arsenocholine) which are relatively nontoxic. However, it has been reported that 5%–12% of the As in freshwater fish is present as iAs (NORIN et al. 1985). To better assess the role of diet in the intake of iAs, additional data are needed on the chemical form of As in food.

β) *Influence of Disease Status on Methylation.* Liver disease has been shown to change capacity for As methylation in humans (BUCHET et al. 1984). Although the cumulative urinary excretion of As in 24 h after i.v. injection of 7.14 μg As^{III}/kg was unaffected by the presence or absence of liver disease, liver disease status did affect the MMA and DMA in urine, resulting in reduced MMA and increased DMA. In individuals without liver disease (n = 13), MMA represented 12.3% ± 2.8% and DMA 23.3% ± 6.4% of the cumulative 24-h urinary As (GEUBEL et al. 1988). In patients with liver cirrhosis injected i.v. with 7.14 μg As^{III}/kg, the proportion of MMA and DMA in cumulative 24-h urine was 4.7% ± 3.3% and 40.4% ± 16.6% (n = 38), respectively. Notably, carbon tetrachloride-induced liver cirrhosis increases the urinary excretion of DMA in male golden Syrian hamsters (TAKAHASHI et al. 1988).

b) Experimental Studies

Experiments conducted on volunteers who received a p.o. dose of carrier-free arsenic (As^V) acid (0.06 ng As) showed that the urinary excretion of As was described by a three-component exponential function (POMROY et al. 1980). A half-life of 2.09 days corresponded to the first component (65.9% of the dose), 9.5 days to the second component (30.4%), and 38.4 days to the third one (3.7%). In a similar experiment (TAM et al. 1979), 50% of the p.o. dose of 10 ng arsenic (As^V) acid was recovered in urine in the first 5 days with proportions of 51% DMA, 21% MMA, and 27% iAs.

Inorganic arsenic, MMA, and DMA were found to be the only metabolic forms of As in urine of volunteers who ingested 125, 250, 500, or 1000 μg sodium As^{III} daily for 5 days (BUCHET et al. 1981a). The amounts of iAs, MMA, and DMA excreted in urine on day 5 (taken separately or together) correlated significantly with daily ingested dose. As noted above, this study provided equivocal evidence that the capacity for methylation of As could be saturated at high levels of iAs intake.

In urine of volunteers who ingested an extract from a seaweed (*Hijikia fusiformis*) containing As as As^V (86%), As^{III} (7%), and DMA (7%), the peak excretion of As^V occurred within 2 h of dosing (YAMAUCHI and YAMAMURA 1979). At the later time points, the concentration of As^{III} in urine exceeded that of As^V, indicating that in vivo reduction of As^V had occurred. MMA content in the urine gradually increased between 4 and 24 h after ingestion. A similar excretion pattern was found for DMA, with the DMA concentration in urine being about twice that of MMA. During 48 h after ingestion, 36% of the As dose was excreted in urine with proportions of As^{V}, As^{III}, MMA, and DMA being 9.7%, 17.5%, 25.3%, and 47.4%, respectively.

In a study of the pattern and extent of urinary excretion of As in healthy adult volunteers during 4 days after p.o. administration of As^{III}, MMA, or DMA (0.5 mg As), the amount of As excreted in the urine represented 46%, 78%, and 75% of the dose, respectively (BUCHET et al. 1981b). About 75% of the As excreted in the urine after ingestion of As^{III} was in the form of MMA and DMA (ratio 1:2); 13% of the MMA appeared methylated as DMA, while DMA was eliminated unchanged in the urine. However, MARAFANTE and associates (1987) have reported the methylation of DMA in humans. During 3 days after ingestion of ^{74}As-DMA (0.1 mg/kg), 4% of the dose was found in the urine in the form of TMAO and 80% was excreted as DMA.

Althouth most studies in experimental species found no metabolic conversion of DMA and TMA, human data suggested that partial demethylation of TMA may occur. While 90% of the ingested amount of TMA (10 μg As/kg) was excreted in the urine in the unmetabolized form, 3%–5% was found in the form of iAs, MMA, and DMA (YAMAUCHI and YAMAMURA 1984b). Notably, these data were obtained as differences between background values determined for each species prior to digestion of TMA and values determined after digestion. Hence, the degree of uncertainty in these calculations cannot easily be assessed.

In conclusion, several significant aspects of the methylation of As in humans can be highlighted. First, MMA is a somewhat more abundant urinary metabolite in humans than in other species. Second, humans methylate DMA to TMA, yielding a species with distinct toxic effects. Third, demethylation of TMA in humans may result in exposure to other iAs and organoAs species. Fourth, the speciation of As in urine may serve as a reliable indicator of occupational exposure to iAs as well as an indicator of the magnitude of exposure to As from environmental sources. Fifth, the concentrations of TMA compounds in urine may be useful indicators of As exposure arising from the consumption of seafood, especially fish and shellfish.

D. Conclusions and Future Research Directions

From exposure in food and water and in occupational and environmental settings, As remains a significant and difficult issue in risk assessment and public health. Despite continued study of the toxic effects of As in humans and other species, there are significant gaps in our knowledge of the metabolism, disposition, and toxicity of this metalloid. The following paragraphs summarize some of these gaps and suggest research which will address these needs.

Because As is a carcinogen in humans, development of a model for As-induced carcinogenesis in an experimental species is of great importance. The availability of an animal model would allow studies of the mechanistic basis of As carcinogenesis and an examination of the possible roles of As as an initiator, promoter, or progressor in the carcinogenic process.

In parallel with studies of As carcinogenicity, additional research is needed on the metabolic fate and toxicity of iAs and organoAs species. The complete speciation of As by valence and form in tissues and excreta of As-exposed humans and experimental animals will provide much-needed data on the capacity for the reduction and oxidation of iAs and for its methylation and on the distribution and retention of all As species. This aspect of the As research is central to a wide range of studies on As metabolism. Hence, the development and validation of improved analytical methods are primary research needs.

Further research is needed to identify the molecular targets for iAs and organoAs in various tissues. The linkage between the reduction of As^{V} by GSH (or by uncharacterized pathways), the formation of unique As^{III}-GSH complexes, and the interactions of As with macromolecular targets remains to be elucidated. Studies of the toxicity of various As species, including As^{III}-GSH complexes and organoAs species, are needed to complement our current understanding of the risk associated with As exposure.

The enzymatic basis of the methylation of iAs also requires further study. Characterization of the methyltransferase(s) which catalyze As methylation will contribute to an understanding of the mechanistic basis of As metabolism and will determine the functional homology between the As methyltransferases and other methyltransferases. The possibility that competition among methyltransferases for cofactors and substrates contributes to the toxicity and carcinogenicity of As merits examination.

Additional studies are needed of the metabolism and toxicity of As in humans. This includes studies of As metabolism in volunteers and epidemiological studies in As-exposed populations. These studies should integrate improved methods for As speciation in tissues and excreta and should test biomarkers of As exposure or effect. Of particular interest in these studies will be estimates of the interindividual variation in capacity for the methylation of As. As described above, interindividual variation in methylation capacity may reflect genotypic variation or may be due to differences in methyl donor availability. Because methyl donor availability may reflect dietary lipotrope content or competition among various methylation processes for a limited methyl donor pool, new epidemiological studies should incorporate measures of dietary composition.

Additional effort should be given to integrating the available data on the metabolism and disposition of As in humans and other species into a physiologically based model for the kinetic behavior of iAs and the organoAs species. Coordination of the research in humans and experimental species could contribute to the rapid development and validation of a robust model for the fate and biological effects of this metalloid.

Acknowledgements. M.S. is a visiting scientist supported by Training Grant T901915 of the U.S. Environmental Protection Agency/University of North Carolina Toxicology Research Program with the Curriculum in Toxicology, University of North Carolina at Chapel Hill. M.D. is a visiting scientist supported by the U.S. Environmental Protection Agency through the Center for Environmental Medicine and Lung Biology (Cooperative Agreement CR817643) at the University of North Carolina at Chapel Hill. We thank our colleagues in the Health Effects Research Laboratory of the U.S. Environmental Protection Agency for valuable discussions on pharmacokinetic, toxicological, and epidemiological aspects of As exposure and Professor Hiroshi Yamauchi, Department of Public Health, St. Marianna Universtiy School of Medicine, Kawasaki, Japan, for useful discussions on arsenic metabolism.

References

Anundi I, Hogberg J, Vahter M (1982) GSH release in bile as influenced by arsenite. FEBS Lett 145:285–288

Bertolero F, Marafante E, Edel Rade J, Pietra R, Sabbioni E (1981) Biotransformation and intracellular binding of arsenic in tissue of rabbits after intraperitoneal administration of ^{74}As labelled arsenite. Toxicology 20:35–44

Braman RS, Foreback CC (1973) Methylated forms of arsenic in the environment. Science 182:1247–1249

Buchanan WD (1962) Toxicity of arsenic compounds. Elsevier, Amsterdam

Buchet JP, Lauwerys R (1985) Study of inorganic arsenic methylation by rat in vitro: relevance for the interpretation of observations in man. Arch Toxicol 57:125–129

Buchet JP, Lauwerys R (1987) Study of factors influencing the in vivo methylation of inorganic arsenic in rats. Toxicol Appl Pharmacol 91:65–74

Buchet JP, Lauwerys R (1988) Role of thiols in the in vitro methylation of inorganic arsenic by rat liver cytosol. Biochem Pharmacol 37:3149–3153

Buchet JP, Lauwerys R, Roels H (1981a) Urinary excretion of inorganic arsenic and its metabolites after repeated ingestion of sodium metaarsenite. Int Arch Occup Environ Health 48:111–118

Buchet JP, Lauwerys R, Roels H (1981b) Comparison of the urinary excretion of arsenic metabolites after a single oral dose of sodium arsenite, monomethylarsonate, or dimethylarsinate in man. Int Arch Occup Environ Health 48:71–79

Buchet JP, Geubel A, Pauwels S, Mahieu P, Lauwerys R (1984) The influence of liver disease on the methylation of arsenite in humans. Arch Toxicol 55:151–154

Cannon JR, Edmonds JS, Francesconi KA, Raston CL, Saunders JB, Skelton BW, White AH (1981) Isolation, crystal structure and synthesis of arsenobetaine a constituent of the western rock lobster Panulirus cygnus, the dusky shark, Carcharhinus obscurus and some samples of human urine. Aust J Chem 34: 787–798

Cebrián ME, Del Razo LM, García-Vargas G (1993) Indicators of susceptibility and damage in arsenic exposed populations. In: Book of abstracts, International Conference on Arsenic Exposure. 28–30 July 1993, New Orleans, LA, p 12

Challenger F (1945) Biological methylation. Chem Revs 36:315–361

Challenger F (1951) Biological methylation. Adv Enzymol 12:429–491

Chen CJ, Chuang YC, Lin TM, Wu HY (1985) Malignant neoplasms among residents of a blackfoot disease-endemic area in Taiwan: high arsenic artesian well water and cancers. Cancer Res 45:5895–5899

Chen CJ, Wu MM, Lee SS, Wang JD, Cheng SH, Wu HY (1988) Atherogenicity and carcinogenicity of high-arsenic artesian well water. Multiple risk factors and related malignant neoplasms of blackfoot disease. Arteriosclerosis 8:452–460
Coulson EJ, Remington RE, Lynch KM (1935) Metabolism in the rat of the naturally occurring arsenic of shrimp as compared with arsenic trioxide. J Nutr 10: 255–270
Cox DP, Alexander M (1973) Production of trimethylarsine gas from various arsenic compounds by three sewage fungi. Bull Environ Contam Toxicol 9:84–88
Crecelius EA (1977) Changes in the chemical speciation of arsenic following ingestion by man. Environ Health Perspect 19:147–150
Dabeka RW, McKenzie AD, Lacroix GMA, Cleroux C, Bowe S, Graham RA, Conacher HBS, Verdier P (1993) Survey of arsenic in total diet food composites and estimation of the dietary intake of arsenic by Canadian adults and children. J AOAC Int 76:14–25
Delnomdedieu M, Basti MM, Otvos JO, Thomas DJ (1993) Transfer of arsenite from glutathione to dithiol: a model of interaction. Chem Res Toxicol 6:598–602
Delnomdedieu M, Basti MM, Otvos JO, Thomas DJ (1994) Reduction of arsenate and dimethylarsinate by glutathione: a magnetic resonance study. Chem Biol Interact (in press)
Doerfler W (1983) DNA methylation and differentiation. Annu Rev Biochem 52:93–124
Doi R (1991) Individual difference of methylmercury metabolism in animals and its significance in methylmercury toxicity. In: Imura N, Clarkson TW (eds) Advances in mercury toxicology. Plenum, New York, p 77
Edmonds JS, Francesconi KA, Cannon JR, Raston CL, Skelton BW, White AH (1977) Isolation, crystal structure and synthesis of arsenobetaine, the arsenical constituent of the western rock lobster Panulirus longipes cygnus George. Tetrahedron Lett 18:1543–1546
Fearon EH, Vogelstein B (1990) A genetic model for colorectal carcinogenesis. Cell 61:757–767
Foa V, Colombi A, Maroni M, Buratti M, Calzaferri G (1984) The speciation of chemical forms of arsenic in the biological monitoring of exposure to inorganic arsenic. Sci Total Environ 34:241–259
Frost DV (1967) Arsenicals in biology. Retrospect and prospect. Fed Proc 26:194–208
Fuentes N, Zambrano F, Rosenmann M (1981) Arsenic contamination: metabolic effects and localization in rats. Comp Biochem Physiol 70C:269–272
Gadd GM (1993) Microbial formation and transformation of organometallic and organometalloid compounds. FEMS Microbiol Rev 11:297–316
Georis B, Cardenas A, Buchet JP, Lauwerys R (1990) Inorganic arsenic methylation by rat tissue slices. Toxicology 63:73–84
Geubel AP, Mairlot MC, Buchet JP, Dive C, Lauwerys R (1988) Abnormal methylation capacity in human liver cirrhosis. Int J Clin Pharmacol Res VIII:117–122
Goodman LS, Gilman A (1941) Drugs used in the chemotherapy of syphilis. I. The pharmacology and clinical toxicology of the anti syphilitic arsenicals – the arsphenamines, mapharsen and tryparasamide. In: The pharmacological basis of therapeutics, 2nd edn. Macmillan, New York, p 946
Gyurasics A, Varga F, Gregus Z (1991a) Effects of arsenicals on biliary excretion of endogenous glutathione and xenobiotics with glutathione-dependent hepatobiliary transport. Biochem Pharmacol 41:937–944
Gyurasics A, Varga F, Gregus Z (1991b) Glutathione-dependent biliary excretion of arsenic. Biochem Pharmacol 42:465–468
Harvey SC (1970) Arsenic. In: Goodman LS, Gilman A (eds) The pharmacological basis of therapeutics, 4th edn. Macmillan, New York, p 958
Hirata M, Hisanaga A, Tanaka A, Ishinishi N (1988) Glutathione and methylation of inorganic arsenic in hamsters. Appl Organomet Chem 2:315–320
Hirata M, Mohri T, Hisanaga A, Ishinishi N (1989) Conversion of arsenite and arsenate to methylarsenic and dimethylarsenic compounds by homogenates

prepared from livers and kidneys of rats and mice. Appl Organomet Chem 3:335–341
Hirata M, Tanaka A, Hisanaga A, Ishinishi N (1990) Effects of glutathione depletion on the acute nephrotoxic potential of arsenite and on arsenic metabolism in hamsters. Toxicol Appl Pharmacol 106:469–481
Hoffman JL (1980) The rate of transmethylation in mouse liver as measured by trapping S-adenosylhomocysteine. Arch Biochem Biophys 205:132–135
Hopenhayn-Rich C, Smith AH, Goeden HM (1993) Human studies do not support the methylation threshold hypothesis for the toxicity of inorganic arsenic. Environ Res 60:161–177
Huang H, Huang CF, Wu DR, Jinn CM, Jan KY (1993) Glutathione as a cellular defense against arsenite toxicity in cultured Chinese hamster ovary cells. Toxicology 79:195–204
Hughes MF, Menache M, Thompson DJ (1994) Dose-dependent disposition of sodium arsenate in mice following acute oral exposure. Fund Appl Toxicol 22:80–89
Hunter FT, Kip AF, Irvine JF Jr (1942) Radioactive tracer studies on arsenic injected as potassium arsenite. J Pharmacol Exp Ther 76:207–220
International Agency for Research on Cancer (1987) Arsenic and arsenic compounds. In: IARC monograph on the evaluation of carcinogenic risks to humans – overall evaluations of carcinogenicity: an update of IARC Monographs 1 to 42, [Suppl 7], Lyon, p 100
Ji G, Silver S (1992) Reduction of arsenate to arsenite by the ArsC protein of the arsenic resistance operon of Staphylococcus aureus plasmid pI258. Proc Natl Acad Sci USA 89:9474–9478
Joint WHO/FAO Expert Committee on Food Additives (1989) Evaluation of certain food additives and contaminants, 33rd report, technical report series 776. WHO, Geneva, p 27
Kurosawa S, Yasuda K, Taguchi M, Yamazaki S, Toda S, Morite M, Uehiro T, Fuga K (1980) Identification of arsenobetaine, a water soluble arsenic compound in muscle and liver of a shark Prionace glauca. Agric Biol Chem 44:1993–1994
Lakso JU, Peoples SA (1975) Methylation of inorganic arsenic by mammals. J Agric Food Chem 23:674–676
Lanz H, Wallace PC, Hamilton JG (1950) The metabolism of arsenic in laboratory animals using As^{74} as a tracer. Univ Calif Publ Pharmacol 2:253–282
Lee T-C, Wei ML, Chang WJ, Ho IC, Lo JF, Jan KY, Huang H (1989) Elevation of glutathione and glutathione S-transferase activity in arsenic resistant Chinese hamster ovary cells. In Vitro Cell Dev Biol 25:442–448
Lerman S, Clarkson TW (1983) The metabolism of arsenite and arsenate by the rat. Fund Appl Toxicol 3:309–314
Luten JB, Riekwel-Booy G, Rauchbaar A (1982) Occurrence of arsenic in plaice (Pleuronectes platessa), nature of organo-arsenic compound present and its excretion by man. Environ Health Perspect 45:165–170
Luten JB, Riekwel-Booy G, Greef JVD, ten Oever de Brauw MC (1983) Identification of arsenobetaine in sole, lemon sole, flounder, dab, crab and shrimps by field desorption and fast bombardment mass spectrometry. Chemosphere 12:131–141
Mahieu P, Buchet J-P, Roels HA, Lauwerys R (1981) The metabolism of arsenic in humans acutely intoxicated by As_2O_3. Its significance for the duration of BAL therapy. Clin Toxicol 18:1067–1075
Mahieu P, Buchet J-P, Lauwerys R (1987) Evolution clinique et biologique d'une intoxication orale aigue par l'anhydride arsenieux et considerations sur l'attitude therapeutique. J Toxicol Clin Exp 7:273–278
Maiorino RM, Aposhian HV (1985) Dimercaptan metal-binding agents influence the biotransformation of arsenite in the rabbit. Toxicol Appl Pharmacol 77:240–250
Marafante E, Vahter M (1984) The effect of methyltransferase inhibition on the metabolism of [^{74}As]arsenite in mice and rabbits. Chem Biol Interact 50:49–57

Marafante E, Vahter M (1986) The effect of dietary and chemically induced methylation deficiency on the metabolism of arsenate in the rabbit. Acta Pharmacol Toxicol 59 [Suppl 7]:35–38
Marafante E, Vahter M, Dencker L (1984) Metabolism of arsenocholine in mice, rats and rabbits. Sci Total Environ 34:223–240
Marafante E, Vahter M, Envall J (1985) The role of methylation in the detoxification of arsenate in the rabbit. Chem Biol Interact 56:225–238
Marafante E, Vahter M, Norin H, Envall J, Sandström M, Christakopoulos A, Ryhage R (1987) Biotransformation of dimethylarsinic acid in mouse, hamster and man. J Appl Toxicol 7:111–117
Mass MJ (1992) Human carcinogenesis by arsenic. Environ Geochem Health 14: 49–54
McBride BC, Wolfe RS (1971) Biosynthesis of dimethylarsine by methanobacterium. Biochemistry 10:4312–4317
McKinney JD (1992) Metabolism and disposition of inorganic arsenic in laboratory animals and humans. Environ Geochem Health 14:43–48
Mercer JF, Livingston J, Hall B, Paynter JA, Begy C, Chandrasekharappa S, Lockhart P, Grimes A, Bhave M, Siemieniak D et al. (1993) Isolation of a candidate gene for Menkes' disease by positional cloning. Nature Genet 3:20–25
Mergeay M (1991) Towards an understanding of the genetics of bacterial metal resistance. Trends Biotech 9:17–24
Michalowsky LA, Jones PA (1989) DNA methylation and differentiation. Environ Health Perspect 80:189–197
Mohri T, Hisanaga A, Ishinishi N (1990) Arsenic intake and excretion by Japanese adults: a 7-day duplicate diet study. Food Chem Toxicol 28:521–529
Moody JP, Williams RT (1964a) The fate of 4-nitrophenylarsonic acid in hens. Food Cosmet Toxicol 2:695–706
Moody JP, Williams RT (1964b) The metabolism of 4-hydroxy-3-nitrophenylarsonic acid in hens. Food Cosmet Toxicol 2:707–715
Murai T, Iwata H, Otoshi T, Endo G, Horiguchi S, Fukushima S (1993) Renal lesions induced in F344/DuCrj rats by 4 weeks oral administration of dimethylarsenic acid. Toxicol Lett 66:53–61
National Research Council (1977) Medical and biologic effects of environmental pollutants. Arsenic. National Academy of Sciences, Washington DC, p 16
Newberne PM, Rogers AE (1986) Labile methyl groups and the promotion of cancer. Annu Rev Nutr 6:407–432
Norin H, Christakopoulos A (1982) Evidence for the presence of arsenobetaine and other organoarsenicals in shrimps. Chemosphere 11:287–298
Norin H, Ryhage R, Christakopoulos A, Sandström (1983) New evidence for the presence of arsenobetaine in shrimps (Pandalus borealis) by use of pyrolysis gas chromatography-atomic absorption spectrometry/mass spectromety. Chemosphere 12:299–315
Norin H, Vahter M, Christakopoulos, Sandstrom M (1985) Concentration of inorganic and total arsenic in fish from industrially polluted water. Chemosphere 14:325–334
Odanaka Y, Matano O, Goto S (1980) Biomethylation of inorganic arsenic by the rat and some laboratory animals. Bull Environ Contam Toxicol 24:452–459
Offergelt JA, Roels H, Buchet JP, Boeckx M, Lauwerys R (1992) Relation between airborne arsenic trioxide and urinary excretion of inorganic arsenic and its methylated metabolites. Br J Ind Med 49:387–393
Overby LR, Frederickson RL (1963) Metabolic stability of radioactive arsanilic acid in chickens. J Agric Food Chem 11:378–381
Overby LR, Frederickson RL (1965) Metabolism of arsanilic acid. II. Localization and type of arsenic excreted and retained by chickens. Toxicol Appl Pharmacol 7:855–867
Overby LR, Straube L (1965) Metabolism of arsanilic acid. I. Metabolic stability of double labeled arsanilic acid in chickens. Toxicol Appl Pharmacol 7:850–854

Peoples SA (1964) Review of arsenical pesticides. In: Woolson EA (ed) Arsenical pesticides. ACS, Washington DC, p 1 (ACS symposium series 7)
Peoples SA (1969) The failure of methanearsonic acid to cross the blood-mammary barrier when administered orally to lactating cows. Fed Proc 28:359 (abstract)
Pershagen G, Bjorklund N-E (1985) On the pulmonary tumorigenicity of arsenic trioxide and calcium arsenate in hamsters. Cancer Lett 27:99–104
Pershagen G, Nordberg G, Bjorklund N-E (1984) Carcinomas of the respiratory tract in hamsters given arsenic trioxide and/or benzo(a)pyrene by the pulmonary route. Environ Res 34:227–241
Petito CT, Beck BD (1991) Evaluation of evidence of nonlinearities in the dose-response curve for arsenic carcinogenesis. Trace Subst Environ Health 24: 143–176
Pomroy C, Charbonneau SM, McCullough RS, Tam GKH (1980) Human retention studies with ^{74}As. Toxicol Appl Pharmacol 53:550–556
Rogers EH, Chernoff N, Kavlock RJ (1981) The teratogenic potential of cacodylic acid in the rat and mouse. Drug Chem Toxicol 4:49–61
Rosen BP, Weigel U, Monticello R, Edwards BPF (1991) Molecular analysis of an anion pump. Arch Biochem Biophys 284:381–385
Rowland IR, Davies MJ (1981) In vitro metabolism of inorganic arsenic by the gastro-intestinal microflora of the rat. J Appl Toxicol 1:278–283
Rudnai P, Borzsonyi M (1980) Carcinogenic effect of arsenic trioxide in trans-placentally and neonatally treated CFLP mice. Nat Sci 2:11–18
Scott N, Hatlelid KM, MacKenzie NE, Carter DE (1993) Reaction of arsenic(III) and arsenic(V) species with glutathione. Chem Res Toxicol 6:102–106
Shibata Y, Yoshinaga J, Morita M (1993) Detection of arsenobetaine in human blood. In: Book of abstracts, 6th international symposium on environmental and industrial arsenic, Kawasaki, Japan, p 12
Silver S, Ji G, Broer S, Dey S, Dou D, Rosen BP (1993a) Orphan enzyme or patriarch of a new tribe: the arsenic resistance ATPase of bacterial plasmids. Mol Microbiol 8:637–642
Silver S, Nucifora G, Phung LT (1993b) Human Menkes X-chromosome disease and the staphylococcal cadmium-resistance ATPase: a remarkable similarity in protein sequence. Mol Microbiol 10:7–12
Smith AH, Hopenhayn-Rich C, Bates MN, Goeden HM, Hertz-Picciotto I, Duggan HM, Wood R, Kosnett MJ, Smyth MT (1992) Cancer risks from arsenic in drinking water. Environ Health Perspect 97:259–267
Stevens JT, DiPasquale LC, Farmer JD (1979) The acute inhalation toxicology of the technical grade organoarsenical herbicides, cacodylic acid and disodium methanearsonic acid: a route comparison. Bull Environ Contam Toxicol 21: 304–311
Takahashi K, Yamauchi H, Yamato N, Yamamura Y (1988) Methylation of arsenic trioxide in hamsters with liver damage induced by long-term administration of carbon tetrachloride. Appl Organomet Chem 2:309–314
Takahashi K, Yamauchi H, Mashiko M, Yamamura Y (1990) Effect of S-adenosyl methionine on methylation of inorganic arsenic. Nippon Eiseigaku Zasshi 45:613–618
Tam GKH, Charbonneau SM, Bryce F, Lacroix G (1978) Separation of arsenic metabolites in dog plasma and urine following intravenous injection of ^{74}As. Anal Biochem 86:505–511
Tam GKH, Charbonneau SM, Bryce F, Pomroy C, Sandi E (1979) Metabolism of inorganic arsenic (^{74}As) in humans following oral ingestion. Toxicol Appl Pharmacol 50:319–322
Tam GKH, Charbonneau SM, Bryce F, Sandi E (1982) Excretion of a single oral dose of fish-arsenic in man. Bull Environ Contam Toxicol 28:669–673
Tezuka M, Hanioka K-I, Yamanaka K, Okada S (1993) Gene damage induced in human alveolar type II (L-132) cells by exposure to dimethylarsinic acid. Biochem Biophys Res Commun 1991:1178–1183

Thomas DJ (1994) Arsenic toxicity in humans: research problems and prospects. Environ Geochem Health (in press)
Thompson DJ (1993) A chemical hypothesis for arsenic methylation in mammals. Chem Biol Interact 88:89–114
Tseng W-P (1989) Blackfoot disease in Taiwan: a 30-year follow-up study. Angiology 40:547–558
Tseng W-P, Chu HM, How SW, Fong JM, Lin CS, Yeh S (1968) Prevalence of skin cancer in an endemic area of chronic arsenicism in Taiwan. J Natl Cancer Inst 40:453–463
US Environmental Protection Agency (1988) Special report on ingested inorganic arsenic-skin cancer; nutritional essentiality. Risk Assessment Forum, EPA/625/3-87/013, Washington DC
Vahter M (1981) Biotransformation of trivalent and pentavalent inorganic arsenic in mice and rats. Environ Res 25:286–293
Vahter M (1986) Environmental and occupational exposure to inorganic arsenic. Acta Pharmacol Toxicol 59 [Suppl 7]:31–34
Vahter M, Envall J (1983) In vivo reduction of arsenate in mice and rabbits. Environ Res 32:14–24
Vahter M, Marafante E (1985) Reduction and binding of arsenate in marmoset monkeys. Arch Toxicol 57:119–124
Vahter M, Marafante E (1987) Effects of low dietary intake of methionine, choline or proteins on the biotransformation of arsenite in the rabbit. Toxicol Lett 37:41–46
Vahter M, Marafante E, Lindgren A, Dencker L (1982) Tissue distribution and subcellular binding of arsenic in marmoset monkeys after injection of ^{74}As-arsenite. Arch Toxicol 51:65–77
Vahter M, Marafante E, Dencker L (1983) Metabolism of arsenobetaine in mice, rats and rabbits. Sci Total Environ 30:197–211
Vulpe C, Levinson B, Whitney S, Packman S, Gitscher J (1993) Isolation of a candidate gene for Menkes disease and evidence that it encodes a copper-transporting ATPase. Nature Genet 3:7–13
Wang H-F, Lee T-C (1993) Glutathione S-transferase pi facilitates the excretion of arsenic from arsenic resistant Chinese hamster ovary cells. Biochem Biophys Res Commun 192:1093–1099
Weinshilboum R (1988) Pharmacogenetics of methylation: Relationship to drug metabolism. Clin Biochem 21:201–210
Wu J, Rosen BP (1993) Metalloregulated expression of the ars operon. J Biol Chem 268:52–58
Wu MM, Kuo TL, Hwang YH, Chen CJ (1989) Dose-response relation between arsenic concentration in well water and mortality from cancer and cardiovascular disease. Am J Epidemiol 130:1123–1132
Yamamura Y (1993) Toxicity and metabolism of alkylarsine. In: Book of abstracts, 6th international symposium on environmental and industrial arsenic. Kawasaki, Japan, p 4
Yamamura Y, Yamauchi H (1980) Arsenic metabolites in hair, blood and urine in workers exposed to arsenic trioxide. Ind Health 18:203–210
Yamanaka K, Hasegawa A, Sawamura R, Okada S (1989) Dimethylated arsenics induce DNA strand breaks in lung via the production of acitve oxygen in mice. Biochem Biophys Res Commun 165:43–50
Yamanaka K, Hoshino H, Sawamura R, Hasegawa A, Okada S (1990) Induction of DNA damage by dimethylarsine, a metabolite of inorganic arsenics, is for the major part likely due to its peroxyl radical. Biochem Biophys Res Commun 168:58–64
Yamanaka K, Hasegawa A, Sawamura R, Okada S (1991) Cellular responses to oxidative damage in lung induced by the administration of dimethylarsinic acid, a metabolite of inorganic arsenics, in mice. Toxicol Appl Pharmacol 108:205–215

Yamanaka K, Tezuka M, Kato K, Hasegawa A, Okada S (1993) Crosslink formation between DNA and nuclear proteins by in vivo exposure of cells to dimethylarsinic acid. Biochem Biophys Res Commun 191:1184–1191
Yamauchi H, Yamamura Y (1979) Urinary inorganic arsenic and methylarsenic excretion following arsenate-rich seaweed ingestion. Jpn J Ind Health 21:47–54
Yamauchi H, Yamamura Y (1984a) Metabolism and excretion of orally administered dimethylarsinic acid in the hamster. Toxicol Appl Pharmacol 74:134–140
Yamauchi H, Yamamura Y (1984b) Metabolism and excretion of orally ingested trimethylarsenic in man. Bull Environ Contam Toxicol 32:682–687
Yamauchi H, Yamamura Y (1985) Metabolism and excretion of orally administered arsenic trioxide in the hamster. Toxicology 34:113–121
Yamauchi H, Yamamura Y (1986) Metabolism and excretion of orally and intraperitoneally administered gallium arsenide in the hamsters. Toxicology 40:237–246
Yamauchi H, Yamato N, Yamamura Y (1988) Metabolism and excretion of orally and intraperitoneally administered methylarsonic acid in the hamster. Bull Environ Contam Toxicol 40:280–286
Yamauchi H, Takahashi K, Mashiko M, Yamamura Y (1989) Biological monitoring of arsenic exposure of gallium arsenide- and inorganic arsenic-exposed workers by determination of inorganic arsenic and its metabolites in urine and hair. Am Ind Hyg Assoc J 50:606–612
Yamauchi H, Kaise T, Takahashi K, Yamamura Y (1990) Toxicity and metabolism of trimethylarsine in mice and hamster. Fund Appl Toxicol 14:399–407
Yamauchi H, Takahashi K, Mashiko M, Saitoh J, Yamamura Y (1992) Intake of different chemical species of dietary arsenic by the Japanese and their blood and urinary arsenic levels. Appl Organomet Chem 6:383–388
Zeisel SH (1988) "Vitamin-like" molecules (A) choline. In: Shils ME, Young VR (eds) Modern nutrition in health and disease, 7th den. Lea and Febiger, Philadelphia, p 440

CHAPTER 19

Bacterial Plasmid-Mediated Resistances to Mercury, Cadmium, and Copper

S. SILVER and M. WALDERHAUG

A. Introduction and Overview: Generalities

Ion homeostasis was undoubtedly one of the earliest attributes of living cells. The maintenance of optimal concentrations of catalytic ions and ionic osmolytes is essential for the conduct of metabolism. Cells, both prokaryotic and eukaryotic, must be especially selective with regard to transition metal ions that cross the cell membranes. Not only must cells accumulate essential metal ions and exclude toxic ions, but they must also limit the influx or stimulate the efflux of essential metal ions so that these ions do not reach toxic levels. Both copper and iron, for example, are essential trace cations that can cause cellular damage if the internal cellular concentration of either cation is too high, or if the storage of these cations becomes deranged.

The mechanisms for achieving metal ion homeostasis depend on the nature of the metal's role in the cell. For metal ions that are constantly needed by cells, a transporter may be constitutively expressed and reside in the membrane for instant use by the cell. Examples of constitutively expressed transporters are the Trk system for K^+ and the Pit system for phosphate, both in *Escherichia coli*. These systems are always expressed, whether environmental levels of substrates are low or high. For metal ions that are needed sparingly (such as iron), or that are only occasionally encountered in the environment (like most toxic heavy metals), tightly regulated ("induced") synthesis of transporter systems is common. For these regulated systems, specific metal sensors (e.g., Fur for Fe^{3+} and MerR for Hg^{2+}) turn on the genetic expression of transporters when needed; for nutrients, they turn off expression in times of excess availability. For toxic metal ions, regulated synthesis of transport systems appears to be the rule.

Chemical coordination of metal ions in living systems is a central question in metal ion homeostasis and is under intense study. The predominant atoms that coordinate with the metals are oxygen, nitrogen, and sulfur. However, sometimes metals can form covalent bonds with carbon (e.g., methylmercury). Oxygen is used to coordinate transition metals for transport in the case of siderophores (e.g., enterochelin and aerobactin), but examples of proteins using oxygen atoms on amino acid side chains to coordinate metals are rare. Sulfur on cysteine or methionine residues is a common coordinating atom. Examples of cysteine coordination with mercury, cadmium, and copper are shown in Sects. B, C, and E, respectively. Alternatively, unionized nitrogen (for example, the N-3 of histidine) is a metal ion coordinating atom. Examples of nitrogen coordination are proposed in Sect. C.II, the Czc transport system.

Gene expression affecting metal ion homeostasis may be regulated indirectly or directly. A sensor–effector form of regulation is an example of an indirect mechanism. A sensor protein, which may be located in the cell membrane, "sees" the metal ion and then communicates with a separate effector protein that might be a DNA-binding repressor protein. This is the mechanism by which the bacterial copper resistance system is regulated (see Sect. E, below). The effector protein may interact with RNA polymerase as well as with DNA directly and cause a change in mRNA expression. For more direct regulation, the metal ion interacts directly with the protein that causes increased or decreased genetic expression. On the other hand, the mercury-responding mercury resistance regulator, MerR, is an example of direct regulation (Sect. B). For both forms of regulation, investigation of the metal-binding domains of the sensing proteins provides examples of ion-binding motifs and promotes insights into the mechanisms of metal ion toxicity and resistance.

B. Mercury and Organomercurial Resistance

I. Introduction

Resistance to the toxic effects of mercury is quite commonly found in bacteria (SILVER and WALDERHAUG 1992; MISRA 1992). Since most bacteria are rarely exposed to toxic levels of mercury, the resistance mechanism is inducible and is frequently found on plasmids and/or transposons. Many of the multi-antibiotic resistance plasmids that are frequently found in clinical collections have determinants of mercury resistance as well. Furthermore, mercury resistance is a consistent component of the chromosomal resistance determinant of MRSA (methicillin-resistant *Staphylococcus aureus*), a current clinical problem and one with no apparent connection to the use of mercurials. If the mercury resistance system is present, the expression of the detoxifying activities is tightly regulated and turned on only when needed. The MerR regulatory protein turns on mRNA synthesis by a positive activator mechanism (for primary references, see SILVER and WALDERHAUG 1992; MISRA 1992).

II. Genetic Organization and Molecular Biology

In all bacterial systems studied, resistance to mercury requires a group of about six genes. Their functions include: (a) regulation so that the system will function only when needed, (b) uptake of extracellular Hg^{2+} in a controlled and bound form to (c) mercuric reductase for conversion of toxic

Hg^{2+} to less toxic and volatile Hg^0, and sometimes (d) the additional enzyme organomercurial lyase, which cleaves the carbon–mercury bond in organomercurials such as phenylmercury and methylmercury (releasing benzene and methane plus Hg^{2+}, the substrate for the mercuric reductase enzyme; WALSH et al. 1988).

1. Operon Structure in Gram-Negative Bacteria

Figure 1A shows the genes and gene functions for the mercury resistance operon from plasmid pDU1358, which contains six genes and includes the *merB* gene for organomercurial lyase. Four sequenced alternative versions of mercury resistance determinants of gram-negative bacteria lack the *merB* gene. All (except for *Thiobacillus*; see next section) start with the *merR* gene, which is transcribed into a separate mRNA that is synthesized in the opposite direction from the major mer mRNA (Fig. 1A). The MerR protein is a positive activator protein. It is unique and is the paradigm of a potential

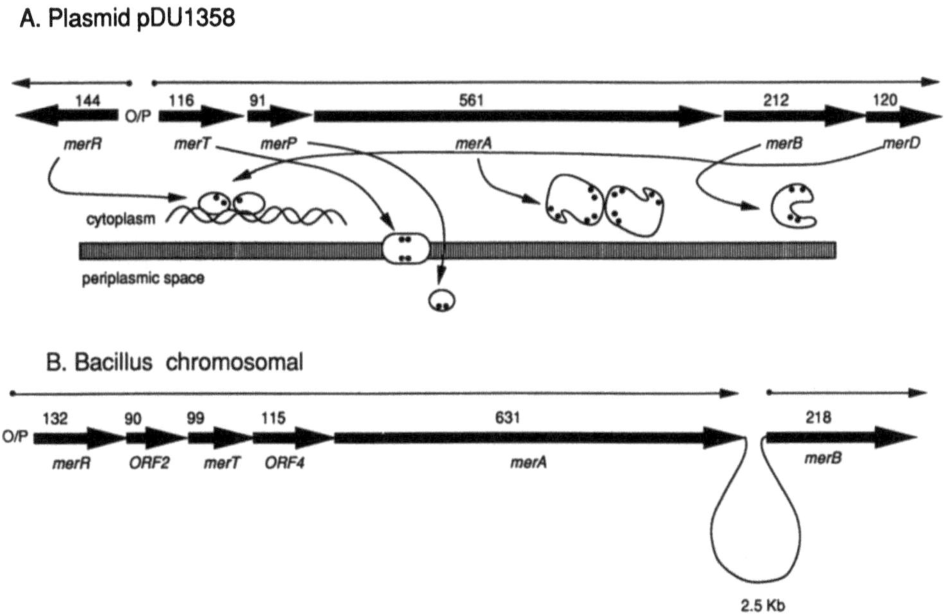

Fig. 1A,B. Genes of the gram-positive and gram-negative bacterial mercury resistance systems. **A** From plasmid pDU1358 from gram-negative bacteria. **B** From the chromosomal determinant of Bacillus. Modified from SILVER and WALDERHAUG (1992). mer genes or open reading frames (ORF) not yet assigned gene mnemonics are indicated by *thick arrows*, corresponding to the transcriptional and translational direction. Lengths are indicated in amino acid (*aa*) equivalents above the genes. *Thin arrows* indicate lengths and directions of mRNA transcripts. Below the plasmid pDU1358 genes, the protein products and their locations are indicated, with *closed dots* representing cysteine residues

new class of activator proteins (O'HALLORAN 1993). MerR binds as a dimer to the operator DNA region, repressing its own synthesis when bound. When Hg^{2+} is added, one Hg^{2+} binds per dimer, utilizing two cysteine residues on one MerR subunit and a different third cysteine on the second MerR subunit to make an unusual $Hg(S-)_3$ coordination complex (O'HALLORAN 1993; SUMMERS 1992). The MerR complex occupies the same position on the operator DNA with or without Hg^{2+}; the binding of Hg^{2+} brings about a localized untwisting of the DNA to a conformation favorable for RNA polymerase binding and initiation of mRNA synthesis.

A single mRNA transcript encodes the remaining five *mer* genes of plasmic pDU1358. The first two genes determine an uptake transport system consisting of a small membrane protein MerT (containing four cysteine residues as potential Hg^{2+}-binding sites) and a still smaller periplasmic protein MerP (with two cysteines). The cysteine pair in the small MerP periplasmic binding protein is closely homologous over a 30 amino acid stretch with the amino-terminal presumed mercury-binding motif in mercuric reductase (for primary references, see SILVER and WALDERHAUG 1992). These Hg^{2+}-binding motifs are in turn homologous to Cd^{2+}-binding motifs at the amino terminus of the Cd^{2+} efflux ATPase (SILVER et al. 1989, 1993b; Sect. C.I below) and the more recently found (presumed) Cu^{2+}-binding motifs in bacterial and animal cell efflux ATPases (SILVER et al. 1993b; Sect. D below). Here we consider only the proposed role of these cysteine pairs in passing toxic Hg^{2+} safely from the cell surface to the active site of the intracellular mercuric reductase.

When we first found this transport system and its genes, we asked why cells would accumulate a toxic cation, instead of just excluding it. However, since transport systems exist in all known versions of the *mer* resistance determinant, we have rationalized this as part of a "bucket brigade" which carries Hg^{2+} (too dangerous to leave loose) across the cell surface to the intracellular mercuric reductase (which must be intracellular, since it requires the high-energy cofactor NADPH). Ionic mercury appears to be coordinated by cysteine residues from the time it enters the periplasmic space (Fig. 1A). At that time the Hg^{2+} ion is coordinated by the cysteine pair on MerP, which is thought to transfer the cation to another cysteine pair on the outer surface of MerT, the membrane-bound transporter (Fig. 1A). MerT has two pairs of cysteine residues, and the model of transport involves one pair at the periplasmic side and another pair at the cytoplasmic side. From MerT, the Hg^{2+} cation is transferred to MerA, which has three pairs of potentially coordinating cysteine pairs per subunit.

After the transport system genes, the *merA* gene encodes the large subunits of the dimeric protein mercuric reductase. SCHIERING et al. (1991) solved the crystal structure for the mercuric reductase protein from the *Bacillus mer* operon (see Sect. B.I.3). The structure (Fig. 2), as anticipated, is very similar to that of the homologous human protein glutathione reductase. The dimer subunits are related by a twofold rotational axis, and

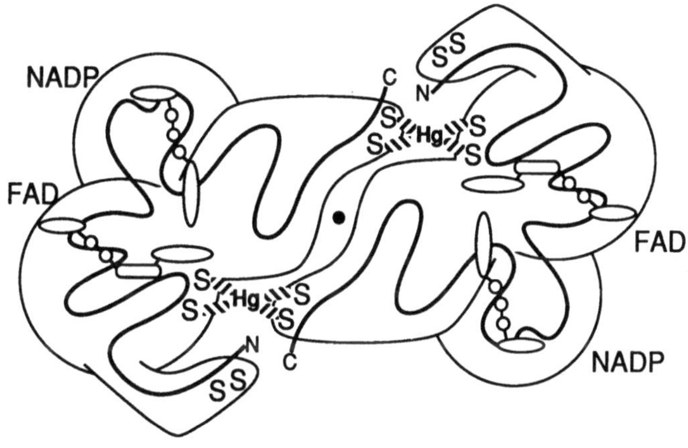

Fig. 2. Structure of mercuric reductase (modeled after the crystal structure of SCHIERING et al. 1991). Hg^{2+} is indicated bound to four cysteine sulfurs (see text). NADPH and FAD are perpendicular to the plane of the page. Amino (*N*) and carboxyl (*C*) termini of the amino acid backbone are indicated, as is the dithiol (*S-S*, although probably reduced cysteines in vivo) near the amino terminus. The *central dot* represents the twofold rotational axis

both cofactors (NADPH and FAD) occur perpendicular to the plane of the structure in Fig. 2. The active site includes an 18 amino acid stretch containing the redox-active dithiol common to all of this family of mercuric reductase, glutathione reductase, and related proteins (SILVER and WALDERHAUG 1992; MISRA 1992). Mercuric reductase also contains a highly conserved stretch at the carboxyl terminus (as shown in Fig. 2; DISTEFANO et al. 1990), raising the possibility that four cysteines (two on each subunit) are involved in coordination of the substrate Hg^{2+}. These four cysteines lie close together in the crystal structure as diagrammed and the carboxyl-terminal region is conserved among all versions of mercuric reductase, but is very different from the carboxyl-terminal (and glutathione-binding) region of glutathione reductase.

An additional feature in Fig. 2 worth noting is the amino-terminal 160 amino acids of mercuric reductase that lacked a fixed position in the crystal and therefore were not part of the solved structure. These contain the sequence that is homologous to MerP and postulated to be a mercury-binding domain. This region is drawn in Fig. 2 as an extension from the protein; perhaps it functions like a baseball mitt that catches Hg^{2+} from the membrane transport proteins and delivers Hg^{2+} to the carboxyl-terminal catalytic binding site, so that, as in the "bucket brigade" model above, Hg^{2+} is never found free within the cell. Mutant strains with the transport system but lacking the MerA detoxification enzyme are "hypersensitive" to mercury salts, as they accumulate Hg^{2+} but cannot get rid of it. After reduction by NADH (via FAD and the active site cysteine pair), metallic Hg^0 is released

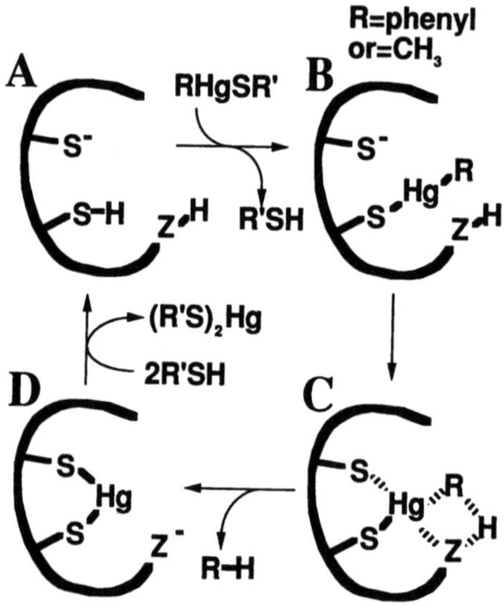

Fig. 3. Reaction mechanism for organomercurial lyase (modified from SILVER 1992, with permission). *A*, Reduced enzyme; *B*, with organomercurial (R-Hg) bound to one cysteine thiol; *C*, R-Hg bound to two thiols and S_E2 attack of proton on R-group; *D*, product Hg^{2+} bound to two cysteine thiols

and may diffuse freely out from the cell. The detoxification mechanism is unusual and interesting in that the toxic cation is first transported into the cytoplasm, rather than being detoxified at the cell surface. This probably is a reflection of the difficulty of transporting reducing equivalents (NADPH) out from the cell. The closed system of detoxification allows the cell to safely handle the toxic Hg^{2+} cation.

Following the *merA* gene for mercuric reductase, the pDU1358 mercuric resistance operon has the *merB* gene that determines the 212-amino acid organomercurial lyase enzyme (Fig. 1), whose reaction mechanism is shown is Fig. 3. Unlike mercuric reductase, no cofactors are involved in organomercurial lyase. Organomercurials, for example methylmercury or phenylmercury (which are of most concern to human toxicology), bind to this small monomeric protein. The reaction mechanism of organomercurial lyase was analyzed by BEGLEY et al. (1986), WALTS and WALSH (1988), and WALSH et al. (1988). The results suggest that a crucial cysteine (probably Cys_{96} or Cys_{117} or Cys_{159} in the plasmid pDU1358 sequence; these three cysteines are conserved in all four sequenced versions of the gene) initially binds the organomercurial substrate (Fig. 3B). An S_E2 mechanism results in both carbon–mercury bond breakage and carbon–hydrogen bond formation. The hydrogen may be donated by a histidine residue ("Z"; perhaps His_{163} or His_{178}, the only two histidine positions conserved in the four sequenced versions of the lyase protein) to convert the "R" group to either benzene from phenylmercury or methane from methylmercury, as mercury associates with a second conserved cysteine residue (Fig. 3C ⇒ D). Then a thiol compound (usually cysteine in vitro, but perhaps glutathione in vivo) releases the Hg^{2+} from the lyase (Fig. 3D ⇒ A), allowing a new reaction cycle to commence. In vivo, the Hg^{2+} may be passed directly from organomercurial lyase to the amino-terminal cysteines of mercuric reductase, again protecting the intracellular environment.

2. The Special Case of Thiobacillus

The mercury resistance system cloned and sequenced from *T. ferrooxidans* is unlike all others to date, in that the genes are not contiguous. Rather there is a mercuric reductase gene located together with only a single transport gene (*merC*; KUSANO et al. 1990). Unlinked elsewhere on the *T. ferrooxidans* chromosome are two additional functional copies of *merC*, each transcribed in the opposite direction (and therefore separately) from the regulatory genes *merR1* and *merR2* (INOUE et al. 1991). Partial (and therefore nonfunctional) additional copies of *merA* are located next to these functional *merC* genes. How this complex pattern arose and how it functions for effective mercury resistance in *Thiobacillus* is unclear, but one can conclude that most (although not all) *mer* operons will follow the pattern of pDU1358 (Fig. 1A).

3. Operon Structure in Gram-Positive Bacteria

The three versions of gram-positive *mer* operon that have been sequenced differ from the gram-negative *mer* operons by the absence of divergent transcription of the regulatory gene *merR* in the opposite direction from the genes for the structural proteins. For the *mer* operons of the *Bacillus* chromosome (Fig. 1B) and of *Staphylococcus aureus* plasmid pI258 (LADDAGA et al. 1987), the basic structure of the operon in the same, with *merR* proximal to the operator/promoter site and the structural genes *merA* and *merB* distal. The poorly defined genes ORF2, *merT* (so named from sequence homology to the *merT* genes of gram-negative bacteria), and ORF4 are all postulated to be involved in mercury transport. The mercuric reductase enzyme product of the *Bacillus merA* is the largest of the eight versions for which the gene has been sequenced and the only one known to date with a duplication of the amino-terminal presumed Hg^{2+}-binding motif (Fig. 2), which lacks a fixed position in the crystal structure solved by SCHIERING et al. (1991). Following *merA* on the *Bacillus* chromosome is approximately 2.5 kb of DNA that has not been sequenced and appears to have no involvement with mercury resistance. This is followed in turn by an isolated *merB* gene for organomercurial lyase. The separate transcript (Fig. 1B) and regulation of this solitary *merB* gene have not been studied. However, the localization of the *merA* and *merB* genes (with a gap between) and even restriction nuclease site patterns of the sequenced *Bacillus* isolate of

WANG et al. (1989) are found in 90% of the *mer* determinants of bacilli isolated from mercury-polluted sediment from Minamata Bay, Japan (K. Nakamura and S. Silver, submitted).

Whereas the overall *mer* operon structure of *S. aureus* plasmid pI258 is basically similar to that of the *Bacillus* system shown is Fig. 1B (except for closely contiguous *merA* and *merB*), the operon of the *Streptomyces lividans* is different from that of the other two gram-positive bacteria by having a divergent operator with *merR* on one side and the structural genes *merA* and *merB* on the other side (SILVER and WALDERHAUG 1992). In contrast to the gram-negative operons, there are structural genes on the same side of the *Streptomyces* operator/promoter site as *merR*.

C. Cadmium and Zinc Resistance

Cadmium and zinc are related transition metals with contrasting biological roles. Zinc is an essential ion (functioning catalytically and structurally in proteins) and probably has a specific transport mechanism for entering all cells, whereas cadmium is a toxic ion with no known biological functions. Cd^{2+} needs to be excluded, if possible, or to be extruded when found inside the cell. Because of the chemical similarity between Zn^{2+} and Cd^{2+}, it is difficult to exclude cadmium specifically from zinc uptake systems. The cell uses an efflux transporter as the strategy for keeping a low cytoplasmic concentration of cadmium.

I. CadA Cd^{2+} ATPase in *Staphylococcus* and *Bacillus*

An ever-increasing range of gram-positive bacteria are being found to share a single mechanism of Cd^{2+} resistance. The Cd^{2+} efflux P-type ATPase was first studied and sequenced in our laboratory (NUCIFORA et al. 1989; SILVER et al. 1989). The CadA Cd^{2+} ATPase of *Staphylococcus* is an outward-directed transport system whose synthesis is induced when resistant cells are exposed to Cd^{2+} (YOON et al. 1991; TSAI et al. 1992; TSAI and LINET 1993). Because the specifics of the CadA structure are considered again in Sect. D for comparisons with mammalian enzymes, some detail is appropriate at this point.

The CadA Cd^{2+} ATPase by sequence homology is a typical P-type ATPase (Fig. 4A; earlier called "E1-E2 ATPases"); it has a phosphorylated intermediate (TSAI and LINET 1993), which is probably an aspartyl-phosphate at position Asp_{415} in an aspartyl-kinase domain highly conserved in this family of proteins, which occur in prokaryote, plant, and animal cells (SILVER et al. 1993b). The CadA ATPase has the entire set of characteristic motifs and conserved residues of P-type ATPases, including a postulated phosphatase domain involving residues 267–270 (as labeled in Fig. 4A) and residues involved in ATP binding, including Lys_{489} and residues 618–622.

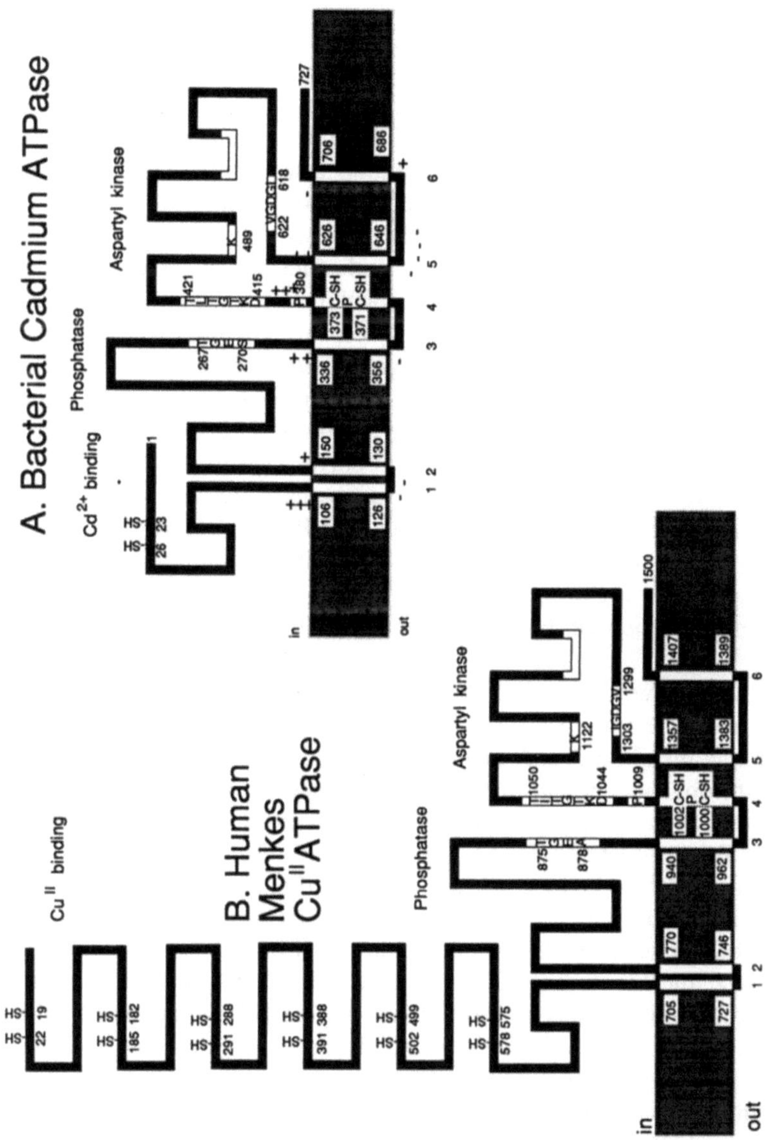

Fig. 4A,B. Comparison of **A** the CadA bacterial Cd^{2+} efflux ATPase and **B** that proposed for the Menkes' syndrome Cu^{2+} ATPase (from SILVER et al. 1993b, with permission)

Our model of the Cd^{2+} ATPase includes six *trans*-membrane α-helical regions, although ODERMATT et al. (1993) are considering an alternative eight *trans*-membrane segments model for this family.

However, SMITH et al. (1993), in a thorough and elegant study of membrane topology of the Mg^{2+} P-type ATPase of *Salmonella typhimurium*, concluded that the amino-terminal 750 amino acids of the Mg^{2+} ATPase (which is homologous to the entire ATPase region of the CadA ATPase) pass across the membrane six times. The evidence for topological folding back and forth across the membrane consisted of measuring enzyme activity of 35 protein fusions with β-lactamase (ampicillin resistance is active only if the β-lactamase fusion is outside or on the cell surface) and β-galactosidase (only allowing growth on lactose if the β-galactosidase domain is within the cytoplasm). SMITH et al. (1993) concluded that models with seven, eight, nine, or 12 transmembrane segments were ruled out and that a model with ten transmembrane segments (which would correspond to six for CadA or Menkes ATPases, plus an additional four in the carboxyl 150 amino acids, after the regions of homology to other bacterial P-type ATPases) was consistent with the results. These results favor the model with six transmembrane segments and argue against the additional two segments toward the amino terminus proposed by VULPE et al. (1993) and ODERMATT et al. (1993).

A proline residue (Pro_{372} in Figs. 4A, 5, in the channel/phosphorylation domain) is conserved in all P-type ATPases. Pro_{372} has been postulated to be within the membrane and involved in conformational movement associated with opening and closing of the membrane channel. This proline is notably flanked by cysteine residues in CadA and some closely related prokaryotic P-type ATPases. Finally, the amino-terminal region of the Cd^{2+} efflux ATPase is thought to contain a Cd^{2+}-binding motif because of the close homology between the 30-amino acid stretch containing a Cys-Xaa-Xaa-Cys motif at residues 23–26 and sequences in the mercury-resistance proteins MerP and MerA (Sect. B.II.1 above). This Cys-Xaa-Xaa-Cys motif in CadA occurs only in this amino-terminal region (Fig. 4); together with the Cys-Pro-Cys motif at residues 371–373, these are the only cysteine residues in the entire 727-amino acid CadA sequence. These cysteines may represent intermediates for binding of Cd^{2+} during transport. It is important to determine whether the movement of Cd^{2+} from the first to the cysteine second site corresponds to the change from the E1 state to the E2 state of the P-type ATPase.

Although the specifics of how the CadA Cd^{2+} ATPase is regulated are not known, experiments with gene fusions have shown that Cd^{2+} can induce the CadA operon (YOON et al. 1991), and biochemical experiments have shown that Cd^{2+} transport activity is found only with membranes from induced cells (TSAI et al. 1992). The product of the second gene in the CadA operon (now called CadC) is required for resistance (YOON and SILVER 1991). This protein shows sequence homology with ArsR, the repressor of

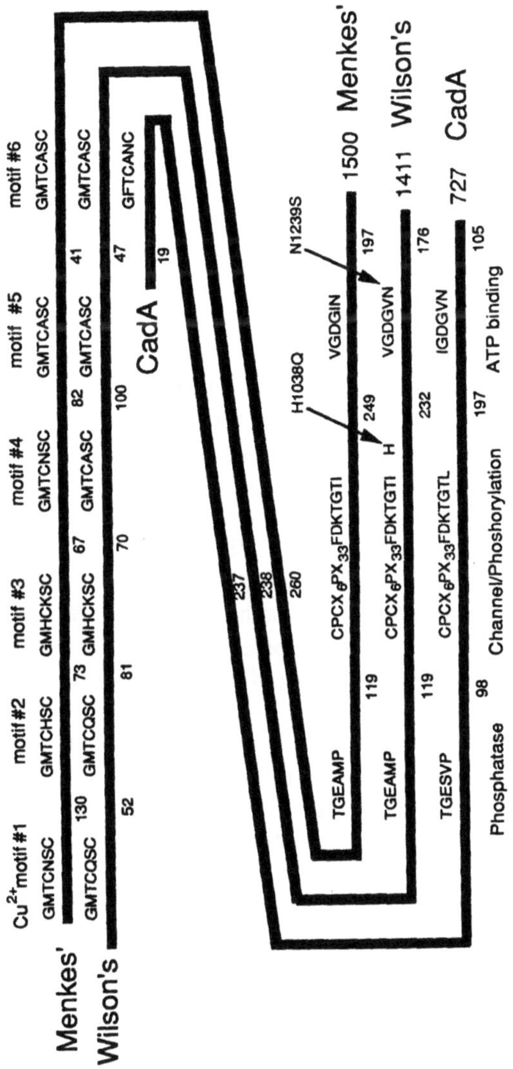

Fig. 5. Alignment of human Menkes' and Wilson's Cu^{2+} P-type ATPase polypeptide sequences with the bacterial Cd^{2+} ATPase (sequence data from VULPE et al. 1993 and BULL et al. 1993)

the arsenic resistance determinant (SILVER and WALDERHAUG 1992). This gene product is the prime candidate for the regulatory protein of the Cd^{2+} ATPase system.

When the CadA P-type ATPase determinant was sequenced (NUCIFORA et al. 1989; SILVER et al. 1989), it was the second bacterial DNA sequence determining a P-type ATPase. Since then, however, the number of available bacterial P-type ATPase sequences has been growing very rapidly (SILVER et al. 1993b), and 16 are currently known. Figure 6 shows an evolutionary "tree," primarily for the CadA "subfamily," and the most closely related other sequences. There are now four probable Cd^{2+} CadA sequences, as shown. The first was from plasmid pI258 of *S. aureus* (NUCIFORA et al. 1989). The next, published by IVEY et al. (1992), comes from a presumed-chromosomal determinant of a soil *Bacillus* isolate. The other two CadA sequences are not published; they come from the 40-kb methicillin-resistance region of the chromosome of a methicillin-resistant *S. aureus* isolate (D.T. Dubin, personal communication) and from an otherwise "cryptic" (or silent) plasmid of the gram-positive human foodborne pathogen *Listeria monocytogenes* (P. Cossart, personal communication).

The staphylococcal pI258 and *Bacillus* CadA sequences are over 80% identical at the amino acid level, but these are only 67%–69% identical with the *Listeria* CadA. These close homologies define the current CadA subfamily, compared with the much lower (24%–34%) identities with the next three sequences listed in Fig. 6. The *Enterococcus hirae* CopA and CopB presumed-Cu^{2+} P-type ATPases (ODERMATT et al. 1992, 1993) were discussed by SILVER et al. (1993b). It is thought that CopA is an uptake ATPase and CopB an efflux ATPase and that the coordinated synthesis and function of these two transport ATPases provide for homeostatic control of cellular copper levels (ODERMATT et al. 1993).

The remaining two sequences in Fig. 6 come from a report by KANAMARU et al. (1993) of two P-type ATPase sequences from a cyanobacterial strain. Since they were obtained by polymerase chain reaction (PCR)-assisted cloning, there was no prior knowledge of possible function associated with these genes. The shorter, 747-amino acid PacS, sequence was noted to be related to CadA (although with only 32% amino acid identities; the *Enterococcus* sequences were not available). The cation specificity of the PacS ATPase is not known, but a mutant strain with the gene disrupted is reported to be somewhat copper hypersensitive. The longer, 926-amino acid PacL ATPase (KANAMARU et al. 1993), which forms the "outlier" in Fig. 6, was noted to be more closely related to eukaryotic Ca^{2+} P-type ATPases. However, these relationships are very weak (other than in the strongly conserved small regions common to all P-type ATPases), and PacL is not particularly closer to still another new presumed-Ca^{2+} ATPase from a different cyanobacterial strain (not shown). Figure 6 is thus intended to summarize the currently known Cd^{2+} branch of bacterial P-type ATPases and to emphasize that the number of P-type ATPase sequences

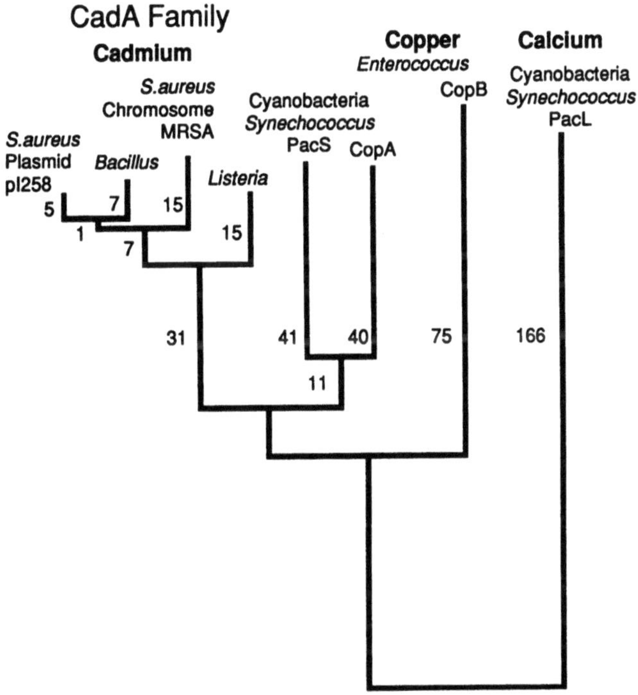

Fig. 6. Alignment tree for selected bacterial divalent-cation translocating P-type ATPases, including the four current CadA cadmium efflux ATPases, three bacterial presumed-copper ATPases (CopA and CopB from Enterococcus and PacS from a cyanobacterium Synechococcus), along with PacL, a possibly calcium ATPase from the same Synechococcus strain. See text for explanations and literature citations

now available precludes putting them all together in a single figure. It is necessary to take a small subclass such as Cd^{2+} or Cu^{2+} ATPases and analyze cation specificity in detail.

II. Czc (Cd^{2+}, Zn^{2+}, and Co^{2+}) and Cnr (Co^{2+} and Ni^{2+}) Resistance Systems of *Alcaligenes*

For *Alcaligenes eutrophus*, two well-defined systems efflux divalent metal cations: the Czc system transports Cd^{2+}, Zn^{2+}, and Co^{2+} out of the cell (NIES 1992a; NIES et al. 1989), and the Cnr system transports Co^{2+} and Ni^{2+} out of the cell (LIESEGANG et al. 1993; COLLARD et al. 1993). Both systems are plasmid based. The Czc and Cnr systems do not (from their sequences) appear to be ATPases (NIES et al. 1989; LIESEGANG et al. 1993), and the Czc system functions as a cation/proton antiporter (NIES 1994). Little is known about the regulation of expression of these systems. The Czc system (Fig. 7) consists of four genes: *czcA* is predicted by sequence to produce a membrane-bound protein; *czcB* also determines a membrane protein; and *czcC*

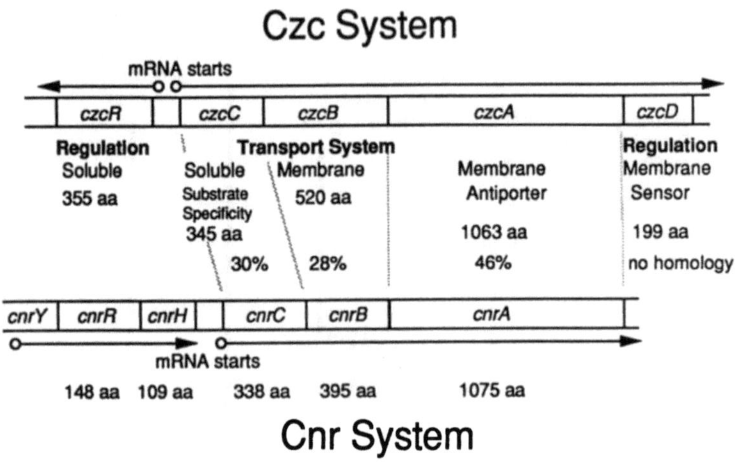

Fig. 7. Comparison of genes and functions of the Alcaligenes czc (cadmium, zinc, and cobalt) and cnr (cobalt and nickel) resistance systems (from SILVER and JI 1994, with permission)

determines a soluble protein (NIES 1994). The gene *czcD* appears to specify a regulatory protein. Another contiguous gene which is divergently transcribed from operon, *czcR*, produces a protein which may work with CzcD to activate expression of the operon (NIES 1992b). Genetic analysis attributes the transport function to *CzcA*, since deletions in the *czcA* gene cause loss of transport (NIES et al. 1989). CzcB and CzcC appear to confer cation specificity to transport.

Surprisingly, the metal-binding domains of the Czc system apparently do not involve cysteine residues. Instead, there appears to be a histidine-based motif for metal ion binding. The sequence Glu-His-His is found once in CzcA and twice in CzcB. His-Xaa-His appears once in CzcB at the carboxyl end and twice in CzcD at the amino-terminal end (NIES et al. 1989; NIES 1992a,b).

The *cnr* system is sufficiently different at the DNA level from *czc* that the two systems do not hybridize with each other on Southern blots. However, recent sequence analysis of the *cnr* genes (Fig. 7) reveals that the three genes *cnrA*, *cnrB*, and *cnrC* are significantly homologous to *czcA*, *czcB*, and *czcC* of the *czc* system, respectively (LIESEGANG et al. 1993). It is likely that the two systems will function similarly. Gene regulation may be quite different, however, since the regulatory proteins of the two systems appear not to be related. The *cnr* system genes *cnrY*, *cnrR*, and *cnrH*, upstream from the structural genes, appear to regulate gene function. *cnrY* has been mutated, resulting in constitutive expression of *cnrABC* (COLLARD et al. 1993). Whether this mutation causes derepression, or unregulated expression, of the structural genes is not known.

D. Relationship Between Human Menkes' and Wilson's ATPases and Bacterial P-Type ATPases

Basically, the products of the genes that are defective in the human diseases Menkes' syndrome and Wilson's disease are remarkably similar to the cadmium P-type ATPase of gram-positive bacteria described above. Since this is the sole chapter in this monograph that concerns bacterial systems and since the "candidate genes" for the Menkes' and Wilson's defects are quite new, we think it appropriate to give a simple (and perhaps premature) picture in this section. Menkes' and Wilson's diseases are the two most familiar hereditary copper diseases of humans (Danks 1989); both have an apparent basis in defects of copper transport. In both cases, the symptoms are complex and result in severe neurological damage, which is lethal for Menkes' syndrome and potentially lethal for Wilson's disease in the absence of copper chelation therapy. Before the defective genes in these two conditions were identified, alternative hypotheses were considered (Danks 1989). However, with the findings of "candidate Menkes'" and "candidate

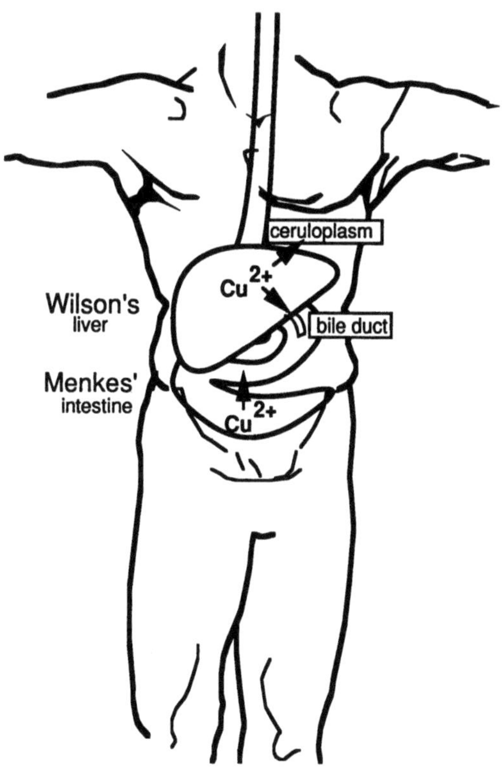

Fig. 8. Proposed primary defects of human copper diseases Menkes' (passage from intestinal mucosa into blood circulation) and Wilson's (passage from liver into the bile and blood circulation)

Wilson's" genes that potentially determine P-type ATPases, it appears that the primary defect in Menkes' syndrome is a block in Cu^{2+} efflux across the upper intestinal mucosa (Fig. 8) and that disease symptoms of Menkes' syndrome result from copper starvation by the tissues with consequent failure to make copper-requiring enzymes (DANKS 1989).

The structure of the gene product disrupted in the cells of Menkes' disease patients was recognized as a P-type ATPase (VULPE et al. 1993) with six probably Cu^{2+} binding motifs (Fig. 4) from protein library searches when the candidate Menkes' cDNAs were first identified and sequenced (CHELLY et al. 1993; MERCER et al. 1993; VULPE et al. 1993). The properties predicted for the copper P-type ATPases appear to adequately explain the primary defects in these diseases. We will briefly describe first the two human diseases and then the predicted properties of the gene products. The basic properties of Menkes' syndrome and the comparison with the bacterial cadmium ATPase were recently described by SILVER et al. (1993b), where the primary references may be found. Most of the following information derives from the current standard chapter on human copper disorders, including Menkes' and Wilson's, by DANKS (1989).

Menkes' syndrome is a human X-chromosome-linked, invariably fatal disease resulting from defective copper transport in basically all tissues (except the liver). Copper is accumulated and subsequent passage is blocked at the level of the intestinal mucosal cells (Fig. 8). We propose that the primary defect in Menkes' syndrome is the blockage of Cu^{2+} efflux across the upper intestinal mucosa (DANKS 1989; Fig. 8), although the gene is also expressed in kidney, brain, and perhaps every cell (MERCER et al. 1994). Therefore the remaining body tissues cannot obtain adequate copper and cannot make copper-containing enzymes, such as lysyl oxidase, superoxide dismutase, and cytochrome oxidase. Without lysyl oxidase, connective tissue collagen and elastin cannot be cross-linked, and all connective tissues are defective.

Menkes' disease children die from neural degeneration or arterial ruptures, which are secondary effects of copper deficiency. Since most cell types express the wild-type gene that is defective in Menkes' syndrome, copper injections are usually not effective treatment for Menkes' disease. Cultured Menkes' fibroblasts are also defective; they show increased copper accumulation and reduced efflux, which give a clue to the primary cause of the disease and allow for a specific radioactive Cu^{2+}-retention diagnostic test. The excessive accumulation of copper by the Menkes' cells, accompanied by the Cu^{2+} sequestration with intracellular metallothionein (which does not occur with wild-type cells), leads to a problem in intracellular copper movement. Consequently, cellular copper is high, while intracellular copper is unavailable for metabolic functions.

It was widely postulated that an intracellular Cu^{2+} carrier protein, whose role might be to carry copper to where it is needed for in corporation into enzymes, is defective. In the absence of this carrier protein, metallothionein synthesis might be induced by low levels of copper, and once metallothionein binds copper, the copper is prevented from reaching the efflux site. However, the alternative hypothesis is a defect in copper efflux (the failure to homeostatically regulate cell copper through efflux). This hypothesis is consistent with the new finding that the Menkes' gene determines a putative P-class ATPase. A mouse model for Menkes' is available, with several mutations at the same X-chromosome locus and with copper deficiency symptoms. The equivalent mouse cDNA has just been sequenced (MERCER et al. 1994), and mutants lacking mRNA ("dappled") or with abnormal-length mRNA ("blotchy") have been identified (MERCER et al. 1994).

Since all copper-dependent enzymes and metabolism are affected, biosynthesis of elastin and collagen is defective. Menkes' victims usually die before 3 years of age, frequently from neurodegenerative defects or connective tissue disorders (DANKS 1989). Identification of the "Menkes' gene" or the "Wilson's gene" will not provide a cure. However, as with other hereditary diseases, early diagnosis gives promise for a major reduction in the frequencies of these diseases.

Wilson's disease is superficially the opposite of Menkes' syndrome: it is a disease of copper overload, not starvation. The gene for Wilson's disease maps to chromosome 13, and "candidate cDNA" for the gene defective in Wilson's disease was recently isolated and sequenced (BULL et al. 1993; TANZI et al. 1993). We propose a simple answer, considering the predicted ATPase structure: that Wilson's disease be considered the "Menkes of the liver" (Figs. 5, 8). The candidate Wilson's disease gene and proposed ATPase product (BULL et al. 1993) are highly homologous to those of Menkes' syndrome (Fig. 5), but appear to function primarily (or with few exceptions) in the liver; the "Menkes' gene" functions only in fetal liver (MERCER et al. 1994).

Wilson's defect primarily affects the liver. In normal humans, the liver is the primary organ of copper accumulation and homeostatic regulation. Copper needed by other organs is *trans*-shipped by the circulating carrier protein ceruloplasmin, made and exported from the liver. Excess copper is accumulated by the liver and excreted by the bile duct from the body (Fig. 8). In Wilson's disease patients, copper accumulates in the liver but both liver-exit pathways, namely serum ceruloplasmin and bile excretion, are defective. Copper overaccumulation leads to liver necrosis and to sudden uncontrolled releases of copper. The overflow is evident elsewhere in the body, in particular in the brain and eyes (where a green–brown ring of copper can be seen on the iris of Wilson's disease patients). Neurological damage is extensive and patients usually die in their teens. Wilson's disease, however, differs significantly from Menkes' syndrome in that effective treatment is available. Copper chelate therapy (penicillamine ingested at about 1 g/day) causes the penicillamine–copper adduct to be rapidly excreted in the urine (about 2 mg copper/day). Once the patient attains normal copper levels, continuing penicillamine treatment or limiting copper uptake by dietary control (along with added zinc in the diet) maintains copper homeostasis for the patient. It seems likely that a normal life expectancy and continued health without symptoms can be achieved for Wilson's disease patients under careful medical monitoring (DANKS 1989).

Two caveats are required with respect to the hypotheses presented here: First, the protein products of the "candidate" Wilson's and Menkes' genes have not yet been seen, but only predicted from candidate cDNA sequences. Second, the biochemical and transport properties of these P-type ATPases have not been measured. However, fibroblasts from both Wilson's and Menkes' disease patients appear to express the basic defects. These fibroblasts accumulate radioactive copper with apparently normal initial uptake kinetics but are defective in copper efflux or exchange. The fibroblasts from Menkes' and Wilson's disease patients both overaccumulate Cu^{2+} and bind Cu^{2+} to metallothionein, but cannot pass on (or efflux) Cu^{2+} appropriately (J. CAMAKARIS, as reported by DANKS 1989).

Given the strong similarities between the presumed protein products of the Menkes' and Wilson's genes diagrammed in Figs. 5 and 9 and the directly established Cd^{2+} ATPase described above, it seems likely that the function of these mammalian ATPases is to pump Cu^{2+} out of the cytoplasm and into either extracellular fluid or intracellular vesicles (there are no experiments to show which), where Cu^{2+}-specific protein binding occurs. Regulation of body copper levels and transport from organ to organ follows directly from the ATPase function. Access to Cu^{2+} enzymes requires proper sequestration and delivery of Cu^{2+} to the apoproteins. Why there are six Cu^{2+} binding motifs in the Menkes' and Wilson's products and usually only one in the CadA Cd^{2+} ATPase is not clear. Catalytic and regulatory distinctions among the six positions seem likely.

E. Copper Transport and Resistance in Bacteria

Copper is an essential metal ion in bacteria as well as in mammals. Bacteria need to keep intracellularly free Cu^{2+} at low concentrations since free Cu^{2+} can facilitate oxygen damage in bacteria as it does in mammalian cells. Two or three rather newly studied copper transport systems are considered here. First, there are the poorly understood CopA and CopB P-type ATPases of *Enterococcus hirae* (ODERMATT et al. 1992, 1993) and the SynA P-type ATPase of *Synechococcus* (SILVER et al. 1993b; PHUNG et al., in preparation) that appear to be the first bacterial examples of Cu^{2+}-transporting ATPases. These proteins have presumed Cu^{2+}-binding motifs (Figs. 4, 9) closely homologous to those of Menkes' and Wilson's proteins. Disruption of CopA led to failure to grow on medium containing very low levels of Cu^{2+} (ODERMATT et al. 1993), and disruption of SynA led to greater resistance to high Cu^{2+} (PHUNG et al., in preparation). These results were both tentatively

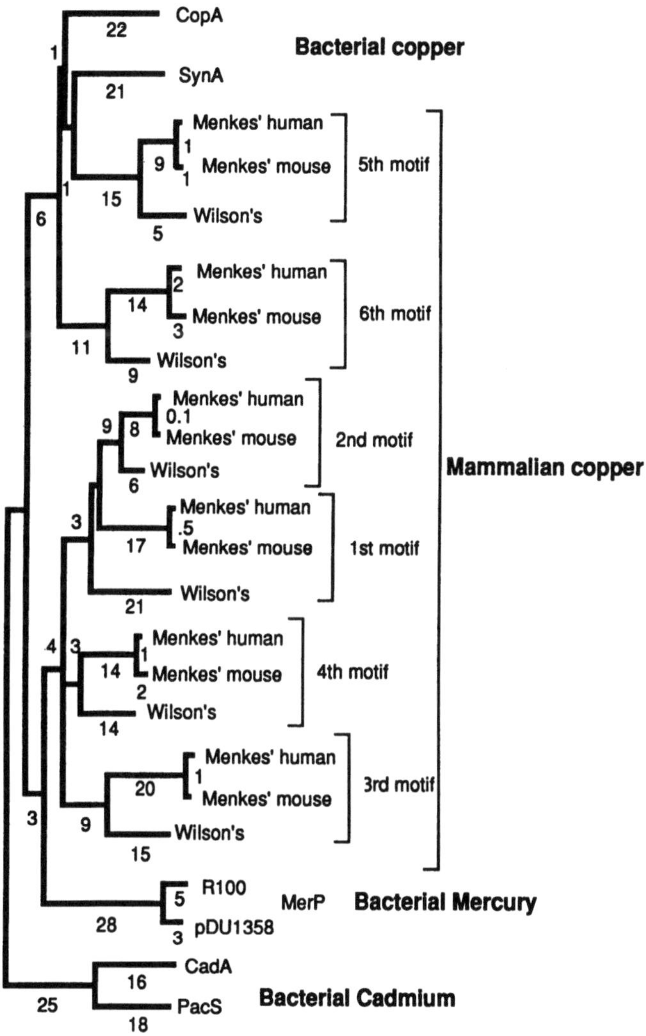

Fig. 9. Alignment tree for metal-binding motifs of the three presumed mammalian copper ATPases, the two bacterial copper ATPases (CopA and SynA), along with bacterial cadmium ATPases (CadA and SynB) and mercury-binding periplasmic proteins, MerP, from plasmids R100 and pDU1358. See text for literature citations

ascribed to roles of the cation transport ATPase in Cu^{2+} uptake. How presumed Cu^{2+}-binding motifs on the cellular interior (Fig. 4) would function in uptake is unclear. CopB, however, which is thought to be a Cu^{2+} efflux ATPase because disruption of the gene leads to copper sensitivity (ODERMATT et al. 1993), lacks the amino-terminal dithiol metal-binding motif, but near its amino terminus CopB has three repeats of a consensus sequence Met-Xaa-His-Xaa-Xaa-Met-Ser-Gly-Met-Xaa-His-Ser that is similar to repeats found in the CopA (or PcoA) proteins of plasmid copper resistance systems (COOKSEY 1993; BROWN et al. 1994; see below).

A series of at least six genes, in which mutations cause increased sensitivity to copper, were identified in *E. coli* (ROUCH 1986) and preliminarily characterized (BROWN et al. 1992, 1994). Of these only one, cutE, has been sequenced (ROGERS et al. 1991). Two preliminary hypotheses for the function of *CutE* are a role in intracellular Cu^{2+} sequestration and movement (ROGERS et al. 1991) and a role in bacterial outer membrane function (GUPTA et al. 1993). Whether subsequently identified genes affecting copper sensitivity levels in *E. coli* will be found to determine P-type ATPases remains to be established.

Resistance to elevated levels of copper is determined by plasmids in *E. coli* and *Pseudomonas* and by chromosomal genes in *Xanthomonas campestris* (SILVER et al. 1993a; BROWN et al. 1992, 1994; MELLANO and COOKSEY 1988; COOKSEY 1993). The plasmid-based copper resistance system in *E. coli* (*pco*) is essentially equivalent to that found in *Pseudomonas* (*cop*) (BROWN et al. 1994; SILVER et al. 1993a; Fig. 10). The *pco* system has been found widely in a range of enteric bacteria (WILLIAMS et al. 1993). The operon is made up of four structural genes: *pcoA* in *E. coli*/*copA* in *Pseudomonas*, *pcoB*/*copB*, *pcoC*/*copC*, and *pcoD*/*copD*. Also present are two regulatory genes: *pcoS*/*copS* and *pcoR*/*copR*. The two regulatory proteins are thought to act as membrane sensor and DNA-binding regulator, respectively. This conclusion is based on sequence homologies to the well-studied two-component ATP-kinase sensor–regulatory systems (PARKINSON and KOFOID 1992). PcoS senses the copper concentration and at some threshold is "kinased" at a conserved histidine residue by ATP. The phosphorylated PcoS protein then activates the inactive PcoR protein by *trans*-phosphorylation of a conserved aspartate residue (SILVER et al. 1993a; BROWN et al. 1994). Active phospho-PcoR protein then stimulates transcription of the *pco* structural genes at operator/promoter region upstream of the structural genes (Fig. 10).

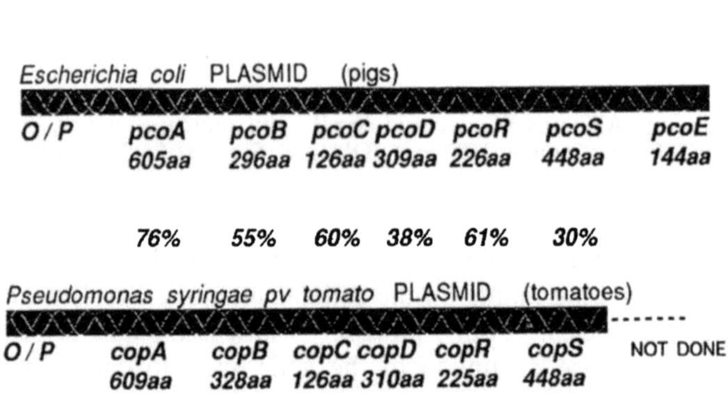

Fig. 10. Genes and products of the Pseudomonas syringae and E. coli copper resistance systems (modified from SILVER and JI 1994, with permission)

The structural gene products in *Pseudomonas* have been isolated and the first three identified. CopA and CopC are periplasmic proteins which turn the cells blue by binding copper ions (CHA and COOKSEY 1991; COOKSEY 1993). The ion binding capacities of the two are 11 and 1 Cu^{2+}, respectively (CHA and COOKSEY 1991). CopB is an outer membrane protein (possibly a porin) and CopD is probably an inner membrane protein, but no phenotype or function has been associated with CopD to date. The mechanism of resistance appears to be the stimulation of periplasmic binding of copper. Cells lacking the CopA periplasmic binding protein are hypersensitive to copper (CHA and COOKSEY 1993). This leads to the suggestion that CopC and CopD function in cellular copper uptake and CopA is essential to prevent overaccumulation (CHA and COOKSEY 1993; COOKSEY 1993).

Since copper is an essential ion for the metabolism of cells, a chromosomal-based uptake system must exist. The *cut* system has been identified in *E. coli*. Little is known about the system since only one of the genes has been sequenced (*cutE*). The *copB* gene in *Enterococcus* that was previously sequenced has now been identified as a *cut* gene. The sequence of the *copB* gene reveals homology to P-type ATPases and may provide a clue to the mechanism of other chromosome-based copper transport systems.

F. Summary and Conclusions

Analysis of the mechanisms of metal ion homeostasis and binding motifs may provide fruitful insights into the problem of metal ion toxicity. The mechanisms by which cells detect, store, and transport metals are the most sensitive components of the cells with regard to metal ions and will be the first systems affected in toxic situations.

We have reviewed the best-known prokaryotic plasmid-based resistance systems from the view of metal ion homeostasis. Plasmid-based systems are, with few exceptions, inducible and are regulated by activators (mercury and copper) or repressors (cadmium) in different cases. The metal-binding motifs used by these systems are frequently localized (e.g., Cys-Xaa-Xaa-Cys, Glu-His-His, and His-Xaa-His), but occasionally coordinating cysteine residues on different subunits interact (as in the MerR regulatory protein and mercuric reductase enzyme).

Prokaryotic systems are beginning to show promise as models for homologous processes in eukaryotic cells. A spectacular example of this occurred recently with the surprising similarities in the metal-binding motifs and overall structure of the bacterial Cd^{2+} efflux ATPase and the putative Cu^{2+} ATPases of human Menkes' syndrome and Wilson's disease. Prokaryotic cells, due to their varied ecological niches, are expected to have more and varied mechanisms for dealing with metal starvation and toxic metal ion onslaught. The range of such bacterial systems increases the probability that prokaryotic systems will have parallels in eukaryotic homeostatic mechanisms. Although eukaryotic cells grow in much more restricted environments, multiple metal ion transporters probably exist for many potentially toxic metal ions. For example, copper regulation in humans requires at least two different Cu^{2+} P-type ATPases, whose locations and physiological functions are different, but whose basic biochemical structure and function are quite similar. This is similar to the situation with P-type Ca^{2+} and Na^{+}/K^{+} ATPases, which also result from several different genes, differentially expressed.

Metal ion homeostasis in prokaryotes has not yet provided a model for regulation in eukaryotes. The mechanisms for gene regulation in prokaryotes are proving to be as variable themselves as the mechanisms of transport. Despite great differences between eukaryotes and prokaryotes, it is still probable that their metal ion homeostasis mechanisms may sometimes be found to be similar. For the present, however, the mechanisms of prokaryotic metal ion regulation sometimes provide first hypotheses for constructing models of eukaryotic metal ion homeostasis.

References

Begley TP, Walts AE, Walsh CT (1986) Mechanistic studies of a protonolytic organomercurial cleaving enzyme: bacterial organomercurial lyase. Biochemistry 25:7192–7200

Brown NL, Rouch DA, Lee BTO (1992) Copper resistance systems in bacteria. Plasmid 27:41–51

Brown NL, Lee BTO, Silver S (1994) Bacterial transport of and resistance to copper. In: Sigel H (ed) Metal ions in biological systems, vol 30. Dekker, New York, pp 405–435

Bull PC, Thomas GR, Rommens JM, Forbes JR, Cox DW (1993) The Wilson disease gene is a putative copper transporting P-type ATPase similar to the Menkes gene. Nature Genet 5:327–337

Cha J-S, Cooksey DA (1991) Copper resistance in Pseudomonas syringae mediated by periplasmic and outer membrane proteins. Proc Natl Acad Sci USS 88: 8915–8919

Cha J-S, Cooksey DA (1993) Copper hypersensitivity and uptake in Pseudomonas syringae containing cloned components of the copper resistance operon. Appl Environ Microbiol 59:1671–1674

Chelly J, Tümer Z, Tonnesen T, Petterson A, Ishikawa-Brush Y, Tommerup N, Horn N, Monaco AP (1993) Isolation of a candidate gene for Menkes disease which encodes for a potential heavy metal binding protein. Nature Genet 3: 14–19

Collard J-M, Provoost A, Taghavi S, Mergeay M (1993) A new type of Alcaligenes eutrophus CH34 zinc resistance generated by mutations affecting regulation of the cnr cobalt-nickel resistance system. J Bacteriol 175:779–784

Cooksey DA (1993) Copper uptake and resistance in bacteria. Mol Microbiol 7:1–5

Danks DM (1989) Disorders of copper transport. In: Scriver CR, Beaudet AL, Sly WS, Valle D (eds) Metabolic basis of inherited disease, 6th edn. McGraw-Hill, New York, pp 1411–1431

Distefano MD, Moore MJ, Walsh CT (1990) Active site of mercuric reductase resides at the subunit interface and requires Cys135 and Cys140 from one subunit and Cys558 and Cys559 from the adjacent subunit: evidence from in vivo and in vitro heterodimer formation. Biochemistry 29:2703–2713

Gupta SD, Gan K, Schmid MB, Wu HC (1993) Characterization of a temperature-sensitive mutant of Salmonella typhimurium defective in apolipoprotein N-acyltransferase. J Biol Chem 268:16551–16556

Inoue C, Sugawara K, Kusano T (1991) The merR regulatory gene in Thiobacillus ferrooxidans is spaced apart from the mer structural genes. Mol Microbiol 5:2707–2718

Ivey DM, Guffanti AA, Shen Z, Kudyan N, Krulwich TA (1992) The CadC gene product of alkaliphilic Bacillus firmus OF4 partially restores Na^+ resistance to an Escherichia coli strain lacking an Na^+/H^+ antiporter (NhaA). J Bacteriol 174:4878–4884

Kanamaru K, Kashiwagi S, Mizuno T (1993) The cyanobacterium, Synechococcus sp. PCC7942, possesses two distinct genes encoding cation-transporting P-type ATPases. FEBS Lett 330:99–104

Kusano T, Ji G, Inoue C, Silver S (1990) Constitutive synthesis of a transport function encoded by the Thiobacillus ferrooxidans merC gene cloned in Escherichia coli. J Bacteriol 172:2688–2692

Laddaga R, Chu L, Misra TK, Silver S (1987) Nucleotide sequence and expression of the mercurial resistance operon from Staphylococcus aureus plasmid pI258. Proc Natl Acad Sci USA 84:5106–5110

Liesegang H, Lemke K, Siddiqui RA, Schlegel H-G (1993) Characterization of the inducible nickel and cobalt resistance determinant cnr from pMOL28 of Alcaligenes eutrophus CH34. J Bacteriol 175:767–778

Mellano MA, Cooksey DA (1988) Nucleotide sequence and organization of copper resistance genes from Pseudomonas syringae pv. tomato. J Bacteriol 170: 2879–2883

Mercer JFB, Livingston J, Hall B, Paynter JA, Begy C, Chandrasekharappa S, Lockhart P, Grimes A, Bhave M, Siemieniak D, Glover TW (1993) Isolation of a partial candidate gene for Menkes' by positional cloning. Nature Genet 3:20–25

Mercer JFB, Grimes A, Ambrosini L, Lockhart P, Paynter JA, Dierick H, Spencer JA, Glover TW (1994) Mutations in the murine homologues of the Menkes' gene in dappled and blotchy mice. Nature Genet (in press)

Misra TK (1992) Bacterial resistances to inorganic mercury salts and organomercurials. Plasmid 27:4–16

Nies DH (1992a) Resistance to cadmium, cobalt, zinc and nickel in microbes. Plasmid 27:17–28

Nies DH (1992b) CzcR and CzcD, gene products affecting regulation of resistance to cobalt, zinc, and cadmium (czc system) in Alcaligenes eutrophus. J Bacteriol 174:8102–8110

Nies DH (1994) Expression of the cobalt, zinc, and cadmium efflux system of Alcaligenes eutrophus in Escherichia coli: CzcABC is a cation-proton antiporter. Eur J Biochem (submitted)

Nies DH, Nies A, Chu L, Silver S (1989) Expression and nucleotide sequence of a plasmid-determined divalent cation efflux system from Alcaligenes eutrophus. Proc Natl Acad Sci USA 86:7351–7355

Nucifora G, Chu L, Misra TK, Silver S (1989) Cadmium resistance from Staphylococcus aureus plasmid pI258 cadA gene results from a cadmium-efflux ATPase. Proc Natl Acad Sci USA 86:3544–3548

Odermatt A, Suter H, Krapf R, Solioz M (1992) An ATPase operon involved in copper resistance by Enterococcus hirae. Ann NY Acad Sci 936:484–486

Odermatt A, Suter H, Krapf R, Solioz M (1993) Primary structure of two P-type ATPases involved in copper homeostasis in Enterococcus hirae. J Biol Chem 268:12775–12779

O'Halloran TV (1993) Transition metals in control of gene expression. Science 261:715–725

Parkinson JS, Kofoid EC (1992) Communication modules in bacterial signaling proteins. Annu Rev Genet 26:71–112

Rogers SD, Bhave MR, Mercer JFB, Camakaris J, Lee BTO (1991) Cloning and characterization of cutE, a gene involved in copper transport in Escherichia coli. J Bacteriol 173:6742–6748
Rouch DA (1986) Plasmid-mediated copper resistance in E. coli. PhD thesis, University of Melbourne, Australia
Schiering N, Kabsch W, Moore MJ, Distefano MD, Walsh CT, Pai EF (1991) Structure of the detoxification catalyst mercuric ion reductase from Bacillus sp. strain RC607. Nature 352:168–172
Silver S (1992) Bacterial heavy metal detoxification and resistance systems. In: Mongkolsuk S, Lovett PS, Trempy J (eds) Biotechnology and environmental science: molecular approaches. Plenum, New York, pp 109–129
Silver S, Ji G (1994) Newer systems for bacterial resistances to toxic heavy metals. Environ Health Perspect 102/4 (in press)
Silver S, Walderhaug W (1992) Gene regulation of plasmid- and chromosome-determined inorganic ion transport in bacteria. Microbiol Rev 56:195–228
Silver S, Nucifora G, Chu L, Misra TK (1989) Bacterial resistance ATPases: primary pumps for exporting toxic cations and anions. Trends Biochem Sci 14:76–80
Silver S, Lee BTO, Brown NL, Cooksey DA (1993a) Bacterial plasmid resistances to copper, cadmium, and zinc. In: Welch AJ, Chapman SK (eds) The chemistry of the copper and zinc triads. Royal Society of Chemistry, London, pp 38–53
Silver S, Nucifora G, Phung LT (1993b) Human Menkes X-chromosome disease and the staphylococcal cadmium-resistance ATPase: a remarkable similarity in protein sequences. Mol Microbiol 10:7–12
Smith DL, Tao T, Maguire ME (1993) Membrane topology of a P-type ATPase. The Mgt magnesium transport protein of Salmonella typhimurium. J Biol Chem 268:22469–22479
Summers AO (1992) Untwist and shout: a heavy metal-responsive transcriptional regulator. J Bacteriol 174:3097–3101
Tanzi RE, Petrukhin K, Chernov I, Pellequer JL et al. (1993) The Wilson disease gene is a copper transporting ATPase with homology to Menkes disease gene. Nature Genet 5:344–350
Tsai K-J, Linet AL (1993) Formation of a phosphorylated enzyme intermediate by the cadA Cd^{2+}-ATPase. Arch Biochem Biophys 305:267–270
Tsai K-J, Yoon KP, Lynn AR (1992) ATP-dependent cadmium transport by the *cadA* cadmium resistance determinant in everted membrane vesicles of Bacillus subtilis. J Bacteriol 174:116–121
Vulpe C, Levinson B, Whitney S, Packman S, Gitschier J (1993) Isolation of a candidate gene for Menkes disease and evidence that it encodes a copper-transporting ATPase. Nature Genet 3:7–13
Walsh CT, Distefano MD, Moore MJ, Shewchuk LM, Verdine GL (1988) Molecular basis of bacterial resistance to organomercurial and inorganic mercuric salts. FASEB J 2:124–130
Walts AE, Walsh CT (1988) Bacterial organomercurial lyase: novel enzymatic protonolysis of organostannanes. J Am Chem Soc 110:1950–1953
Wang Y, Moore M, Levinson HS, Silver S, Walsh C, Mahler I (1989) Nucleotide sequence of a chromosomal mercury resistance determinant from a Bacillus sp. with broad-spectrum mercury-resistance. J Bacteriol 171:83–92
Williams JR, Morgan A, Rouch DA, Brown NL, Lee BTO (1993) Copper-resistant enteric bacteria from United Kingdom and Australian piggeries. Appl Environ Microbiol 59:2531–2537
Yoon KP, Silver S (1991) A second gene in the Staphylococcus aureus cadA cadmium resistance determinant of plasmid pI258. J Bacteriol 173:7636–7642
Yoon KP, Misra TK, Silver S (1991) Regulation of the cadA cadmium resistance determinant of Staphylococcus aureus plasmid pI258. J Bacteriol 173:7643–7649

Subject Index

Springer-Verlag and the Environment

We at Springer-Verlag firmly believe that an international science publisher has a special obligation to the environment, and our corporate policies consistently reflect this conviction.

We also expect our business partners – paper mills, printers, packaging manufacturers, etc. – to commit themselves to using environmentally friendly materials and production processes.

The paper in this book is made from low- or no-chlorine pulp and is acid free, in conformance with international standards for paper permanency.

MIX
Papier aus verantwortungsvollen Quellen
Paper from responsible sources
FSC® C105338

If you have any concerns about our products,
you can contact us on
ProductSafety@springernature.com

In case Publisher is established outside the EU,
the EU authorized representative is:
Springer Nature Customer Service Center GmbH
Europaplatz 3, 69115 Heidelberg, Germany

Printed by Libri Plureos GmbH
in Hamburg, Germany